GAUGE THEORIES IN PARTICLE PHYSICS

A PRACTICAL INTRODUCTION

2nd edition

Graduate Student Series in Physics
Other books in the series

Weak Interactions
D Bailin

Collective Effects in Solids and Liquids
N H March and M Parrinello

Hadron Interactions
P D B Collins and A D Martin

Superfluidity and Superconductivity
D R Tilley and J Tilley

Introduction to Gauge Field Theory
D Bailin and A Love

Mechanics of Deformable Media
A B Bhatia and R N Singh

Supersymmetry, Superfields and Supergravity
Prem P Srivastava

The Physics of Structurally Disordered Matter
N E Cusack

GRADUATE STUDENT SERIES IN PHYSICS

Series Editor: Professor Douglas F Brewer, M.A., D.Phil.
Professor of Experimental Physics, University of Sussex

GAUGE THEORIES IN PARTICLE PHYSICS

A PRACTICAL INTRODUCTION

2nd edition

IAN J R AITCHISON
Department of Theoretical Physics
University of Oxford

ANTHONY J G HEY
Department of Electronics and Computer Science
University of Southampton

ADAM HILGER, BRISTOL AND PHILADELPHIA

British Library Cataloguing in Publication Data

Aitchison, I. J. R. (Ian Johnston Rhind)
Gauge theories in particle physics.—
2nd ed.
1. Elementary particles. Gauge theories
I. Title II. Hey, Anthony J. G. III. Series
539.7′21

ISBN 0-85274-329-7 (hbk)
ISBN 0-85274-328-9 (pbk)
ISSN 0261-7242 (Graduate Student Series in Physics)

US Library of Congress Cataloging-in-Publication Data
Aitchison, Ian Johnston Rhind, 1936–
Gauge theories in particle physics: a practical introduction
Ian J. R. Aitchison and Anthony J. G. Hey, – 2nd ed.
550p. 23cm.– (Graduate student series in physics)
Bibliography: 8p.
Includes index.
ISBN 0-85274-329-7 ISBN 0-85274-328-9 (pbk.):
1. Gauge fields (Physics) 2. Particles (Nuclear physics) 3. Weak interactions (Nuclear physics) 4. Quantum electrodynamics. 5. Feynman diagrams. I. Hey, Anthony J. G. II. Title. III. Series.
QC793.3–bA34 1989
539.721–dc19 88-19162
CIP

Published under the Adam Hilger imprint by
IOP Publishing Ltd
Techno House, Redcliffe Way, Bristol BS1 6NX, England
242 Cherry Street, Philadelphia, PA 19106, USA

Typeset by KEYTEC, Bridport, Dorset
Printed in Great Britain by Butler and Tanner Ltd, Frome

PREFACE TO THE SECOND EDITION

In the Preface to the first edition of this book, published in 1982, we wrote: 'This book is intended to be a *practical* introduction to the calculation of many quantities of physical interest in particle physics. Today, this means a guide to *Feynman diagrams* appropriate to *gauge field theories*, and to the calculational techniques based on these diagrams. Our aim is to provide the minimum essential equipment required for a reasonable understanding of gauge theories that are being tested by contemporary experiments in high-energy physics'. We were strongly motivated by a desire to make the predictions of gauge theories *accessible* to a wider audience, since the literature, up to that point, had been generally very technical. Three essential features of our presentation were determined by the 'accessibility' requirement, as we then saw it: we dispensed with the machinery of Lagrangian-based quantum field theory and followed instead an intuitive approach (using wavefunctions) based on Feynman's ideas; we devoted the first third of the book to the original and best-understood gauge theory (quantum electrodynamics), introducing many calculational techniques in this relatively familiar context and generalising them later to the strong and weak gauge theories; and we tried, where possible, to provide physical arguments and analogies to illuminate difficult formal and conceptual points.

In the six years that have elapsed since then, the situation in particle physics has changed very significantly in several respects. Almost as soon as our book appeared, of course, a series of major discoveries was reported from the UA1 and UA2 groups using the $\bar{p}p$ collider at CERN. These experiments spectacularly confirmed the main predictions of the Glashow–Salam–Weinberg theory of weak interactions and also provided further support for QCD. They have been followed by higher-precision experiments, with the result that now the permanence of 'the standard model' is not in serious doubt: future, more ambitious, theories will have to include it, in some suitable limit or approximation.

A second noteworthy feature has been, in fact, the failure of any single such 'unified theory' to emerge as the undisputably most promising way to go 'beyond the standard model'. Grand unification, technicolour, supersymmetry, supergravity and strings all have exciting things to say—but more experimental clues, such as will be provided (one hopes) by the next generation of machines, are badly needed. In this context, the standard model—in both its theoretical and experimental aspects—appears as a relatively self-contained and consolidated chapter of physics. It makes sense to 'draw a line' at this point and await news from the TeV region.

The experimental discoveries, and the resulting rather different theoretical perspective, both indicated the necessity for a major revision of this book, while remaining faithful to the original aims, motivation and approach. Most obviously, we felt that it was essential to include considerable discussion of the recent experimental findings. A major aim of the original text was to enable readers to work through simple calculations, in the three gauge theories of the standard model, to the point of getting a meaningful result that could be compared with experiment. Now that so many experimental results are available, the value of exhibiting the confrontation between theory and experiment is enormous. Thus, whereas our first edition had but five diagrams showing experimental data, the present edition has over fifty more.

On the theoretical side we have also made major changes. Indeed, we have virtually turned the first edition inside out. Whereas the gauge principle and the discussion of QED as a gauge theory only made their appearance in Chapter 8 of the first edition, we now introduce these ideas very early on, in Chapter 2. We adopt, in fact, a more deductive approach generally. Instead of concentrating on a heuristic development of QED, by analogy with which other theories were built, we now take the 'gauge' view of QED to be the most appropriate logical starting point, given the centrality of this concept to the standard model.

Furthermore, we have reversed ourselves on the crucial issue of whether to introduce the methods of quantum field theory or not. Perhaps there is some irony in deciding thus, at the very moment when string theory offers the promise of superseding quantum field theory. Nevertheless, the success of the standard model does constitute a remarkable vindication of the quantum field concept, comprehending as it does, in one conceptual framework (albeit with subtle variations), three of the known forces of nature.

Indeed, we are now convinced that the basic physics of quantum fields—the heart of the twentieth century answer to all those ancient questions concerning the nature of matter and force—should be far more readily accessible than it is. We have therefore tried to provide a self-contained elementary introduction to quantum field theory (in Chapter 4), stressing wherever possible physical models and analogies, while also developing the formalism to a useful calculational stage. This allows us to discuss in some detail (in Chapter 5) just why it is that the electromagnetic interaction between two charged particles can be regarded as due to the 'exchange' of photons. Lacking a formalism for describing emission and absorption of quanta, the treatment of this crucial point in the heuristic approach of our first edition was, as we then admitted, something of a hand-wave. It is, of course, central to the physical interpretation of the covariant Feynman graph amplitudes.

Lagrangian-based quantum field theory also allows us to repair another admitted deficiency of the first edition—the missing link between the conservation of a quantity such as electric charge and the corresponding symmetry: see §4.6 and Chapter 8. We are also able to improve our discussion of spontaneous symmetry breakdown (Chapter 13).

However, we have quite definitely *not* abandoned wavefunctions! It seems to us possible to introduce just such a limited amount of the formalism of quantum field theory as will allow us to gain crucial *physical* insights which are obtainable in no other way. For the rest—i.e. most of the actual calculations—we rely as before on the wavefunction approach, since the full apparatus of field theoretic perturbation theory is beyond our scope. We also follow the previous and obvious plan of developing the easiest gauge theory (QED) first.

The last major change is the addition of a (long) chapter on QCD—a theory which received very little attention in the first edition. As in other parts of the book, our aim has been to develop the theory in as simple a way as possible, from first principles, to the point of confrontation with experiment. This is a tall order, and the later sections (§§9.5 and 9.6) may be omitted at first reading. Of course, we are discussing here *perturbative* QCD: we have included as one of our 'Four last things' (now Chapter 15) a brief introduction to the lattice approach to field theory, including the non-perturbative aspects of QCD.

The first edition *did* turn out to be quite accessible—at least, it was read by a wide spectrum of scientists from different disciplines. No doubt this was largely because at that time little else existed at a comparably 'low' level. We hope very much that we have not compromised this accessibility—which remains a fundamental aim—by including all these various bits of additional material. We feel that the experimental data should be no barrier—but rather, an aid to understanding. And we would like to think that a grasp of the essentials of quantum field theory will, in the end, make comprehension easier and be stimulating in itself. But we have also added one further, wholly new, chapter—namely the first, which provides a brief survey of the 'facts' of particle physics and of the theoretical framework that will be used to analyse them. Here we have tried to set the standard model in the context of the conceptual developments of the last hundred years, hoping thereby to offer a smoother entry to the whole subject for non-specialists.

In the present edition, therefore, we begin in Part I with a brief review of the salient experimental facts, which is then followed by an extended presentation of the various theoretical concepts that will be used to comprehend them. This represents a marked departure from the approach of the first edition, in which we aimed to reach 'real calculations' for QED as quickly as possible, postponing deeper conceptual matters until later in the book. The present structure seems more logical, but there is of course a price to pay: a quarter of the book is gone before a cross section is calculated. Nevertheless, we hope that readers (and their instructors) will feel that the deeper conceptual base provided by Part I makes this wait worthwhile. Calculations begin in earnest in Part II, which is the QED material from the first edition essentially unchanged. In Part III we present non-Abelian gauge theories and our review of QCD. Part IV contains weak interaction phenomenology, and Part V develops the electroweak theory: the major changes here are the (limited) use of field theory and the inclusion of much experimental data. Finally, Part VI takes the reader beyond tree graphs, beyond perturbation theory and (a little) beyond the standard model. The total additions are considerable, and to match the increased girth of the text, Appendix A has been considerably enlarged.

Acknowledgments

We reproduce below our expressions of thanks, taken from the Preface to the first edition, to the many people who helped us then; their efforts are of course preserved in the bulk of the present edition:

> I J R A has enjoyed the privilege of having an office in the same corridor as John C Taylor, Chris Llewellyn Smith, and latterly Graham Ross, from all of whom he has learned a great deal. The latter two very kindly read over a draft of Chapter 12 [now 15], making useful comments. Likewise, Robin Stinchcombe made helpful suggestions on an early draft of Chapter 9 [now 13]. Caroline Fraser contributed much friendly support and encouragement, as well as perceptive criticism of early drafts of many chapters. Finally, I J R A acknowledges with sincere thanks the generosity of the R T French Company, Rochester, New York, which supported him fully during a year's leave at the University of Rochester (1978–79), throughout which, in the absence of all formal teaching commitments, he was able to make at least some progress in repairing the gaps in his knowledge. A J G H wishes to thank his colleagues at Southampton for their interest and comments, and especially Garry McEwen, who actively encouraged the writing of a book such as this. Our joint thanks are also due to Jeffery Mandula and Denis Nicole for some incisive comments on a draft of Chapter 9; to Jack Paton, who, in addition to commenting on Chapter 9 when it was in a formative stage, read the entire completed manuscript, and made many extremely valuable suggestions; to Jan Ambjorn and Richard Hughes, for carefully reading and commenting on Chapters 8 (now 2 and 8), 9 and 12 in proof; and to David Bailin and Bob

Kelly, each of whom detected one very bad confusion. We are also grateful to several generations of students, both at Southampton and Oxford, and at summer schools, who lived through the genesis of this project, and supplied many constructive comments. We trust that we have now sufficiently insured ourselves against blame for any residual errors; it remains only to thank Susan Makin for cheerfully transforming a long and complicated manuscript into an accurate typescript, and Jessie Hey for assistance in preparing the index.

We are also grateful to those who drew our attention to a number of minor slips and misprints in the first edition, which we were able to correct in the 1983 reprint. In bringing the section on the status of QED up to date (now §6.10), we had the benefit of helpful correspondence with Professor T Kinoshita.

The chapters which are new in the present edition owe much to various people. The germ of Chapter 4 was a crash course on 'Feynman rules from quantum field theory', consisting of just one lecture, given by A J G H at the School for Young High Energy Physicists at the Rutherford Appleton Laboratory in 1984 and 1985. Important inspiration for the approach came from CERN Summer School lectures by John Bell on 'Experimental quantum field theory'. Roger Cashmore, then the director of these RAL Schools, encouraged this innovation, and also made room in the programme for an additional lecture by A J G H on 'Lattice gauge theory for plumbers', part of which forms the basis of the last section of Chapter 15. A J G H's place in the RAL School was taken over by I J R A in 1986. John Dainton, the new director, enthusiastically encouraged the more wholesale corruption of his innocent charges with the doctrines of quantum field theory, and Chapter 4 is part of the result. We benefited enormously from the expert assistance of many people in producing successive drafts of Chapter 9, and of the new sections of Chapter 14. These were started while I J R A was on leave at CERN during the year 1985–86, and he is grateful to the University of Oxford, and to Worcester College, for granting that leave of absence, and to Maurice Jacob for his warm hospitality in the CERN Theory Division. This stay at CERN provided a unique opportunity to get to grips with the new data, through contact with the people who had obtained them. In particular, Luigi DiLella most generously made available many of his original notes, transparencies and review materials, which were of superb pedagogic quality. Tony Weidberg made a careful reading of Chapters 9 and 14, as did Harry Cheung (then a research student, and an ideal guinea pig). Likewise, Steve Geer and Bill Scott provided useful input on this material. Mike Pennington, also, commented on Chapter 9—parts of which draw important ideas from his excellent article in *Reports on Progress in Physics*, **46**, 393 (1983). We have tried to give a balanced treatment of the major discoveries, but we are aware that strong feelings are sometimes aroused by accounts of 'who did what'. Despite the above-named assistance, the responsibility for what appears herein is ours: we apologise if we have inadvertently created any false or misleading impressions, and will be prepared to listen to politely worded corrections.

Some particular acknowledgements remain. Euan Squires, for Adam Hilger, read right through the entire manuscript at a late stage, and detected a considerable number of infelicities, and worse. We are very grateful to him. A J G H thanks Gary McEwen, tireless as ever, and colleagues and students in the Southampton Group for helpful comments on sections of the manuscript. We owe most special thanks to George Emmons, Principal of the Merchant Navy College, Greenhithe, Kent. Not a particle theorist himself, but possessing an exuberant and avid interest in this sort of stuff, he has lived with the second edition almost as long as we have. He gave wise suggestions for qualitative changes in early drafts, and with meticulous thoroughness checked details of equations in later ones. Going far beyond all reasonable demands of friendship, he even joined us in checking the proofs. We owe him a great debt. All this said, however, we are only too well aware that errors will persist: we should be most grateful to have them brought to our attention.

Ian J R Aitchison
Anthony J G Hey
May 1988

CONTENTS

PART III: NON-ABELIAN GAUGE THEORY AND QCD

PART IV: NOT QUITE A THEORY: PHENOMENOLOGY OF WEAK INTERACTIONS

PART V: THEORY OF ELECTROWEAK INTERACTIONS

PART VI: BEYOND THE TREES

PART I

EXPERIMENTAL AND THEORETICAL ELEMENTS

There are therefore Agents in Nature able to make the Particles of Bodies stick together by very strong Attractions. And it is the Business of experimental Philosophy to find them out.

Isaac Newton, *Optics*, from the thirty-first (and last) 'query' at the end of the book (page 394 of the reprinted fourth edition, G Bell and Sons, Ltd, London, 1931).

1

QUARKS AND LEPTONS, AND THE FORCES BETWEEN THEM

1.1 The standard model

The traditional goal of particle physics, in response to Newton's challenge, has been to identify what appear to be structureless units of matter, and to understand the nature of the forces acting between them. Stated thus, the enterprise has a two-fold aspect: matter on the one hand, forces on the other. The expectation is that the smallest units of matter should interact in the simplest way: or, that there is a deep connection between the basic units of matter and the basic forces. The joint matter/force nature of the enquiry is perfectly illustrated by Thomson's discovery of the electron and Maxwell's theory of the electromagnetic field, which together mark the birth of modern particle physics. The electron was recognised both as the 'particle of electricity'—or as we might now say, as an elementary source of the electromagnetic field, with its motion constituting an electromagnetic current—and also as an important constituent of matter. In retrospect, the story of particle physics over the subsequent one hundred years or so has consisted in the discovery and study of two new (non-electromagnetic) forces—the *weak* and the *strong* forces—and in the search for 'electron-figures' to serve both as constituents of the new layers of matter which were uncovered (first nuclei, and then hadrons), and also as sources of the new force fields. This effort has recently culminated in decisive progress: the identification of a collection of matter units which are indeed analogous to the electron, and the partial, but highly convincing, experimental verification of theories of the associated strong and weak force fields, which incorporate and generalise in a beautiful way the original electron/electromagnetic field relationship. These theories are collectively (and rather modestly, if uninformatively) called 'the standard model', to which this book is intended as an elementary introduction.

In brief, the picture is as follows. There are two types of matter units: *leptons* and *quarks*. Both have spin $\frac{1}{2}$ (in units of $\hbar$) and are structureless at the smallest distances currently probed by the highest-energy accelerators. The leptons are direct generalisations of the electron, the term denoting particles which, if charged, interact both electromagnetically and weakly, and, if neutral, only weakly. The involvement of the electron in weak, as well as electromagnetic, interactions became clear quite early on, with the study of nuclear β decay—though it took many years before it was conclusively established that the electron emitted in this process was invariably accompanied by a neutral leptonic partner, the antineutrino $\bar{\nu}_e$. By contrast, the quarks—which are the fermionic constituents of hadrons, and thence of nuclei—are more like 'strongly interacting analogues' of the electron, since they interact via all three interactions, strong, electromagnetic and weak. This, indeed, is the basis for distinguishing between these two types of matter unit, a distinction which will presumably disappear if (see §15.1) it eventually proves possible to unify all three types of force. At all events, the weak and electromagnetic interactions of both quarks and leptons are described in a (partially) unified way by the *electroweak theory* (see Chapters 13 and 14), which is a generalisation of *quantum electrodynamics* or *QED* (Chapters 5 and 6); while the strong interactions of quarks are described by *quantum chromodynamics* or *QCD* (Chapters 8 and 9), which is also analogous to QED. Both quarks and leptons act as sources of the electroweak field, while quarks alone act as the sources for the QCD field; we may speak of weak and strong 'charges' and 'currents', by analogy with similar concepts in QED.

According to this picture, the observed 'levels of structure'—atomic, nuclear and hadronic—are built up in a series of stages. Quarks interacting via QCD bind to form hadrons, such as neutrons, protons and pions. It has to be admitted at once that the details of this process are not yet understood, owing to the calculational difficulties of dealing with a strongly interacting quantum field theory. We shall return to this point at the end of the book (see §15.4). Residual interquark forces are then thought to give rise to the rather complicated observed forces between neutrons and protons, which are responsible for binding these *nucleons* to form nuclei. This stage is also not well understood, but is regarded as somewhat analogous to the way in which complicated interatomic forces are ultimately due to residual electromagnetic interactions between the constituents—for of course it is these latter interactions which bind electrons and nuclei into atoms. The reasons why, despite these gaps in the argument, we do nevertheless believe in QCD will be discussed in Chapter 9. As regards the weak force, it appears that, as its name already indicates, it is not strong enough to produce any binding effects among the systems studied so far.

The reader will certainly have noticed that the most venerable force of all—gravity—is conspicuously absent from our story. It will continue to be so. In practical terms this is quite reasonable, since its effect is very many orders of magnitude smaller than even the weak force, at least until the interparticle separation reaches distances far smaller than we shall be discussing. Conceptually, also, gravity still seems to be somewhat distinct from the other forces, which, as we have already indicated, are encouragingly similar. There are no particular fermionic sources carrying 'gravity charges': it seems that *all* matter gravitates. This of course was a motivation for Einstein's geometrical approach to gravity. Despite remarkable recent developments in the theory of *strings* (Green *et al* 1987), it is fair to say that the vision of the unification of all the forces, which possessed Einstein, is still some way from realisation. To include any further discussion of these ideas would take us far beyond our stated aim.

This book is not intended as a completely self-contained textbook on particle physics, which would survey the broad range of observed phenomena and outline the main steps by which the above picture has come to be accepted. For this we must refer the reader to other sources (e.g. Perkins 1987). However, in the remainder of this chapter we shall offer a brief summary of the salient experimental facts which have established this current view of matter and forces, together with an introduction to the conceptual framework needed to understand it.

1.2 Levels of structure: from atoms to quarks

What kind of experiments are done to probe the structure of microscopically small systems? We may distinguish three basic types: *scattering*, *spectroscopy* and *break-up* experiments, the exploitation of which constitutes a common theme in the chain of discovery: atom $\rightarrow$ nucleus $\rightarrow$ hadron $\rightarrow$ quark.

1.2.1 Atoms $\rightarrow$ nucleus

Around the turn of the last century, the experiments of Thomson and others established the charge and mass of electrons. In particular, the charge was exactly equal in magnitude to that of a singly ionised atom. Thus part, at least, of the atom had been 'broken off'. Then, in 1909, Geiger and Marsden, of the University of Manchester in England, reported an unusual experimental finding. Working on a programme initiated by Rutherford, they were measuring how α particles are reflected by thin metal foils. In perhaps the understatement of the century, they remarked that 'It seems surprising that some of the α particles, as the experiment shows, can be turned within a layer of 6×10^{-5} cm of gold through an angle of 90°, and even more' (Geiger and Marsden 1909).

Many years later, Rutherford, in a famous phrase, said that when he became aware of this result he was as amazed as if he had seen a shell bounce back from hitting a sheet of paper. It took him two years to find the explanation. The atom was known to be electrically neutral, and to contain negatively charged electrons of mass very much less than that of the atom. The question was: how is the heavy, positively charged matter distributed? It seems to have been generally thought of as being uniformly spread over the atomic volume, and a specific model of this type was proposed by J J Thomson in 1910. We shall call this the 'soft' model. In such a model, Rutherford (1911) noted, α particles are deflected very little in any single scattering process, so that large deflections can occur only as a result, as he put it, 'of a multitude of small scatterings by the atoms of matter traversed'. However, 'a simple calculation based on the theory of probability shows that the chance of an α particle being deflected through 90° is vanishingly small'.

Rutherford went on: 'It seems reasonable to suppose that the deflection through a large angle is due to a single atomic encounter, for the chance of a second encounter of a kind to produce a large deflection must in most cases be exceedingly small. A simple calculation shows that the atom must be a seat of an intense electric field in order to produce such a large deflection at a single encounter'.

Thus was the nuclear atom born, in which the atom contained a 'hard' constituent—the small, massive, positively charged nucleus. In more precise terms, according to the soft model the distribution of scattered particles falls off exponentially with the angle of deflection; the departure of the observed distribution from this exponential form is the signal for hard scattering. Geiger and Marsden found that about one in 20 000 α particles was turned through 90° or more in passing through a thin gold foil; by contrast, a 'soft' model would predict a fraction of something like 1 in 10^{3500}! (Eisberg 1961, ch 4).

Rutherford went much further than this qualitative explanation of course. He calculated the angular distribution expected from his nuclear model, obtaining the famous $\sin^{-4}\phi/2$ law, where ϕ is the scattering angle. Two years later, Geiger and Marsden (1913) published a second paper which was a long and detailed verification of this prediction and of other features of the scattering. They recorded observed scintillations versus angle in the form of a table, and commented: 'In columns IV and VI the ratios of the numbers of scintillations to $1/\sin^4\phi/2$ are entered. It will be seen that in both sets the values are approximately constant'. Column IV was scattering from silver, column VI was gold. It is slightly curious that they chose not to plot a graph of their angular distribution, but we have taken the liberty of doing it for them: figure 1.1 records some points from their data for silver.

At almost the same time, Bohr (1913)—then also in Rutherford's group at Manchester—proposed his famous model for the *dynamics* of the nuclear atom, based on an inspired blend of classical mechanics (which Rutherford had also used in deriving his angular distribution)

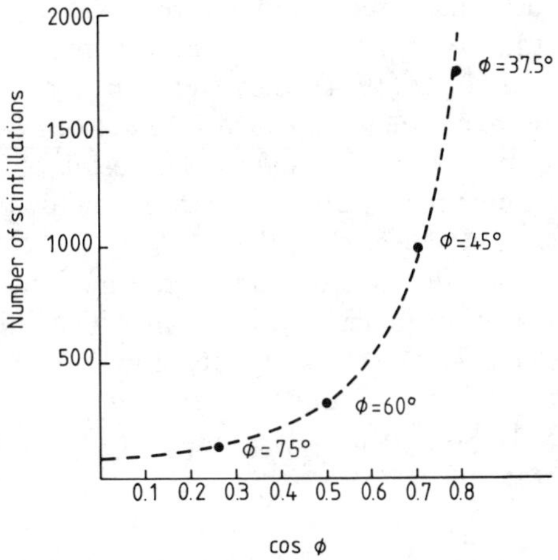

Figure 1.1 Four of the data points reported by Geiger and Marsden (1913), replotted here as the number of scintillations (for scattering of α particles by a silver foil) versus the cosine of the scattering angle ϕ. The curve is the Rutherford prediction $\sin^{-4}\phi/2$, normalised to the data point at $\phi = 60°$.

and early quantum theory. This succeeded in giving an excellent account of the *spectroscopy* of hydrogen, as it was then known. Figure 1.2(*a*) shows the gross structure of the hydrogen energy levels; the ground state is marked at zero energy, and the threshold for ionising the electron is about 13.6 eV. Also shown in figure 1.2(*b*) are some of the low-lying levels of the two-electron atom, helium.

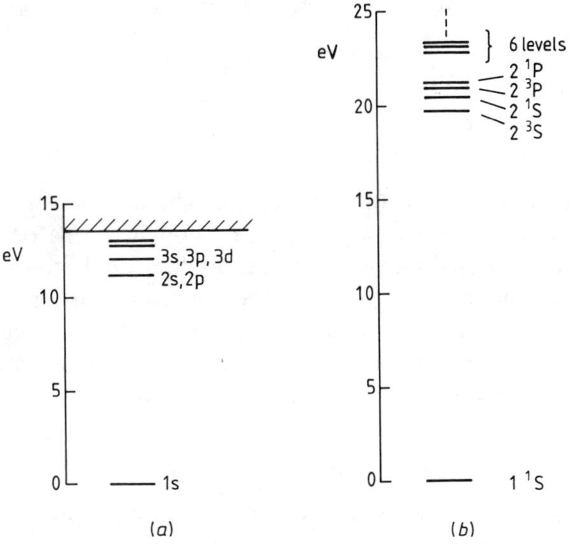

Figure 1.2 (*a*) Gross structure of hydrogen energy levels. (*b*) Gross structure of low-lying levels of helium.

Evidence for such quantised energy levels had of course originally come from the study of emission spectra. Beginning with the experiments of Franck and Hertz (1914), they were further confirmed in a new type of scattering experiment, in which electrons were *inelastically* scattered from atoms. Figure 1.3 shows the result of a typical experiment, the relative number of scattered electrons being plotted versus the energy of the scattering electron. The peak at 200 eV corresponds to elastic scattering, since this was the beam energy. The peak just below 180 eV corresponds to an inelastic process in which the beam electrons have delivered some 21 eV of energy to the helium atoms, thereby exciting them to the first group of excited levels (unresolved in figure 1.3). Similarly, the peak at about 177 eV corresponds to the excitation of the states near 23 eV.

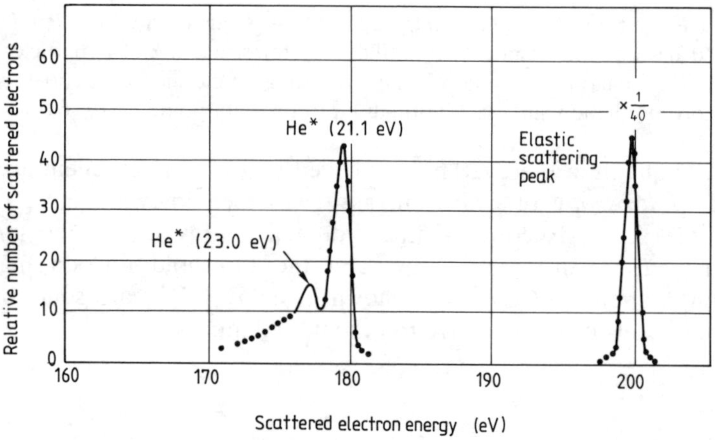

Figure 1.3 Elastic and inelastic electron scattering from helium atoms, taken from figure 1-13 of French and Taylor (1978).

With the development of quantum mechanics in the 1920s, such spectra began to be quantitatively understood in terms of the internal excitations (vibrations and rotations) of the light constituents, the electrons. The spatial distribution of the electrons was also mapped out by x-ray scattering, and extended typically over sizes of the order of a few angstroms. The uncertainty principle gives an estimate for the typical electron momenta when they are confined to such a linear domain:

$$p \sim \hbar/10^{-10}\,\mathrm{m} \sim 2\,\mathrm{keV}/c$$

so that (with $mc^2 \approx 0.5$ MeV for the electron) the ratio $v/c \sim 4 \times 10^{-3}$. The gross features of atomic structure were thus well described by the non-relativistic quantum mechanics of pointlike electrons interacting with each other, and with a pointlike nucleus, via Coulomb forces.

1.2.2 Nuclei → nucleon

But is the nucleus really pointlike? As accelerator technology developed, it became possible to get beams of much higher energy or momentum. From the de Broglie relation $\lambda = h/p$ this meant that the resolving power of the beam became much finer, and *deviations* from the simple Rutherford formula for charged particle scattering (which assumed that both α particle and nucleus were pointlike) began to be observed. The most precise series of experiments was done by Hofstadter and collaborators at SLAC in the 1950s (Hofstadter 1963). They used an electron beam, to avoid the complication of the poorly understood strong interactions between α particles and nuclei. Figure 1.4 shows how the angular distribution of electrons scattered *elastically* from gold at an energy of 126 MeV falls *below* the 'point-nucleus' prediction. Qualitatively this is due to wave mechanical diffraction effects over the finite volume of the nucleus. The observed

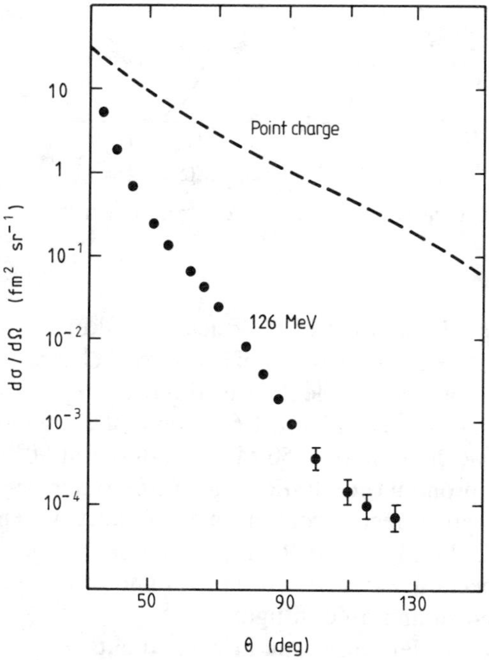

Figure 1.4 Angular distribution of 126 MeV electrons elastically scattered from gold, taken from Hofstadter (1963). Data points are full circles; the broken curve is the 'point-nucleus' prediction.

distribution is a product of two factors: the first is the distribution corresponding to scattering from a single pointlike target (generalising Rutherford's formula to include corrections for relativistic kinematics, quantum mechanics, electron spin, target recoil and so on), while the second is called the *form factor*, which is characteristic of the spatial extension of the target's charge density (Perkins 1987, §6.2; see also §6.6 below). Figure 1.5 shows the charge density of $^{12}_{6}C$ extracted (Hofstadter 1963) from such elastic electron scattering data. It is clear that this nucleus has a charge radius of about 1–2 fm (1 fm = 10^{-15} m); in heavier nuclei of mass number A the radius goes roughly as $A^{1/3}$ fm.

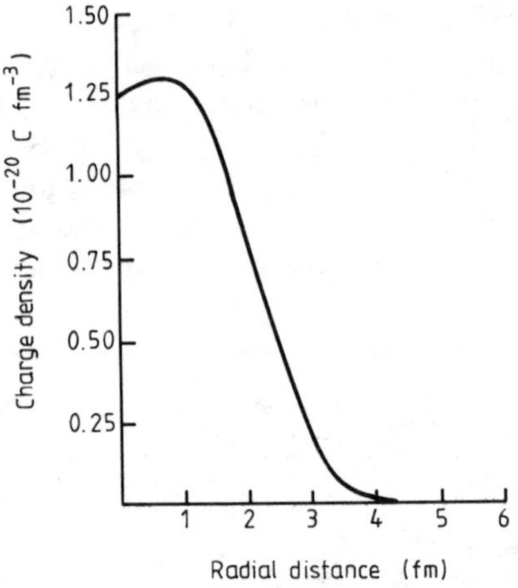

Figure 1.5 Charge density of $^{12}_{6}C$, extracted from electron scattering data (Hofstadter 1963).

The finite spatial extension of the nucleus is already an indication that it is not structureless. As with atoms, *inelastic* electron scattering from nuclei reveals that the nucleus can be excited into a sequence of quantised energy states. Figure 1.6 shows the energy spectrum of electrons with incident energy 150 MeV scattered at 90° from carbon. In addition to the pronounced elastic peak, three other peaks can be seen, corresponding (after recoil corrections) to electron energy losses of about 4.4 MeV, 7.7 MeV and 9.6 MeV. These energies correspond to the first three excited states of carbon as found in independent spectroscopic studies and summarised in figure 1.7.

It is clear that nuclei must contain constituents distributed over a size of order a few fermis, whose internal quantised motions lead to the observed nuclear spectra involving energy differences of order a few MeV. Chadwick's discovery of the neutron in 1932 established that, to a good approximation, the constituents of nuclei are neutrons and protons—generically called *nucleons*. Since the neutron is electrically neutral, it was clear that a novel force had to be operating over the

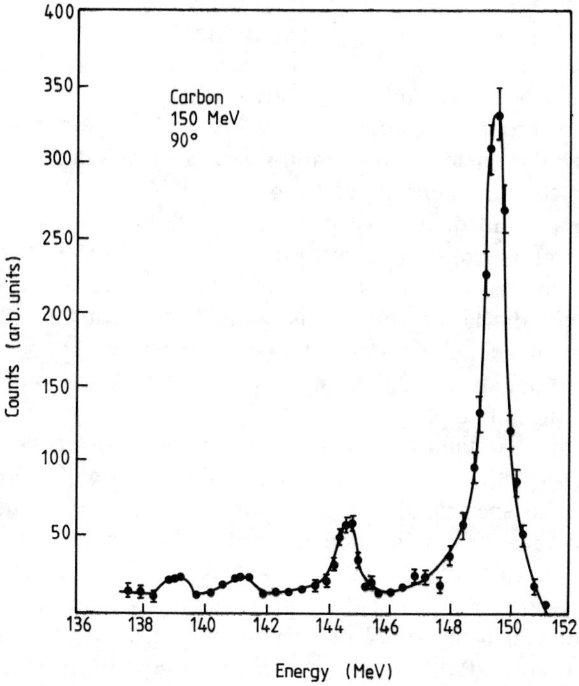

Figure 1.6 Elastic and inelastic scattering from carbon nuclei (Fregeau and Hofstadter 1955).

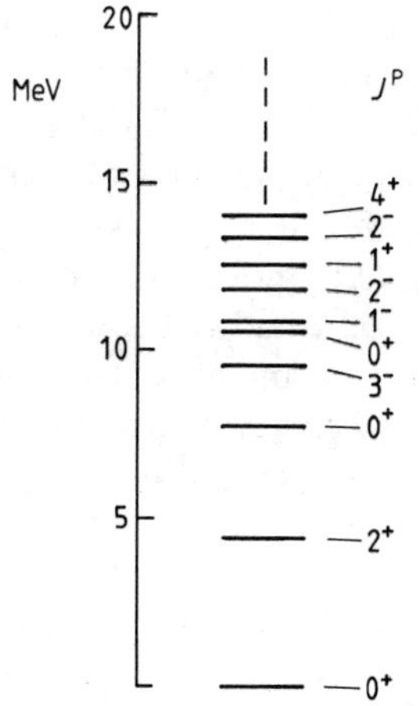

Figure 1.7 Energy levels of $^{12}_{6}C$.

range of nuclear dimensions in order to bind nucleons into nuclei. It must also be considerably stronger than the electromagnetic force, since it has to counterbalance the typical 'uncertainty' energy, which is now of order

$$\frac{1}{2M}\left(\frac{\hbar}{R}\right)^2 \sim 20 \text{ MeV}$$

for M = nucleon mass and $R \sim 1$ fm; the (repulsive) potential energy between two protons at this separation is about ten times smaller. Nevertheless, the nucleons move non-relativistically since 20 MeV is well below the nucleon rest energy of about 940 MeV.

A most important qualitative point now emerges: the typical scales of *size* and *energy* are quite different for atoms and nuclei. The excitation energies in atoms are in general insufficiently great to excite the nucleus: hence the latter appears as a small inert core inside the atom, and only when excited by appropriately higher-energy beams does it reveal its internal structure. The nuclear degrees of freedom are 'frozen' on the scale of atomic physics.

One further detail of nuclear spectra is of concern to us. Figures 1.8(a)–1.8(c) show how, after allowing for Coulomb energy differences, the spectra of nuclei which differ from each other by the replacement of a proton by a neutron (or vice versa) contain striking similarities. In the case of figure 1.8(a) we have examples of 'mirror nuclei', in which a nucleus with Z protons and N neutrons is compared with another consisting of N protons and Z neutrons. In figure 1.8(b) the lowest (ground) state of ^{18}F corresponds to a state which is excluded, by the Pauli principle, for the neighbouring nuclei in which a spare n–p pair is replaced by a p–p or n–n pair; otherwise the levels of these three 'analogue' nuclei match very well. Figure 1.8(c) shows near coincidences between the energy levels of four nuclei, with mass number 21. $^{21}_{10}$Ne and $^{21}_{11}$Na are a mirror pair, but beyond a certain excitation energy their levels also correspond to states in neighbouring nuclei, with $Z = 9$ and $Z = 12$. A pair of nuclei with corresponding levels, as in figure 1.8(a), is said to form a *doublet*; three such nuclei, as in figure 1.8(b), form a *triplet*; and four, as in figure 1.8(c), a *quartet*. Even without the benefit of Chadwick's discovery, these figures would surely indicate that nuclei are made of two basic objects, one charged and one neutral, *the forces between which are very nearly independent of their charge*. We will make use of just this sort of argument in the following section, and we will return to the mathematics of such *charge multiplets* in Chapter 8.

1.2.3 Nucleons → quarks

Are nucleons pointlike? The answer, in broad outline, is a remarkable repetition of the pictures just sketched for atoms and nuclei. First, elastic scattering experiments of electrons from nucleons by Hofstadter and co-workers in the 1960s (Hofstadter 1963) revealed that the proton has a well defined form factor, indicating an approximately exponential distribution of charge with a root mean square radius of about 0.8 fm. The magnetic moments of nucleons also have a spatial distribution; when scaled by their total (integrated) values μ_p and μ_n, the proton and neutron magnetic moment distributions fall on very nearly the same curve as the proton charge distribution (Perkins 1987, §6.5). These distributions have been well measured down to distances of order 0.04 fm, corresponding to a momentum transfer to the nucleon of order 5 GeV/c.

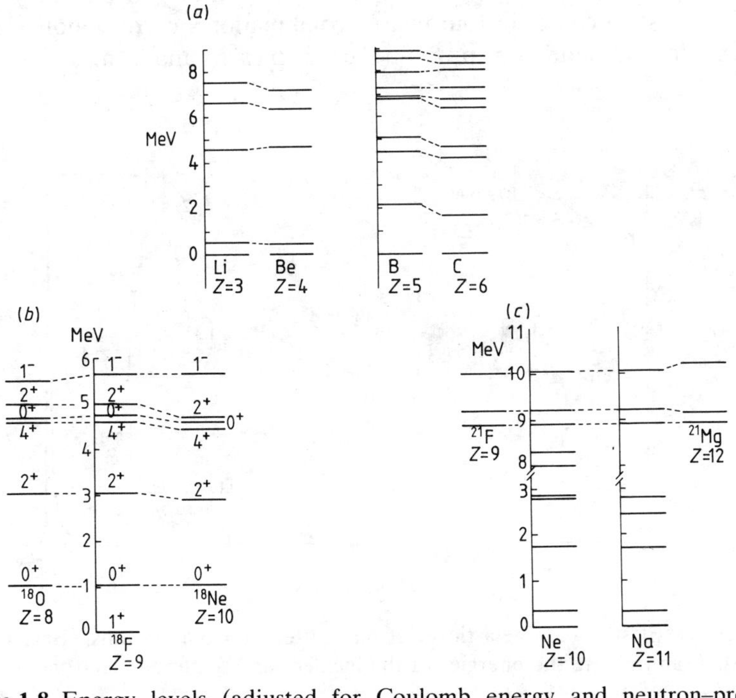

Figure 1.8 Energy levels (adjusted for Coulomb energy and neutron–proton mass differences) of nuclei of the same mass number but different charge, showing (*a*) 'mirror' doublets, (*b*) triplets and (*c*) doublets and quartets.

As with both the atom and the nucleus, the clearly defined spatial extension measured by elastic electron scattering already indicates that the nucleon is not structureless. Further information bearing on nucleon structure can be obtained, as in the atomic and nuclear cases, by studying inelastic electron scattering to see if there are signs of a *nucleon spectroscopy*, which could be interpreted in terms of internal motions of constituents. Figure 1.9 shows the energy spectrum of electrons of incident energy 4.879 GeV scattered at 10° from protons. Once again we see the usual large elastic peak (shifted from 4.879 GeV by recoil effects), but there is also clear evidence for other peaks which presumably correspond to excitations of the recoiling system. However, the detailed interpretation of figure 1.9 is a complicated story: not only are there, in some cases, several excited recoil states contributing to a 'single' peak, but even the apparently featureless regions of the curve conceal considerable structure. Only the first of the peaks beyond the elastic peak, that at $E' \approx 4.2$ GeV (which corresponds to an energy of about 1.23 GeV for the recoiling system), has a simple interpretation. It corresponds exactly to a long-established resonant state observed in π–N scattering and denoted by the symbol Δ (Perkins 1987, §4.9). This particular state comes in four charge combinations, corresponding to the accessible πN channels π^+p, π^+n (π^0p), π^-p (π^0n) and π^-n.

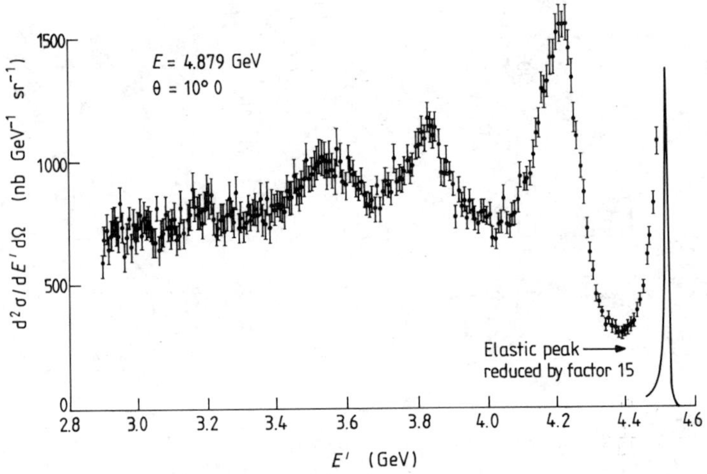

Figure 1.9 Elastic and inelastic electron scattering from protons (Bartel *et al* 1968). E and E' are the energies of the incident and scattered electrons.

The detailed deciphering of spectra such as figure 1.9 involved much careful experimentation in the field of *baryon spectroscopy* in the 1960s and 1970s. The result of all the analysis revealed an elaborate level sequence, paralleling in a remarkable way the spectra established for atoms and, even more so, nuclei. The result is partially summarised in figures 1.10(a) and 1.10(b). One series of levels, figure 1.10(a), comes in *two* charge combinations (charged and neutral) and is built on the proton and neutron as ground states. The two nearby states around 1.5 GeV correspond to the peak at $E' \approx 3.8$ GeV in figure 1.9 (but the latter figure shows no trace of the level at about 1.4 GeV in figure 1.10(*a*)!). This sequence is strikingly similar to the *doublet* sequences we saw in nuclear physics, figure 1.8(*a*). Another series, figure 1.10(*b*), comes in *four* charge combinations and is built on the Δ's as lowest levels. This one recalls the *quartets* of figure 1.8(*c*). It is noteworthy that these are all the observed nucleon-like excitations—for example, no triply charged states are seen.

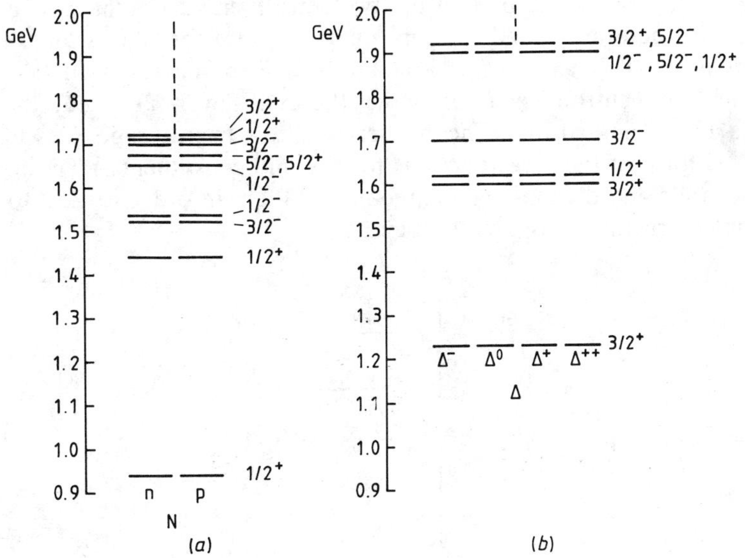

Figure 1.10 Baryon energy levels: (*a*) doublets (N); (*b*) quartets (Δ).

There are also mesons. Originally predicted by Yukawa (1935) as the quantum of the short-range nuclear force field (see §1.3 below), the pion was eventually discovered in 1947. In the 1960s it too turned out to be merely the ground state in a sequence of further excited states, forming charge triplets as shown in figure 1.11 (cf figure 1.8(*b*)).

Unfortunately this simple world of neutrons, protons and pions, together with their excited states, is not all there is to life. We shall briefly describe the additional complications in §1.4 below; see also Perkins (1987). But we have presented enough selective evidence to make our main point: the conclusion seems to be irresistible that we are looking, once again, at the internal excitations of the constituents of composite systems. A specific proposal along these lines was made independently in 1964 by Gell-Mann (1964) and Zweig (1964). Though based on somewhat different (and much more fragmentary) evidence, their suggestion has turned out to be essentially correct. They proposed that the nucleon-like states are made of three spin-$\frac{1}{2}$ constituents called (by Gell-Mann) quarks, while the mesons are quark–antiquark bound states. Such states are collectively called *hadrons*. As in the nuclear case, the simplest interpretation of the hadronic charge multiplets displayed in figures 1.10 and 1.11 is that the states are built out of two types of constituent, which differ by one unit of charge. However, the proton itself, for example, has unit charge—hence it would appear that its constituents must be fractionally charged. The most economical assignment turns out to be that the two constituents have charges $\frac{2}{3}e$ and $-\frac{1}{3}e$, and these are called the 'up' quark (u) and the 'down' quark (d) respectively. The series of proton levels is then the states of the uud system, the neutron levels are udd, the Δ^{++} levels are uuu, the π^+ levels are u$\bar{\text{d}}$ and so on. (The Δ^+ series differs from the p series in the detailed form of the wavefunctions for the uud constituents.) The forces between u's and d's must be charge-independent in order to lead to the multiplet structure of figures 1.10 and 1.11.

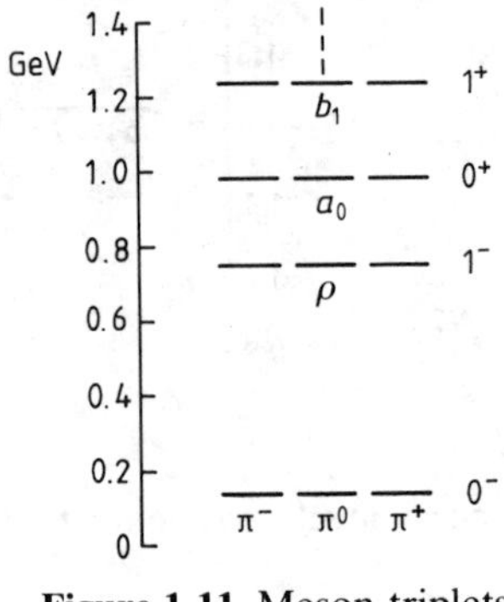

Figure 1.11 Meson triplets.

We may at this point recall that typical energy level differences in nuclei are measured in MeV. Thus in the vast majority of nuclear phenomena, the neutrons and protons remain in their unexcited ground states, hadronic excitations being typically of order 100 MeV: the hadronic degrees of freedom are largely 'frozen' in nuclear physics, just as the nuclear ones are frozen in atomic physics. The hadronic level is therefore effectively separated from the nuclear one, though in neither size nor energy scales is the separation as marked as that between nuclei and atoms. Nuclear physics thus becomes a semi-autonomous discipline.

Despite the 'obviousness' (in retrospect) of the quark hypothesis as developed thus far, it is a fact that—with some notable exceptions, for example Dalitz (1965, 1967)—most physicists were reluctant to accept it until the mid-1970s. We shall begin to appreciate some of the reasons for urging caution as we proceed. Nevertheless, a surprisingly simple 'shell model' approach is capable of giving an excellent description of all the known hadronic spectra, in terms of the (qqq) and ($q\bar{q}$) picture (Isgur and Karl 1983). The problem, as we shall ultimately see (§15.4), is how to relate this to QCD!

The nucleon spectroscopy data are indirect evidence for constituents. Is there any way of 'seeing' the quarks directly? Here we naturally think of the original Rutherford-type experiment, in which the existence of the nuclear constituent of the atom was inferred from the anomalously large number of particles scattered at large angles. To probe the structure of nucleons in this way, it is not convenient to use α particles, since the interpretation of the scattering data is then complicated by the fact that the α's interact strongly with the nucleon. Instead, electrons are used, which interact via the well understood electromagnetic force. Figure 1.12 shows the result of extending the inelastic electron scattering measurements (shown in figure 1.9 for $\theta = 10°$) to larger scattering angles. We know that the height of the elastic peak will fall rapidly due to the exponential fall-off of the nucleon form factor (recall figures 1.4 and 1.5 for the nuclear case). The same is true for the other distinct peaks, indicating that the well defined excited nucleon states have definite spatial extension. But in the region of largest energy transfer, something dramatically new happens: the curve does *not* fall as the angle increases! This is illustrated in figure 1.12. In other words, for large enough energy transfer, the electrons bounce backwards quite happily, just as Geiger and Marsden's α particles did, suggesting that the nucleon contains 'hard' constituents. In Chapter 7 we shall analyse the *deep inelastic scattering* experiments with the proper technical detail, and show how they do imply the existence of spin-$\frac{1}{2}$ hadronic constituents, which are pointlike down to currently probed scales and which possess the other attributes of the quarks hypothesised from the hadronic spectroscopy.

Despite mounting evidence, of the type described, from both scattering and spectroscopy during the 1960s and early 1970s, many physicists at that time continued to regard quarks more as useful devices for systematising a mass of complicated data than as genuine items of physical reality. One reason for their scepticism we must now confront, for it constitutes a major new twist in the story of the layers of matter. Gell-Mann ended his 1964 paper with the remark: 'A search for stable quarks of charge $-\frac{1}{3}$ or $+\frac{2}{3}$ and/or stable di-quarks of charge $-\frac{2}{3}$ or $+\frac{1}{3}$ or $+\frac{4}{3}$ at the highest energy accelerators would help to reassure us of the non-existence of real quarks'.

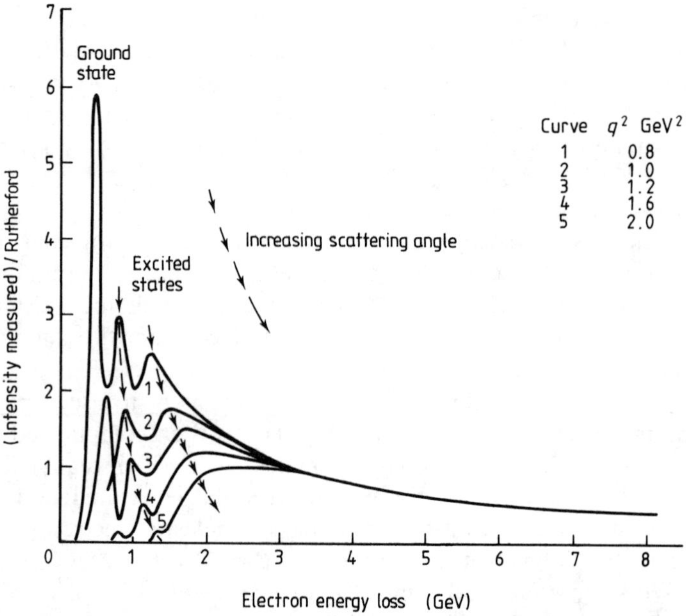

Figure 1.12 Inelastic electron–proton scattering cross section as a function of energy transfer ν, for different values of q^2, the invariant square of the 4-momentum transfer; q^2 increases as the scattering angle increases. (From Perkins D 1972 *Introduction to High Energy Physics* 1st edn, courtesy Addison-Wesley Publishing Company.)

Indeed, with one possible exception (La Rue *et al* 1977, 1981), this 'reassurance' has been handsomely provided! *Unlike* the constituents of atoms and nuclei, quarks have *not* been observed as free isolated particles. When hadrons of the highest energies currently available are smashed into each other, what is observed downstream is only lots more hadrons, not fractionally charged quarks. The explanation for this novel behaviour of quarks—that they appear perfectly real when clustered into hadrons, but are not to be seen in isolation—is now believed to lie in the nature of the interquark force (QCD). We shall briefly touch on this force again in §1.3, and take it up in more detail in Chapters 9 and 15. The consensus at present is that QCD does imply '*confinement*' of quarks—that is, they do not exist as isolated single particles, only as groups confined to hadronic volumes.

We end this section on 'levels' by presenting more experimental data which, besides substantiating the existence of quarks, also indicate a remarkable similarity between them and leptons, suggesting that quarks and leptons—if not 'fundamental'—are at least to be thought of as the natural elements of a common level of structure.

1.2.4 Quarks and leptons

Most sceptics were finally converted to real quarks by the 'November Revolution' of 1974: the discovery of a new series of mesonic spectra, the ψ/J particles (Aubert *et al* 1974, Augustin *et al* 1974). Beyond reasonable doubt, these levels had the quantum numbers characteristic of fermion–antifermion states. Indeed, the ψ/J spectrum is very well described in terms of $c\bar{c}$ states, where 'c' is a new type of quark, called the 'charm' quark (see §1.4). The $c\bar{c}$ system is now called 'charmonium', by analogy with the e^-e^+ system called positronium. This analogy is far more than superficial, as figure 1.13 shows: the energy scales differ by a factor of 10^8, but the similarity is truly remarkable. It is all the more surprising when we recall that the e^+e^- interact electromagnetically to form positronium, whereas it is the strong (QCD) interaction between the $c\bar{c}$ pair that forms charmonium.

It is already clear from this example that, in discussing possible similarities between quarks and leptons, we cannot avoid mention of the forces that act between them. Indeed, all talk of structure for *matter* units must inevitably be linked to some theory of the *forces* by which they are probed (cf our remarks in §1.2.3 about using electrons rather than α particles to probe nucleon structure). But we cannot discuss everything simultaneously, and forces will get their turn in the next section. We continue with further evidence for quarks in hadrons, and for a striking similarity between the strong (QCD) and electromagnetic forces.

Let us return to the 'Rutherford programme' once again. We have seen that the relatively high cross section for large-angle, large-energy-transfer electron scattering from nucleons provided evidence for 'hard' constituents. What if we collide two nucleons together? With a 'soft' model of the nucleon we should expect the number of the observed reaction products to fall off dramatically as a function of their angle to the beam direction. On a 'hard' model we should see prominent events at wide angles, corresponding to collisions between the constituents. But what exactly are these 'reaction products' and 'events'? The result of a hard collision between, say, a quark from each nucleon would normally be to scatter the quarks to wide angles—but we have seen that the quarks are *not* observed as free isolated particles! Instead, by a process which is not understood in detail, the two emerging hard-scattered quarks get converted into two roughly collimated 'jets' of *hadrons*. The existence and angular distributions of these jets provide further (in-direct) evidence relating to quarks and their interactions, as will be discussed in some detail in Chapters 7 and 9.

At this point we content ourselves with a trailer. Hard scattering events in hadronic collisions first showed up really clearly in experiments

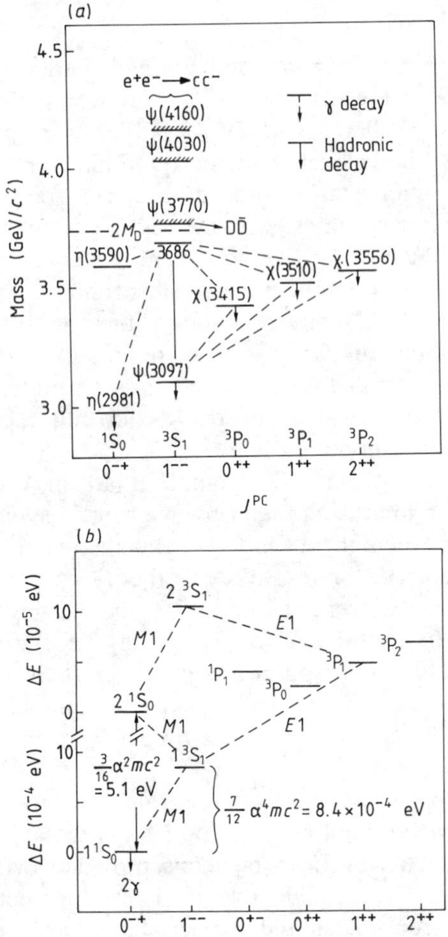

Figure 1.13 Energy levels of (*a*) charmonium ($c\bar{c}$) and (*b*) positronium (e^+e^-). (From Perkins D 1987 *Introduction to High Energy Physics* 3rd edn, courtesy Addison-Wesley Publishing Company.)

done at the CERN $\bar{p}p$ collider, since it provided the largest energies and momentum transfers then available (early 1980s) and thus was able to probe the smallest distance scales within the hadrons. Figure 1.14 shows the cross section for hadronic jets produced at wide angles to the $\bar{p}$–p beams of the CERN collider, plotted against the total transverse energy carried by the jets. Without entering into details (see §9.4), it is clear that large transverse energy corresponds to large scattering angles. The broken line on the figure is the prediction for a 'soft' model of p and $\bar{p}$, falling away exponentially as in Thomson's atomic model. The data clearly stand out many orders of magnitude greater than this prediction at larger angles, just as in the experiments of Geiger and Marsden.

Furthermore, when the actual angular distribution of two jet events is plotted, it follows almost exactly the Rutherford curve! Figure 1.15 shows this two-jet angular distribution, together with the Geiger–Marsden data of figure 1.1, suitably rescaled (the scattering angle is now called θ). Perhaps the most remarkable thing about figure 1.15 is the way in which the angular distribution of the two jets, which is a consequence of the primary *strong* interaction between the constituents, is so similar to the Geiger–Marsden distribution—which of course is just 'Rutherford scattering' determined by the *electromagnetic* interaction between the nuclei. This result provides dramatic confirmation of a deep similarity between strong and electromagnetic forces, as already hinted at in figure 1.13. Indeed, the curve 'leading-order QCD' is the appropriate generalisation to the strong interaction case of Rutherford's calculation, as we shall see explicitly in Chapter 9, and it is indistinguishable on this plot from the Rutherford form $\sin^{-4}\theta/2$!

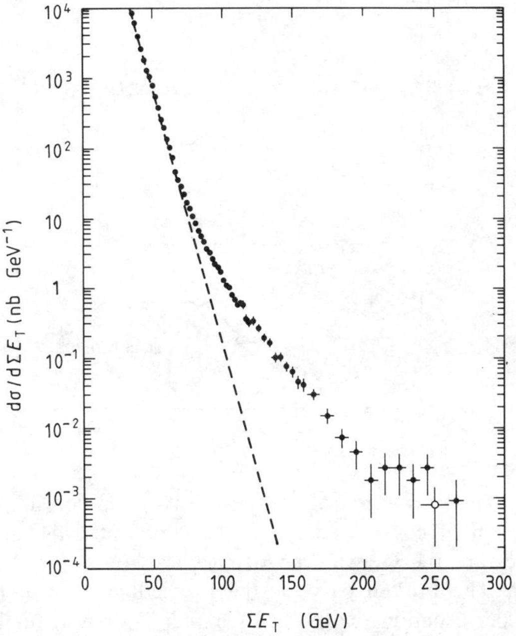

Figure 1.14 Distribution of events versus total transverse energy ΣE_T as observed in the UA2 central detector (DiLella 1986).

Clear evidence for hadronic jets, associated with primary quark processes, was actually first found in a different class of experiments, which is now very common in high-energy physics but which we have so far not mentioned—the head-on collision between an electron e^- and a positron e^+ (the charmonium spectroscopy of figure 1.13 is best studied in such experiments). This particle–antiparticle pair can annihilate, the available energy rematerialising in many forms. In particular a $q\bar{q}$ pair may be formed, possessing very high total energy (say 100 GeV) but concentrated in a very small volume. As the q and $\bar{q}$ fly apart from the interaction region, it seems that the 'confining' forces of QCD come into play, converting them (as in the $\bar{p}p$ case) into two *hadronic* jets. Figure 1.16 shows such a two-jet event in e^+e^- annihilation. When the angular distribution of such hadronic jets is studied, a very simple result is found: it is exactly what is predicted for e^+e^- annihilation into two pointlike spin-$\frac{1}{2}$ quarks! This will be shown in §7.5 below. The hadronic jets seem to remember very well the distribution imprinted on them by the primitive $q\bar{q}$ pair.

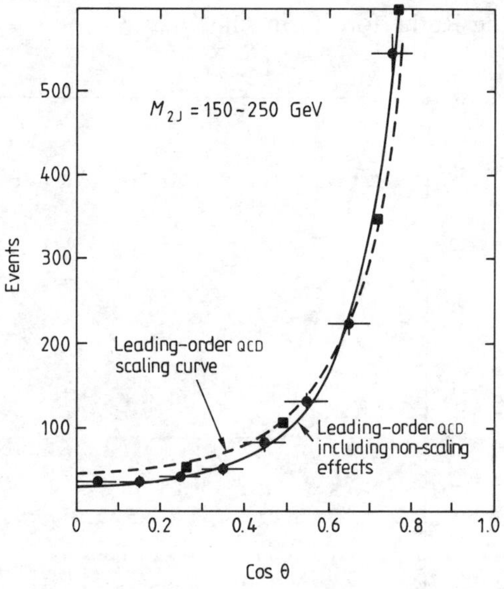

Figure 1.15 Angular distribution of two-jet events in $\bar{p}p$ collisions (Arnison *et al* 1985) as a function of $\cos\theta$, where θ is the CMS scattering angle. The points shown as squares are the Geiger–Marsden values from figure 1.1, scaled by an overall constant. The broken curve is the prediction of QCD, obtained in the lowest order of perturbation theory (see Chapter 9); it is virtually indistinguishable from the Rutherford shape $\sin^{-4}\theta/2$. The full curve includes corrections (Chapter 9).

More evidence that quarks and leptons appear to be on an equal footing can be found in weak interactions, but the discussion is too technical for inclusion here and will be postponed until Chapter 10. Indeed it is time we introduced the leading ideas relevant to the other half of the matter/force story.

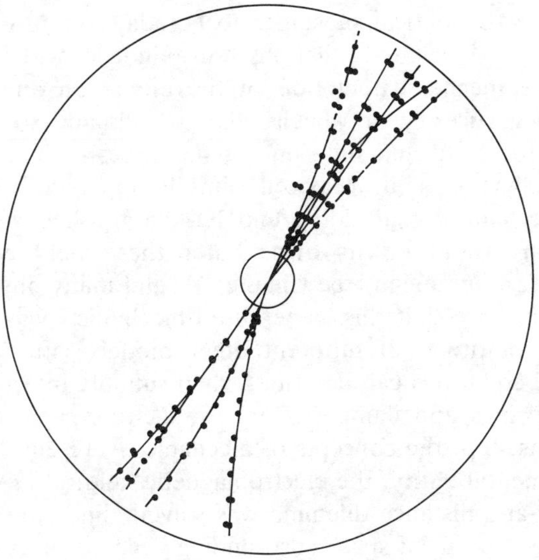

Figure 1.16 Two-jet event from the TASSO detector at the e^+e^- storage ring PETRA.

1.3 The theoretical framework: matter and force in relativistic quantum field theory

In the world of the classical physicist, matter and force were clearly separated. The nature of matter was intuitive, based on everyday macroscopic experience; force, on the other hand, was more problematical. Contact forces between bodies were easy to understand, but forces which seemed capable of acting at a distance caused difficulties. 'That gravity should be innate, inherent and essential to matter, so that one body can act upon another at a distance, through a vacuum, without the mediation of anything else, by and through which action and force may be conveyed from one to the other, is to me so great an absurdity, that I believe no man who has in philosophical matters a competent faculty of thinking can ever fall into it' (letter from Newton to Bentley). Newton could find no satisfactory mechanism, or physical model, for the transmission of the gravitational force between two distant bodies; but his dynamical equations provided a powerful predictive framework, given the (unexplained) gravitational force law; and this eventually satisfied most people.

The nineteenth century saw the precise formulation of the more intricate force laws of electromagnetism. Here too the distaste for action-at-a-distance theories led to numerous mechanical or fluid mechanical models of the way electromagnetic forces (and light) are transmitted. Maxwell made brilliant use of such models as he struggled to give physical and mathematical substance to Faraday's empirical ideas about lines of force. Maxwell's equations were indeed widely regarded as describing the mechanical motion of the ether—an amazing medium, composed of vortices, gear wheels, idler wheels and so on. But in his 1864 paper, the third and final one of the series on lines of force and the electromagnetic field, Maxwell himself appeared ready to throw away the mechanical scaffolding and let the finished structure of the *field equations* stand on its own. Later these field equations were derived from a Lagrangian (see Chapter 4), and many physicists came to agree with Poincaré that this 'generalised mechanics' was more satisfactory than a multitude of different ether models; after all, the same mathematical equations can describe, when suitably interpreted, systems of masses, springs and dampers, or of inductors, capacitors and resistors. With this step, the concepts of mechanics were enlarged to include a new fundamental entity, the electromagnetic field.

The action-at-a-distance dilemma was solved, since the electromagnetic field permeates all of space surrounding charged or magnetic bodies, responds locally to them, and itself acts on other distant bodies, propagating the action to them at the speed of light: for Maxwell's theory, besides unifying electricity and magnetism, also predicted the existence of electromagnetic waves which should travel with the speed of light, as was confirmed by Hertz in 1888. Indeed, light *was* a form of electromagnetic wave.

As we have seen, electrons were isolated, around 1900, as the 'particles of electricity'—the prototype of all 'charged or magnetic bodies'. It seems almost to be implied by the local field concept, and the desire to avoid action at a distance, that the fundamental carriers of electricity should themselves be pointlike, so that the field does not, for example, have to interact with different parts of an electron simultaneously. Thus the pointlike nature of elementary matter units seems intuitively to be tied to the local nature of the force field via which they interact.

Very soon after these triumphs of classical physics, however, another world began to make its appearance—the quantum one. First the photoelectric effect and then—much later—the Compton effect showed unmistakably that these electromagnetic *waves* somehow also had a *particle*-like aspect, the photon. At about the same time, the intuitive understanding of the nature of matter began to fail as well: supposedly *particle*-like things, like electrons, displayed *wave*-like properties (interference and diffraction). Thus the conceptual distinction between matter and forces, or between particle and field, was no longer so clear. On the one hand, electromagnetic forces, treated in terms of fields, now had a particle aspect; and on the other hand, particles now had a wave-like or field aspect. 'Electrons', writes Feynman (1965) at the beginning of Volume 3 of his *Lectures on Physics*, 'behave just like light'.

How can we build a theory of electrons and photons which does justice to all the 'pointlike', 'local', 'wave/particle' ideas just discussed? Consider the apparently quite simple process of spontaneous decay of an excited atomic state in which a photon is emitted:

$$A^* \to A + \gamma. \tag{1.1}$$

Ordinary non-relativistic quantum mechanics cannot provide a first-principles account of this process, because the degrees of freedom it normally discusses are those of the '*matter*' units alone—that is, in this example, the electronic degrees of freedom. However, it is clear that something has changed radically in the *field* degrees of freedom. On the left-hand side, the matter is in an excited state and the electromagnetic field is somehow not manifest; on the right, the matter has made a transition to a lower-energy state and the energy difference has gone into creating a quantum of electromagnetic radiation. In Chapter 4 we shall try to give an elementary and self-contained introduction to the quantum theory of fields. We shall see that a field—which has infinitely many degrees of freedom—can be thought of as somewhat analogous to a vibrating solid (which has merely a very large number). The field can exist in well defined states of definite quantised energy—just like the energy levels of systems of quantum particles. In the process (1.1) the electromagnetic field was originally in its ground state, and was raised finally to an excited state by the transfer of energy from the matter degrees of freedom. The excited field state is described quantum mechanically by the presence of the photon: in the ground state there are no photons.

These ideas will be formalised in Chapter 4. For the present we will be satisfied to give a pictorial representation of this process, which will ultimately be correlated to actual formulae. We know that the atom is *not* elementary, and has spatial extension. The electromagnetic field cannot therefore respond locally to the whole atom, but rather (it is presumed) to one single pointlike source, such as an electron, at a time. Thus the basic process operative in (1.1) may be represented as in figure 1.17. The crucial idea captured by this figure is that the electromagnetic quantum materialises at one single point, which is the position of the electron at a certain instant.

Figure 1.17 cannot be the full story of the process (1.1) however. It is easy to check that such a transition $e^- \to e^- + \gamma$ is impossible for a free electron, since energy and momentum cannot be conserved (problem 1.1(a)). But of course the electron is, in reality, bound inside the atom, and the *change* in the binding energy provides the energy to excite the electromagnetic field and so produce the photon.

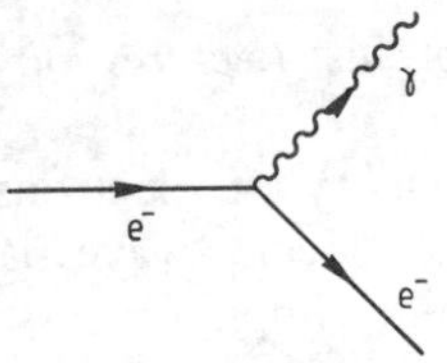

Figure 1.17 Electron radiating a photon by local single quantum emission.

The wave nature of matter has already provided us with one motivation for trying to treat, say, electrons and photons on a roughly equal footing. Another appears as soon as we consider the implications of relativity. As we have already seen, in order to probe matter at the smallest distance scales, experiments must be performed at the highest available energies. At energies much higher than the rest mass energies of the particles involved, the number of particles is not constant, since particle production processes are common, through the conversion of energy into matter via the celebrated $E = mc^2$ relation. For example, the electron beams discussed in the previous sections had energies many orders of magnitude greater than the rest mass energy of the electron (~ 0.5 MeV$/c^2$). Thus the *matter* degrees of freedom also change, in a way which is quite similar to the appearance (or disappearance) of field quanta such as the photon. It seems, therefore, that we must also regard 'matter' particles as quanta of excitation of quantum fields: in this way electrons and photons are described in an analogous way (though there are certainly significant differences—see Chapter 4).

A single electron cannot materialise on its own, because—so it seems—electric charge must always be conserved. But provided energy and momentum conservation laws are satisfied, *pairs* of electrons and positrons—or other particle–antiparticle pairs—can always be produced. The fundamental process by which this occurs is again assumed to be *local* and is represented by figure 1.18. As in the case of figure 1.17, however, energy and momentum cannot be conserved for figure 1.18 as it stands (problem 1.1(b)). Instead it can be regarded as part of a larger process such as figure 1.19 (pair production by γ rays in the vicinity of a recoiling atom) or figure 1.20 (nuclear de-excitation by internal conversion to an e^+e^- pair).

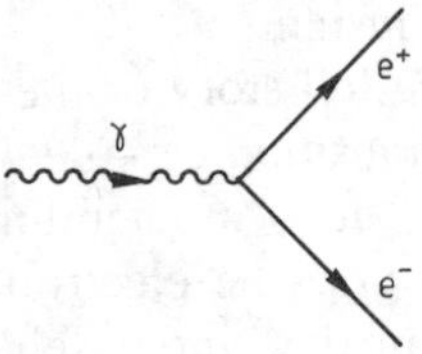

Figure 1.18 Dissociation of a (virtual) photon into an e^+e^- pair.

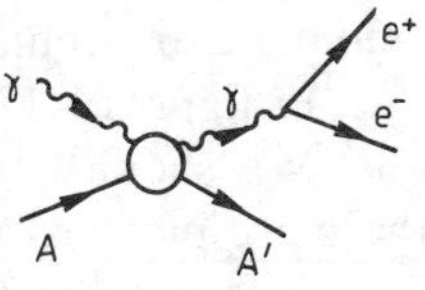

Figure 1.19 Pair production process.

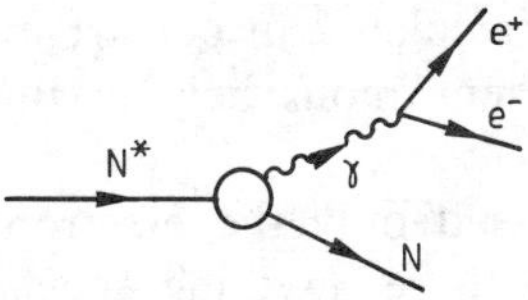

Figure 1.20 Internal conversion.

Although figure 1.17 cannot occur on its own, it is possible to imagine a process in which a single electron *emits* a photon, which is then *absorbed* by a second electron, as shown in figure 1.21. Why should this be allowed while figure 1.17 is not? The answer is provided by a fundamental feature of quantum mechanics. As will be discussed for figure 1.21 explicitly in §5.6, the rules of quantum mechanical perturbation theory (see Appendix A, equation (A.84)) show that quantum amplitudes contain contributions, in general, from subprocesses involving 'virtual' transitions; that is, transitions involving states with energies *different* from the initial (or final) state energy. The only requirement is that the initial and final energies must be the same. Thus the subprocesses $e \rightarrow e + \gamma$ and $e + \gamma \rightarrow e$ can occur as 'virtual' transitions, as in figure 1.21, provided energy is conserved overall. In figure 1.21, then, the two electrons exchange a variable amount of energy and momentum, which is 'carried' from one to the other by the photon. Since, in general, changes of momentum are interpreted as being due to forces, we will infer that the electrons have exerted a force on each other. This interpretation of *force* as being due to *the exchange of a field quantum* is fundamental, and will be discussed in detail for the electromagnetic case in §§5.6 and 5.8.2.

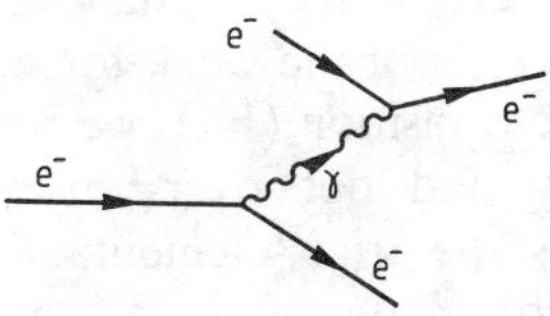

Figure 1.21 Electromagnetic scattering via γ exchange.

This is an appropriate moment to remark that these suggestive diagrams which we have drawn in figures 1.17–1.21 are called 'Feynman diagrams', after their inventor. We should stress that there is far more to Feynman diagrams than the mere pictorial representation of a process (however helpful—or perhaps misleading—this might be). The real power of these diagrams lies in the fact that, as we shall see for the first time in Chapter 5, there is a precise and quantitative correspondence between a given diagram and a specific mathematical expression for a quantum amplitude, as calculated in perturbation theory. The diagrams are thus 'visual aids' of enormous help in thinking about complicated formulae.

So far we have discussed only the electromagnetic force in general quantum field terms. Is there anything special about the structure of electromagnetic theory itself? Indeed there is. As we shall see in Chapter 2, it is a theory which is very strongly constrained by *symmetry* considerations. The particular symmetry is called gauge invariance—and thus it is a *gauge theory*.

What about the other forces? Nuclear β decay proceeds by a process quite analogous to that shown in figure 1.20. The emitted e^- is accompanied by a $\bar{\nu}_e$ instead of by an e^+, as shown in figure 1.22. However, there are many differences of detail between this process and the electromagnetic one of figure 1.20. The field quantum in this case is the W^-, which 'carries' the weak force as the photon carries the electromagnetic one. However, unlike the photon, the W^- is charged (W^+ also exists, mediating positron emission for example), and it is also very massive ($m_W \simeq 82\ \mathrm{GeV}/c^2$). The second property of the W is actually the real reason for the apparent weakness of the weak force. The process shown in figure 1.22 is *highly* 'virtual', since in the subprocess $N(A, Z) \rightarrow W^- + N'(A, Z+1)$ the difference in energy between the initial and final nuclei is of the order of only a few MeV, while the rest mass energy of the W is about 82 GeV! Formula (A.84) shows that the quantum mechanical amplitude for such a virtual two-step process is inversely proportional to the energy difference between the 'intermediate' state and the initial or final state. In the present case the intermediate state includes a W, and this energy difference is huge on a nuclear scale. Hence the amplitude is strongly suppressed, and the rate for the decay is very slow, so that the force causing it appears to be very weak. We shall return in a moment to the question of the 'true' strength of the weak force.

Just as with the atomic transition (1.1), we believe that the weak force field has to act locally and not over the whole nuclear, or even nucleonic, volume. Thus the true 'elementary' process contributing to figure 1.22, or to neutron β decay, is the one shown in figure 1.23. Positron emission occurs via the process shown in figure 1.24. Weak scattering processes analogous to figure 1.21 also exist, for example figure 1.25.

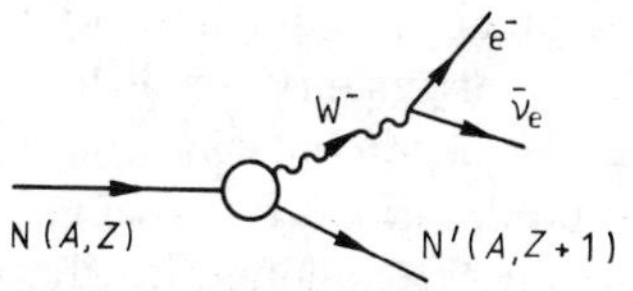

Figure 1.22 Nuclear β decay.

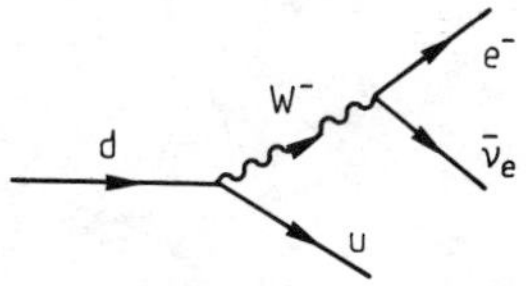

Figure 1.23 Quark β^- decay.

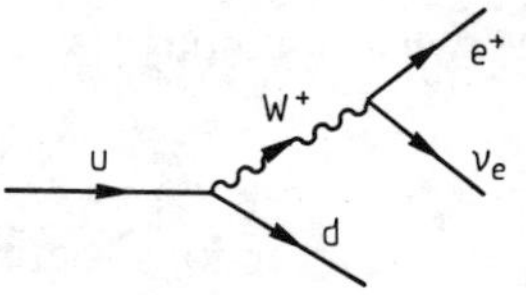

Figure 1.24 Quark β^+ decay.

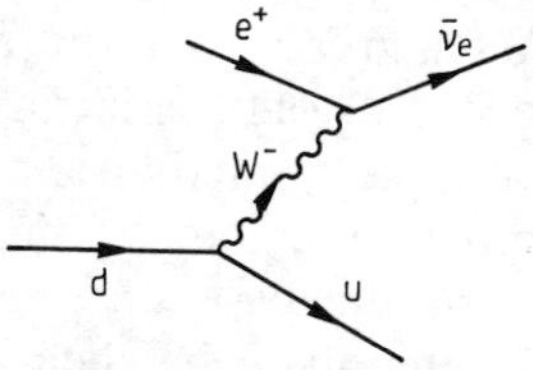

Figure 1.25 Weak scattering via W exchange.

A further type of weak process is also possible, which is mediated by a neutral particle, the Z^0 (mass 93 GeV/c^2), as shown in figure 1.26 for example. A fuller discussion of weak interaction phenomenology is deferred until Chapters 10–12, and the electroweak theory is developed in Chapters 13 and 14.

Despite its apparent 'weakness', the weak force does betray some similarities with electromagnetism. Although the masses of the W and Z quanta are very far from equal to that of the photon, the *spins* of all three quanta are 1. Furthermore, the weak interactions, via the W and Z particles, of quarks and leptons are extremely similar—much as the electromagnetic charges of an e^+ and a proton are, indeed, identical. This property—known as 'universality'—will be discussed in Chapter 11. Chapter 12 provides further arguments indicating that the weak interactions must also, in fact, be described by a gauge theory—but in this case it is of a subtle kind, as will be explained in Chapters 13 and 14. Finally, as we shall also see in Chapter 14, the intrinsic strengths of the weak and electromagnetic forces are actually very similar. All these features are built into the 'electroweak theory'.

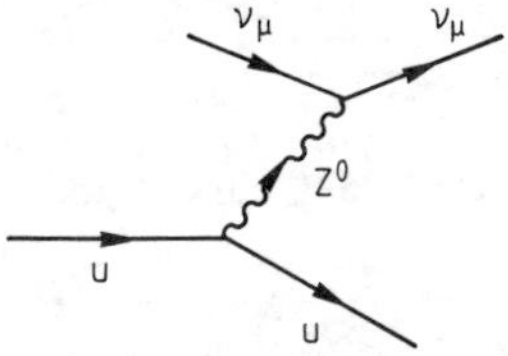

Figure 1.26 Weak scattering via Z^0 exchange.

There remain the strong interactions between quarks. As already remarked in §1.3 (cf figures 1.13 and 1.15), these too appear to be in some way similar to QED. The strong quark interaction (QCD) proceeds by the exchange of force quanta called gluons, which have spin 1 and are believed to be massless. To this extent, gluons are more like photons than are the W's and Z's, but the precise form of the QCD interaction, though analogous to QED, is intrinsically more complicated. QCD will be discussed in Chapters 8 and 9, and for the moment we only want to mention the nature of the complicating factor, which relates to the concept of 'charge'. In the case of electromagnetism, the force between two particles is determined by their charges, which are reckoned in terms of simple multiples of one basic unit of charge (including zero, of course). In QCD the notion of 'charge' becomes considerably extended. The attribute analogous to charge is called 'colour', and it comes in more than one variety (as the name implies). As far as quarks are concerned, each one exists in three 'colour varieties', for example u_r (red up quark), u_b (blue up) and u_g (green up). The QCD force is sensitive to this *colour* label. Indeed, the gluons effectively 'carry colour' back and forth between the quarks, as shown in the basic gluon exchange process of figure 1.27.

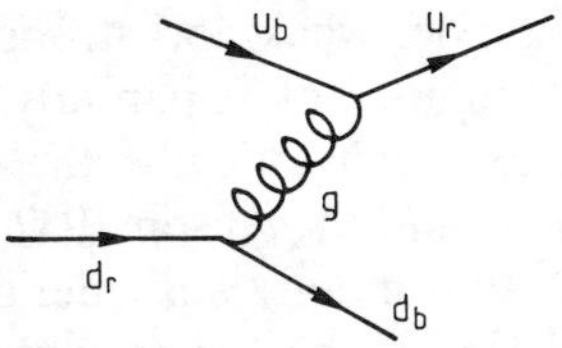

Figure 1.27 Strong scattering via gluon exchange.

Because the gluons carry colour, they can interact with themselves. This is perhaps the single most important difference between QCD and QED: in the latter, the photons—being without electromagnetic charge—do not have the ordinary electromagnetic interaction. It is believed that gluons too are confined by QCD, so that—like quarks—they are not seen as isolated free particles. Nevertheless, they also seem to 'hadronise' after being produced in a primitive short-distance collision process, as we saw happened in the case of q's and $\bar{q}$'s. Such 'gluon jets' provide indirect evidence for the existence and properties of gluons, as we shall see in Chapter 9.

A relatively simple overall picture has therefore begun to emerge. The quarks are the basic source particles of QCD, and they bind together to make hadrons. Like the electron and neutrino, quarks have spin $\frac{1}{2}$. The weak interactions involve *pairs* of quarks and leptons, the *quark doublet* (u, d) and the *lepton doublet* (ν_e, e^-). These appear to be sources for the weak force. This family of four fermions $\{(\nu_e, e^-), (u, d)\}$ constitutes our basic 'matter' set, and actually suffices to account (in principle) for nearly all aggregated matter, from atoms to stars and galaxies. (The corresponding antiparticles are not regarded as new species.) On the 'force' side, the gluons are the force quanta of QCD, and the $W^\pm$ and Z^0 particles are those of weak interactions; all these quanta, like the photon, have spin 1. The weak and strong force fields are generalisations of electromagnetism; all three are examples of gauge theories. (We may note, parenthetically, that gravity is still the odd one out from this viewpoint, since the force quantum in that case—the 'graviton'—is presumed to have spin 2.)

In the following three chapters our aim will be to establish the mathematical formalism—involving quantum mechanics, relativity and field theory—which is necessary to put quantitative calculations alongside all the foregoing words. Before that, a few more words—concerning some unfortunate relatives.

1.4 Unwanted complications

We have presented, in broad outline, an account of the evidence that the fermion family $\{(\nu_e, e^-), (u, d)\}$ contains the 'electron-figures' appropriate to electromagnetic, weak and strong forces. Most curiously, however, a considerable number of apparently rather redundant 'relatives' of this family *also* exist. Historically, the first of the others to be found was the muon (Street and Stevenson 1937, Anderson and Neddermeyer 1937). At first, the muon was identified with the particle postulated by Yukawa (1935) as the field quantum of the 'nuclear force field', the exchange of which between two nucleons would account for their interaction. Yukawa predicted the mass of the exchanged quantum by relating it, qualitatively, to the range of the associated force—a connection which is quite general and of fundamental importance, as we shall discuss further in Chapter 5 (§5.8.2). Yukawa's prediction for the mass was in the range 100–200 MeV†, and the muon was found to have a mass of some 105.7 MeV. Nevertheless, experiments by Conversi *et al* (1947) established that the muon did not interact strongly, as would be required for a quantum of the (strong) nuclear force field. The muon is therefore a lepton. Indeed, it is an 'electron-figure' in an only too literal way: it seems, to all intents and purposes, to behave in exactly the same way as an electron, interacting only electromagnetically and weakly, with basic interaction strengths identical to those of an electron. No one has found a convincing explanation for this duplication—or a 'use' for the muon in the scheme of things, unless muon-induced fusion (Bracci and Fiorentini 1982, Jones 1986) turns out to solve our energy problems.

What of Yukawa's quantum? Also in 1947, Lattes *et al* found evidence for a particle somewhat heavier than the muon and decaying into it. These heavier particles—pions of mass ~ 140 MeV—were found to interact strongly (Perkins 1947, Occhialini and Powell 1947), and accordingly were identified with Yukawa's force carrier. As will have been apparent from the preceding sections, we do not now regard the nuclear force field as a fundamental one, and correspondingly the pion is not a 'pointlike quantum' but a composite hadron. Nevertheless, it is still true that the long-range part of the nucleon–nucleon interaction is usefully described in terms of pion exchange (see §5.8.2 below).

The decay of the pion to the muon has a lifetime of the order 10^{-8} s, characteristic (for such an energy release) of a weak interaction. The emitted muon is accompanied, as in β decay, by a neutrino: for example

$$\pi^- \rightarrow \mu^- + \bar{\nu}_\mu. \tag{1.2}$$

Are these (anti-) neutrinos the same as the ones accompanying the e^- in β decay? An important experiment by Danby *et al* (1962) provided evidence that they are not. Consider the β decay process

$$\mathrm{n} \rightarrow \mathrm{p} + \mathrm{e}^- + \bar{\nu}_\mathrm{e}. \tag{1.3}$$

†Henceforth we shall quote masses in energy units, setting $c = 1$ (see Appendix B).

In a manipulation which may be unfamiliar, but which should become less so as the reader progresses further with us into relativisitic quantum theory, we may infer from the existence of (1.3) the related process

$$e^+ + n \to p + \bar{\nu}_e \tag{1.4}$$

in which we have transferred an *electron* from the *final* state into the *initial* state, converting it into a positron. All the required conservation laws are satisfied by this transaction: figure 1.25 shows a contribution to (1.4) at the fundamental level. We can also infer the existence of the inverse process

$$\bar{\nu}_e + p \to e^+ + n. \tag{1.5}$$

Consider now the $\bar{\nu}_\mu$'s emitted by the π^-'s in flight, according to (1.2). Danby *et al* (1962) found that they produced μ^+'s via

$$\bar{\nu}_\mu + p \to \mu^+ + n \tag{1.6}$$

but not e^+'s via

$$\bar{\nu}_\mu + p \to e^+ + n \quad (\textit{not}\text{ observed}). \tag{1.7}$$

Of course, the 'non-observation' means, in practice, an upper limit on the cross section. However, accepting (1.5)–(1.7), we must conclude that the electron and its partner neutrino on the one hand, and the muon and its neutrino on the other, appear to lead (at least to a good approximation) quite separate lives, and do not mix with each other. Indeed *something* seems to be stopping the μ^- from decaying into the e^- via the process

$$\mu^- \to e^- + \gamma \quad (\textit{not}\text{ observed}). \tag{1.8}$$

This is shown in figure 1.28, which is analogous to figure 1.17. Whereas figure 1.17 is forbidden to occur as a real process by energy–momentum conservation (problem 1.1(a)), there is plenty of energy available for figure 1.28—indeed, given the strong *similarity* between μ^- and e^-, it should confidently be expected to occur! In the same way, only the *pair* production processes (cf figure 1.19)

$$\gamma + A \to e^+ + e^- + A' \tag{1.9}$$

and

$$\gamma + A \to \mu^+ + \mu^- + A' \tag{1.10}$$

are observed, never the 'mixed' one

$$\gamma + A \to \mu^+ + e^- + A' \quad (\textit{not}\text{ observed}) \tag{1.11}$$

indicating the absence of the basic interaction of figure 1.29.

The upshot of this is that we can attach 'electron-like' and 'muon-like' labels to the respective neutrinos, and systematise the observations by the rule that these two types of lepton do not mix. Thus the β decay of the muon itself is

$$\mu^- \to e^- + \bar{\nu}_e + \nu_\mu \tag{1.12}$$

in which the muon number, carried initially by the μ^-, is preserved by the final ν_μ, and the electron numbers of e^- and $\bar{\nu}_e$ cancel each other out.

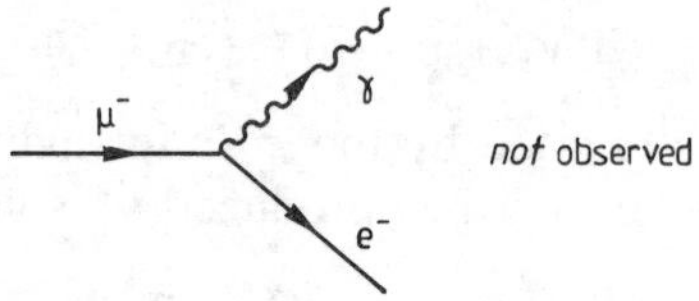

Figure 1.28 Radiative μ decay (*not* observed).

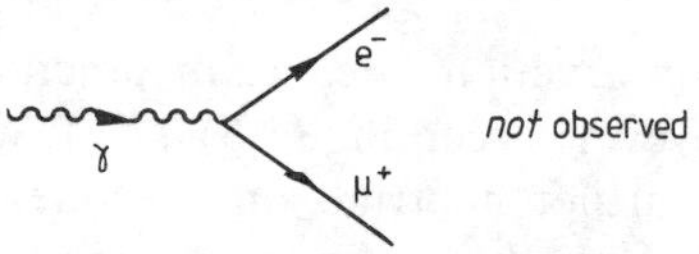

Figure 1.29 Dissociation of a (virtual) photon into a μ^+e^- 'pair' (*not* observed).

We therefore have to accept the existence of a second, separate, lepton doublet (ν_μ, μ^-). Are there more quarks too? For the first family—or as it is called, *generation*—of fermions $\{(\nu_e, e^-), (u, d)\}$ there is already a slight suggestion of 'parallelism' between the lepton and the quark doublets, in that the two members of each doublet have charges differing by one unit. Hopes of an eventual unification of all forces would also tend to imply (see Chapter 15) some kind of symmetry between leptons and quarks. Indeed, there *is* a further quark doublet (c, s), where the 'c' refers to the 'charm' quark (charge $\frac{2}{3}e$) (mentioned in §1.2.4) and the 's' to the 'strange' quark (charge $-\frac{1}{3}e$)—exactly the same charges as the u and d quarks. This new quark doublet completes a second fermion generation $\{(\nu_\mu, \mu^-), (c, s)\}$.

The quark attributes 'charm' and 'strangeness' are somewhat analogous to the empirical quantum numbers already introduced for leptons. The necessity for the strangeness quantum number became clear in the 1950s. In cloud (and, later, bubble) chambers the occurrence of events of the type

$$\pi^- + \mathrm{p} \rightarrow \mathrm{x}^0 + \mathrm{X}^0, \quad \mathrm{X}^0 \rightarrow \pi^- + \mathrm{p}, \quad \mathrm{x}^0 \rightarrow \pi^+ + \pi^- \tag{1.13}$$

was well established. Figure 1.30 shows an example of such a process. The two broken lines on the reconstruction (figure 1.30(*b*)) represent the neutral particles x^0 and X^0 which leave no ionising track in figure 1.30(*a*). They are detected via their decay modes $\mathrm{x}^0 \rightarrow \pi^+ + \pi^-$, $\mathrm{X}^0 \rightarrow \pi^- + \mathrm{p}$. By measuring the momenta of these decay products, the masses of the x^0 and X^0 particles were found to be some 500 MeV and 1100 MeV respectively. One curious feature of these new states was that they were invariably produced in association with each other in such collisions, never singly: for example, the apparently quite possible (indeed, on energy grounds, more favourable) processes

$$\left.\begin{aligned} \pi^- + \mathrm{p} &\rightarrow \mathrm{x}^0 + \mathrm{n} \\ \pi^- + \mathrm{p} &\rightarrow \pi^0 + \mathrm{X}^0 \end{aligned}\right\} \quad (\textit{not}\ \text{observed}) \tag{1.14}$$

were not observed. This is reminiscent of the fact that the *pair* production processes (1.9) and (1.10) are observed, while the 'mixed' process (1.11) is not. Perhaps a pair of new quarks is being created in (1.13), one of which ends up in the x^0 and the other in the X^0. However, there is a major difference between (1.13) and (1.9) or (1.10): the cross section for (1.13) indicated that it was proceeding via the strong interaction, not the electromagnetic one. Nevertheless, we might imagine a mechanism such as figure 1.31 to be responsible, in which the $\bar{\mathrm{u}}$ quark from the π^- *annihilates* with a u quark from the p and produces, via a strong interaction mediated by a gluon, a new pair of quarks, $\mathrm{s}\bar{\mathrm{s}}$. The outgoing x^0 is then a meson with composition ($\mathrm{d}\bar{\mathrm{s}}$), while X^0 is a baryon (sud). Of course, we cannot take figure 1.31 too literally: for example, it totally fails to explain how the s and $\bar{\mathrm{s}}$ become bound into hadrons in the final state. There will, presumably, be all kinds of complicated processes, typically involving many gluons, to produce these bindings. Still, it does embody what is now believed to be the reason for the associated production phenomenon: the quarks evidently carry labels—collectively called 'flavours'—which the strong gluon force field does not mix; that is, processes such as $\mathrm{s} \rightarrow \mathrm{d} + \mathrm{g}$ do not occur. 'Strangeness' (S) is such a label, and the s quark is assigned $S = -1$. The reason for the minus sign is historical: the meson ($\mathrm{d}\bar{\mathrm{s}}$) is here the K^0, assigned $S = +1$ in the empirical Gell-Mann (1953) and Nishijima (1955) classification, which followed the 'associated production' hypothesis of Pais (1952). The X^0 baryon is the Λ^0, assigned strangeness -1.

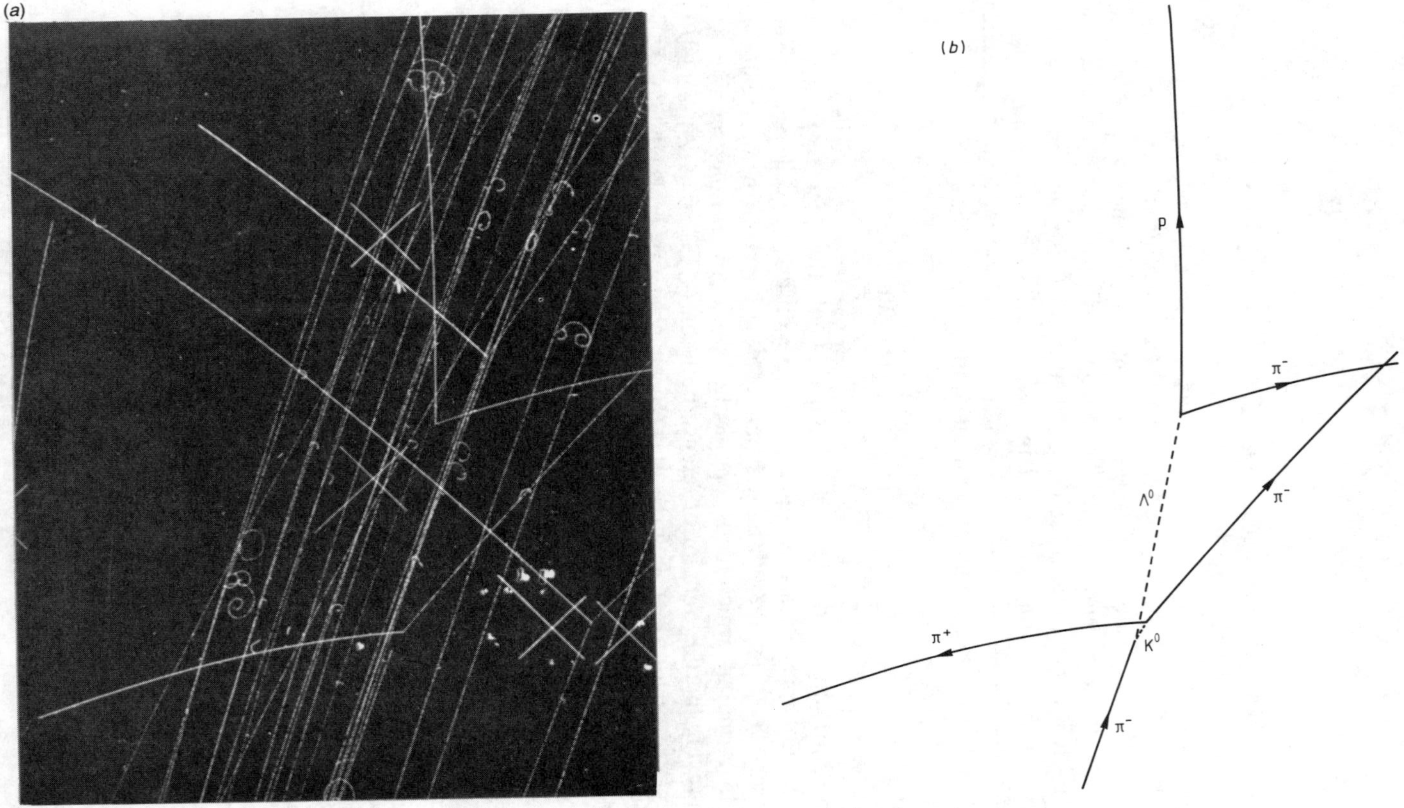

Figure 1.30 Associated production of strange particles.

We can now construct a variety of strange meson states $(q\bar{s})$, $(\bar{q}s)$, and strange baryon states (qqs); these are the various K mesons and hyperons (Λ and Σ particles), together with their excited states—a 'strange' spectroscopy. The strangeness quantum number is conserved in all strong interactions.

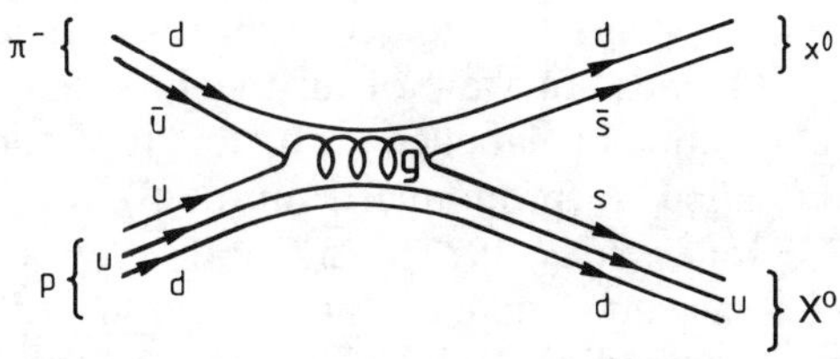

Figure 1.31 Simple quark mechanism for associated production.

Why was it called 'strange'? The answer is apparent from (1.13): although the K^0 and Λ^0 were invariably *produced* in association, they *decayed* quite happily into ordinary known particles which carry $S = 0$! The resolution of this puzzle came with the recognition that the decay processes had lifetimes characteristic of weak interactions, and the idea (novel at the time) that the quantum number S might be conserved for strong interactions but violated for weak interactions: that is, the weak interactions *do* mix $S = 0$ and $S \neq 0$ states. (Note that this implies that reactions (1.14) do occur in principle, but with a very small cross section.) At the quark level, a typical strangeness-changing transition is shown in figure 1.32, which includes the flavour-mixing subprocess $s \to W^- + u$. If the pair (ud) are added as 'spectators' throughout, figure 1.32 will become a mechanism for $\Lambda^0 \to e^-\bar{\nu}_e p$, the (strangeness-changing) β decay of the Λ particle. We shall discuss strangeness-changing weak interactions in more detail in Chapter 11.

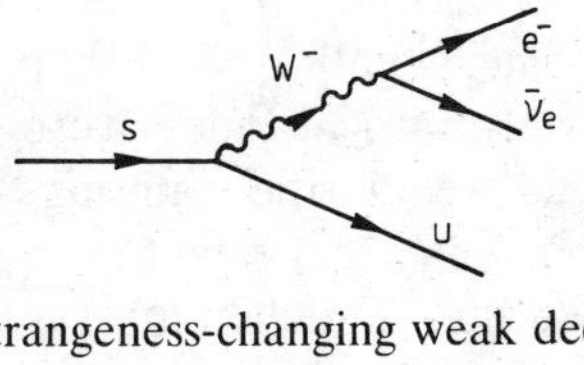

Figure 1.32 Strangeness-changing weak decay of s quark.

'Charm' is another such flavour quantum number, conserved in strong interactions but violated in weak. The meson system $(c\bar{c})$—'charmonium'—has already been mentioned in §1.2 above. This has no net charm, but charmed mesons of the type ($q\bar{c}$) or ($\bar{q}c$) have been identified, as have charmed baryons. We shall have more to say about charm in weak processes in Chapter 11.

This is still not the end of the 'replication of generations'. In 1975 Perl *et al* (1975) discovered evidence for a third 'electron analogue', the τ^- with a mass of 1.78 GeV. Its decay characteristics (in $\tau \rightarrow \mu\nu\bar{\nu}_\mu$, $\tau \rightarrow e\nu\bar{\nu}_e$) indicate the existence of the expected electron partner, ν_τ. The (ν_τ, τ^-) doublet appears to interact in all respects identically to the (ν_e, e^-) and (ν_μ, μ^-) ones. In view of all the foregoing, one expects a corresponding quark doublet, labelled (t, b) for 'top' and 'bottom'. ($b\bar{b}$) states, generically called Y (pronounced 'upsilon'), were found in 1977 (Herb *et al* 1977, Innes *et al* 1977), and both they and the $c\bar{c}$ states continue to be extensively studied. Further details of the $c\bar{c}$ and $b\bar{b}$ spectroscopy are contained in Perkins (1987). At the time of writing there is no firm evidence for the expected t state (charge $+\frac{2}{3}$).

We have been purposely vague about the quark masses. Since quarks do not, apparently, exist as free isolated particles, it is not straightforward to decide how to define their mass. Nevertheless, good descriptions of at least the heavier hadron states (such as $c\bar{c}$ and $b\bar{b}$) are obtained by supposing that the major contribution to the hadronic mass is given by an effective quark mass contributing additively. Thus from the ($c\bar{c}$) states of mass $\sim$ 3 GeV, $m_c \sim 1.5$ GeV, and from the ($b\bar{b}$) ones at $\sim$ 10 GeV, $m_b \sim$ 5GeV, in very rough terms. As with the leptons, the quarks of successive generations appear to get heavier. Estimates of the top quark mass range from 40 to 50 GeV and upwards.

It has to be admitted that our final collection of supposedly basic fermions seems decidedly uneconomical—and there may be more generations ahead (we don't know why there are three, so why shouldn't there be even more?). It is certainly tempting to think of the sequence e^- (mass $\sim$ 0.5 MeV), μ^-($\sim$ 106 MeV) and τ^- ($\sim$ 1.8 GeV), say, as some kind of new spectroscopy. Yet no one has managed to make a convincing model of this. For one thing, all these states have the same spin, which is certainly unlike a conventional excitation spectrum. For another, no γ transitions between the states occur, though this would normally be expected. The situation is perhaps analogous to the early days of atomic physics. Empirical rules were formulated in order to systematise the observed regularities among transitions. These rules were couched in terms of empirical quantum numbers, which only later became understood in terms of such things as angular momentum and parity, with the advent of quantum mechanics. With 'electron-ness' and other lepton quantum numbers, and with 'strangeness' and the other quark flavours, we are still at the purely empirical stage: there is no dynamical understanding of these quantities. Perhaps this is an indication that some quite new dynamical theory will ultimately be needed.

This brings us to our final point, and prepares the way for the following chapter. We must emphasise what appears to be a crucial distinction between attributes such as electromagnetic charge and colour on the one hand, and flavour labels on the other. The former have a direct dynamical significance, whereas the latter do not. In the case of electric charge, for example, this means simply that a particle carrying this property responds in a definite way to the presence of an electromagnetic field, and itself creates such a field. No such force fields are known for any of the lepton numbers, or for the quark flavours; these labels are at present purely empirical classification devices, without dynamical significance. With this point cleared, it is time to leave the cataloguing of particles and begin our study of the forces that act between them, starting with the electromagnetic interaction.

Summary

Matter and forces.

The standard model: leptons and quarks, and the strong, electromagnetic and weak forces.

Levels of matter structure. Scattering and spectroscopy for atoms, nuclei and nucleons.

Evidence implying the existence of quarks.

Forces in quantum field theory. Force quanta and matter quanta.

Forces and the exchange of quanta.

Generations. Flavours.

Flavours versus dynamical quantum numbers.

Problems

1.1 (a) Show that the process (in free space)

$$e^- \rightarrow e^- + \gamma$$

is forbidden by energy–momentum conservation.

(b) Show, similarly, that the process

$$\gamma \rightarrow e^+ + e^-$$

is forbidden in free space.

1.2 It is a remarkable fact (to be commented on further in Chapter 9) that the observed hadronic states are extremely limited as to possible quantum numbers. For example, no doubly charged meson (e.g. 'π^{++}') has ever been discovered. Such a state can *not* be made out of $q\bar{q}$ combinations of known quarks—and hence its absence confirms the simple quark model picture. Suggest

(a) another mesonic state that can *not* be made out of known $q\bar{q}$ combinations, and

(b) a possible baryonic state that can *not* be made out of known qqq combinations;

and for each state in (a), (b), give a reaction in which your proposed state could, if it existed, be produced while respecting standard strong interaction conservation laws (e.g. $\pi^+ p \to$ 'π^{++}' + n).

Examine the evidence for the existence of baryons with $S = +1$.

2

ELECTROMAGNETISM AS A GAUGE THEORY

2.1 Introduction

The previous chapter reviewed briefly some of the main reasons for thinking that the simplest constituents of matter are quarks and leptons. As we have seen, both quarks and leptons appear to be pointlike down to the smallest distance scales currently attainable by the highest-energy particle accelerators. We must now introduce the main concern of this book—namely, the nature of the forces between these 'elementary particles'.

One of the relevant forces—electromagnetism—has been well understood in its classical guise for many years. Over a century ago, Faraday, Maxwell and others developed the theory of the electromagnetic interaction, culminating in Maxwell's paper of 1864 (Maxwell 1864). Today Maxwell's theory still stands—unlike Newton's 'classical mechanics' which was shown by Einstein to require modification at relativistic velocities—speeds approaching the velocity of light. Moreover, Maxwell's electromagnetism, when suitably married to quantum mechanics, gives us, in '*quantum electrodynamics*', or QED, what we call in Part II 'the best theory we have'. As we shall see in Chapter 6, this theory is in truly remarkable agreement with experiment. As we have already indicated, the theories of the weak and strong forces included in the standard model are generalisations of QED, and promise to be as successful as that theory. QED has therefore become the paradigmatic theory.

From today's perspective, the crucial thing about electromagnetism is that it is a theory in which the *dynamics* (i.e. the behaviour of the forces) is intimately related to a *symmetry* principle. In the everyday world, a symmetry operation is something that can be done to an object that leaves the object looking the same after the operation as before. By extension, we may consider mathematical operations—or 'transformations'—applied to the objects in our theory such that the physical laws look the same after the operations as they did before. Such transformations are usually called *invariances* of the laws. Familiar examples are, for instance, the translation and rotation invariance of all fundamental laws: Newton's laws of motion remain valid whether or not we translate or rotate a system of interacting particles. But of course—precisely because they do apply to all laws, classical or quantum—these two invariances have no special connection with any particular force law. Instead, they constrain the form of the allowed laws to a considerable extent, but by no means uniquely determine them. Nevertheless, this line of argument leads one to speculate whether it might in fact be possible to impose further types of symmetry constraints so that the forms of the force laws *are* essentially determined. This would then be one possible answer to the question: why are the force laws the way they are? (Ultimately of course this only replaces one question by another!)

In this chapter we shall discuss electromagnetism from this point of view. This is not the historical route to the theory, but it is the one which generalises to the other two interactions. This is why we believe it important to present the central ideas of this approach in the familiar context of electromagnetism at this early stage.

A distinction that is vital to the understanding of all these interactions is that between a *global* invariance and a *local* invariance. In a global invariance the same transformation is carried out at all space–time points: it has an 'everywhere simultaneously' character. In a local invariance different transformations are carried out at different individual space–time points. In general, as we shall see, a theory that is globally invariant will not be invariant under locally varying transformations. However, by introducing new force fields that interact with the original particles in the theory in a specific way, and which also transform in a particular way under the local transformations, a sort of local invariance can be restored. We will see all these things more clearly when we go into more detail, but the important conceptual point to be grasped is this: one may view these special force fields and their interactions as existing in order to permit certain local invariances to be true. The particular local invariance relevant to electromagnetism is the well known *gauge invariance* of Maxwell's equations: in the quantum form of the theory this property is directly related to an invariance under local *phase transformations* of the quantum fields. A generalised form of this phase invariance also underlies the theories of the weak and strong interactions. For this reason they are all known as 'gauge theories'.

2.2 The Maxwell equations: current conservation

We begin by considering the basic laws of classical electromagnetism, the Maxwell equations. We use a system of units (Heaviside–Lorentz) which is convenient in particle physics (see Appendix C). Before Maxwell's work these laws were (in free space)

$$\nabla \cdot \boldsymbol{E} = \rho_{\text{em}} \quad \text{(Gauss's law)} \tag{2.1}$$

$$\nabla \times \boldsymbol{E} = -\partial \boldsymbol{B}/\partial t \quad \text{(Faraday–Lenz laws)} \tag{2.2}$$

$$\nabla \cdot \boldsymbol{B} = 0 \quad \text{(no magnetic charges)} \tag{2.3}$$

and, for steady currents,

$$\nabla \times \boldsymbol{B} = \boldsymbol{j}_{\text{em}} \quad \text{(Ampère's law).} \tag{2.4}$$

Maxwell noticed that taking the divergence of this last equation leads to conflict with the continuity equation for electric charge

$$\partial \rho_{\text{em}}/\partial t + \nabla \cdot \boldsymbol{j}_{\text{em}} = 0. \tag{2.5}$$

Since

$$\nabla \cdot (\nabla \times \boldsymbol{B}) = 0 \tag{2.6}$$

from (2.4) there follows the result

$$\nabla \cdot \boldsymbol{j}_{\text{em}} = 0. \tag{2.7}$$

This can only be true in situations where the charge density is constant in time. For the general case, Maxwell modified Ampère's law to read

$$\nabla \times \boldsymbol{B} = \boldsymbol{j}_{\text{em}} + \partial \boldsymbol{E}/\partial t \tag{2.8}$$

which is now consistent with (2.5). Equations (2.1)–(2.3), together with (2.8), constitute Maxwell's equations in free space.

The vitally important continuity equation (2.5) states that the rate of decrease of charge in any arbitrary volume Ω is due precisely and only to the flux of current out of its surface; that is, no net charge can be created or destroyed in Ω. Since Ω can be made as small as we please, this means that *electric charge must be locally conserved*: a process in which charge is created at one point and destroyed at a distant one is not allowed, despite the fact that it conserves charge overall, or 'globally'. The ultimate reason for this is that the global form of charge conservation would necessitate the instantaneous propagation of signals (such as 'now, create a positron over there'), and this conflicts with special relativity—a theory which, historically, flowered from the soil of electrodynamics. The extra term introduced by Maxwell—the 'electric displacement current'—owes its place in the dynamical equations to a local conservation requirement.

We remark at this point that we have just introduced another local/global distinction, similar to that discussed above in connection with invariances. In this case the distinction applies to a conservation law—since invariances are related to conservation laws in both classical and quantum mechanics, we should perhaps not be too surprised by this. However, as with invariances, conservation laws—such as charge conservation in electromagnetism—play a central role in gauge theories in that they are closely related to the dynamics. The point is simply illustrated by asking how we could measure the charge of a newly created subatomic particle X. There are two conceptually different ways:

(i) We could arrange for X to be created in a reaction such as

$$\mathrm{A} + \mathrm{B} \rightarrow \mathrm{C} + \mathrm{D} + \mathrm{X}$$

where the charges of A, B, C and D are already known. In this case we can use *charge conservation* to determine the charge of X.

(ii) We could see how particle X responded to known electromagnetic fields. This uses *dynamics* to determine the charge of X.

Either way gives the same answer: it is the conserved charge which determines the particle's response to the field. By contrast, there are several other conservation laws that seem to hold in particle physics, such as lepton number and baryon number, that apparently have no dynamical counterpart (cf the remarks at the end of §1.4). To determine the baryon number of a newly produced particle, we have to use B conservation and tot up the total baryon number on either side of the reaction. As far as we know there is no baryonic force field.

Thus gauge theories are characterised by a close interrelation between *three* conceptual elements: symmetries, conservation laws and dynamics. In fact, it is now widely believed that the *only* exact quantum number conservation laws are those which have an associated gauge theory force field—see comment (i) in §2.6 below. Thus one might suspect that baryon number is not absolutely conserved—as is indeed the case in proposed unified gauge theories of the strong, weak and electromagnetic interactions. In the above discussion we have briefly touched on the connection between two pairs of these three elements: symmetries ↔ dynamics, and conservation laws ↔ dynamics. The precise way in which the remaining link is made—between the symmetry of electromagnetic gauge invariance and the conservation law of charge—is more technical. We will discuss this connection with the help of simple ideas from quantum field theory in Chapter 4. For the present we continue with our study of the Maxwell equations and, in particular, of the gauge invariance they exhibit.

2.3 The Maxwell equations: gauge invariance

In classical electromagnetism, and especially in quantum mechanics, it is convenient to introduce the vector potential $A_\mu(x)$ in place of the fields $\boldsymbol{E}$ and $\boldsymbol{B}$. We write

$$\boldsymbol{B} = \boldsymbol{\nabla} \times \boldsymbol{A} \tag{2.9}$$

$$\boldsymbol{E} = -\boldsymbol{\nabla} V - \partial \boldsymbol{A}/\partial t \tag{2.10}$$

which defines the 3-vector potential $\boldsymbol{A}$ and the scalar potential V. With these definitions, equations (2.2) and (2.3) are then automatically satisfied.

The origin of gauge invariance in classical electromagnetism lies in the fact that the potentials $\boldsymbol{A}$ and V are not unique for given physical fields $\boldsymbol{E}$ and $\boldsymbol{B}$. The transformations which $\boldsymbol{A}$ and V may undergo while preserving $\boldsymbol{E}$ and $\boldsymbol{B}$ (and hence the Maxwell equations) unchanged are called gauge transformations, and the associated invariance of the Maxwell equations is called gauge invariance.

What are these transformations? Clearly $\boldsymbol{A}$ can be changed by

$$\boldsymbol{A} \rightarrow \boldsymbol{A}' = \boldsymbol{A} + \boldsymbol{\nabla}\chi \tag{2.11}$$

where χ is an arbitrary function, with no change in $\boldsymbol{B}$ since curl grad $\equiv 0$. To preserve $\boldsymbol{E}$, V must then change simultaneously by

$$V \rightarrow V' = V - \partial\chi/\partial t. \tag{2.12}$$

These transformations can be combined into a single compact equation by introducing the 4-vector potential†

$$A^{\mu} \equiv (V, \boldsymbol{A}) \tag{2.13}$$

and noting (from problem 2.1) that the differential operators $(\partial/\partial t, -\boldsymbol{\nabla})$ form the components of a 4-vector ∂^{μ}.

A gauge transformation is then specified by

$$\boxed{A^{\mu} \rightarrow A'^{\mu} = A^{\mu} - \partial^{\mu}\chi.} \tag{2.14}$$

The Maxwell equations can also be written in a manifestly covariant form† using the 4-current $j^{\mu}_{\text{em}}(x)$ given by

$$j^{\mu}_{\text{em}} = (\rho_{\text{em}}, \boldsymbol{j}_{\text{em}}) \tag{2.15}$$

in terms of which the continuity equation takes the form (problem 2.1)

$$\partial_{\mu} j^{\mu}_{\text{em}} = 0. \tag{2.16}$$

The Maxwell equations (2.1) and (2.8) then become (problem 2.2)

$$\partial_{\mu} F^{\mu\nu} = j^{\nu}_{\text{em}} \tag{2.17}$$

where we have defined the field strength tensor

$$F^{\mu\nu} \equiv \partial^{\mu} A^{\nu} - \partial^{\nu} A^{\mu}. \tag{2.18}$$

Since $F^{\mu\nu}$ is obviously invariant under the gauge transformation

$$A^{\mu} \rightarrow A'^{\mu} = A^{\mu} - \partial^{\mu}\chi \qquad (2.14)$$

the Maxwell equations in this form are manifestly gauge invariant. The

†See Appendix A.2 for relativistic notation.

'covariant field equations' satisfied by A^μ then follow from equations (2.17) and (2.18):

$$\Box A^\nu - \partial^\nu(\partial_\mu A^\mu) = j^\nu_{\text{em}}. \tag{2.19}$$

Since gauge transformations turn out to be of central importance in the quantum theory of electromagnetism, it would be nice to have some insight into why Maxwell's equations are gauge invariant. The all-important 'fourth' equation (2.8) was inferred by Maxwell from local charge conservation, as expressed by the continuity equation

$$\partial_\mu j^\mu_{\text{em}} = 0. \tag{2.16}$$

The field equation

$$\partial_\mu F^{\mu\nu} = j^\nu_{\text{em}} \tag{2.17}$$

then of course automatically embodies (2.16). The mathematical reason it does so is that $F^{\mu\nu}$ is a four-dimensional kind of 'curl'

$$F^{\mu\nu} = \partial^\mu A^\nu - \partial^\nu A^\mu \tag{2.18}$$

which is obviously unchanged by a gauge transformation

$$A^\mu \rightarrow A'^\mu = A^\mu - \partial^\mu \chi. \tag{2.14}$$

Hence there is the suggestion that the gauge invariance is related in some way to charge conservation. However, the connection is not so simple. Wigner (1949) has given a simple argument to show that the principle that no physical quantity can depend on the absolute value of the electrostatic potential, when combined with energy conservation, implies the conservation of charge. Wigner's argument relates charge (and energy) conservation to an invariance under transformation of the electrostatic potential by a constant: charge conservation alone does not seem to require the more general space–time-dependent transformation of gauge invariance.

Changing the value of the electrostatic potential by a constant amount is an example of what we have called a *global* transformation (since the change in the potential is the same everywhere). Invariance under this global transformation is related to a conservation law: that of charge. But this global invariance is not sufficient to generate the full Maxwellian dynamics. However, as remarked by 't Hooft (1980), one can regard equations (2.11) and (2.12) as expressing the fact that a *local* change in the electrostatic potential V (the $\partial\chi/\partial t$ term in (2.12)) can be compensated—in the sense of leaving the Maxwell equations unchanged—by a corresponding local change in the magnetic vector potential $\boldsymbol{A}$. Thus by including magnetic effects, the global invariance under a change of V by a constant can be extended to a local invariance (which is a much more restrictive condition to satisfy). Hence there is the beginning of a suggestion that one might almost 'derive' the complete Maxwell equations, which unify electricity and magnetism, from the requirement that the theory be expressed in terms of potentials in such a way as to be invariant under local (gauge) transformations on those potentials. Certainly special relativity must play a role too: this also links electricity and magnetism, via the magnetic effects of charges as seen by an observer moving relative to them. If a 4-vector potential A^{μ} is postulated, and it is then demanded that the theory involve it only in a way which is insensitive to local changes of the form (2.14), one is led naturally to the idea that the physical fields enter only via the quantity $F^{\mu\nu}$, which is invariant under (2.14). From this, one might conjecture the field equation on grounds of Lorentz covariance.

It goes without saying that this is certainly not a 'proof' or 'derivation' of the Maxwell equations. Nevertheless, the idea that *dynamics* (in this case, the complete interconnection of electric and magnetic effects) may be intimately related to a *local invariance requirement* (in this case, electromagnetic gauge invariance) turns out to be a fruitful one. As indicated in §2.1 above, it is generally the case that, when a certain global invariance is generalised to a local one, the existence of a new 'compensating' field is entailed, interacting in a specified way. The first example of a dynamical theory 'derived' from a local invariance requirement seems to be the theory of Yang and Mills (1954) (see also Shaw 1955), from whose paper a crucial quotation appears at the start of Part III of this book. Their work was extended by Utiyama (1956), who developed a general formalism for such compensating fields. As we have said, these types of dynamical theories, based on local invariance principles, are called gauge theories.

It is a remarkable fact that all the interactions currently regarded as fundamental are of precisely this type. We have briefly discussed the Maxwell equations in this light, and will continue with (quantum) electrodynamics in the following two sections. Another example, but one which we shall not pursue in this book, is that of general relativity (the theory of the gravitational interaction). Utiyama (1956) showed that this theory could be arrived at by generalising the global (space–time-independent) coordinate transformations of special relativity to local ones; as with electromagnetism, the more restrictive local invariance requirements entailed the existence of a new field—the gravitational one—with an (almost) prescribed form of interaction. The two other fundamental interactions—the strong interaction between quarks, and the weak interaction between quarks and leptons—also seem to be described by gauge theories (of essentially the Yang–Mills type), as we shall see in detail in Parts III and V of this book.

In order to proceed further, we must now discuss how such ideas are incorporated into quantum mechanics.

2.4 Gauge invariance in quantum mechanics

The Lorentz force law for a non-relativistic particle of charge q moving with velocity $\boldsymbol{v}$ under the influence of both electric and magnetic fields is

$$\boldsymbol{F} = q\boldsymbol{E} + q\boldsymbol{v} \times \boldsymbol{B}. \tag{2.20}$$

It may be derived, via Hamilton's equations, from the classical Hamiltonian†

$$H = (1/2m)(\boldsymbol{p} - q\boldsymbol{A})^2 + qV. \tag{2.21}$$

The Schrödinger equation for such a particle in an electromagnetic field is

$$\left(\frac{1}{2m}(-\mathrm{i}\boldsymbol{\nabla} - q\boldsymbol{A})^2 + qV\right)\psi(\boldsymbol{x}, t) = \mathrm{i}\frac{\partial\psi(\boldsymbol{x}, t)}{\partial t} \tag{2.22}$$

which is obtained from the classical Hamiltonian by the usual prescription, $\boldsymbol{p} \to -\mathrm{i}\boldsymbol{\nabla}$, for Schrödinger's wave mechanics ($\hbar = 1$). Notice the appearance of the operator combinations

$$\boxed{\begin{aligned} \boldsymbol{D} &\equiv \boldsymbol{\nabla} - \mathrm{i}q\boldsymbol{A} \\ D^0 &\equiv \partial/\partial t + \mathrm{i}qV \end{aligned}} \tag{2.23}$$

in place of $\boldsymbol{\nabla}$ and $\partial/\partial t$, in going from the free particle Schrödinger equation to the electromagnetic field case.

The solution of $\psi(\boldsymbol{x}, t)$ of the Schrödinger equation (2.22) describes completely the state of the particle moving under the influence of the potentials V, $\boldsymbol{A}$. However, these potentials are not unique, as we have already seen: they can be changed by a gauge transformation

$$\boldsymbol{A} \to \boldsymbol{A}' = \boldsymbol{A} + \boldsymbol{\nabla}\chi \tag{2.11}$$

$$V \to V' = V - \partial\chi/\partial t \tag{2.12}$$

and the Maxwell equations for the fields $\boldsymbol{E}$ and $\boldsymbol{B}$ will remain the same. This immediately raises a serious question: if we carry out such a change of potentials in equation (2.22), will the solution $\psi'(\boldsymbol{x}, t)$ of the resulting equation

$$\left(\frac{1}{2m}(-\mathrm{i}\boldsymbol{\nabla} - q\boldsymbol{A}')^2 + qV'\right)\psi'(\boldsymbol{x}, t) = \mathrm{i}\frac{\partial\psi'(\boldsymbol{x}, t)}{\partial t} \tag{2.22a}$$

describe the same physics as the solution $\psi(\boldsymbol{x}, t)$ of equation (2.22)? If it does, we shall be able to assume the validity of Maxwell's theory for the

†We set $\hbar = c = 1$ throughout (see Appendix B).

quantum world; if not, some modification will be necessary, since the gauge symmetry possessed by the Maxwell equations will be violated in the quantum theory.

Since we know the relations (2.11) and (2.12) between $\boldsymbol{A}, V$ and $\boldsymbol{A}', V'$, we can actually find out what $\psi'(\boldsymbol{x}, t)$ must be in order that equation (2.22a) be consistent with (2.22). We shall state the answer and then verify it; then we shall discuss the physical interpretation. The required $\psi'(\boldsymbol{x}, t)$ is

$$\psi'(\boldsymbol{x}, t) = \exp[\mathrm{i}q\chi(\boldsymbol{x}, t)]\psi(\boldsymbol{x}, t) \tag{2.24}$$

where χ is the same space–time-dependent function as appears in equations (2.11) and (2.12). To verify this we consider

$$\begin{aligned}(-\mathrm{i}\boldsymbol{\nabla} - q\boldsymbol{A}')\,\psi' &= [-\mathrm{i}\boldsymbol{\nabla} - q\boldsymbol{A} - q(\boldsymbol{\nabla}\chi)]\,[\exp(\mathrm{i}q\chi)\psi] \\ &= q(\boldsymbol{\nabla}\chi)\exp(\mathrm{i}q\chi)\psi + \exp(\mathrm{i}q\chi)\cdot(-\mathrm{i}\boldsymbol{\nabla}\psi) \\ &\quad + \exp(\mathrm{i}q\chi)\cdot(-q\boldsymbol{A}\psi) - q(\boldsymbol{\nabla}\chi)\exp(\mathrm{i}q\chi)\psi.\end{aligned} \tag{2.25}$$

The first and last terms cancel leaving the result

$$(-\mathrm{i}\boldsymbol{\nabla} - q\boldsymbol{A}')\psi' = \exp(\mathrm{i}q\chi)\cdot(-\mathrm{i}\boldsymbol{\nabla}-q\boldsymbol{A})\psi \tag{2.26}$$

which may be written using equation (2.23) as

$$(-\mathrm{i}\boldsymbol{D}'\psi') = \exp(\mathrm{i}q\chi)\cdot(-\mathrm{i}\boldsymbol{D}\psi). \tag{2.27a}$$

Thus, although the space–time-dependent phase factor feels the action of the gradient operator $\boldsymbol{\nabla}$, it 'passes through' the combined operator $\boldsymbol{D}'$ and converts it into $\boldsymbol{D}$: in fact, comparing equations (2.24) and (2.27a), we see that $\boldsymbol{D}'\psi'$ bears to $\boldsymbol{D}\psi$ exactly the same relation as ψ' bears to ψ. In just the same way we find (cf equation 2.23))

$$(\mathrm{i}D'^0\psi') = \exp(\mathrm{i}q\chi)\cdot(\mathrm{i}D^0\psi) \tag{2.27b}$$

where we have used equation (2.12) for V'. Once again, $D'^0\psi'$ is simply related to $D^0\psi$. Repeating the operation which led to equation (2.27a), we find

$$\begin{aligned}(1/2m)\,(-\mathrm{i}\boldsymbol{D}')^2\psi' &= \exp(\mathrm{i}q\chi)\cdot(1/2m)\,(-\mathrm{i}\boldsymbol{D})^2\psi \\ &= \exp(\mathrm{i}q\chi)\cdot\mathrm{i}D^0\psi \quad \text{(using equation (2.22))} \\ &= \mathrm{i}D'^0\psi' \quad \text{(using equation (2.27b))}.\end{aligned} \tag{2.28}$$

Equation (2.28) is just (2.22a) written in the D notation of equation (2.23), so we have verified that (2.24) is the correct relationship between ψ' and ψ to ensure consistency between equations (2.22) and (2.22a).

Do ψ and ψ' describe the same physics? The answer is yes, but it is not quite trivial. It is certainly obvious that the probability densities $|\psi|^2$

and $|\psi'|^2$ are equal, since in fact ψ and ψ' in equation (2.24) are related by a *phase* transformation. However, we can be interested in other observables involving the derivative operators $\boldsymbol{\nabla}$ or $\partial/\partial t$—for example, the current, which is essentially $\psi^*(\boldsymbol{\nabla}\psi) - (\boldsymbol{\nabla}\psi)^*\psi$. It is easy to check that this is *not* invariant under (2.24), because the phase $\chi(\boldsymbol{x}, t)$ is $\boldsymbol{x}$-dependent. But equations (2.27a) and (2.27b) show us what we must do to construct *gauge invariant currents*: namely, we must replace $\boldsymbol{\nabla}$ by $\boldsymbol{D}$ (and in general also $\partial/\partial t$ by D^0) since then

$$\psi^{*\prime}(\boldsymbol{D}'\psi') = \psi^* \exp(-\mathrm{i}q\chi)\cdot\exp(\mathrm{i}q\chi)\cdot(\boldsymbol{D}\psi) = \psi^*\boldsymbol{D}\psi \qquad (2.29)$$

for example. Thus the identity of the physics described by ψ and ψ' is indeed ensured.

We summarise these important considerations by the statement that the gauge invariance of Maxwell's equations remains an invariance in quantum mechanics provided we make the combined transformation

$$\boxed{\begin{aligned} \boldsymbol{A} &\rightarrow \boldsymbol{A}' = \boldsymbol{A} + \boldsymbol{\nabla}\chi \\ V &\rightarrow V' = V - \partial\chi/\partial t \\ \psi &\rightarrow \psi' = \exp(\mathrm{i}q\chi)\psi \end{aligned}} \qquad (2.30)$$

on the potentials and on the wavefunction.

The Schrödinger equation is non-relativistic, but the Maxwell equations are of course fully relativistic. One might therefore suspect that the prescriptions discovered here are actually true relativistically as well, and this is indeed the case. We shall introduce the spin-0 and spin-$\frac{1}{2}$ relativistic wave equations in Chapter 3. For the present we note that (2.23) can be written in manifestly covariant form as

$$\boxed{D^\mu \equiv \partial^\mu + \mathrm{i}qA^\mu} \qquad (2.31)$$

in terms of which (2.27a) and (2.27b) become

$$-\mathrm{i}D'^\mu\psi' = \exp(\mathrm{i}q\chi)\cdot(-\mathrm{i}D^\mu\psi). \qquad (2.32)$$

It follows that any wave equation involving the operator ∂^μ can be made gauge invariant under the combined transformation

$$A^\mu \rightarrow A'^\mu = A^\mu - \partial^\mu\chi$$
$$\psi \rightarrow \psi' = \exp(\mathrm{i}q\chi)\psi$$

if ∂^μ is replaced by D^μ. In fact, we seem to have a very simple prescription for obtaining the wave equation for a particle in the presence of an electromagnetic field from the corresponding *free particle* wave equation: make the replacement

$$\boxed{\partial^\mu \to D^\mu \equiv \partial^\mu + iqA^\mu.} \tag{2.33}$$

In the following section this will be seen to be the basis of the so-called 'gauge principle' whereby, in accordance with the idea advanced in the previous sections, the form of *interaction* is determined by the insistence on (local) gauge invariance.

One final remark: this new kind of derivative

$$D^\mu \equiv \partial^\mu + iqA^\mu \tag{2.31}$$

turns out to be of fundamental importance—it will be the operator which generalises from the (Abelian) phase symmetry of QED (see comment (iii) of §2.6) to the (non-Abelian) phase symmetries of our weak and strong interaction theories. It is called the *'covariant derivative'*.

2.5 The argument reversed: the gauge principle

In the preceding section, we took it as *known* that the Schrödinger equation, for example, for a charged particle in an electromagnetic field, has the form

$$[(1/2m)(-i\boldsymbol{\nabla} - q\boldsymbol{A})^2 + qV]\psi = i\partial\psi/\partial t. \tag{2.22}$$

We then checked its gauge invariance under the combined transformation

$$\begin{aligned} \boldsymbol{A} &\to \boldsymbol{A}' = \boldsymbol{A} + \boldsymbol{\nabla}\chi \\ V &\to V' = V - \partial\chi/\partial t \\ \psi &\to \psi' = \exp(iq\chi)\psi. \end{aligned} \tag{2.30}$$

We now want to reverse the argument: we shall start by demanding that our theory is invariant under the *space–time-dependent phase transformation*

$$\psi(\boldsymbol{x}, t) \to \psi'(\boldsymbol{x}, t) = \exp[iq\chi(\boldsymbol{x}, t)]\psi(\boldsymbol{x}, t). \tag{2.34}$$

We shall demonstrate that such a phase invariance is not possible for a free theory, but rather requires an *interacting* theory, involving a (4-vector) field whose interactions with the charged particle are precisely determined, and which undergoes the transformation

$$\boldsymbol{A} \to \boldsymbol{A}' = \boldsymbol{A} + \boldsymbol{\nabla}\chi \tag{2.11}$$

$$V \to V' = V - \partial\chi/\partial t \tag{2.12}$$

when $\psi \rightarrow \psi'$. The demand of this type of phase invariance will then have dictated the form of the interaction—this is the basis of the *gauge principle*.

We therefore focus attention on the phase of the wavefunction. The absolute phase of a wavefunction in quantum mechanics cannot be measured; only relative phases are measurable, via some sort of interference experiment. A simple example is provided by the diffraction of particles by a two-slit system. Downstream from the slits, the wavefunction is a coherent superposition of two components, one originating from each slit: symbolically,

$$\psi = \psi_1 + \psi_2. \tag{2.35}$$

The probability distribution $|\psi|^2$ will then involve, in addition to the separate intensities $|\psi_1|^2$ and $|\psi_2|^2$, the *interference* term

$$2\mathrm{Re}(\psi_1^* \psi_2) = 2|\psi_1||\psi_2|\cos\delta$$

where δ $(= \delta_1 - \delta_2)$ is the *phase difference* between components ψ_1 and ψ_2. The familiar pattern of alternating intensity maxima and minima is then attributed to variation in the phase difference δ. Where the components are in phase, the interference is constructive and $|\psi|^2$ has a maximum; where they are out of phase, it is destructive and $|\psi|^2$ has a minimum. It is clear that if the individual phases δ_1 and δ_2 are each shifted by the same amount, there will be no observable consequences, since only the phase difference δ enters.

The situation in which a wavefunction can be changed in a certain way without leading to any observable effects is precisely what is entailed by a symmetry or invariance principle in quantum mechanics. In the case under discussion, the invariance is that of a constant overall change in phase. In performing calculations it is necessary to make some definite choice of phase; that is, to adopt a 'phase convention'. The actual value chosen is irrelevant, as is guaranteed by the invariance principle, but some choice has to be made.

Invariance under a constant change in phase is an example of a *global* invariance, according to the terminology introduced in the previous section. We make this point quite explicit by writing out the transformation as

$$\boxed{\begin{aligned} \psi \rightarrow \psi' &= \mathrm{e}^{\mathrm{i}\alpha}\psi \\ \alpha &= \text{constant} \end{aligned}} \quad \text{global phase invariance.} \tag{2.36}$$

That α in (2.36) is a constant, the same for all space–time points, expresses the fact that once a phase convention (choice of α) has been made at one space–time point, the same convention must be adopted at all other points. Thus in the two-slit experiment we are not free to make a *local* change of phase: for example, as discussed by 't Hooft (1980), inserting a half-wave plate behind just one of the slits will certainly have observable consequences.

There is a sense in which this may seem an unnatural state of affairs (cf the quotation from Yang and Mills cited at the start of Part III). Once a phase convention has been adopted at one space–time point, the same convention must be adopted at all other ones: the half-wave plate must extend instantaneously across all of space, or not at all. Following this line of thought, one might then be led to 'explore the possibility' of requiring invariance under *local* phase transformations; that is, independent choices of phase convention at each space–time point. By itself, the foregoing is not a compelling motivation for such a step. However, as we pointed out in §2.3, such a move from a global to a local invariance is apparently of crucial significance in classical electromagnetism and general relativity, and seems now to provide the key to an understanding of elementary particle interactions. Let us see, then, where the demand of '*local* phase invariance'

$$\boxed{\psi(\boldsymbol{x},t) \rightarrow \psi'(\boldsymbol{x},t) = \exp[\mathrm{i}\alpha(\boldsymbol{x},t)]\psi(\boldsymbol{x},t)} \quad \text{local phase invariance} \qquad (2.37)$$

leads us.

There is immediately a problem: this is *not* an invariance of the free particle Schrödinger equation or of any relativistic wave equation! For example, if the original wavefunction $\psi(\boldsymbol{x}, t)$ satisfied the free particle Schrödinger equation

$$(1/2m)(-\mathrm{i}\boldsymbol{\nabla})^2\psi(\boldsymbol{x}, t) = \mathrm{i}\partial\psi(\boldsymbol{x}, t)/\partial t$$

then the wavefunction ψ', given by the local phase transformation above, will not, since both $\boldsymbol{\nabla}$ and $\partial/\partial t$ now act on $\alpha(\boldsymbol{x}, t)$ in the phase factor. Thus local phase invariance is not an invariance of the free particle wave equation. If we wish to satisfy the demands of local phase invariance, we are obliged to modify the free particle Schrödinger equation into something for which there is a local phase invariance. But this modified equation will no longer describe a free particle: in other words, the freedom to alter the phase of a charged particle's wavefunction locally is only possible if some kind of force field is introduced in which the particle moves. In more physical terms, the invariance will now be manifested in the inability to distinguish observationally between the effect of making a local change in phase convention and the effect of some new field in which the particle moves.

What kind of field will this be? In fact, we know immediately what the answer is, since the local phase transformation

$$\psi \rightarrow \psi' = \exp[i\alpha(\boldsymbol{x}, t)]\psi \tag{2.37}$$

with $\alpha = q\chi$ is just the phase transformation associated with electromagnetic gauge invariance! Thus we must modify the Schrödinger equation

$$(1/2m)(-i\nabla)^2\psi = i\partial\psi/\partial t \tag{2.38}$$

to

$$(1/2m)(-i\nabla - q\boldsymbol{A})^2\psi = (i\partial/\partial t - qV)\psi$$

and satisfy the local phase invariance

$$\psi \rightarrow \psi' = \exp[i\alpha(\boldsymbol{x}, t)]\psi$$

by demanding that $\boldsymbol{A}$ and V transform by

$$\begin{aligned} \boldsymbol{A} &\rightarrow \boldsymbol{A}' = \boldsymbol{A} + q^{-1}\nabla\alpha \\ V &\rightarrow V' = V - q^{-1}\partial\alpha/\partial t \end{aligned} \tag{2.39}$$

when $\psi \rightarrow \psi'$. But the modified wave equation is of course precisely the Schrödinger equation describing the interaction of the charged particle with the electromagnetic field described by $\boldsymbol{A}$ and V.

In a covariant treatment, $\boldsymbol{A}$ and V will be regarded as parts of a 4-vector A^μ, just as $-\nabla$ and $\partial/\partial t$ are parts of ∂^μ. Thus the presence of the vector field A^μ, interacting in a 'universal' prescribed way with any particle of charge q, is dictated by local phase invariance. A vector field such as A^μ, introduced in order to guarantee local phase invariance, is called a 'gauge field'. The principle that the interaction should be so dictated by the phase (or gauge) invariance is called the *gauge principle*: it allows us to write down the wave equation for the interaction directly from the free particle equation†. As before, the method clearly generalises to the four-dimensional case.

2.6 Comments on the gauge principle in electromagnetism

(i) A properly sceptical reader may have detected an important sleight of hand in the discussion above. Where exactly did the electromagnetic charge appear from? The trouble with our argument as so far presented is that we could have defined fields $\boldsymbol{A}$ and V so that they coupled equally to all particles—instead we smuggled in a factor q.

Actually we can do a bit better than this. We can use the fact that the electromagnetic charge is absolutely conserved, to claim that there can

†Actually, the electromagnetic interaction is uniquely specified by this procedure only for particles of spin 0 and spin $\frac{1}{2}$; see §8.4 for the case of spin 1.

be no quantum mechanical interference between states of different charge q. Hence different phase changes are allowed within each 'sector' of definite q:

$$\psi' = \exp(\mathrm{i}q\chi)\psi \tag{2.40}$$

let us say. When this becomes a local transformation, $\chi \to \chi(\boldsymbol{x}, t)$, we shall need to cancel a term $q\boldsymbol{\nabla}\chi$, which will imply the presence of a '$-q\boldsymbol{A}$' term, as required. Note that such an argument is only possible for an *absolutely* conserved quantum number q—otherwise we cannot split up the states of the system into non-communicating sectors specified by different values of q. Reversing this line of reasoning, a conservation law such as baryon number conservation, with no related gauge field, would therefore now be suspected of not being absolutely conserved.

We still have not tied down why q is the electromagnetic charge and not some other absolutely conserved quantum number. A proper discussion of the reasons for identifying A^{μ} with the electromagnetic potential and q with the particle's charge will be given in Chapter 4, with the help of quantum field theory.

(ii) Accepting these identifications, we note that the form of the interaction contains but one parameter, the electromagnetic charge q of the particle in question. It is the *same* whatever the type of particle with charge q, whether it be lepton, hadron, nucleus, ion, atom, etc. Precisely this type of 'universality' is present in the weak couplings of quarks and leptons, as we shall see in Chapter 11. This strongly suggests that some form of gauge principle must be at work in generating weak interactions as well. The associated symmetry or conservation law is, however, of a very subtle kind, as we shall discuss in Chapter 13. Incidentally, although all particles of a given charge q interact electromagnetically in a universal way, there is nothing at all in the preceding argument to indicate why, in nature, the charges of observed particles are all integer multiples of one basic charge. We shall return to this problem of charge quantisation in §15.2.

(iii) Returning to point (i), we may wish that we had not had to introduce the absolute conservation of charge as a separate axiom. As remarked earlier, at the end of §2.2, we should like to relate that conservation law to the symmetry involved, namely invariance under (2.37). It is worth looking at the nature of this symmetry in a little more detail. It is not a symmetry which—as in the case of translation and rotation invariances for instance—involved changes in the space–time coordinates $\boldsymbol{x}$ and t. Instead, it operates on the *real and imaginary parts of the wavefunction*. Let us write

$$\psi = \psi_{\mathrm{R}} + \mathrm{i}\psi_{\mathrm{I}}. \tag{2.41}$$

Then

$$\psi' = e^{i\alpha}\psi = \psi_R' + i\psi_I' \tag{2.42}$$

can be written as

$$\begin{aligned} \psi_R' &= (\cos\alpha)\psi_R - (\sin\alpha)\psi_I \\ \psi_I' &= (\sin\alpha)\psi_R + (\cos\alpha)\psi_I \end{aligned} \tag{2.43}$$

from which we can see that it is indeed a kind of 'rotation', but in the ψ_R–ψ_I plane, whose 'coordinates' are the real and imaginary parts of the wavefunction. We call this plane an *internal* space, and the associated symmetry an *internal symmetry*. Thus our phase invariance can be looked upon as a kind of internal space rotational invariance.

We can imagine doing two successive such transformations

$$\psi \to \psi' \to \psi'' \tag{2.44}$$

where

$$\psi'' = e^{i\beta}\psi' \tag{2.45}$$

and so

$$\psi'' = e^{i(\alpha+\beta)}\psi = e^{i\delta}\psi \tag{2.46}$$

with $\delta = \alpha + \beta$. This is a transformation of the same form as the original one. The set of all such transformations forms what mathematicians call a *group*, in this case U(1), meaning the group of all unitary one-dimensional matrices. A unitary matrix $\mathbf{U}$ is one such that

$$\mathbf{U}\mathbf{U}^\dagger = \mathbf{U}^\dagger\mathbf{U} = \mathbf{I} \tag{2.47}$$

where $\mathbf{I}$ is the identity and † denotes the Hermitian conjugate. A one-dimensional matrix is of course a single number—in this case, a complex number. Condition (2.47) limits this to being a simple phase: the set of phase factors of the form $e^{i\alpha}$, where α is any real number, form the elements of a U(1) group. These are just the factors that enter into our gauge (or phase) transformations for wavefunctions. Thus we say that the electromagnetic gauge group is U(1).

The transformations of the U(1) group have the simple property that it does not matter in what order they are performed: referring to (2.44)–(2.46), we would have got the same final answer if we had done the β 'rotation' first and then the α one, instead of the other way around; this is because, of course,

$$\exp(i\alpha)\cdot\exp(i\beta) = \exp[i(\alpha+\beta)] = \exp(i\beta)\cdot\exp(i\alpha).$$

Mathematicians call U(1) an *Abelian* group: different transformations commute. We shall see later (in Chapters 9 and 14) that the 'internal' symmetry spaces relevant to the strong and weak gauge invariances are not so simple. The 'rotations' in these cases are more like full three-dimensional rotations of real space, rather than the two-dimensional rotation of (2.43). We know that, in general, such real space rotations do *not* commute, and the same will be true of the strong and weak rotations. Their gauge groups are called *non-Abelian*.

Once again, we shall have to wait until Chapter 4 before understanding how the symmetry represented by (2.42) is really related to the conservation law of charge.

(iv) The attentive reader may have picked up one further loose end. The vector potential $\boldsymbol{A}$ is related to the magnetic field $\boldsymbol{B}$ by

$$\boldsymbol{B} = \boldsymbol{\nabla} \times \boldsymbol{A}. \tag{2.9}$$

Thus if $\boldsymbol{A}$ has the special form

$$\boldsymbol{A} = \boldsymbol{\nabla} f \tag{2.48}$$

$\boldsymbol{B}$ will vanish. The question we must answer, therefore, is: how do we know that the $\boldsymbol{A}$ field introduced by our gauge principle is not of the form (2.48), leading to a trivial theory ($\boldsymbol{B} = \boldsymbol{0}$)? The answer to this question will lead us on a very worthwhile detour.

The Schrödinger equation with $\boldsymbol{\nabla} f$ as the vector potential is

$$(1/2m)(-\mathrm{i}\boldsymbol{\nabla} - q\boldsymbol{\nabla} f)^2\psi = E\psi. \tag{2.49}$$

We can write the formal solution to this equation as

$$\psi = \exp\left(\mathrm{i}q\int_{-\infty}^{\boldsymbol{x}} \boldsymbol{\nabla} f\cdot \mathrm{d}\boldsymbol{l}\right)\cdot\psi(f = 0) \tag{2.50}$$

which may be checked by using the fact that

$$\frac{\partial}{\partial a}\int^{a} f(t)\mathrm{d}t = f(a). \tag{2.51}$$

The notation $\psi(f = 0)$ means just the free particle solution with $f = 0$; the line integral is taken along an arbitrary path ending in the point $\boldsymbol{x}$. But we have

$$\mathrm{d}f = \frac{\partial f}{\partial x}\mathrm{d}x + \frac{\partial f}{\partial y}\mathrm{d}y + \frac{\partial f}{\partial z}\mathrm{d}z \equiv \boldsymbol{\nabla} f\cdot \mathrm{d}\boldsymbol{l}. \tag{2.52}$$

Hence the line integral can be done trivially and the solution becomes

$$\psi = \exp[\mathrm{i}q(f(\boldsymbol{x}) - f(-\infty))]\cdot\psi(f = 0). \tag{2.53}$$

We say that the phase factor introduced by the (in reality, field-free) vector potential $\boldsymbol{A} = \boldsymbol{\nabla} f$ is *integrable*: the effect of this particular $\boldsymbol{A}$ is merely to multiply the free particle solution by an $\boldsymbol{x}$-dependent phase (apart from a trivial constant phase). Since this $\boldsymbol{A}$ should give no real electromagnetic effect, we must hope that such a change in the wavefunction is also somehow harmless. Indeed, Dirac showed (Dirac

1981, pp 92–3) that such a phase factor corresponds merely to a redefinition of the momentum operator $\hat{\boldsymbol{p}}$. The essential point is that (in one dimension, say) $\hat{p}$ is defined ultimately by the commutator ($\hbar = 1$)

$$[\hat{x}, \hat{p}] = \mathrm{i}. \tag{2.54}$$

Certainly the familiar choice

$$\hat{p} = -\mathrm{i}\frac{\partial}{\partial x} \tag{2.55}$$

satisfies this commutation relation. But we can also add any function of x to $\hat{p}$, and this modified $\hat{p}$ will still be satisfactory since x commutes with any function of x. More detailed considerations by Dirac showed that this arbitrary function must actually have the form $\partial F/\partial x$, where F is arbitrary. Thus

$$\hat{p}' = -\mathrm{i}\frac{\partial}{\partial x} + \frac{\partial F}{\partial x} \tag{2.56}$$

is an acceptable momentum operator. Consider then the quantum mechanics defined by the wavefunctions $\psi(f = 0)$ and the momentum operator $\hat{p} = -\mathrm{i}\partial/\partial x$. Under the unitary transformation

$$\psi(f = 0) \rightarrow \mathrm{e}^{\mathrm{i}qf(x)}\psi(f = 0) \tag{2.57}$$

$\hat{p}$ will be transformed to

$$\hat{p} \rightarrow \mathrm{e}^{\mathrm{i}qf(x)}\hat{p}\,\mathrm{e}^{-\mathrm{i}qf(x)}. \tag{2.58}$$

But the right-hand side of this equation is just $\hat{p} - q\partial f/\partial x$ (problem 2.3), which is an equally acceptable momentum operator, identifying qf with the F of Dirac.

What of the physically interesting case in which $\boldsymbol{A}$ is *not* of the form $\boldsymbol{\nabla} f$? The equation is now

$$(1/2m)(-\mathrm{i}\boldsymbol{\nabla} - q\boldsymbol{A})^2\psi = E\psi \tag{2.59}$$

to which the solution is

$$\psi = \exp\left(\mathrm{i}q\int_{-\infty}^{\boldsymbol{x}} \boldsymbol{A}\cdot\mathrm{d}\boldsymbol{l}\right)\cdot\psi(\boldsymbol{A} = 0). \tag{2.60}$$

The line integral can now not be done so trivially: one says that the $\boldsymbol{A}$-field has produced a *non-integrable phase factor*. There is more to this terminology than the mere question of whether the integral is easy to do. The crucial point is that the integral now depends on the *path followed* in reaching the point $\boldsymbol{x}$, whereas the integrable phase factor in (2.50) depends only on the end-points of the integral, not on the path joining them.

Consider two paths C_1 and C_2 (figure 2.1) from $-\infty$ to the point $\boldsymbol{x}$. The difference in the two line integrals is the integral over a *closed* curve C, which can be evaluated by Stokes' theorem:

$$\int_{C_1}^{x} \boldsymbol{A}\cdot\mathrm{d}\boldsymbol{l} - \int_{C_2}^{x} \boldsymbol{A}\cdot\mathrm{d}\boldsymbol{l} = \oint_C \boldsymbol{A}\cdot\mathrm{d}\boldsymbol{l} = \int\int_S \boldsymbol{\nabla}\times\boldsymbol{A}\cdot\mathrm{d}\boldsymbol{S} = \int\int_S \boldsymbol{B}\cdot\mathrm{d}\boldsymbol{S} \quad (2.61)$$

where S is any surface spanning the curve C. In this form we see that if $\boldsymbol{A} = \boldsymbol{\nabla} f$, then indeed the line integrals over C_1 and C_2 are equal since curl grad $\equiv 0$, but if $\boldsymbol{B} = \boldsymbol{\nabla}\times\boldsymbol{A}$ is not zero, the difference between the integrals is determined by the enclosed flux of $\boldsymbol{B}$.

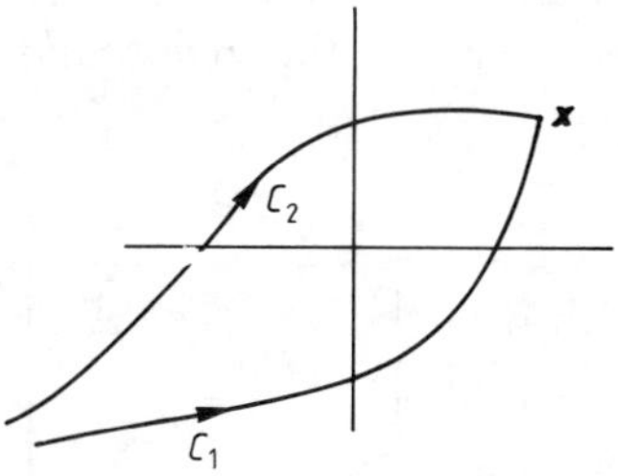

Figure 2.1 Two paths C_1 and C_2 (in two dimensions for simplicity) from $-\infty$ to the point $\boldsymbol{x}$.

The above analysis turns out to imply the existence of a remarkable phenomenon—the Aharonov–Bohm effect, named after its discoverers (Aharonov and Bohm 1959). Suppose we go back to our two-slit experiment of §2.5, only this time we imagine that a long thin solenoid is inserted between the slits, so that the components ψ_1 and ψ_2 of the split beam pass one on each side of the solenoid (figure 2.2). After passing round the solenoid, the beams are recombined, and the resulting interference pattern is observed downstream. At any point $\boldsymbol{x}$ of the pattern, the phase of the ψ_1 and ψ_2 components will be modified—relative to the $\boldsymbol{B} = \boldsymbol{0}$ case—by factors of the form (2.60). These factors depend on the respective paths, which are different for the two components ψ_1 and ψ_2. Thus the phase difference between these components, which determines the interference pattern, will involve the $\boldsymbol{B}$-dependent factor (2.61). Thus, even though the field $\boldsymbol{B}$ is essentially totally contained within the solenoid, and the beams themselves have passed through $\boldsymbol{B} = \boldsymbol{0}$ regions only, there is nevertheless an observable effect on the pattern provided $\boldsymbol{B} \neq \boldsymbol{0}$! This effect—a shift in the pattern as $\boldsymbol{B}$ varies—was first confirmed experimentally by Chambers (1960), soon after its prediction by Aharonov and Bohm. It was anticipated in work by Ehrenburg and Siday (1949); further references and discussion are contained in Berry (1984).

(v) In conclusion, we must emphasise that there is ultimately no compelling logic for the vital leap to a local phase invariance from a global one. The latter is, by itself, both necessary and sufficient in quantum field theory to guarantee local charge conservation. Nevertheless, the gauge principle—deriving interactions from the requirement of local phase invariance—is so simple, beautiful and powerful (and apparently successful) that it has taken command of elementary particle physics. In later parts of the book we shall consider generalisations of the electromagnetic gauge principle. It will be important always to bear in mind that any attempt to base theories of non-electromagnetic interactions on some kind of gauge principle can only make sense if there is an exact symmetry involved. The reason for this will only become clear when we consider the *renormalisability* of our theories (Chapter 6).

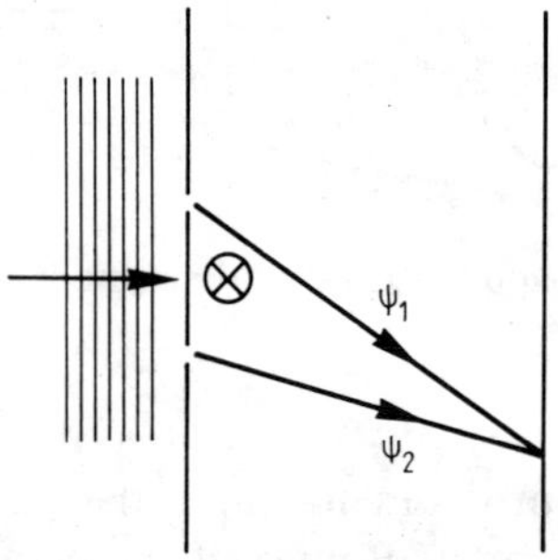

Figure 2.2 The Aharonov–Bohm effect.

Summary

Maxwell's equations and the local conservation of charge.

Electromagnetic potentials A^{μ} and gauge transformations.

Field strength tensor $F^{\mu\nu}$.

Lorentz covariant and gauge invariant forms of Maxwell's equations, using A^{μ}.

Gauge invariance in quantum mechanics requires wavefunction to change by space–time-dependent phase factor when A^{μ} changes by gauge transformation.

The gauge principle for generating interactions (electromagnetic case): local and global invariances.

U(1) transformations.

Non-integrable phase factor and the Aharonov–Bohm effect.

Problems

2.1 (*a*) A Lorentz transformation in the x^1 direction is given by

$$t' = \gamma(t - vx^1),\ x'^1 = \gamma(-vt + x^1)$$
$$x'^2 = x^2,\ x'^3 = x^3$$

where $\gamma = (1 - v^2)^{-1/2}$ and $c = 1$. Write down the inverse of this transformation (i.e. express $(t,\ x^1)$ in terms of $(t',\ x'^1)$), and use the 'chain rule' of partial differentiation to show that, under the Lorentz transformation, the two quantities $(\partial/\partial t,\ -\partial/\partial x^1)$ transform in the same way as $(t,\ x^1)$.

[The general result is that the four-component quantity $(\partial/\partial t, -\partial/\partial x^1, -\partial/\partial x^2, -\partial/\partial x^3) \equiv (\partial/\partial t, -\boldsymbol{\nabla})$ transforms in the same way as (t, x^1, x^2, x^3). Four-component quantities transforming this way are said to be 'contravariant four vectors', and are written with an upper four-vector index; thus $(\partial/\partial t, -\boldsymbol{\nabla}) \equiv \partial^\mu$. Upper indices can be lowered by using the metric tensor $g_{\mu\nu}$, see Appendix A.2, which reverses the signs of the spatial components. Thus $\partial^\mu = (\partial/\partial t, \partial/\partial x_1, \partial/\partial x_2, \partial/\partial x_3)$. Similarly the four quantities $(\partial/\partial t, \boldsymbol{\nabla}) = (\partial/\partial t, \partial/\partial x^1, \partial/\partial x^2, \partial/\partial x^3)$ transform as $(t, -x^1, -x^2, -x^3)$ and are a 'covariant four-vector', denoted, by ∂_μ.]

(*b*) Check that equation (2.5) can be written as (2.16).

2.2 How many independent components does the field strength $F^{\mu\nu}$ have? Express each component in terms of electric and magnetic field components. Hence verify that equation (2.17) correctly reproduces both equations (2.1) and (2.8).

2.3 Verify the result

$$\mathrm{e}^{\mathrm{i}qf(x)}\hat{p}\ \mathrm{e}^{-\mathrm{i}qf(x)} = \hat{p} - q\frac{\partial f}{\partial x}.$$

3

THE KLEIN–GORDON AND DIRAC WAVE EQUATIONS, AND THE INTERPRETATION OF THEIR NEGATIVE-ENERGY SOLUTIONS

What we call the beginning is often the end
And to make an end is to make a beginning.
The end is where we start from.

from *Little Gidding* by T S Eliot

3.1 Introduction

It is clear that the non-relativistic Schrödinger equation is quite inadequate to analyse the results of experiments performed at energies far higher than the rest mass energies of the particles involved. The first purpose of this chapter, therefore, is to introduce, as an inescapable preliminary to all subsequent material, the two simplest relativistic wave equations with which we shall be principally concerned—the Klein–Gordon (spin 0) and Dirac (spin $\frac{1}{2}$) equations. We shall be brief, as fuller accounts are available elsewhere (Bjorken and Drell 1964, Aitchison 1972).

There is one important matter that we shall, however, discuss in some detail—the treatment of *antiparticles* within such a 'one-particle wavefunction' type of quantum mechanics. As we noted in §1.3, particle creation—usually in the form of particle–antiparticle pairs—is an essential feature of high-energy physics. The traditional wave equation formalism—even if relativistic in form—would seem inappropriate, since one introduces one wavefunction for each 'particle' degree of freedom, and the number of such particles is supposed not to change. The natural formalism would seem to be one which embodied the possibility of particle creation and destruction from the beginning—and this will indeed be provided in the following chapter on quantum field theory.

However, it turns out that there *is* a way of handling particle creation and annihilation, even within the traditional 'wave equation' approach. This is provided by Feynman's interpretation of the *negative-energy solutions* of a wave equation as describing *antiparticles*, which will be discussed in §3.4. There are at least two reasons for pursuing this approach instead of passing directly to quantum field theory. The first is that the basic physical ideas are of permanent importance and provide useful additional physical insight into the somewhat more abstract field theory formalism. The second is that we shall actually *use* rather little of the full apparatus of Lagrangian-based quantum field theory. In fact, we shall only rarely consider anything beyond the lowest-order approximation for the quantum amplitudes, when they are expressed as a power series in the interaction strength. For this limited purpose, ordinary quantum mechanical perturbation theory gives us expressions for the amplitudes in terms of the interaction potentials and the free particle wavefunctions—and we can get a long way on this basis alone, provided of course we make the potentials and wavefunctions appropriately relativistic (in §3.6 we shall see how this is done for electromagnetism).

The fact that so much can be done on the basis of quantum *mechanics* rather than quantum *field theory* is a fortunate circumstance, since the full intricacies of the latter are truly formidable. Undoubtedly they have to be mastered by anyone wishing to be a professional particle theorist. But we are not writing for such intending professionals. Rather, our aim is to provide relatively quick access, for a wider readership, to the important concepts and simplest applications of gauge theories. The bulk of the calculational sections of the book will therefore be based on the quantum mechanical, rather than the field theoretic, approach. But we shall appeal to the latter for certain fundamental conceptual points (for example, the connection between phase transformations and charge conservation). We shall also be able to motivate in detail the physical interpretation of the 'Feynman diagrams' which we shall use so freely, and which otherwise easily degenerates into mumbo-jumbo.

3.2 The Klein–Gordon equation

The non-relativistic Schrödinger equation may be put into correspondence with the non-relativistic energy–momentum relation

$$E = \boldsymbol{p}^2/2m \tag{3.1}$$

by means of the operator replacements†

$$E \to \mathrm{i}\partial/\partial t \tag{3.2}$$

$$\boldsymbol{p} \to -\mathrm{i}\boldsymbol{\nabla} \tag{3.3}$$

these differential operators being understood to act on the Schrödinger wavefunction.

For a relativistic wave equation we must start with the correct relativistic energy–momentum relation. Energy and momentum appear

†Recall that $\hbar = c = 1$ throughout (see Appendix B).

as the 'time' and 'space' components of the momentum 4-vector

$$p^\mu = (E, \boldsymbol{p}) \tag{3.4}$$

which satisfy the mass-shell condition

$$p^2 = p_\mu p^\mu = E^2 - \boldsymbol{p}^2 = m^2. \tag{3.5}$$

Since energy and momentum are merely different components of a 4-vector, an attempt to base a relativistic theory on the relation

$$E = +(\boldsymbol{p}^2 + m^2)^{1/2} \tag{3.6}$$

is unattractive, as well as having obvious difficulties in interpretation for the square root operator. Schrödinger, before settling for the less ambitious non-relativistic Schrödinger equation, and later Klein and Gordon, attempted to build relativistic quantum mechanics (RQM) from the squared energy relation

$$E^2 = \boldsymbol{p}^2 + m^2. \tag{3.7}$$

Using the operator replacements for E and $\boldsymbol{p}$ we are led to

$$-\partial^2\phi/\partial t^2 = (-\boldsymbol{\nabla}^2 + m^2)\phi \tag{3.8}$$

which is the Klein–Gordon equation (KG equation). We consider the case of a one-component scalar wavefunction $\phi(\boldsymbol{x}, t)$: one expects this to be appropriate for the description of spin-0 bosons.

3.2.1 Solutions in coordinate space

In terms of the D'Alembertian operator

$$\square \equiv \partial_\mu \partial^\mu = \frac{\partial^2}{\partial t^2} - \boldsymbol{\nabla}^2 \tag{3.9}$$

the KG equation reads

$$(\square + m^2)\phi(\boldsymbol{x},t) = 0. \tag{3.10}$$

Let us look for a plane wave solution of the form

$$\phi(\boldsymbol{x}, t) = N\mathrm{e}^{-\mathrm{i}Et+\mathrm{i}\boldsymbol{p}\cdot\boldsymbol{x}} = N\mathrm{e}^{-\mathrm{i}p\cdot x} \tag{3.11}$$

where we have written the exponent in suggestive 4-vector scalar product notation

$$p\cdot x = p_\mu x^\mu = Et - \boldsymbol{p}\cdot\boldsymbol{x} \tag{3.12}$$

and N is a normalisation factor which we will determine later. In order that this wavefunction be a solution of the KG equation, we find by direct substitution that E must be related to $\boldsymbol{p}$ by the condition

$$E^2 = \boldsymbol{p}^2 + m^2. \tag{3.13}$$

This looks harmless enough, but it actually implies that for a given 3-momentum $\boldsymbol{p}$ there are in fact *two* possible solutions for the energy, namely

$$E = \pm\, (\boldsymbol{p}^2 + m^2)^{1/2}. \tag{3.14}$$

As Schrödinger and others quickly found, it is not possible to ignore these negative solutions without obtaining inconsistencies. What then do these negative-energy solutions mean?

3.2.2 *Probability current*

In exactly the same way as for the non-relativistic Schrödinger equation, it is possible to derive a conservation law for a 'probability current' of the KG equation. We have

$$\frac{\partial^2 \phi}{\partial t^2} - \boldsymbol{\nabla}^2\phi + m^2\phi = 0 \tag{3.15}$$

and by multiplying this equation by ϕ^*, and subtracting ϕ times the complex conjugate of equation (3.15), one obtains, after some manipulation (see problem 3.1), the result

$$\frac{\partial \rho}{\partial t} + \boldsymbol{\nabla}\cdot\boldsymbol{j} = 0 \tag{3.16}$$

where

$$\rho = \mathrm{i}\left[\phi^*\frac{\partial \phi}{\partial t} - \left(\frac{\partial \phi^*}{\partial t}\right)\phi\right] \tag{3.17}$$

and

$$\boldsymbol{j} = \mathrm{i}^{-1}[\phi^*\boldsymbol{\nabla}\phi - (\boldsymbol{\nabla}\phi^*)\phi] \tag{3.18}$$

(the derivatives $(\partial_\mu\phi^*)$ act only within the bracket). In explicit 4-vector notation this conservation condition reads (cf (A.62))

$$\partial^\mu j_\mu = 0 \tag{3.19}$$

with

$$j^\mu \equiv (\rho, \boldsymbol{j}) = \mathrm{i}[\phi^*\partial^\mu\phi - (\partial^\mu\phi^*)\phi]. \tag{3.20}$$

The spatial current $\boldsymbol{j}$ is identical in form to the Schrödinger current, but for the KG case the 'probability density' now contains time derivatives since the KG equation is second order in $\partial/\partial t$. This means that ρ is not constrained to be positive-definite—so how can ρ represent a probability density? We can see this problem explicitly for the plane wave solutions

$$\phi = N\mathrm{e}^{-\mathrm{i}Et + \mathrm{i}\boldsymbol{p}\cdot\boldsymbol{x}} \tag{3.21}$$

which give (problem 3.1)

$$\rho = 2|N|^2 E \tag{3.22}$$

and E can be positive or negative; that is, the sign of ρ is the sign of energy.

Historically, this problem of negative probabilities coupled with that of negative energies led to the abandonment of the KG equation. For the moment we follow history—although we shall soon come back and rescue the KG equation.

3.3 The Dirac equation

In the case of the KG equation it is clear why the problem arose:

(a) In constructing a wave equation in close correspondence with the squared energy–momentum relation

$$E^2 = \boldsymbol{p}^2 + m^2$$

we immediately allowed negative–energy solutions.

(b) The KG equation has a $\partial^2/\partial t^2$ term: this leads to a continuity equation with a 'probability density' containing $\partial/\partial t$, and hence to negative probabilities.

Dirac approached these problems in his characteristically direct way. In order to obtain a positive-definite probability density $\rho \geqslant 0$, he required an equation linear in $\partial/\partial t$. Then, for relativistic covariance (see below), the equation must also be linear in $\boldsymbol{\nabla}$. He postulated the equation (Dirac 1928)

$$\begin{aligned} \mathrm{i}\frac{\partial\psi(\boldsymbol{x},t)}{\partial t} &= \left[-\mathrm{i}\left(\alpha_1\frac{\partial}{\partial x^1}+\alpha_2\frac{\partial}{\partial x^2}+\alpha_3\frac{\partial}{\partial x^3}\right)+\beta m\right]\psi(\boldsymbol{x},t) \\ &= (-\mathrm{i}\boldsymbol{\alpha}\cdot\boldsymbol{\nabla}+\beta m)\psi(\boldsymbol{x},t). \end{aligned} \tag{3.23}$$

What are the α's and β? To find the conditions on the α's and β, consider what we require of a relativistic wave equation:

(a) the correct relativistic relation between E and $\boldsymbol{p}$,

$$E = +(\boldsymbol{p}^2 + m^2)^{1/2};$$

(b) the equation should be covariant under Lorentz transformations.

We shall not consider the second requirement here. In Appendix A, the concept of covariance is illustrated by the example of the Pauli equation for a non-relativistic spin degree of freedom:

$$\boldsymbol{\sigma}\cdot\boldsymbol{B}\chi = E\chi. \tag{3.24}$$

Here χ is a two-component spinor and $\boldsymbol{\sigma}\cdot\boldsymbol{B}$ and E are scalars under rotations. For the Dirac case we will write equations in terms of 4-vectors, scalars and so on, and assume without proof the corresponding transformation properties. Proofs may be found in Bjorken and Drell (1964) and Aitchison (1972).

To solve requirement (a), Dirac in fact demanded that his wavefunction ψ satisfy, in addition, a KG-type condition

$$-\partial^2\psi/\partial t^2 = (-\nabla^2 + m^2)\psi. \tag{3.25}$$

We note with hindsight that we have once more opened the door to negative-energy solutions: Dirac's remarkable achievement was to turn this apparent defect into one of the triumphs of theoretical physics!

We can now derive conditions on $\boldsymbol{\alpha}$ and β. We have

$$\mathrm{i}\partial\psi/\partial t = (-\mathrm{i}\boldsymbol{\alpha}\cdot\nabla + \beta m)\psi \tag{3.26}$$

and so, squaring the operator on both sides,

$$\begin{aligned}\left(\mathrm{i}\frac{\partial}{\partial t}\right)^2\psi &= (-\mathrm{i}\boldsymbol{\alpha}\cdot\nabla + \beta m)(-\mathrm{i}\boldsymbol{\alpha}\cdot\nabla + \beta m)\psi\\ &= -\sum_{i=1}^{3}\alpha_i^2\frac{\partial^2\psi}{(\partial x^i)^2} - \sum_{\substack{i,j=1\\ i>j}}^{3}(\alpha_i\alpha_j + \alpha_j\alpha_i)\frac{\partial^2\psi}{\partial x^i\partial x^j}\\ &\quad - \mathrm{i}m\sum_{i=1}^{3}(\alpha_i\beta + \beta\alpha_i)\frac{\partial\psi}{\partial x^i} + \beta^2 m^2\psi.\end{aligned} \tag{3.27}$$

But by our assumption that ψ also satisfies the KG condition, we must have

$$\left(\mathrm{i}\frac{\partial}{\partial t}\right)^2\psi = -\sum_{i=1}^{3}\frac{\partial^2\psi}{(\partial x^i)^2} + m^2\psi. \tag{3.28}$$

It is thus evident that the α's and β cannot be ordinary, classical, commuting quantities. Instead, they must satisfy the following *anticommutation relations* in order to eliminate the unwanted terms on the right-hand side of equation (3.27):

$$\alpha_i\beta + \beta\alpha_i = 0, \quad i = 1, 2, 3 \tag{3.29a}$$

$$\alpha_i\alpha_j + \alpha_j\alpha_i = 0, \quad i, j = 1, 2, 3; i \neq j. \tag{3.29b}$$

In addition we require

$$\alpha_i^2 = \beta^2 = 1. \tag{3.29c}$$

Remembering the appearance of the Pauli matrices, with their anticommutation relations, for non-relativistic spin-$\frac{1}{2}$ particles, it is natural to interpret $\boldsymbol{\alpha}$ and β as matrices acting on a column vector (spinor) ψ. One therefore expects the Dirac equation to represent a particle with $\frac{1}{2}$-odd-integral spin. It is not difficult to prove (Bjorken and Drell 1964,

Aitchison 1972) that the smallest possible dimension of the matrices for which the Dirac conditions can be satisfied is 4 × 4. The conventional choice for the α's and β is

$$\alpha_i = \begin{pmatrix} \mathbf{0} & \sigma_i \\ \sigma_i & \mathbf{0} \end{pmatrix}, \qquad \beta = \begin{pmatrix} \mathbf{1} & \mathbf{0} \\ \mathbf{0} & -\mathbf{1} \end{pmatrix} \tag{3.30}$$

where we have written these 4 × 4 matrices in 2 × 2 'block diagonal' form, and the σ_i's are the usual 2 × 2 Pauli matrices and $\mathbf{1}$ is the 2 × 2 unit matrix. Readers not familiar with this labour-saving form should verify, both by using the corresponding explicit 4 × 4 matrices, such as

$$\alpha_1 = \begin{pmatrix} 0 & 0 & 0 & 1 \\ 0 & 0 & 1 & 0 \\ 0 & 1 & 0 & 0 \\ 1 & 0 & 0 & 0 \end{pmatrix} \tag{3.31}$$

and so on, and by the block diagonal form, that this choice does indeed satisfy the required conditions. These are

$$\{\alpha_i, \beta\} = 0 \tag{3.32a}$$

$$\{\alpha_i, \alpha_j\} = 2\delta_{ij}\mathbf{1} \tag{3.32b}$$

$$\beta^2 = \mathbf{1} \tag{3.32c}$$

where $\{\mathbf{A}, \mathbf{B}\}$ is the anticommutator of the two matrices, $\mathbf{AB} + \mathbf{BA}$, and $\mathbf{1}$ is here the 4 × 4 unit matrix. Notice that this choice of $\boldsymbol{\alpha}$ and β is not unique. In fact, all matrices related to these by any unitary 4 × 4 matrix $\mathbf{U}$ (which thus preserves the anticommutation relations) are allowed:

$$\alpha_i' = \mathbf{U}\alpha_i\mathbf{U}^{-1} \tag{3.33a}$$

$$\beta' = \mathbf{U}\beta\mathbf{U}^{-1}. \tag{3.33b}$$

For the most part we shall use the standard representation above, but in some cases it is convenient (but not necessary) to use other representations in which β is not diagonal.

3.3.1 Free particle solutions

Since the Dirac Hamiltonian now involves 4 × 4 matrices, it is clear that we must interpret the Dirac wavefunction ψ as a four-component column vector—the so-called Dirac spinor. Let us look at the explicit form of the free particle solutions. By analogy with the non-relativistic case, where we may write a general spin-$\frac{1}{2}$ spinor in terms of factorised solutions

$$\psi = \chi\cdot(\text{plane wave}) \tag{3.34}$$

where χ is a two-component spinor, we therefore look for solutions of the Dirac equation of the form

$$\psi = \omega e^{-ip \cdot x} \tag{3.35}$$

where ω is a four-component spinor and $e^{-ip\cdot x}$, with $p^\mu = (E, \boldsymbol{p})$, is the plane wave solution. We substitute this into the Dirac equation

$$i\partial\psi/\partial t = (-i\boldsymbol{\alpha}\cdot\boldsymbol{\nabla} + \beta m)\psi \tag{3.26}$$

using the explicit $\boldsymbol{\alpha}$ and β matrices. In order to use the 2×2 block form, it is conventional (and convenient) to split up the spinor ω into two two-component spinors ϕ and χ:

$$\omega = \begin{pmatrix} \phi \\ \chi \end{pmatrix}. \tag{3.36}$$

We obtain the matrix equation (see problem 3.3)

$$E\begin{pmatrix} \phi \\ \chi \end{pmatrix} = \begin{pmatrix} m\mathbf{1} & \boldsymbol{\sigma}\cdot\boldsymbol{p} \\ \boldsymbol{\sigma}\cdot\boldsymbol{p} & -m\mathbf{1} \end{pmatrix} \begin{pmatrix} \phi \\ \chi \end{pmatrix} \tag{3.37}$$

representing two coupled equations for ϕ and χ:

$$(E - m)\phi = \boldsymbol{\sigma}\cdot\boldsymbol{p}\chi \tag{3.38}$$

and

$$(E + m)\chi = \boldsymbol{\sigma}\cdot\boldsymbol{p}\phi. \tag{3.39}$$

Solving for χ from (3.39), the general four-component spinor may be written

$$\omega = \begin{pmatrix} \phi \\ \dfrac{\boldsymbol{\sigma}\cdot\boldsymbol{p}}{E+m}\phi \end{pmatrix}. \tag{3.40}$$

What is the relation between E and $\boldsymbol{p}$ for this to be a solution of the Dirac equation? If we substitute χ from (3.39) into (3.38) and remember that (problem 3.2)

$$(\boldsymbol{\sigma}\cdot\boldsymbol{p})^2 = \boldsymbol{p}^2\mathbf{1} \tag{3.41}$$

we find that

$$(E - m)(E + m)\phi = \boldsymbol{p}^2\phi \tag{3.42}$$

for any ϕ. Hence we arrive at the same result as for the KG equation in that for a given value of $\boldsymbol{p}$, two values of E are allowed:

$$E = \pm (\boldsymbol{p}^2 + m^2)^{1/2} \tag{3.14}$$

i.e. positive *and* negative energy states are still admitted.

The Dirac equation does not therefore solve this problem. What about the probability current?

3.3.2 Probability current for the Dirac equation

Consider the following quantity which we denote (suggestively) by ρ:

$$\rho = \psi^\dagger(x)\psi(x). \tag{3.43}$$

Here $\psi^\dagger$ is the Hermitian conjugate row vector of the column vector ψ. In terms of components

$$\rho = (\psi_1^*, \psi_2^*, \psi_3^*, \psi_4^*)\begin{pmatrix}\psi_1\\ \psi_2\\ \psi_3\\ \psi_4\end{pmatrix} \tag{3.44}$$

so

$$\rho = \sum_{a=1}^{4} |\psi_a|^2 > 0 \tag{3.45}$$

and we see that ρ is a scalar density which is explicitly positive-definite. This is one property we require of a probability density; in addition, we require a conservation law, coming from the Dirac equation, and a corresponding probability current. In fact (see problem 3.4) we can demonstrate using the Dirac equation

$$\mathrm{i}\partial\psi/\partial t = (-\mathrm{i}\boldsymbol{\alpha}\cdot\boldsymbol{\nabla} + \beta m)\psi \tag{3.26}$$

and its Hermitian conjugate

$$-\mathrm{i}\partial\psi^\dagger/\partial t = \psi^\dagger(+\mathrm{i}\boldsymbol{\alpha}\cdot\overleftarrow{\boldsymbol{\nabla}} + \beta m) \tag{3.46}$$

that there is a conservation law of the required form

$$\partial\rho/\partial t + \boldsymbol{\nabla}\cdot\boldsymbol{j} = 0. \tag{3.47}$$

The notation $\psi^\dagger\overleftarrow{\boldsymbol{\nabla}}$ requires some comment: it is shorthand for three row matrices

$$\psi^\dagger\overleftarrow{\boldsymbol{\nabla}}_x \equiv \partial\psi^\dagger/\partial x, \quad \text{etc}$$

(recall that $\psi^\dagger$ is a row matrix).

The probability current for the above equation is

$$\boldsymbol{j}(x) = \psi^\dagger(x)\boldsymbol{\alpha}\psi(x) \tag{3.48}$$

representing a 3-vector with components

$$(\psi^\dagger\alpha_1\psi, \psi^\dagger\alpha_2\psi, \psi^\dagger\alpha_3\psi). \tag{3.49}$$

Thus the interpretation of

$$j^\mu = (\rho, \boldsymbol{j}) \tag{3.50}$$

as a probability current is acceptable, unlike the current for the KG equation. Again this is as we might have anticipated.

3.4 The Dirac and Feynman interpretations of negative-energy solutions

The story so far may be summarised by

Klein–Gordon equation { negative probabilities; negative energies

Dirac equation { positive probabilities; negative energies.

The historical route proceeds via the Dirac equation and Dirac's brilliant reinterpretation of the negative-energy solutions. For a free Dirac electron the available positive and negative energy levels are symmetric about $E = 0$: the situation is sketched in figure 3.1.

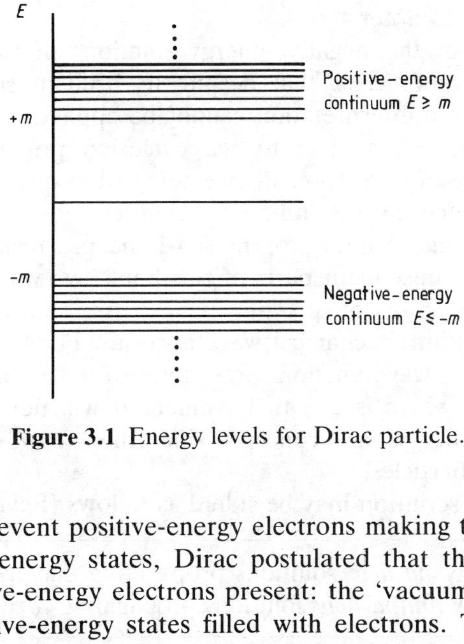

Figure 3.1 Energy levels for Dirac particle.

In order to prevent positive-energy electrons making transitions to the lower, negative-energy states, Dirac postulated that the normal empty state—no positive-energy electrons present: the 'vacuum state'—consists of all the negative-energy states filled with electrons. The Pauli exclusion principle then forbids any positive-energy electrons from falling into these lower energy levels. The 'vacuum' now has infinite negative charge and energy, but since all observations represent *finite* fluctuations in energy and charge with respect to this vacuum, this leads to an acceptable theory. For example, if one negative-energy electron is absent from the Dirac sea, we have a 'hole' relative to the normal vacuum:

$$\text{energy of 'hole'} = -(E_{\text{neg}}) \rightarrow \text{positive energy}$$

$$\text{charge of 'hole'} = -(q_{\text{e}}) \rightarrow \text{positive charge.}$$

Thus the *absence* of a negative-energy electron is equivalent to the *presence* of a positive-energy positively charged version of the electron, that is a positron. This prediction of 'antiparticles' is one of the greatest achievements of theoretical physics†: Carl Anderson received the Nobel Prize for his discovery of the positron in 1932 (Anderson 1932).

In this way it proved possible to obtain sensible results from the Dirac equation and its negative-energy solutions. It is clear, however, that the theory is no longer a real 'single-particle' theory, since we can excite the infinite sea of negative-energy electrons that constitute the normal 'empty state'. For example, if we excite a negative-energy electron to a positive-energy state, we have in the final state a positive-energy electron plus a positive-energy positron 'hole' in the vacuum: this corresponds physically to the process of e^+e^- pair creation. Thus this way of dealing with the negative-energy problem for fermions leads us directly to the need for a quantum field theory. We shall have more to say about this in Chapter 4.

But what about the negative-energy solutions of the Klein–Gordon equation? It is quite clear that despite its brilliant success for spin-$\frac{1}{2}$ particles, the Dirac interpretation cannot be applied to spin-0 particles, since bosons are not subject to the exclusion principle. Actually, a consistent picture of *both* cases does emerge from quantum field theory, as we shall sketch in the following chapter. But, as stated in the introduction to this chapter, for most of the practical calculations we shall consider in subsequent parts of this book, we want to short-circuit the elaborate apparatus of quantum field theory and use only the ideas of ordinary quantum mechanical wavefunctions. For this reason we now describe a purely 'wavefunction' prescription for handling the negative-energy solutions which is due to Feynman. It will not make use of the Dirac negative-energy 'sea' concept—since then it would obviously not work for spin-0 particles.

Feynman's prescription may be stated as follows (Feynman 1962):

negative-energy *particle* solutions propagating *backward* in time ≡ positive-energy *antiparticle* solutions propagating *forward* in time.

†At that time, this was not universally recognised. For example, Pauli (1933) wrote: 'Dirac has tried to identify holes with anti-electrons . . . we do not believe that this explanation can be seriously considered'.

Let us now attempt to make this plausible. Consider the example of the scattering of a particle, say a π^+, by a potential in second order of perturbation theory. In non-relativistic quantum mechanics (NQRM), we can draw a 'space–time' plot of the particle's trajectory (figure 3.2). According to Feynman, in RQM we must also allow for the possibility that particles may be scattered 'backwards in time' (this will be made more precise as we go along). Thus in RQM, in addition to the trajectory above, one must also allow the trajectory shown in figure 3.3. We interpret this second picture using Feynman's prescription, allowing positive-energy π^+ solutions to propagate only forward in time, and negative-energy solutions only backward in time. The π^+ travelling backward from t_2 to t_1 must therefore be assigned a negative energy; it is then equivalent, according to the prescription, to a π^- of positive energy travelling forward from t_1 to t_2. We therefore arrive at the physical process represented by the diagram shown in figure 3.4.

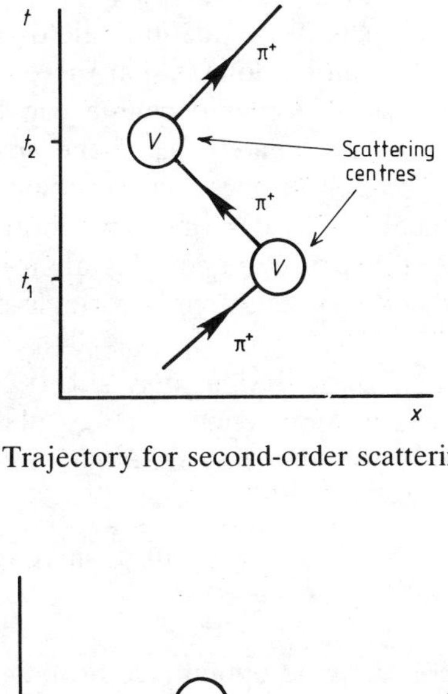

Figure 3.2 Trajectory for second-order scattering in NRQM.

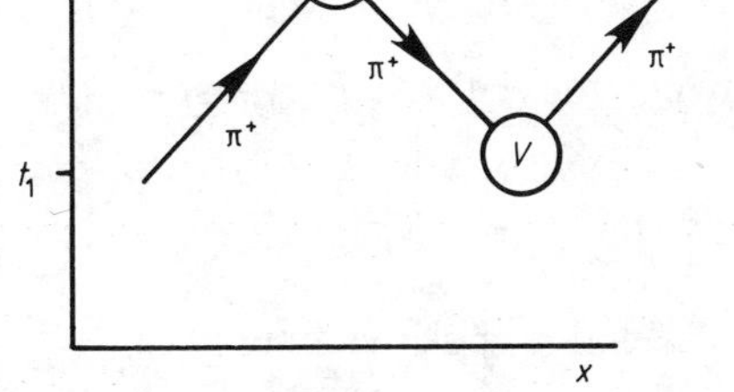

Figure 3.3 Additional trajectory for second-order scattering in RQM.

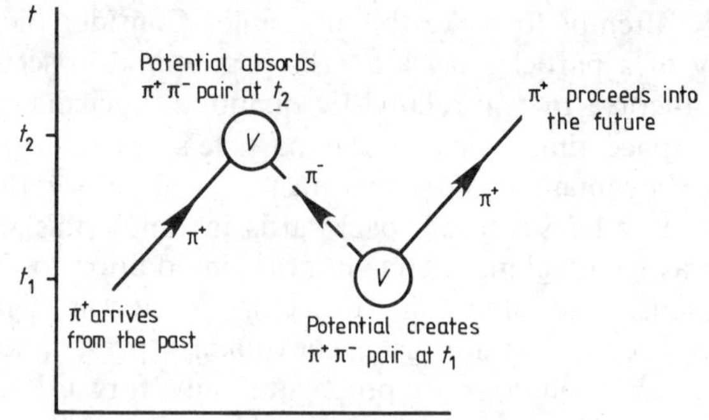

Figure 3.4 Feynman reinterpretation of figure 3.3.

The basic point, then, is that the *negative-energy* solutions of a wave equation describing a certain *particle* will be used to describe processes involving *positive-energy antiparticles*, after suitably interchanging initial and final states. In this way, many-particle situations (like the pair creation process above) can be handled by a single-particle wavefunction formalism, and the complexity of quantum field theory avoided. It is clear that the reason this can be done is that there is no physical process whereby a single particle of non-zero charge can be created by itself: only a pair or pairs can be created. Thus the space–time trajectories with which we are concerned are all continuous. All conceivable processes can be successfully described in terms of *either* particle trajectories alone *or* antiparticle trajectories alone—in either case (and by convention one of course uses the first) a single-particle wavefunction suffices.

The following further considerations may also be helpful. The electromagnetic current for a positive-energy π^+ is plausibly given by the probability current for positive-energy solutions multiplied by the charge $Q\ (= +e)$:

$$j^\mu_{\text{em}}(\pi^+) = (+e) \times (\text{probability current for positive-energy } \pi^+ \text{ solutions}). \tag{3.51}$$

For the plane wave solutions we obtain (see problem 3.1)

$$j^\mu_{\text{em}}(\pi^+) = (+e)2|N|^2[+(\boldsymbol{p}^2 + m^2)^{1/2}, \boldsymbol{p}]. \tag{3.52}$$

What about the current for the π^-? For positive-energy π^- particles we expect

$$j^\mu_{\text{em}}(\pi^-) = \underset{\substack{\uparrow \\ \text{charge}}}{(-e)}2|N|^2[\underset{\substack{\uparrow \\ \text{positive energy}}}{+}(\boldsymbol{p}^2 + m^2)^{1/2}, \boldsymbol{p}] \tag{3.53}$$

and it is evident that this may be rewritten as

$$j^{\mu}_{em}(\pi^-) = (+e)2|N|^2[-(\boldsymbol{p}^2 + m^2)^{1/2}, -\boldsymbol{p}]. \tag{3.54}$$

This is just $j^{\mu}_{em}(\pi^+)$ with *negative* 4-momentum.

Can we push the connection further? Consider what happens to the total charge and energy of a system A which emits a π^- of positive energy (figure 3.5). The energy of A decreases by E, and the charge of A decreases by $(-e)$ which is equivalent to the charge of A having increased by $+e$. This increase of charge could equally well be caused by the absorption of a π^+, but to make this absorption process *equivalent* to the emission process, the π^+ will have to have negative energy (since A loses energy in the emission). Thus we arrive at the equivalence (shown pictorially in figure 3.6):

> the changes of energy and charge of a system for the *emission* of a *positive*-energy π^- are the same as for the *absorption* of a *negative*-energy π^+.

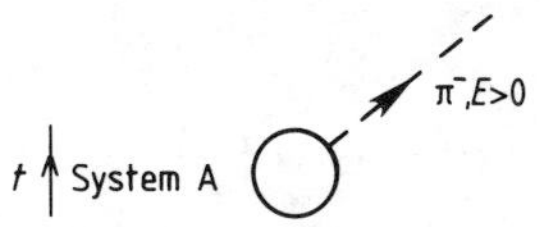

Figure 3.5 π^- emission process.

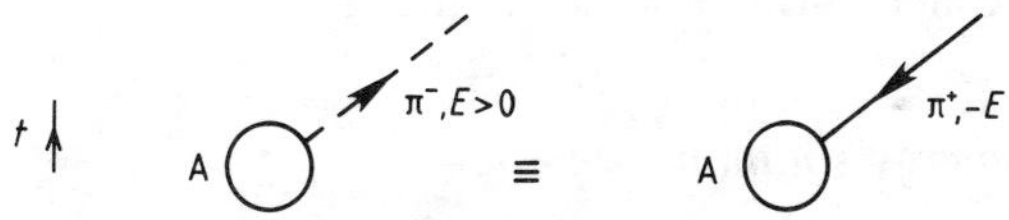

Figure 3.6 Feynman reinterpretation of figure 3.5.

Generalising from charge to all other relevant quantum numbers, and from energy to 4-momentum, we are led to the following hypothesis:

> the emission (absorption) of an antiparticle of 4-momentum p^{μ} is physically equivalent to the absorption (emission) of a particle of 4-momentum $-p^{\mu}$.

In other words, to go from a process involving antiparticles to one involving only particles, change the signs of the 4-momenta of all antiparticles and reverse the role of 'entry' and 'exit' states. It is in this sense that we understand the 'backwards in time' prescription.

It should be clear that the basic physical idea in the above interpretation is not limited to bosons. Let us now follow it through for the spin-$\frac{1}{2}$ case too.

3.5 Feynman interpretation of the negative-energy solutions of the Dirac equation

We return to the Dirac equation

$$\mathrm{i}\partial\psi/\partial t = (-\mathrm{i}\boldsymbol{\alpha}\cdot\boldsymbol{\nabla} + \beta m)\psi \tag{3.26}$$

and its plane wave solutions

$$\psi = \omega \mathrm{e}^{-\mathrm{i}p\cdot x} \tag{3.35}$$

where

$$p^\mu = (p^0, \boldsymbol{p}) \tag{3.55}$$

and ω is the four-component Dirac spinor. We decompose this in the usual way and obtain two coupled equations for the two two-component spinors ϕ and χ:

$$\omega = \begin{pmatrix} \phi \\ \chi \end{pmatrix} \tag{3.36}$$

$$p^0 \begin{pmatrix} \phi \\ \chi \end{pmatrix} = \begin{pmatrix} m\mathbf{1} & \boldsymbol{\sigma}\cdot\boldsymbol{p} \\ \boldsymbol{\sigma}\cdot\boldsymbol{p} & -m\mathbf{1} \end{pmatrix} \begin{pmatrix} \phi \\ \chi \end{pmatrix}. \tag{3.56}$$

We first find explicit forms for the solutions.

3.5.1 Positive-energy solutions

For these

$$p^0 = +(\boldsymbol{p}^2 + m^2)^{1/2} \equiv E > 0. \tag{3.57}$$

What are the solutions at rest? We have then

$$p^0 = +m, \qquad \boldsymbol{p} = 0$$

and our coupled equations become

$$m \begin{pmatrix} \phi \\ \chi \end{pmatrix} = \begin{pmatrix} m\mathbf{1} & 0 \\ 0 & -m\mathbf{1} \end{pmatrix} \begin{pmatrix} \phi \\ \chi \end{pmatrix} \tag{3.58}$$

leading to the result

$$\chi = 0. \tag{3.59}$$

We therefore write

$$\omega(p^0 = +m) = \begin{pmatrix} \phi \\ 0 \end{pmatrix} \tag{3.60}$$

and clearly there are two independent spin states for a positive-energy electron at rest:

$$\phi^1 = \begin{pmatrix} 1 \\ 0 \end{pmatrix} \quad \text{spin up} \tag{3.61}$$

$$\phi^2 = \begin{pmatrix} 0 \\ 1 \end{pmatrix} \quad \text{spin down} \tag{3.62}$$

indicating that the Dirac equation indeed describes a spin-$\frac{1}{2}$ particle.

For non-zero momentum

$$\boldsymbol{p} \neq 0, \qquad E = +(\boldsymbol{p}^2 + m^2)^{1/2} = p^0 \tag{3.63}$$

we eliminate χ and obtain positive-energy spinors in the form

$$\omega^{1,2} = \begin{pmatrix} \phi^{1,2} \\ \dfrac{\boldsymbol{\sigma}\cdot\boldsymbol{p}}{E+m}\phi^{1,2} \end{pmatrix}. \tag{3.64}$$

3.5.2 *Negative-energy spinors*

Now we look for spinors appropriate to the solution

$$p^0 = -(\boldsymbol{p}^2 + m^2)^{1/2} \equiv -E \tag{3.65}$$

(E is always defined to be positive). What are the appropriate solutions at rest? We have now

$$p^0 = -m, \qquad \boldsymbol{p} = 0 \tag{3.66}$$

and

$$-m\begin{pmatrix} \phi \\ \chi \end{pmatrix} = \begin{pmatrix} m\mathbf{1} & 0 \\ 0 & -m\mathbf{1} \end{pmatrix}\begin{pmatrix} \phi \\ \chi \end{pmatrix} \tag{3.67}$$

leading to

$$\phi = 0. \tag{3.68}$$

Thus the two independent negative-energy solutions at rest are just

$$\omega(p^0 = -m) = \begin{pmatrix} 0 \\ \chi \end{pmatrix}. \tag{3.69}$$

The solution for finite momentum $+\boldsymbol{p}$, i.e. for 4-momentum $p^\mu = (-E, \boldsymbol{p})$, is then

$$\omega = \begin{pmatrix} \dfrac{-\boldsymbol{\sigma}\cdot\boldsymbol{p}}{E+m}\chi \\ \chi \end{pmatrix}. \tag{3.70}$$

However, in order to use our antiparticle correspondence

$$\text{positive-energy solutions} \sim \mathrm{e}^{-\mathrm{i}p\cdot x} \tag{3.71a}$$

$$\text{negative-energy solutions} \sim \mathrm{e}^{+\mathrm{i}p\cdot x} \tag{3.71b}$$

we need instead the solution for $p^\mu \to -p^\mu$, i.e. negative-energy solutions with $p^\mu = (-E, -\boldsymbol{p})$. We therefore define

$$\omega^{3,4} = \begin{pmatrix} \dfrac{\boldsymbol{\sigma}\cdot\boldsymbol{p}}{E+m}\chi^{1,2} \\ \chi^{1,2} \end{pmatrix}. \tag{3.72}$$

There is one more subtlety. As in our arguments for charge and energy in the previous section, we expect the *absence* of 'spin up' (along an axis defined in the rest frame) to be equivalent to the *presence* of 'spin down'. In other words,

negative-energy solutions with spin-$\downarrow$

$$\equiv \text{positive-energy solutions with spin-}\uparrow. \tag{3.73}$$

For this reason we choose

$$\chi^1 = \begin{pmatrix} 0 \\ 1 \end{pmatrix}, \qquad \chi^2 = \begin{pmatrix} 1 \\ 0 \end{pmatrix}. \tag{3.74}$$

The labels in the negative-energy spinor, E, $\boldsymbol{p}$, 1 and 2, then refer to the *physical antiparticle values*. We now discuss in detail the negative- and positive-energy connection.

3.5.3 *Feynman interpretation of negative-energy solutions*

There is a difference between the KG and Dirac cases in that the Dirac equation was explicitly designed to yield a probability density and current independent of the sign of the energy:

$$\left.\begin{aligned} \rho &= \psi^\dagger\psi \\ \boldsymbol{j} &= \psi^\dagger\boldsymbol{\alpha}\psi \end{aligned}\right\} \quad \frac{\partial\rho}{\partial t} + \boldsymbol{\nabla}\cdot\boldsymbol{j} = 0. \tag{3.75}$$

Thus for any solution of the form

$$\psi = \omega\phi(\boldsymbol{x}, t) \tag{3.76}$$

we have

$$\rho = \omega^\dagger \omega |\phi(\boldsymbol{x}, t)|^2 \tag{3.77a}$$

$$\boldsymbol{j} = \omega^\dagger \boldsymbol{\alpha} \omega |\phi(\boldsymbol{x}, t)|^2 \tag{3.77b}$$

and $\rho \geqslant 0$ always. We nevertheless want to set up a correspondence so that positive-energy solutions describe *electrons* and negative-energy solutions describe *positrons*, if we reverse the sense of incoming and outgoing waves. For the KG case this was straightforward, since the probability current was proportional to the 4-momentum:

$$j^\mu(\text{KG}) \sim p^\mu. \tag{3.78}$$

We were therefore able to set up the correspondence for the electromagnetic current of π^+ and π^-:

$$\pi^+: \quad j^\mu_{\text{em}} \sim ep^\mu \qquad \text{positive-energy } \pi^+ \tag{3.79a}$$

$$\pi^-: \quad j^\mu_{\text{em}} \sim (-e)p^\mu \qquad \text{positive-energy } \pi^- \tag{3.79b}$$

$$\equiv (+e)(-p^\mu) \quad \text{negative-energy } \pi^+. \tag{3.79c}$$

This simple connection does not hold for the Dirac case since $\rho \geqslant 0$ for both signs of the energy. It is still possible to set up the correspondence, but now an extra minus sign must be inserted 'by hand' whenever we have a negative-energy fermion in the final state. We shall see in Chapter 6 that this will ensure that the e^+ scattering matrix element has the opposite sign for e^- scattering, as we require physically. We therefore state the Feynman hypothesis for fermions:

> the invariant amplitude for the emission (absorption) of an antifermion of 4-momentum p^μ and spin projection s_z in the rest frame is equal to the amplitude (minus the amplitude) for the absorption (emission) of a fermion of 4-momentum $-p^\mu$ and spin projection $-s_z$ in the rest frame.

3.6 Inclusion of electromagnetic interactions via the gauge principle

Having set up the relativistic spin-0 and spin-$\frac{1}{2}$ free particle wave equations, we are now in a position to use the machinery developed in Chapter 2, in order to include electromagnetic interactions. All we have to do is make the replacement

$$\partial^\mu \to D^\mu \equiv \partial^\mu + \mathrm{i}qA^\mu \tag{2.33}$$

for a particle of charge q. For the spin-0 KG equation (3.1) we obtain, after some rearrangement (problem 3.6),

$$(\Box + m^2)\,\phi = -iq\,(\partial_\mu A^\mu + A^\mu \partial_\mu)\,\phi + q^2 A^2 \phi \tag{3.80}$$

$$\equiv -\hat{V}_{\mathrm{KG}}\phi. \tag{3.81}$$

Note that the potential $\hat{V}_{\mathrm{KG}}$ contains the differential operator ∂_μ; the sign of $\hat{V}_{\mathrm{KG}}$ is a convention chosen so as to maintain the same relative sign between $\boldsymbol{\nabla}^2$ and $\hat{V}$ as in the Schrödinger equation. For the spin-$\frac{1}{2}$ Dirac equation (3.26) the same manoeuvre produces

$$\mathrm{i}\frac{\partial\psi}{\partial t} = (-\mathrm{i}\boldsymbol{\alpha}\cdot\boldsymbol{\nabla} + \beta m + \hat{V}_{\mathrm{D}})\psi \tag{3.82}$$

where

$$\hat{V}_{\mathrm{D}} = qA^0\mathbf{1} - q\boldsymbol{\alpha}\cdot\boldsymbol{A}. \tag{3.83}$$

We note that $\hat{V}_{\mathrm{D}}$ is a 4×4 matrix

$$\hat{V}_{\mathrm{D}} = q\begin{pmatrix} A^0 & -\boldsymbol{\sigma}\cdot\boldsymbol{A} \\ -\boldsymbol{\sigma}\cdot\boldsymbol{A} & A^0 \end{pmatrix} \tag{3.84}$$

which acts on the four-component Dirac spinor. Thus the electromagnetic interactions of spin-0 and spin-$\frac{1}{2}$ particles are completely determined by the gauge principle. (The position is *not* so simple for spin-1 particles, as we shall see in Chapter 8.)

Summary

Klein–Gordon equation introduced and abandoned because of problems with negative-energy solutions and negative probabilities.

Dirac equation overcomes negative probabilities but still allows negative energies. Dirac particles described by four-component spinor. Dirac matrices obey anticommutation relations.

Dirac sea reinterprets negative-energy solutions of Dirac equation in terms of antiparticles.

Feynman prescription allows negative-energy solutions of both Dirac and Klein–Gordon equations to be reinterpreted.

Inclusion of electromagnetism via the gauge principle.

Problems

3.1 (a) In natural units $\hbar = c = 1$ and with $2m = 1$, the Schrödinger equation may be written as

$$-\boldsymbol{\nabla}^2\psi + V\psi - \mathrm{i}\partial\psi/\partial t = 0.$$

Multiply this equation from the left by ψ^* and multiply the complex conjugate of this equation by ψ (assume V is real). Subtract the two equations and show that your answer may be written in the form of a continuity equation

$$\partial\rho/\partial t + \nabla\cdot\boldsymbol{j} = 0$$

where $\rho = \psi^*\psi$ and $\boldsymbol{j} = \mathrm{i}^{-1}\,[\psi^*(\nabla\psi) - (\nabla\psi^*)\psi]$.

(b) Perform the same operations for the Klein–Gordon equation and derive the corresponding 'probability' density and current. Show also that for a free particle solution

$$\phi = N\mathrm{e}^{-\mathrm{i}p\cdot x}$$

with $p^\mu = (E, \boldsymbol{p})$, the probability current $j^\mu = (\rho, \boldsymbol{j})$ is proportional to p^μ.

3.2 (a) Using the explicit forms for the 2×2 Pauli matrices, verify the commutation (square brackets) and anticommutation (braces) relations†

$$[\sigma_i, \sigma_j] = 2\mathrm{i}\varepsilon_{ijk}\sigma_k, \qquad \{\sigma_i, \sigma_j\} = 2\delta_{ij}\mathbf{1}$$

where ε_{ijk} is the usual antisymmetric tensor

$$\varepsilon_{ijk} = \begin{cases} +1 & \text{for an even permutation of 123} \\ -1 & \text{for an odd permutation of 123} \\ 0 & \text{if two or more indices are the same} \end{cases}$$

δ_{ij} is the usual Kronecker delta and $\mathbf{1}$ is the 2×2 unit matrix. Hence show that

$$\sigma_i\sigma_j = \delta_{ij}\mathbf{1} + \mathrm{i}\varepsilon_{ijk}\sigma_k.$$

(b) Use this last identity to prove the result

$$(\boldsymbol{\sigma}\cdot\boldsymbol{a})(\boldsymbol{\sigma}\cdot\boldsymbol{b}) = \boldsymbol{a}\cdot\boldsymbol{b}\mathbf{1} + \mathrm{i}\boldsymbol{\sigma}\cdot\boldsymbol{a}\times\boldsymbol{b}.$$

Using the explicit 2×2 form for

$$\boldsymbol{\sigma}\cdot\boldsymbol{p} = \begin{pmatrix} p_z & p_x - \mathrm{i}p_y \\ p_x + \mathrm{i}p_y & -p_z \end{pmatrix}$$

show that

$$(\boldsymbol{\sigma}\cdot\boldsymbol{p})^2 = \boldsymbol{p}^2\mathbf{1}.$$

3.3 For free particle solutions of the Dirac equation

$$\psi = \omega\mathrm{e}^{-\mathrm{i}p\cdot x}$$

the four-component spinor ω may be written in terms of two two-component spinors

†Note summation convention: see footnote on p 510.

$$\omega = \begin{pmatrix} \phi \\ \chi \end{pmatrix}.$$

From the Dirac equation for ψ

$$\mathrm{i}\partial\psi/\partial t = (-\mathrm{i}\boldsymbol{\alpha}\cdot\boldsymbol{\nabla} + \beta m)\psi$$

using the explicit forms for the Dirac matrices

$$\boldsymbol{\alpha} = \begin{pmatrix} \mathbf{0} & \boldsymbol{\sigma} \\ \boldsymbol{\sigma} & \mathbf{0} \end{pmatrix}, \qquad \beta = \begin{pmatrix} \mathbf{1} & \mathbf{0} \\ \mathbf{0} & -\mathbf{1} \end{pmatrix}$$

show that ϕ and χ satisfy the coupled equations

$$(E - m)\phi = \boldsymbol{\sigma}\cdot\boldsymbol{p}\chi, \qquad (E + m)\chi = \boldsymbol{\sigma}\cdot\boldsymbol{p}\phi$$

where $p^\mu = (E, \boldsymbol{p})$.

3.4 (a) Verify explicitly that the matrices $\boldsymbol{\alpha}$ and β of problem 3.3 satisfy the Dirac anticommutation relations. Defining the four 'γ matrices'

$$\gamma^\mu = (\gamma^0, \boldsymbol{\gamma})$$

where $\gamma^0 = \beta$ and $\boldsymbol{\gamma} = \beta\boldsymbol{\alpha}$, show that the Dirac equation can be written in the form $(\mathrm{i}\gamma^\mu\partial_\mu - m)\psi = 0$. Find the anticommutation relations of the γ matrices.

(b) Define the conjugate spinor

$$\bar{\psi}(x) = \psi^\dagger(x)\gamma^0$$

and use the result above to find the equation satisfied by $\bar{\psi}$ in γ matrix notation.

(c) The Dirac probability current may be written as

$$j^\mu = \bar{\psi}(x)\gamma^\mu\psi(x).$$

Show that it satisfies the conservation law

$$\partial_\mu j^\mu(x) = 0.$$

3.5 As will be discussed further in §6.1 below, it is conventional to define positive-energy spinors $u(p, s)$ which differ from (3.64) by a factor:

$$u(p, s) = (E + m)^{1/2} \begin{pmatrix} \phi^s \\ \dfrac{\boldsymbol{\sigma}\cdot\boldsymbol{p}}{E + m}\phi^s \end{pmatrix} \qquad s = 1, 2.$$

Verify that these satisfy $u^\dagger u = 2E$.

In a similar way, negative-energy spinors $v(p, s)$ are defined by

$$v(p, s) = (E + m)^{1/2} \begin{pmatrix} \dfrac{\boldsymbol{\sigma}\cdot\boldsymbol{p}}{E + m}\chi^s \\ \chi^s \end{pmatrix} \qquad s = 1, 2$$

Verify that $v^\dagger v = 2E$.

3.6 Using the KG equation together with the replacement $\partial^\mu \to \partial^\mu + iqA^\mu$, find the form of the potential $\hat{V}_{KG}$ in the resulting equation

$$(\Box + m^2)\,\phi = -\hat{V}_{KG}\phi$$

in terms of A^μ.

It was a wonderful world my father told me about.

You might wonder what he got out of it all. I went to MIT. I went to Princeton. I came home, and he said, "Now you've got a science education. I have always wanted to know something that I have never understood; and so, my son, I want you to explain it to me." I said yes.

He said, "I understand that they say that light is emitted from an atom when it goes from one state to another, from an excited state to a state of lower energy."

I said, "That's right."

"And light is a kind of particle, a photon, I think they call it."

"Yes."

"So if the photon comes out of the atom when it goes from the excited to the lower state, the photon must have been in the atom in the excited state."

I said, "Well, no."

He said, "Well, how do you look at it so you can think of a particle photon coming out without it having been in there in the excited state?"

I thought a few minutes, and I said, "I'm sorry; I don't know. I can't explain it to you."

He was very disappointed after all these years and years of trying to teach me something, that it came out with such poor results.

R P Feynman, *The Physics Teacher*, vol 7, No 6, September 1969

All the fifty years of conscious brooding have brought me no closer to the answer to the question, 'What are light quanta?' Of course today every rascal thinks he knows the answer, but he is deluding himself.

A Einstein (1951)

Quoted in 'Einstein's researches on the nature of light'
E Wolf (1979), *Optics News*, vol 5, No 1, page 39.

I never satisfy myself until I can make a mechanical model of a thing. If I can make a mechanical model I can understand it. As long as I cannot make a mechanical model all the way through I cannot understand; and that is why I cannot get the electromagnetic theory.

Sir William Thomson, Lord Kelvin, 1884 *Notes of Lectures on Molecular Dynamics and the Wave Theory of Light delivered at the Johns Hopkins University, Baltimore*, stenographic report by A S Hathaway (Baltimore: Johns Hopkins University) Lecture XX, pp 270–1.

4

QUANTUM FIELD THEORY

In this chapter we shall give an elementary introduction to quantum field theory, which is the best 'language' currently available for the description of the fundamental theories of matter and force. Even so long after Maxwell's theory of the (classical) electromagnetic field, the concept of a 'disembodied' field is not an easy one; and we are going to have to add the complications of quantum mechanics to it. In such a situation, it is helpful to have some physical model in mind. For most of us, this still means a mechanical model. Thus in the following two sections we begin by considering a mechanical model for a quantum field. At the end, we shall—like Maxwell—throw away the 'mechanism' and have simply quantum field theory, the new fundamental reality which has no need of any further underlying framework. Section 4.1 describes this programme qualitatively; §4.2 presents a more complete formalism.

4.1 The quantum field: (i) descriptive

Mechanical systems are usefully characterised by the number of *degrees of freedom* (DF) they possess: thus a one-dimensional pendulum has one DF, two coupled one-dimensional pendulums have two DF—which may be taken to be their angular displacements, for example. A scalar field $\phi(\boldsymbol{x}, t)$ corresponds to a system with an infinite number of DF, since at each continuously varying point $\boldsymbol{x}$ an independent 'displacement' $\phi(\boldsymbol{x}, t)$, which also varies with time, has to be determined. Thus quantum field theory involves two major mathematical steps: the description of continuous systems (fields) which have infinitely many degrees of freedom, and the application of *quantum* theory to such systems. These two aspects are clearly separable. It is certainly easier to begin by considering systems with a discrete—but possibly very large—number of degrees of freedom, for example a solid. We shall treat such systems first classically and then quantum mechanically. Then, returning to the classical case, we shall allow the number of degrees of freedom to become infinite, so that the system corresponds to a classical field. Finally, we shall apply quantum mechanics directly to fields.

We begin by considering a rather small solid—one that has only two atoms free to move. The atoms, each of mass m, are connected by a string, and each is also connected to a fixed support by a similar string (figure 4.1(a)); all the strings are under tension T. We consider small transverse vibrations of the atoms (figure 4.1(b)), and we call $q_r(t)$ ($r = 1, 2$) the transverse displacements. We are interested in the total energy E of this system. According to classical mechanics, this is equal to the sum of the kinetic energies $\frac{1}{2}m\dot{q}_r^2$ of each atom, together with a potential energy V which can be calculated as follows. Referring to figure 4.1(b), when atom 1 is displaced by q_1, it experiences a restoring force

$$F_1 = T \sin \alpha - T \sin \beta \tag{4.1}$$

assuming a constant tension T along the string. For small displacements q_1 and q_2 we have

$$\begin{aligned} \sin \alpha &= q_1/(l^2 + q_1^2)^{1/2} \approx q_1/l \\ \sin \beta &= (q_2 - q_1)/[l^2 + (q_2 - q_1)^2]^{1/2} \approx (q_2 - q_1)/l. \end{aligned} \tag{4.2}$$

Thus the restoring force on particle 1 is, in this approximation,

$$F_1 = k(2q_1 - q_2) \tag{4.3}$$

with $k = T/l$. Similarly, the restoring force on particle 2 is

$$F_2 = k(2q_2 - q_1) \tag{4.4}$$

and the equations of motion are

$$m\ddot{q}_1 = -\, k(2q_1 - q_2) \tag{4.5}$$

$$m\ddot{q}_2 = -\, k(2q_2 - q_1). \tag{4.6}$$

The potential energy is then determined (up to an irrelevant constant) by the requirement that (4.5) and (4.6) are of the form

$$m\ddot{q}_1 = -\, \partial V/\partial q_1 \tag{4.7}$$

$$m\ddot{q}_2 = -\, \partial V/\partial q_2. \tag{4.8}$$

Thus we deduce that

$$V = k(q_1^2 + q_2^2 - q_1 q_2). \tag{4.9}$$

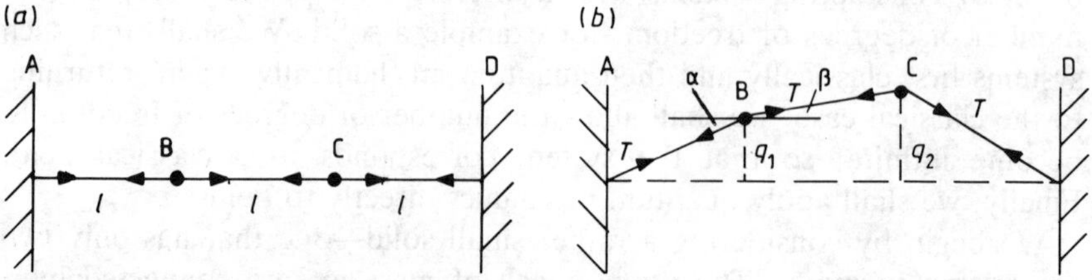

Figure 4.1 A vibrating system with two degrees of freedom: (a) two mass points at rest, with the strings under tension; (b) a small transverse displacement.

Equations (4.5) and (4.6) form a pair of *coupled* differential equations, i.e. the right-hand side of the q_1 equation depends on q_2 as well as q_1, and similarly for the q_2 equation. This 'mathematical' coupling has its origin in the term $-kq_1q_2$ in V, which corresponds to the 'physical' coupling of the string BC connecting the two atoms. If this coupling were absent, equations (4.5) and (4.6) would describe two independent (uncoupled) harmonic oscillators, each of frequency $(2k/m)^{1/2}$. With this coupling, the solutions of (4.5) and (4.6) are not quite so obvious. However, a simple mathematical step makes the equations much easier. Suppose we add the two equations so as to obtain

$$m(\ddot{q}_1 + \ddot{q}_2) = -k(q_1 + q_2) \tag{4.10}$$

and subtract them to obtain

$$m(\ddot{q}_1 - \ddot{q}_2) = -3k(q_1 - q_2). \tag{4.11}$$

A remarkable thing has happened: the two *combinations* $q_1 + q_2$ and $q_1 - q_2$ of the original coordinates satisfy *uncoupled* equations—which are of course very easy to solve. The combination $q_1 + q_2$ oscillates with frequency $\omega_1 = (k/m)^{1/2}$, while $q_1 - q_2$ oscillates with frequency $\omega_2 = (3k/m)^{1/2}$.

Let us introduce

$$Q_1 = (q_1 + q_2)/\sqrt{2}, \qquad Q_2 = (q_1 - q_2)/\sqrt{2} \tag{4.12}$$

(the $\sqrt{2}$'s are for later convenience). Then the solutions of (4.10) and (4.11) are

$$Q_1(t) = A \cos \omega_1 t + B \sin \omega_1 t \tag{4.13}$$

$$Q_2(t) = C \cos \omega_2 t + D \sin \omega_2 t. \tag{4.14}$$

Suppose that the initial conditions are such that

$$q_1(0) = q_2(0) = a, \qquad \dot{q}_1(0) = \dot{q}_2(0) = 0; \tag{4.15}$$

i.e. the atoms are released from rest, at equal transverse displacements a. In terms of the Q_r's, the conditions (4.15) are

$$\begin{aligned} Q_2(0) &= \dot{Q}_2(0) = 0 \\ Q_1(0) &= \sqrt{2}\, a, \qquad \dot{Q}_1(0) = 0. \end{aligned} \tag{4.16}$$

Thus from (4.13) and (4.14) we find that the complete solution, for these initial conditions, is

$$Q_1(t) = \sqrt{2}\, a \cos \omega_1 t \tag{4.17a}$$

$$Q_2(t) = 0. \tag{4.17b}$$

We see from (4.17b) that the motion is such that $q_1 = q_2$ throughout, and from (4.17a) that the system vibrates with a single definite frequency ω_1. A form of motion in which the system as a whole moves with a definite frequency is called a '*normal mode*', or simply a 'mode' for short. Figure 4.2(*a*) shows two 'snapshot' configurations of our two-atom system when it is oscillating in the mode characterised by $q_1 = q_2$. In this mode only $Q_1(t)$ changes; $Q_2(t)$ is always zero. Another mode also exists in which $q_1 = -q_2$ at all times: here $Q_1(t)$ is zero and Q_2 oscillates with frequency ω_2. Figure 4.2(*b*) shows two snapshots of the atoms when they are vibrating in this second mode. The coordinate combinations Q_1, Q_2, in terms of which this 'single-frequency motion' occurs, are called 'normal mode coordinates', or '*normal coordinates*' for short.

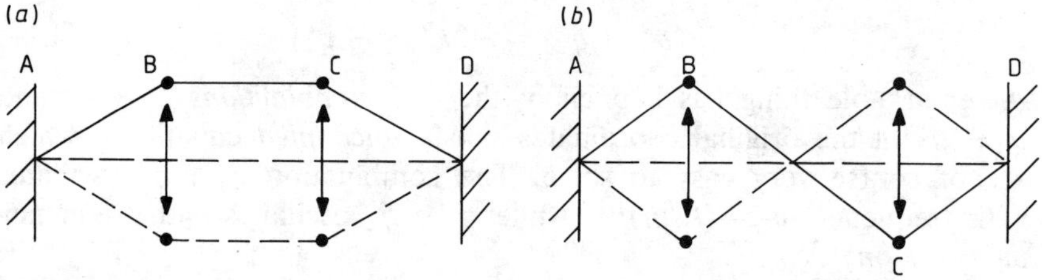

Figure 4.2 Motion in the two normal modes: (*a*) frequency ω_1; (*b*) frequency ω_2.

In general, the initial conditions will not be such that the motion is a pure mode; both $Q_1(t)$ and $Q_2(t)$ will be non-zero. From (4.12) we have

$$q_1(t) = [Q_1(t) + Q_2(t)]/\sqrt{2} \tag{4.18}$$

and

$$q_2(t) = [Q_1(t) - Q_2(t)]/\sqrt{2} \tag{4.19}$$

so that q_1 and q_2 are expressed as a sum of two terms oscillating with frequencies ω_1 and ω_2. We say the system is in 'a superposition of modes'. Nevertheless, the mode idea is still very important as regards the total energy of the system, as we shall now see. The kinetic energy can be written in terms of the mode coordinates Q_r as

$$T = \tfrac{1}{2}m\dot{Q}_1^2 + \tfrac{1}{2}m\dot{Q}_2^2 \tag{4.20}$$

while the potential energy becomes

$$V = \tfrac{1}{2}m\omega_1^2Q_1^2 + \tfrac{1}{2}m\omega_2^2Q_2^2 \equiv V(Q_1, Q_2). \tag{4.21}$$

The total energy is therefore

$$E = [\tfrac{1}{2}m\dot{Q}_1^2 + \tfrac{1}{2}m\omega_1^2 Q_1^2] + [\tfrac{1}{2}m\dot{Q}_2^2 + \tfrac{1}{2}m\omega_2^2 Q_2^2]. \tag{4.22}$$

This equation shows that, when written in terms of the normal coordinates, the total energy contains *no* coupling terms of the form Q_1Q_2; indeed, the energy has the remarkable form of a simple sum of two independent uncoupled oscillators, one with characteristic frequency ω_1, the other with frequency ω_2. The energy (4.22) has exactly the form appropriate to a system of two *non-interacting* 'things', each executing simple harmonic motion: the 'things' are actually the two modes. Modes do not interact, whereas the original atoms do! (Though we should add the caution that this is only true here because we ignored higher-than-quadratic terms in $V(q_1, q_2)$; such higher-order 'anharmonic' corrections will produce couplings between the modes: see §4.3.) Of course, this decoupling in the expression for the total energy is reflected in the decoupling of the equations of motion for the Q variables:

$$m\ddot{Q}_r = -\frac{\partial V(Q_1, Q_2)}{\partial Q_r}, \quad r = 1, 2. \tag{4.23}$$

The system under discussion had just two DF. We began by describing it in terms of the obvious DF, the physical displacements of the two atoms q_1 and q_2. But we have learned that it is very illuminating to describe it in terms of the normal coordinate combinations Q_1 and Q_2. The normal coordinates are really the relevant degrees of freedom. Of course, for just two particles, the choice between the q_r's and the Q_r's may seem rather academic; but the important point—and the reason for going through these simple manipulations in detail—is that the basic idea of the normal mode, and of normal coordinates, generalises immediately to the much less trivial N-atom problem (and also to the field problem). For N atoms there are (for one-dimensional displacements) N DF, and if we take them to be the actual atomic displacements, the total energy will be

$$E = \sum_{r=1}^{N} \tfrac{1}{2}m\dot{q}_r^2 + V(q_1, q_2, \ldots, q_N) \tag{4.24}$$

which includes all the couplings between the atoms. We assume, as before, that the q_r's are small enough so that only quadratic terms need be kept in V; then by a linear transformation of variables (generalising (4.12))

$$Q_r = \sum_{s=1}^{N} a_{rs} q_s \tag{4.25}$$

it is possible to write E as a sum of N separate terms, just as in (4.22):

$$E = \sum_{r=1}^{N} [\tfrac{1}{2}m\dot{Q}_r^2 + \tfrac{1}{2}m\omega_r^2 Q_r^2]. \tag{4.26}$$

The Q_r's are the normal coordinates and the ω_r's are the normal frequencies, and there are N of them. If only one of the Q_r's is non-zero, the N atoms are moving in a single mode. The fact that the total energy in (4.26) is a *sum* of N single-mode energies allows us to say that our N-atom solid *behaves as if it consisted of N separate and free harmonic oscillators*—which, however, are *not* to be identified with the coordinates of the original atoms. Once again, and now much more crucially, it is the *mode coordinates* that are the relevant DF rather than those of the original particles.

The second stage in our programme is to treat such systems quantum mechanically, as we should certainly have to for a real solid. It is still true that—if the potential energy is a quadratic function of the displacements—the transformation (4.25) allows us to write the total energy as a sum of N mode energies, each of which has the form of a harmonic oscillator. Now, however, these oscillators obey the laws of quantum mechanics, so that each mode oscillator exists only in certain definite states, whose energy eigenvalues are quantised. For each mode of frequency ω_r, the allowed energy values are

$$\varepsilon_r = (n_r + \tfrac{1}{2})\hbar\omega_r \tag{4.27}$$

where n_r is a positive integer or zero. This is in sharp contrast to the classical case, of course, in which arbitrary values are allowed for the oscillator energies. The total energy eigenvalue then has the form

$$E = \sum_{r=1}^{N}(n_r + \tfrac{1}{2})\hbar\omega_r. \tag{4.28}$$

The frequencies ω_r are determined by the interatomic forces and are common to both the classical and quantum descriptions; in quantum theory, though, *the states of definite energy of the vibrating N-body system are characterised by the values of a set of integers* (n_1, n_2, . . ., n_N), *which determine the energies of each mode oscillator.*

For each mode oscillator, $\hbar\omega_r$ measures the quantum of vibrational energy; the energy of an allowed mode state is determined uniquely by the number n_r of such quanta of energy in the state. We now make a profound reinterpretation of this result (first given, almost *en passant*, by Born, Heisenberg and Jordan (1926) in one of the earliest papers on quantum mechanics). We forget about the original N DF $q_1, q_2, \ldots, q_N$ and the original N 'atoms', which indeed are only remembered in (4.28) via the fact that there are N different mode frequencies ω_r. Instead, we concentrate on the *quanta* and treat *them* as the 'things' which really determine the behaviour of our quantum system. We say that 'in a state with energy $(n_r + \frac{1}{2})\hbar\omega_r$ there are n_r quanta present'. For the state characterised by $(n_1, n_2, \ldots, n_N)$ there are n_1 quanta of mode 1 (frequency ω_1), n_2 of mode 2, . . ., and n_N of mode N. Note particularly that although the number of modes N is (so far) finite, the values of the n_r's are unrestricted, except insofar as the total energy is fixed. Thus we are moving from a 'fixed number' picture (N DF) to a 'variable number' picture (the n_r's restricted only by the total energy constraint (4.28)). In the case of a real solid, these quanta of vibrational energy are called *phonons*. We summarise the point we have reached by the important statement that *a phonon is an elementary quantum of vibrational excitation*.

Now we take one step backward in order, afterwards, to take two steps forward. We return to the classical mechanical model with N harmonically interacting degrees of freedom. It is possible to imagine increasing the number N to infinity, and decreasing the interatomic spacing a to zero, in such a way that the product Na stays finite, say $Na = L$. We then have a classical *continuous* system—for example, a string of length L. (We stay in one dimension for simplicity.) The transverse vibrations of this string are now described by a *field* $\phi(x, t)$, where at each point x of the string $\phi(x, t)$ measures the displacement from equilibrium, at time t, of a small element of string around the point x. Thus we have passed from a system described by a discrete number of degrees of freedom, $q_r(t)$ or $Q_r(t)$, to one described by a continuous degree of freedom, the displacement field $\phi(x, t)$. The discrete suffix r has become the continuous argument x—and to prepare for later abstraction, we have denoted the displacement by $\phi(x, t)$ rather than, say, $q(x, t)$.

In the continuous problem the analogue of the small-displacement assumption, which limited the potential energy in the discrete case to quadratic powers, implies that $\phi(x, t)$ obeys the wave equation

$$\frac{1}{c^2}\frac{\partial^2\phi(x, t)}{\partial t^2} = \frac{\partial^2\phi(x, t)}{\partial x^2} \tag{4.29}$$

where c is the wave propagation velocity. Again, we consider first the classical treatment of this system. Out aim is to find, for this continuous field problem, the analogue of the normal coordinates—or in physical terms, the *modes* of vibration—which were so helpful in the discrete case. Fortunately, the string's modes are very familiar. By imposing suitable boundary conditions at each end of the string, we determine the allowed wavelengths of waves travelling along the string. Suppose, for simplicity, that the string is stretched between $x = 0$ and $x = L$. This constrains $\phi(x, t)$ to vanish at these end points. A suitable form for $\phi(x, t)$ which does this is

$$\phi^{(r)}(x, t) = A_r(t)\sin\left(\frac{r\pi x}{L}\right) \tag{4.30}$$

where $r = 1, 2, 3, \ldots$, which expresses the fact that an exact number of half-wavelengths must fit onto the interval $(0, L)$. Inserting (4.30) into (4.29), we find

$$\ddot{A}_r = -\omega_r^2 A_r \tag{4.31}$$

where

$$\omega_r^2 = r^2\pi^2 c^2/L^2. \tag{4.32}$$

Thus the amplitude $A_r(t)$ of the particular waveform (4.30) executes simple harmonic motion with frequency ω_r. Each motion of the string which has a definite wavelength also has a definite frequency; it is therefore precisely a mode. Figure 4.3(*a*) shows two snapshots of the string when it is oscillating in the mode for which $r = 1$, and figure 4.3(*b*) shows the same for the mode $r = 2$; these may be compared with figures 4.2(*a*) and 4.2(*b*). Just as in the discrete case, the general motion of the string is a superposition of modes

$$\phi(x, t) = \sum_{r=1}^{\infty} A_r(t) \sin\left(\frac{r\pi x}{L}\right); \tag{4.33}$$

in short, a Fourier series!

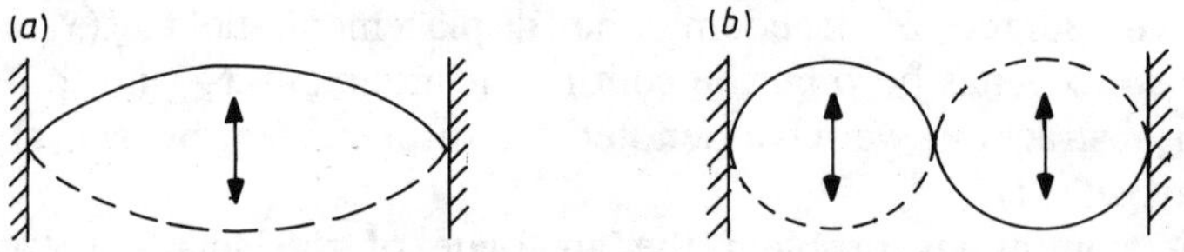

Figure 4.3 String motion in two normal modes: (*a*) $r = 1$ in equation (4.30); (*b*) $r = 2$.

We must now examine the total energy of the vibrating string, which we expect to be greatly simplified by the use of the mode concept. The total energy is the continuous analogue of the discrete summation in (4.24), namely the integral

$$E = \int_0^L \left[\frac{1}{2}\rho\left(\frac{\partial\phi}{\partial t}\right)^2 + \frac{1}{2}\rho c^2\left(\frac{\partial\phi}{\partial x}\right)^2\right] \mathrm{d}x \tag{4.34}$$

where the first term is the kinetic energy and the second is the potential energy (ρ is the mass per unit length of the string, assumed constant). Inserting (4.33) into (4.34), and using the orthonormality of the sine functions on the interval $(0, L)$, one obtains (problem 4.1) the crucial result

$$E = (L/2)\sum_{r=1}^{\infty}[\tfrac{1}{2}\rho\dot{A}_r^2 + \tfrac{1}{2}\rho\omega_r^2 A_r^2]. \tag{4.35}$$

Indeed, just as in the discrete case, the total energy of the string can be written as a sum of individual mode energies. We note that *the Fourier* *amplitude A_r acts as a normal coordinate*. Comparing (4.35) with (4.26), we see that the string behaves exactly like a system of independent uncoupled oscillators, the only difference being that now there are an infinite number of them, corresponding to the infinite number of DF in the continuous field $\phi(x, t)$. The normal coordinates $A_r(t)$ are, for many purposes, a much more relevant set of DF than the original displacements $\phi(x, t)$.

The final step is to apply quantum mechanics to this classical field system. Once again, the total energy is equivalent to that of a sum of (infinitely many) mode oscillators, each of which has to be quantised. The total energy eigenvalue has the form (4.28), except that now the sum extends to infinity:

$$E = \sum_{r=1}^{\infty} (n_r + \tfrac{1}{2})\hbar\omega_r. \tag{4.36}$$

The excited states of the quantised field $\phi(x, t)$ are characterised by saying how many phonons of each frequency are present; the ground state has no phonons at all. We remark that as $L \rightarrow \infty$, the mode sum in (4.35) or (4.36) will be replaced by an integral over a continuous frequency variable.

We have now completed, in outline, the programme introduced above, ending up with the quantisation of a 'mechanical' field system. All of the foregoing, it must be clearly emphasised, is absolutely basic to modern solid state physics. The essential idea—quantising independent modes—can be applied to an enormous variety of 'oscillations'. In all cases the crucial concept is the elementary excitation—the mode quantum. Thus we have plasmons (quanta of plasma oscillations), magnons (magnetic oscillations), . . ., as well as phonons (vibrational oscillations). All this is securely anchored in the physics of many-body systems.

Now we come to the use of these ideas as an *analogy*, to help us understand the (presumably non-mechanical) quantum fields with which we shall actually be concerned in this book—for example, the electromagnetic field. Consider a region of space containing electromagnetic fields. These fields obey (a three-dimensional version of) the wave equation (4.29), with c now standing for the speed of light. By imposing suitable boundary conditions, the total electromagnetic energy in any region of space can be written as a sum of mode energies. Each mode has the form of an oscillator, whose amplitude is (see (4.30)) the Fourier component of the wave, for a given wavelength. These oscillators are each quantised. Their quanta are called photons. Thus, *a photon is an elementary quantum of excitation of the electromagnetic field*.

So far the only kind of 'particle' we have in our relativistic quantum field theoretic world is the photon. What about the electron, say? Well, recalling Feynman again, 'There is one lucky break, however—electrons behave just like light'. In other words, we shall regard an electron also as an elementary quantum excitation of an 'electron field'. What is 'waving' to supply the vibrations for this electron field? We do not answer this question just as we didn't for the photon. We *postulate* a relativistic quantum field for the electron which obeys some suitable wave equation—in this case, for non-interacting electrons, the Dirac equation. The field is expanded as a sum of Fourier components, as with the electromagnetic field. Each component behaves as an independent oscillator degree of freedom (and there are, of course, an infinite number of them); the quanta of these oscillators are electrons.

Actually the above, though correctly expressing the basic idea, omits one crucial factor, which makes it almost fraudulently oversimplified. There is of course one very big difference between photons and electrons. The former are *bosons* and the latter *fermions*; photons have spin angular momentum of one (in units of $\hbar$), electrons of one half. It is very difficult, if not downright impossible, to construct any mechanical model at all which has fermionic excitations. Phonons have spin 1, in fact, corresponding to three states of polarisation of the corresponding vibrational waves. But 'phonons' carrying spin $\frac{1}{2}$ are hard to come by. No matter, you may say, Maxwell has weaned us away from jelly, so we shall be grown up and boldly postulate the electron field as a basic *thing*.

Certainly this is what we do. But we also know that fermionic particles, like electrons, have to obey an exclusion principle: no two identical fermions can have the same quantum numbers. Some of the consequences of this for our field–particle view of electrons will be discussed in later sections. For the moment we merely register the fact that the world seems to consist of two types of basic field: 'force fields', whose quanta are bosons (like the photons for the electromagnetic force field); and what we might call 'matter fields', whose quanta are fermions (like the electrons). To construct the world, then, one has to postulate a set of matter fields (e.g. quarks and leptons) and a set of force fields (e.g. the electroweak fields, the strong force fields and the gravitational fields), and explain how they are all to interact with each other. That is what we shall be doing in subsequent chapters, with the aid of the gauge principle.

4.2 The quantum field: (ii) Lagrange–Hamilton formulation

4.2.1 The action principle: Lagrangian particle mechanics

We must now make the foregoing qualitative picture more mathematically precise. It is clear that we would like a formalism capable of treating, within a single overall framework, the mechanics of both fields and particles, in both classical and quantum aspects. Remarkably enough, such a framework does exist (and was developed long before quantum theory): Hamilton's principle of *least action*, with the 'action' defined in terms of a *Lagrangian*. We strongly recommend the reader with no prior acquaintance with this profound approach to physical laws to read Chapter 19 of Volume 2 of Feynman's *Lectures on Physics* (Feynman 1964).

The least action approach differs radically from the more familiar one which can conveniently be called 'Newtonian'. Consider the simplest case, that of classical particle mechanics. In the Newtonian approach, equations of motion are postulated which involve forces as the essential physical input; from these, the trajectories of the particles can be calculated. In the least action approach, equations of motion are *not* postulated as basic, and the primacy of forces yields to that of *potentials*. The path by which a particle actually travels is determined by the postulate (or principle) that it has to follow that particular path, out of infinitely many possible ones, for which a certain quantity—the *action*—is minimised. The action S is defined by

$$S = \int_{t_1}^{t_2} L(q(t), \dot{q}(t))\mathrm{d}t \qquad (4.37)$$

where $q(t)$ is the position of the particle as a function of time, $\dot{q}(t)$ is its velocity and the all-important function L is the Lagrangian. Given L as an explicit function of the variables $q(t)$ and $\dot{q}(t)$, we can imagine evaluating S for all sorts of possible $q(t)$'s, starting at time t_1 and ending at time t_2. We can draw these different *possible trajectories* on a q versus t diagram as in figure 4.4. For each path we evaluate S: the *actual* path is the one for which S is smallest, by hypothesis.

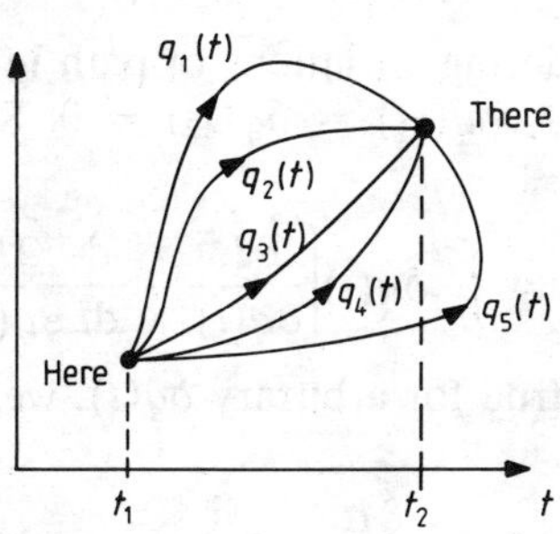

Figure 4.4 Possible space–time trajectories from 'here' ($q(t_1)$) to 'there' ($q(t_2)$).

But what is L? In simple cases (as we shall verify below) L is just T–V, the difference of kinetic and potential energies. Thus for a single particle in a potential V,

$$L = \tfrac{1}{2}m\dot{x}^2 - V(x). \tag{4.38}$$

Knowing $V(x)$, we can try and put the 'action principle' into action. However, how can we set about finding which trajectory minimises S? It is quite interesting to play with some simple specific examples and actually calculate S for several 'fictitious' trajectories—i.e. ones that we know from the Newtonian approach are *not* followed by the particle—and try and get a feeling for what the actual trajectory that minimises S might be like (of course it is the Newtonian one—see problem 4.2). But clearly this is not a practical answer to the general problem of finding the $q(t)$ that minimises S. Actually, we can solve this problem by calculus.

Our problem is something like the familiar one of finding the point t_0 at which a certain function $f(t)$ has a stationary value. In the present case, however, the function S is not just a simple function of t—rather it is a function of the entire set of points $q(t)$. It is a *function of the function* $q(t)$, or a '*functional*' of $q(t)$. We want to know what particular '$q_0(t)$' minimises S.

By analogy with the single-variable case, we consider a small variation $\delta q(t)$ in the path from $q(t_1)$ to $q(t_2)$. At the minimum, the change δS corresponding to a change δq must vanish. This change in the action is given by

$$\delta S = \int_{t_1}^{t_2} \left(\frac{\partial L}{\partial q(t)} \delta q(t) + \frac{\partial L}{\partial \dot{q}(t)} \delta \dot{q}(t) \right) \mathrm{d}t. \tag{4.39}$$

Using $\delta \dot{q}(t) = \mathrm{d}(\delta q(t))/\mathrm{d}t$ and integrating the second term by parts yields

$$\delta S = \int_{t_1}^{t_2} \delta q(t) \left[\frac{\partial L}{\partial q(t)} - \frac{\mathrm{d}}{\mathrm{d}t} \frac{\partial L}{\partial \dot{q}(t)} \right] \mathrm{d}t + \left[\frac{\partial L}{\partial \dot{q}(t)} \delta q(t) \right]_{t_1}^{t_2}. \tag{4.40}$$

Since we are considering variations of path in which all trajectories start at t_1 and end at t_2, $\delta q(t_1) = \delta q(t_2) = 0$. So the condition that S be stationary is

$$\delta S = \int_{t_1}^{t_2} \delta q(t) \left[\frac{\partial L}{\partial q(t)} - \frac{\mathrm{d}}{\mathrm{d}t} \frac{\partial L}{\partial \dot{q}(t)} \right] \mathrm{d}t = 0. \tag{4.41}$$

Since this must be true for arbitrary $\delta q(t)$, we must have

$$\boxed{\frac{\partial L}{\partial q(t)} - \frac{\mathrm{d}}{\mathrm{d}t} \frac{\partial L}{\partial \dot{q}(t)} = 0.} \tag{4.42}$$

This is the celebrated Euler–Lagrange equation of motion. Its solution gives the '$q_0(t)$' which the particle actually follows.

We can see how this works for the simple case (4.38) where q is the coordinate x. We have immediately

$$\partial L/\partial \dot{x} = m\dot{x} = p \tag{4.43}$$

and

$$\partial L/\partial x = -\,\partial V/\partial x = F \tag{4.44}$$

where p and F are respectively the momentum and the force of the Newtonian approach. The Euler–Lagrange equation then reads

$$F = \mathrm{d}p/\mathrm{d}t; \tag{4.45}$$

precisely the Newtonian equation of motion. For the special case of a harmonic oscillator (obviously fundamental to the quantum field idea, as §4.1 should have made clear), we have

$$L = \tfrac{1}{2}m\dot{x}^2 - \tfrac{1}{2}m\omega^2 x^2 \tag{4.46}$$

which can be immediately generalised to N independent oscillators (see §4.1) via

$$L = \sum_{r=1}^{N}(\tfrac{1}{2}m\dot{Q}_r^2 - \tfrac{1}{2}m\omega_r^2 Q_r^2). \tag{4.47}$$

We now consider the passage to quantum mechanics.

4.2.2 Quantum mechanics à la Heisenberg–Lagrange–Hamilton

It seems likely that a particularly direct correspondence between the quantum and the classical cases will be obtained if we use the Heisenberg formulation of quantum mechanics (see Appendix A). In the Schrödinger picture, dynamical variables such as position x are independent of time, and the time dependence is carried by the wavefunction. Thus we seem to have nothing like the $q(t)$'s. However, one can always do a unitary transformation to the Heisenberg formulation, in which the wavefunction is fixed and the dynamical variables change with time. This is what we want in order to parallel the Lagrangian quantities $q(t)$. But of course there is one fundamental difference between quantum mechanics and classical mechanics: in the former, the dynamical variables are operators which in general do not commute. In particular, the fundamental commutator states that ($\hbar = 1$)

$$[\hat{q}(t), \hat{p}(t)] = \mathrm{i} \tag{4.48}$$

where the ^ indicates the operator character of the quantity. Here $\hat{p}$ is defined by the generalisation of (4.43):

$$\hat{p} \equiv \partial \hat{L}/\partial \dot{\hat{q}}. \tag{4.49}$$

In this formulation of quantum mechanics we do not have the

Schrödinger-type equation of motion. Instead we have the Heisenberg equation of motion

$$\dot{\hat{A}} = -\mathrm{i}[\hat{A}, \hat{H}] \tag{4.50}$$

where the *Hamiltonian* operator $\hat{H}$ is defined in terms of the Lagrangian operator $\hat{L}$ by

$$\hat{H} = \hat{p}\dot{\hat{q}} - \hat{L} \tag{4.51}$$

and $\hat{A}$ is any dynamical observable. For example, in the oscillator case

$$\hat{L} = \tfrac{1}{2}m\dot{\hat{q}}^2 - \tfrac{1}{2}m\omega^2\hat{q}^2 \tag{4.52}$$

$$\hat{p} = m\dot{\hat{q}} \tag{4.53}$$

and

$$\hat{H} = (1/2m)\hat{p}^2 + \tfrac{1}{2}m\omega^2\hat{q}^2 \tag{4.54}$$

which is the total energy. Notice that $\hat{p}$, obtained from the Lagrangian using (4.49), had better be consistent with the Heisenberg equation of motion for the operator $\hat{A} = \hat{q}$. The Heisenberg equation of motion for $\hat{A} = \hat{p}$ leads to

$$\dot{\hat{p}} = -m\omega^2\hat{q} \tag{4.55}$$

which is an operator form of Newton's law for the harmonic oscillator. Using the expression for $\hat{p}$ (4.53), we find

$$\ddot{\hat{q}} = -\omega^2\hat{q}. \tag{4.56}$$

Now, although this looks like the familiar classical equation of motion for the position of the oscillator—and recovering it via the Lagrangian formalism is encouraging—we must be very careful to appreciate that this is an equation stating how an *operator* evolves with time. Where the quantum particle will actually be *found* is an entirely different matter. By sandwiching (4.56) between wavefunctions, we can at once see that the *average* position of the particle will follow the classical trajectory (remember that wavefunctions are independent of time in the Heisenberg formulation). But *fluctuations* about this trajectory will certainly occur: a quantum particle does not follow a ray-like classical trajectory. Come to think of it, neither does a photon . . ..

4.2.3 Interlude: the quantum oscillator

As we saw in §4.1, we need to know the energy spectrum and associated states of a quantum harmonic oscillator. This is a standard problem, but there is one particular way of solving it—the 'operator' approach due to Dirac (1981, Chapter 6)—that is so crucial to all subsequent development that we include a discussion here in the body of the text.

For the oscillator Hamiltonian

$$\hat{H} = (1/2m)\hat{p}^2 + \tfrac{1}{2}m\omega^2\hat{q}^2 \tag{4.54}$$

if $\hat{p}$ and $\hat{q}$ were not operators, we could attempt to factorise the Hamiltonian in the form '$(q + \mathrm{i}p)(q - \mathrm{i}p)$' (apart from factors of $2m$ and ω). In the quantum case, in which $\hat{p}$ and $\hat{q}$ do not commute, it still turns out to be very helpful to introduce such combinations. If we define the operator

$$\hat{a} = \frac{1}{\sqrt{2}}\left(\sqrt{m\omega}\,\hat{q} + \frac{\mathrm{i}}{\sqrt{m\omega}}\hat{p}\right) \tag{4.57}$$

and its Hermitian conjugate

$$\hat{a}^\dagger = \frac{1}{\sqrt{2}}\left(\sqrt{m\omega}\,\hat{q} - \frac{\mathrm{i}}{\sqrt{m\omega}}\hat{p}\right) \tag{4.58}$$

the Hamiltonian may be written (see problem 4.4)

$$\hat{H} = \tfrac{1}{2}(\hat{a}^\dagger\hat{a} + \hat{a}\hat{a}^\dagger)\omega = (\hat{a}^\dagger\hat{a} + \tfrac{1}{2})\omega. \tag{4.59}$$

The second form for $\hat{H}$ may be obtained from the first using the commutation relation between $\hat{a}$ and $\hat{a}^\dagger$

$$[\hat{a}, \hat{a}^\dagger] = 1 \tag{4.60}$$

derived using the fundamental commutator between $\hat{p}$ and $\hat{q}$. Using this basic commutator (4.60) and our expression for $\hat{H}$, (4.59), one can prove the relations (see problem 4.4)

$$\begin{aligned}[\hat{H}, \hat{a}] &= -\omega\hat{a}\\ [\hat{H}, \hat{a}^\dagger] &= \omega\hat{a}^\dagger.\end{aligned} \tag{4.61}$$

Consider now a state $|n\rangle$ which is an eigenstate of $\hat{H}$ with energy E_n:

$$\hat{H}|n\rangle = E_n|n\rangle. \tag{4.62}$$

Using this definition and the commutators (4.61), we can find the energy of the states $(\hat{a}^\dagger|n\rangle)$ and $(\hat{a}|n\rangle)$. We find

$$\hat{H}(\hat{a}^\dagger|n\rangle) = (E_n + \omega)(\hat{a}^\dagger|n\rangle) \tag{4.63}$$

$$\hat{H}(\hat{a}\,|n\rangle) = (E_n - \omega)(\hat{a}\,|n\rangle). \tag{4.64}$$

Thus the operators $\hat{a}^\dagger$ and $\hat{a}$ respectively raise and lower the energy of $|n\rangle$ by one unit of ω ($\hbar = 1$). Now since $\hat{H} \sim \hat{p}^2 + \hat{q}^2$ with $\hat{p}$ and $\hat{q}$ Hermitian, we can prove that $\langle\psi|\hat{H}|\psi\rangle$ is positive-definite for any state $|\psi\rangle$. Thus the operator $\hat{a}$ cannot lower the energy indefinitely—there must exist a lowest state $|0\rangle$ such that

$$\hat{a}|0\rangle = 0. \tag{4.65}$$

This defines the lowest-energy state of the system: its energy is

$$\hat{H}|0\rangle = \tfrac{1}{2}\omega|0\rangle \tag{4.66}$$

the 'zero-point energy' of the quantum oscillator. The first excited state is

$$|1\rangle = \hat{a}^{\dagger}|0\rangle$$

with energy $(1 + \frac{1}{2})\omega$. The nth state has energy $(n + \frac{1}{2})\omega$ and is proportional to $(\hat{a}^{\dagger})^n|0\rangle$. To obtain a normalisation

$$\langle n|n\rangle = 1$$

the correct normalisation factor can be shown to be (problem 4.4)

$$|n\rangle = \frac{1}{\sqrt{n!}}(\hat{a}^{\dagger})^n|0\rangle. \tag{4.67}$$

Returning to the eigenvalue equation for $\hat{H}$, we have arrived at the result

$$\hat{H}|n\rangle = (\hat{a}^{\dagger}\hat{a} + \tfrac{1}{2})\omega|n\rangle = (n + \tfrac{1}{2})\omega|n\rangle \tag{4.68}$$

so that the state $|n\rangle$ defined by (4.67) is an eigenstate of the *number operator* $\hat{n} = \hat{a}^{\dagger}\hat{a}$, with integer eigenvalue n:

$$\hat{n}|n\rangle = n|n\rangle. \tag{4.69}$$

It is straightforward to generalise all the foregoing to a system whose Lagrangian is a sum of N independent oscillators, as in (4.47):

$$\hat{L} = \sum_{r=1}^{N}(\tfrac{1}{2}m\dot{\hat{q}}_r^2 - \tfrac{1}{2}m\omega_r^2\hat{q}_r^2). \tag{4.70}$$

The required generalisation of the basic commutation relations (4.48) is

$$\begin{aligned} [\hat{q}_r, \hat{p}_s] &= \mathrm{i}\delta_{rs} \\ [\hat{q}_r, \hat{q}_s] &= [\hat{p}_r, \hat{p}_s] = 0 \end{aligned} \tag{4.71}$$

since the different oscillators labelled by the index r or s are all independent. The Hamiltonian is (cf (4.26))

$$\hat{H} = \sum_{r=1}^{N}[(1/2m)\hat{p}_r^2 + \tfrac{1}{2}m\omega_r^2\hat{q}_r^2] \tag{4.72}$$

$$= \sum_{r=1}^{N}(\hat{a}_r^{\dagger}\hat{a}_r + \tfrac{1}{2})\omega_r \tag{4.73}$$

with $\hat{a}_r$ and $\hat{a}_r^{\dagger}$ defined via the analogues of (4.57) and (4.58). Since the eigenvalues of *each* number operator $\hat{n}_r = \hat{a}_r^{\dagger}\hat{a}_r$ are n_r, by the previous results, the eigenvalues of $\hat{H}$ indeed have the form (4.28),

$$E = \sum_{r=1}^{N}(n_r + \tfrac{1}{2})\omega_r. \tag{4.28}$$

The corresponding eigenstates are products $|n_1\rangle|n_2\rangle \ldots |n_N\rangle$ of N individual oscillator eigenstates, where $|n_r\rangle$ contains n_r quanta of excitation, of frequency ω_r; the product state is usually abbreviated to $|n_1, n_2, \ldots, n_N\rangle$. In the ground state of the system, each individual oscillator is unexcited: this state is $|0, 0, \ldots, 0\rangle$, which is abbreviated to $|0\rangle$, where it is understood that

$$\hat{a}_r|0\rangle = 0 \quad \text{for all } r. \tag{4.74}$$

The operators $\hat{a}_r^\dagger$ *create oscillator quanta*; the operators $\hat{a}_r$ *destroy oscillator quanta*.

4.2.4 Lagrange–Hamilton classical field mechanics

We now consider how to use the Lagrange–Hamilton approach for a *field*, starting again with the classical case and limiting ourselves to one dimension to start with.

As explained in the previous section, we shall have in mind the $N \to \infty$ limit of the N degrees of freedom case:

$$\{q_r(t);\ r = 1, 2, \ldots, N\} \xrightarrow[N\to\infty]{} \phi(x, t) \tag{4.75}$$

where x is now a continuous variable labelling the displacement of the 'string' (to picture a concrete system, see figure 4.5). At each point x we have an independent degree of freedom $\phi(x, t)$—thus the *field* system has a 'continuous infinity' of degrees of freedom. We now formulate everything in terms of a Lagrangian density $\mathscr{L}$:

$$S = \int \mathrm{d}t\, L \tag{4.76}$$

where (in one dimension)

$$L = \int \mathrm{d}x \mathscr{L}. \tag{4.77}$$

Equation (4.75) suggests that ϕ has dimension of [length], and since in the discrete case $L = T - V$, $\mathscr{L}$ has dimension [energy/length]. (In general $\mathscr{L}$ has dimension [energy/volume].)

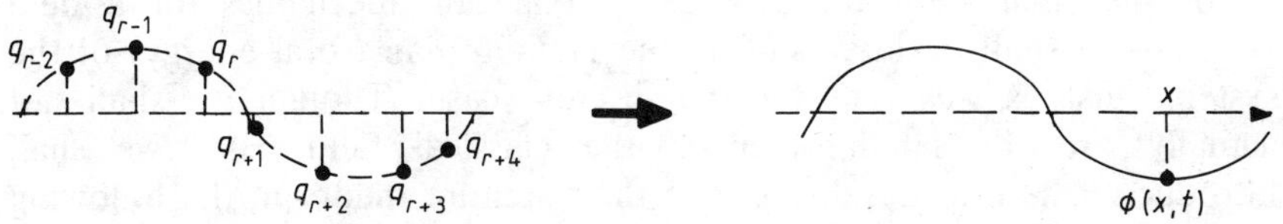

Figure 4.5 The passage from a large number of discrete degrees of freedom (mass points) to a continuous degree of freedom (field).

A new feature arises because ϕ is now a continuous function of x, so that $\mathcal{L}$ can depend on $\partial\phi/\partial x$ as well as on ϕ and $\dot{\phi} = \partial\phi/\partial t$: $\mathcal{L} = \mathcal{L}(\phi, \partial\phi/\partial x, \dot{\phi})$.

As before, we postulate the same fundamental principle

$$\delta S = 0 \tag{4.78}$$

meaning that the dynamics of the field ϕ is governed by minimising S. This time the total variation is given by

$$\delta S = \int \mathrm{d}t \int \left[\frac{\partial \mathcal{L}}{\partial \phi}\delta\phi + \frac{\partial \mathcal{L}}{\partial(\partial\phi/\partial x)}\delta\left(\frac{\partial\phi}{\partial x}\right) + \frac{\partial \mathcal{L}}{\partial \dot{\phi}}\delta\dot{\phi}\right]\mathrm{d}x. \tag{4.79}$$

Integrating the $\delta\dot{\phi}$ by parts in t, and the $\delta(\partial\phi/\partial x)$ by parts in x, and discarding the resulting 'surface' terms, we obtain

$$\delta S = \int \mathrm{d}t \int \mathrm{d}x\,\delta\phi\left[\frac{\partial \mathcal{L}}{\partial \phi} - \frac{\partial}{\partial x}\left(\frac{\partial \mathcal{L}}{\partial(\partial\phi/\partial x)}\right) - \frac{\partial}{\partial t}\left(\frac{\partial \mathcal{L}}{\partial \dot{\phi}}\right)\right] \tag{4.80}$$

and $\delta S = 0$ yields the Euler–Lagrange *field* equations

$$\frac{\partial \mathcal{L}}{\partial \phi} - \frac{\partial}{\partial x}\left(\frac{\partial \mathcal{L}}{\partial(\partial\phi/\partial x)}\right) - \frac{\partial}{\partial t}\left(\frac{\partial \mathcal{L}}{\partial \dot{\phi}}\right) = 0. \tag{4.81}$$

The generalisation to three dimensions is

$$\boxed{\frac{\partial \mathcal{L}}{\partial \phi} - \boldsymbol{\nabla}\cdot\left(\frac{\partial \mathcal{L}}{\partial(\boldsymbol{\nabla}\phi)}\right) - \frac{\partial}{\partial t}\left(\frac{\partial \mathcal{L}}{\partial \dot{\phi}}\right) = 0.} \tag{4.82}$$

As an example, consider

$$\mathcal{L} = \frac{1}{2}\rho\left(\frac{\partial\phi}{\partial t}\right)^2 - \frac{1}{2}\rho c^2\left(\frac{\partial\phi}{\partial x}\right)^2 \tag{4.83}$$

where the factors ρ (mass density) and c (a velocity) have been introduced to get the dimension of $\mathcal{L}$ right. Inserting this into the Euler–Lagrange field equation (4.81), we obtain

$$\frac{\partial^2\phi}{\partial x^2} - \frac{1}{c^2}\frac{\partial^2\phi}{\partial t^2} = 0 \tag{4.84}$$

which is precisely the wave equation, now obtained via the Euler–Lagrange field equations.

For the final step—the passage to quantum mechanics for a field system—we shall be interested in the Hamiltonian (total energy) of the system, just as we were for the discrete case. Though we shall not actually *use* the Hamiltonian in the classical field case, we shall introduce it here, generalising it to the quantum theory in the following section. We recall that Hamiltonian mechanics is formulated in terms of coordinate variables ('q') and momentum variables ('p'), rather than the

q and $\dot{q}$ of Lagrangian mechanics. In the continuum (field) case, the Hamiltonian H is written as the integral of a density $\mathcal{H}$ (we remain in one dimension)

$$H = \int \mathrm{d}x\mathcal{H} \tag{4.85}$$

while the coordinates $q_r(t)$ become the 'coordinate field' $\phi(x, t)$. The question is what is the corresponding 'momentum field'?

The answer to this is provided by a continuum version of the generalised momentum derived from the Lagrangian approach (cf equation (4.43))

$$p = \partial L/\partial \dot{q}. \tag{4.86}$$

We define a 'momentum field' $\pi(x, t)$—technically called the 'momentum canonically conjugate to ϕ'—by

$$\pi(x, t) = \partial \mathcal{L}/\partial \dot{\phi}(x, t) \tag{4.87}$$

where $\mathcal{L}$ is now the Lagrangian density. Note that π has dimensions of a momentum density. In the classical particle mechanics case we define the Hamiltonian by

$$H(p, q) = p\dot{q} - L.$$

Here we define a Hamiltonian density $\mathcal{H}$ by

$$\mathcal{H}(\phi, \pi) = \pi(x, t)\dot{\phi}(x, t) - \mathcal{L}. \tag{4.88}$$

Let us see how this works for the 1-D string with $\mathcal{L}$ given by

$$\mathcal{L} = \frac{1}{2}\rho\left(\frac{\partial \phi}{\partial t}\right)^2 - \frac{1}{2}\rho c^2\left(\frac{\partial \phi}{\partial x}\right)^2. \tag{4.83}$$

We have

$$\pi(x, t) = \rho \partial\phi/\partial t \tag{4.89}$$

and

$$\begin{aligned}\mathcal{H} &= \frac{1}{\rho}\pi^2 - \frac{1}{2}\left[\frac{1}{\rho}\pi^2 - \rho c^2\left(\frac{\partial \phi}{\partial x}\right)^2\right] \\ &= \frac{1}{2}\left[\frac{1}{\rho}\pi^2 + \rho c^2\left(\frac{\partial \phi}{\partial x}\right)^2\right]\end{aligned} \tag{4.90}$$

so that

$$H = \int\left[\frac{1}{2\rho}\pi^2(x, t) + \frac{1}{2}\rho c^2\left(\frac{\partial \phi(x, t)}{\partial x}\right)^2\right]\mathrm{d}x. \tag{4.91}$$

This has exactly the form we expect (see (4.34)), thus verifying the plausibility of the above prescriptions.

4.2.5 *Lagrange–Hamilton quantum field mechanics*

We must now *quantise* the classical field formalism: *a priori*, it is not at all obvious how to set about this task. Indeed, the fact that this formalism indicates a definite answer to this problem is the real reason for developing the quite elaborate Lagrange–Hamilton field formalism in the first place. We have introduced the 'coordinate-like' field $\phi(x, t)$ and also the 'momentum-like' field $\pi(x, t)$. To pass to the quantised version of the field theory, we mimic the procedure followed in the discrete case and promote both of the quantities ϕ and π to operators $\hat{\phi}$ and $\hat{\pi}$. As usual, the distinctive feature of quantum theory is the non-commutativity of certain basic quantities in the theory—for example, the fundamental commutator ($\hbar = 1$)

$$[\hat{q}_r(t), \hat{p}_s(t)] = \mathrm{i}\delta_{rs} \tag{4.92}$$

of the discrete case. Thus we expect that the operators $\hat{\phi}$ and $\hat{\pi}$ will obey some commutation relation which is a continuum generalisation of (4.92). The commutator will be of the form $[\hat{\phi}(x, t), \hat{\pi}(y, t)]$, since—recalling figure 4.5—the discrete index r or s becomes the continuous variable x or y; we also note that (4.92) is between operators at equal times. The continuum generalisation of the δ_{rs} symbol is the Dirac δ function, $\delta(x - y)$, with the properties

$$\int_{-\infty}^{\infty} \delta(x)\mathrm{d}x = 1 \tag{4.93}$$

$$\int_{-\infty}^{\infty} \delta(x - y)f(x)\mathrm{d}x = f(y) \tag{4.94}$$

for all reasonable functions f (see Appendix A). Thus the fundamental commutator of quantum field theory is taken to be

$$[\hat{\phi}(x, t), \hat{\pi}(y, t)] = \mathrm{i}\delta(x - y) \tag{4.95}$$

in the one-dimensional case, with obvious generalisation to the three-dimensional case via the symbol $\delta^3(\boldsymbol{x} - \boldsymbol{y})$. Remembering that we have set $\hbar = 1$, it is straightforward to check that the dimensions are consistent on both sides. Variables $\hat{\phi}$ and $\hat{\pi}$ obeying such a commutation relation are said to be 'conjugate' to each other.

Let us now proceed to explore the effect of this fundamental commutator assumption, for the case of the Lagrangian which yielded the wave equation via the Euler–Lagrange equations:

$$\hat{\mathcal{L}} = \frac{1}{2}\rho\left(\frac{\partial\hat{\phi}}{\partial t}\right)^2 - \frac{1}{2}\rho c^2\left(\frac{\partial\hat{\phi}}{\partial x}\right)^2. \tag{4.83}$$

If we replace c and ρ by 1, we can think of it as a highly simplified (spin-0, one-dimensional) version of the wave equation satisfied by the electromagnetic potential A^{μ}. We then have

$$\hat{\pi}(x, t) = \dot{\hat{\phi}}(x, t) \tag{4.96}$$

and the Hamiltonian density is

$$\hat{\mathcal{H}} = \hat{\pi}\dot{\hat{\phi}} - \hat{\mathcal{L}} = \tfrac{1}{2}\hat{\pi}^2 + \frac{1}{2}\left(\frac{\partial\hat{\phi}}{\partial x}\right)^2. \tag{4.97}$$

The total Hamiltonian is

$$\hat{H} = \int \hat{\mathcal{H}}\,\mathrm{d}x = \int \frac{1}{2}\left[\hat{\pi}^2 + \left(\frac{\partial\hat{\phi}}{\partial x}\right)^2\right]\mathrm{d}x. \tag{4.98}$$

In this form it is not immediately clear how to find the eigenvalues and eigenstates of $\hat{H}$. However, it is exactly at this point that all our preliminary work on *normal modes* comes into its own. If we can write the Hamiltonian as some kind of sum over independent oscillators—i.e. modes—we shall know how to proceed. But we learned that for the classical 'string' field considered in §4.1, the mode expansion was simply a Fourier expansion. In the present case we do not particularly want to consider a field defined on a finite interval $(0, L)$ (though sometimes one does 'put the field in a box' in this way); we shall effectively let $L \to \infty$, so that the Fourier series is replaced by a Fourier integral, and standing waves are replaced by travelling waves. For the classical field obeying the wave equation (4.29) there are plane wave solutions

$$\phi(x, t) \propto \mathrm{e}^{\mathrm{i}kx - \mathrm{i}\omega t} \tag{4.99}$$

where $(c = 1)$

$$\omega = k. \tag{4.100}$$

The general field may be Fourier expanded in terms of these solutions:

$$\phi(x, t) = \int_{-\infty}^{\infty} \frac{\mathrm{d}k}{2\pi\cdot 2\omega}[a(k)\mathrm{e}^{\mathrm{i}kx - \mathrm{i}\omega t} + a^*(k)\mathrm{e}^{-\mathrm{i}kx + \mathrm{i}\omega t}] \tag{4.101}$$

where we have required ϕ to be real. (The rather fussy factors $(2\pi\cdot 2\omega)^{-1}$ are purely conventional, and determine the normalisation of the expansion coefficients a, a^*; they have been chosen so that the normalisation of the quantum field $\hat{\phi}(x, t)$ will be consistent with the normalisation adopted for wavefunctions.) Similarly, the 'momentum field' $\pi = \dot{\phi}$ is expanded as

$$\pi(x, t) = \int_{-\infty}^{\infty} \frac{\mathrm{d}k}{2\pi\cdot 2\omega}(-\mathrm{i}\omega)[a(k)\mathrm{e}^{\mathrm{i}kx - \mathrm{i}\omega t} - a^*(k)\mathrm{e}^{-\mathrm{i}kx + \mathrm{i}\omega t}]. \tag{4.102}$$

We quantise these mode expressions by promoting $\phi \to \hat{\phi}$, $\pi \to \hat{\pi}$ and assuming the commutator (4.95). Thus we write

$$\hat{\phi}(x, t) = \int_{-\infty}^{\infty} \frac{\mathrm{d}k}{2\pi\cdot 2\omega}[\hat{a}(k)\mathrm{e}^{\mathrm{i}kx - \mathrm{i}\omega t} + \hat{a}^\dagger(k)\mathrm{e}^{-\mathrm{i}kx + \mathrm{i}\omega t}] \tag{4.103}$$

and similarly for $\hat{\pi}$. The commutator (4.95) now *determines* the commutators of the *mode operators* $\hat{a}$ and $\hat{a}^\dagger$:

$$\begin{aligned}[\hat{a}(k), \hat{a}^\dagger(k')] &= 2\pi\cdot 2\omega\delta(k-k') \\ [\hat{a}(k), \hat{a}(k')] &= [\hat{a}^\dagger(k), \hat{a}^\dagger(k')] = 0\end{aligned} \tag{4.104}$$

as shown in problem 4.6. These are the desired continuum analogues of the *discrete* oscillator commutation relations

$$\begin{aligned}[\hat{a}_r, \hat{a}_s^\dagger] &= \delta_{rs} \\ [\hat{a}_r, \hat{a}_s] &= [\hat{a}_r^\dagger, \hat{a}_s^\dagger] = 0.\end{aligned} \tag{4.105}$$

The precise factors in front of the δ function depend on the normalisation choice made in the expansion of $\hat{\phi}$, (4.103).

The form of the $\hat{a}$, $\hat{a}^\dagger$ commutation relations (4.104) already suggests that the $\hat{a}(k)$ and $\hat{a}^\dagger(k)$ operators are precisely the single-quantum destruction and creation operators for the continuum problem. To verify this interpretation and find the eigenvalues of $\hat{H}$, we now insert the expansions for $\hat{\phi}$ and $\hat{\pi}$ into $\hat{H}$. One finds the remarkable result (problem 4.7)

$$\hat{H} = \int \frac{\mathrm{d}k}{2\pi\cdot 2\omega}\{\tfrac{1}{2}[\hat{a}^\dagger(k)\hat{a}(k) + \hat{a}(k)\hat{a}^\dagger(k)]\omega\}. \tag{4.106}$$

Comparing this result with the single-oscillator result

$$\hat{H} = \tfrac{1}{2}(\hat{a}^\dagger\hat{a} + \hat{a}\hat{a}^\dagger)\omega \tag{4.59}$$

shows that, as anticipated in §4.1, each classical mode of the 'string' field can be quantised, and behaves like a separate oscillator coordinate, with its own frequency $\omega = k$. The operator $\hat{a}^\dagger(k)$ creates, and $\hat{a}(k)$ destroys, a quantum of the k mode. The factor $(2\pi\cdot 2\omega)^{-1}$ in $\hat{H}$ arises from our normalisation choice.

What is the energy spectrum? We expect the ground state to be determined by the continuum analogue of

$$\hat{a}_r|0\rangle = 0 \quad \text{for all } r; \tag{4.74}$$

namely

$$\hat{a}(k)|0\rangle = 0 \quad \text{for all } k. \tag{4.107}$$

However, there is a problem with this. We can write $\hat{H}$ as

$$\hat{H} = \int \frac{\mathrm{d}k}{2\pi\cdot 2\omega}\hat{a}^\dagger(k)\hat{a}(k)\omega + \int \frac{\mathrm{d}k}{2\pi\cdot 2\omega}\cdot\tfrac{1}{2}[\hat{a}(k), \hat{a}^\dagger(k)]\omega. \tag{4.108}$$

Now consider $\hat{H}|0\rangle$: we see from the definition of the vacuum (4.107) that the first term will give zero as expected—but the second term is infinite, since the commutation relation (4.104) produces the infinite quantity '$\delta(0)$' as $k \to k'$; moreover, the k integral diverges.

This term is obviously the continuum analogue of the zero-point energy $\frac{1}{2}\omega$—but because there are infinitely many oscillators, it is infinite. The conventional ploy is to argue that only energy *differences*, relative to a conveniently defined ground state, really matter—so that we may discard the infinite constant in (4.108). Then the ground state $|0\rangle$ has energy zero, by definition, and the eigenvalues of $\hat{H}$ are of the form

$$\int \frac{\mathrm{d}k}{2\pi \cdot 2\omega} n(k)\omega \tag{4.109}$$

where $n(k)$ is the number of quanta (counted by the number operator $\hat{a}^\dagger(k)\hat{a}(k)$) of energy $\omega = k$. For each definite k, and hence ω, the spectrum is like that of the simple harmonic oscillator.

Any desired state in which excitation quanta are present can be formed by the appropriate application of $\hat{a}^\dagger(k)$ operators to the ground state $|0\rangle$. For example, a two-quantum state containing one quantum of momentum k_1 and another of momentum k_2 may be written

$$|k_1, k_2\rangle = \hat{a}^\dagger(k_1)\hat{a}^\dagger(k_2)|0\rangle. \tag{4.110}$$

A general state will contain an arbitrary number of quanta.

Once again, and this time more formally, we have completed the programme outlined in §4.1, ending up with the 'quantisation' of a classical field $\phi(x, t)$, as exemplified in the basic expression (4.103), together with the interpretation of the operators $\hat{a}(k)$ and $\hat{a}^\dagger(k)$ as destruction and creation operators for mode quanta. We have, at least implicitly, still retained up to this point the 'mechanical model'—thus our field $\hat{\phi}(x, t)$ could describe the quantised vibrations of a (one-dimensional) solid in the Debye (continuum) approximation. The quanta are then phonons, and $|0\rangle$ is the ground state of the solid. We now throw away the mechanical props and embrace the unadorned quantum field theory! We do not ask *what* is waving, we simply postulate a field—such as ϕ—and quantise it. *Its quanta of excitation are what we call particles*—for example, photons in the electromagnetic case. We note that in the field operator $\hat{\phi}$ of (4.103), those terms which destroy quanta go with the factor $\mathrm{e}^{-\mathrm{i}\omega t}$, while those which create quanta go with $\mathrm{e}^{+\mathrm{i}\omega t}$. This choice is deliberate and goes with the 'absorption' and 'emission' factors $\mathrm{e}^{\pm \mathrm{i}\omega t}$ of ordinary time-dependent perturbation theory (Appendix A, equations (A.23) and (A.24)).

We end this long section with some further remarks about the physical interpretation of our quantum field $\hat{\phi}$. First, note that this formalism encompasses *both* the wave *and* the particle aspect of matter and radiation. The former is evident from the plane *wave* expansion functions in the expansion of $\hat{\phi}$, (4.103), which in turn originate from the fact that ϕ—and hence also $\hat{\phi}$—was taken to obey the wave equation (4.29). The latter follows from the discrete nature of the energy spectrum and the associated operators $\hat{a}$, $\hat{a}^\dagger$ which refer to individual quanta, i.e. *particles*.

Next, we may ask: what is the meaning of the ground state $|0\rangle$ for a quantum field? It is undoubtedly the state with $n(k) = 0$ for all k, i.e. the state with no quanta in it—and hence no *particles* in it, on our new interpretation. It is therefore the vacuum! As we shall see later, this understanding of the vacuum as the ground state of a field system is fundamental to much of modern particle physics—for example, to quark confinement and to the generation of mass for the weak vector bosons. Note that although we discarded the overall (infinite) constant in $\hat{H}$, differences in zero-point energies *can* be detected; for example, in the Casimir effect (Casimir 1948, Kitchener and Prosser 1957, Sparnaay 1958). These and other aspects of the quantum field theory vacuum are discussed in Aitchison (1985).

Thirdly, consider the two-particle state (4.110): $|k_1, k_2\rangle = \hat{a}^\dagger(k_1)\hat{a}^\dagger(k_2)|0\rangle$. Since the $\hat{a}^\dagger$ operators commute, (4.104), this state is symmetric under the interchange $k_1 \leftrightarrow k_2$. This is an inevitable feature of the formalism as so far developed—there is no possible way of distinguishing one quantum of energy from another, and we expect the two-quantum state to be indifferent to the order in which the quanta are put in it. However, this has an important implication for the *particle* interpretation: since the state is symmetric under interchange of the particle labels k_1 and k_2, it must describe identical *bosons*. How the formalism is modified in order to describe the antisymmetric states required for two fermionic quanta will be discussed in §4.4.

Finally, the reader may well wonder how to connect the quantum field theory formalism to ordinary 'wavefunction' quantum mechanics. The ability to see this connection will be important in subsequent chapters and it is, indeed, quite simple. Suppose we form a state containing one quantum of the $\hat{\phi}$ field, with momentum k':

$$|k'\rangle = \hat{a}^\dagger(k')|0\rangle.$$

Now consider the amplitude $\langle 0|\hat{\phi}(x, t)|k'\rangle$. We can expand this out as

$$\langle 0|\hat{\phi}(x, t)|k'\rangle = \langle 0|\int \frac{\mathrm{d}k}{2\pi\cdot 2\omega}[\hat{a}(k)\mathrm{e}^{ikx-i\omega t} + \hat{a}^\dagger(k)\mathrm{e}^{-ikx+i\omega t}]\hat{a}^\dagger(k')|0\rangle. \tag{4.111}$$

The '$\hat{a}^\dagger\hat{a}^\dagger$' term will give zero since $\langle 0|\hat{a}^\dagger = 0$. For the other term we use the commutation relations (4.104) to write it as

$$\langle 0|\int \frac{\mathrm{d}k}{2\pi\cdot 2\omega}[\hat{a}^\dagger(k')\hat{a}(k) + 2\pi\cdot 2\omega\delta(k-k')]\mathrm{e}^{ikx-i\omega t}|0\rangle = \mathrm{e}^{ik'x-i\omega' t} \tag{4.112}$$

using the vacuum condition once again, and integrating over the δ function using the property (4.94) which sets $k = k'$ and hence $\omega = \omega'$.

The vacuum is normalised to unity, $\langle 0|0\rangle = 1$. The result is just the plane wave *wavefunction* for a particle in the state $|k'\rangle$! Thus we discover that the vacuum to one-particle matrix elements of the field operators are just the familiar wavefunctions of single-particle quantum mechanics. In this connection we can explain some common terminology. The path to quantum field theory that we have followed is sometimes called 'second quantisation'—ordinary single particle quantum mechanics being the first-quantised version of the theory.

4.3 Interactions in quantum field theory

We have so far only considered free—i.e. non-interacting—quantum fields. The fact that they are non-interacting is evident in a number of ways. The mode expansion for the field $\hat{\phi}$, (4.103), is written in terms of the (free) plane wave solutions of the associated wave equation. Also, the Hamiltonian turned out to be just the sum of individual oscillator Hamiltonians for each mode frequency, (4.106). The energies of the quanta add up—they are non-interacting quanta. Finally, since the Hamiltonian $\hat{H}$ is just a sum of number operators

$$\hat{n}(k) = \hat{a}^{\dagger}(k)\hat{a}(k) \tag{4.113}$$

it is obvious that each such operator commutes with $\hat{H}$ and is therefore a constant of the motion. Thus two waves, each with one excitation quantum, travelling towards each other will pass smoothly through each other and emerge unscathed on the other side—they will not interact at all.

How can we get the mode quanta to interact? If we return to our discussion of classical mechanical systems in §4.1, we see that the crucial step in arriving at the 'sum over oscillators' form for the energy was the assumption that the potential energy was quadratic in the small displacements q_r. We expect that the 'modes will interact' when we go *beyond this harmonic approximation*. The same is true in the continuous (wave or field) case. In the derivation of the appropriate wave equation you will find that somewhere an approximation like $\tan\phi \approx \phi$ or $\sin\phi \approx \phi$ is made. This linearises the equation, and solutions to linear equations can be linearly superposed to make new solutions. If we retain higher powers of ϕ, such as ϕ^3, the resulting non-linear equation has solutions that cannot be obtained by superposing two independent solutions. Thus two waves travelling towards each other will not just pass smoothly through each other: various forms of interaction and distortion of the original waveforms will occur.

What happens when we quantise such anharmonic systems? To gain some idea of the new features that emerge, consider just one 'anharmonic' oscillator with Hamiltonian

$$\widehat{H} = (1/2m)\hat{p}^2 + \tfrac{1}{2}m\omega^2\hat{q}^2 + \lambda\,\hat{q}^3. \tag{4.114}$$

In terms of the $\hat{a}$ and $\hat{a}^\dagger$ combinations this becomes

$$\widehat{H} = \tfrac{1}{2}(\hat{a}^\dagger\hat{a} + \hat{a}\hat{a}^\dagger)\omega + \frac{\lambda}{(2m\omega)^{3/2}}(\hat{a}+\hat{a}^\dagger)^3 \tag{4.115}$$

$$\equiv \widehat{H}_0 + \widehat{H}' \tag{4.116}$$

where $\widehat{H}_0$ is our previous free oscillator Hamiltonian. The algebraic tricks we used to find the spectrum of $\widehat{H}_0$ do *not* work for this new $\widehat{H}$ because of the addition of the $\widehat{H}'$ interaction term. In particular, although $\widehat{H}_0$ commutes with the number operator $\hat{a}^\dagger\hat{a}$, $\widehat{H}'$ does not. Therefore, whatever the eigenstates of $\widehat{H}$ are, they will not in general have a definite number of '$\widehat{H}_0$ quanta'. In fact, we cannot find an exact algebraic solution to this new eigenvalue problem, and we must resort to perturbation theory or to numerical methods.

The perturbative solution to this problem treats $\widehat{H}'$ as a perturbation and expands the true eigenstates of $\widehat{H}$ in terms of the eigenstates of $\widehat{H}_0$:

$$|\bar{r}\rangle = \sum_n c_{rn}|n\rangle. \tag{4.117}$$

From this expansion we see that, as expected, the true eigenstates $|\bar{r}\rangle$ will 'contain different numbers of $\widehat{H}_0$ quanta': $|c_{rn}|^2$ is the probability of finding n '$\widehat{H}_0$ quanta' in the state $|\bar{r}\rangle$. Perturbation theory now proceeds by expanding the coefficients c_{rn} and exact energy eigenvalues $\overline{E}_r$ in power series in the strength λ of the perturbation. For example, the exact energy eigenvalue has the expansion

$$\overline{E}_r = E_r^{(0)} + \lambda E_r^{(1)} + \lambda^2 E_r^{(2)} + \ldots \tag{4.118}$$

where

$$\widehat{H}_0|r\rangle = E_r^{(0)}|r\rangle \tag{4.119}$$

and

$$E_r^{(1)} = \langle r|\widehat{H}'|r\rangle \tag{4.120}$$

$$E_r^{(2)} = \sum_{s\neq r}\frac{\langle r|\widehat{H}'|s\rangle\langle s|\widehat{H}'|r\rangle}{E_r^{(0)} - E_s^{(0)}}. \tag{4.121}$$

To evaluate the second-order shift in energy, we therefore need to consider matrix elements of the form

$$\langle s|(\hat{a} + \hat{a}^\dagger)^3|r\rangle. \tag{4.122}$$

Keeping careful track of the order of the $\hat{a}$ and $\hat{a}^\dagger$ operators, we can evaluate these matrix elements and find, in this case, that there are non-zero matrix elements for states $\langle s| = \langle r+3|$, $\langle r+1|$, $\langle r-1|$ and $\langle r-3|$.

What about the quantum mechanics of two coupled non-linear oscillators? In the same way, the general state is assumed to be a superposition

$$|\bar{r}\rangle = \sum_{n_1,n_2} c_{r,n_1n_2}|n_1\rangle|n_2\rangle \tag{4.123}$$

of states of arbitrary numbers of the quanta of the unperturbed oscillators $\hat{H}_{0(1)}$ and $\hat{H}_{0(2)}$. States of the unperturbed system contain definite numbers n_1 and n_2, say, of the '1' and '2' quanta. Perturbation calculations of the interacting system will involve matrix elements connecting such $|n_1\rangle|n_2\rangle$ states to states $|n_1'\rangle|n_2'\rangle$ with different numbers of these quanta.

All this can be summarised by the remark that the typical feature of quantised interacting modes is that we need to consider processes in which the numbers of the different mode quanta are not constants of the motion. This is, of course, exactly what happens when we have collisions between high-energy particles. When far apart, the particles are indeed free and are just the mode quanta of some quantised fields. But, when they interact, we must expect to see changes in the numbers of quanta, and can envisage processes in which a different number of quanta emerge finally as free particles from the number that originally collided. From the quantum mechanical examples we have discussed, we expect that these interactions will be produced by terms like $\hat{\phi}^3$ or $\hat{\phi}^4$, since the free—'harmonic'—case has $\hat{\phi}^2$, analogous to $\hat{q}^2$ in the quantum mechanics example. Such terms arise in the solid state phonon application precisely from anharmonic corrections involving the atomic displacements. These terms lead to non-trivial phonon–phonon scattering, the treatment of which forms the basis of the quantum theory of thermal resistivity. In the quantum field theory case, when we have generalised the formalism to fermions and photons, the non-linear interaction terms will produce e^-e^-scattering, $q\bar{q}$ annihilation and so on. We now turn to the quantum field theory of a massive Klein–Gordon field.

4.4. The quantum field theory of the Klein–Gordon field

In the previous sections we have shown how quantum mechanics may be married to field theory. Now, for the Klein–Gordon field, we must incorporate the demands of relativity. In the Lagrangian approach this is very easy to do. The classical Euler–Lagrange equation

$$\frac{\partial \mathcal{L}}{\partial \phi} - \boldsymbol{\nabla}\cdot\frac{\partial \mathcal{L}}{\partial(\boldsymbol{\nabla}\phi)} - \frac{\partial}{\partial t}\left(\frac{\partial \mathcal{L}}{\partial \dot{\phi}}\right) = 0 \tag{4.82}$$

may immediately be rewritten in relativistically invariant form

$$\frac{\partial \mathcal{L}}{\partial \phi} - \partial_\mu \left(\frac{\partial \mathcal{L}}{\partial(\partial_\mu \phi)} \right) = 0 \tag{4.124}$$

where $\partial_\mu = \partial/\partial x^\mu$. Similarly, the action

$$S = \int \mathrm{d}t \int \mathrm{d}^3x\, \mathcal{L} = \int \mathrm{d}^4x\, \mathcal{L} \tag{4.125}$$

will be relativistically invariant if $\mathcal{L}$ is, since the volume element d^4x is invariant. Thus, to construct a relativistic field theory, we have to construct an invariant density $\mathcal{L}$ and use the covariant Euler–Lagrange equation above. Thus our previous string Lagrangian

$$\mathcal{L} = \frac{1}{2}\rho\left(\frac{\partial \phi}{\partial t}\right)^2 - \frac{1}{2}\rho c^2 \left(\frac{\partial \phi}{\partial x}\right)^2 \tag{4.83}$$

with $\rho = c = 1$ generalises to

$$\mathcal{L} = \tfrac{1}{2}(\partial_\mu \phi)\cdot(\partial^\mu \phi) \tag{4.126}$$

and produces the wave equation

$$\partial_\mu \partial^\mu \phi = \left(\frac{\partial^2}{\partial t^2} - \boldsymbol{\nabla}^2\right)\phi = 0. \tag{4.127}$$

All of this goes through when the fields are quantised.

This invariant Lagrangian describes a massless particle. What Lagrangian will give us the Klein–Gordon equation? The answer is a simple generalisation of the above:

$$\mathcal{L} = \tfrac{1}{2}\partial_\mu \phi \partial^\mu \phi - \tfrac{1}{2}m^2\phi^2. \tag{4.128}$$

The plane wave solutions of the field equation—now the KG equation—have frequencies given by

$$\omega^2 = \boldsymbol{k}^2 + m^2 \tag{4.129}$$

which is the correct energy–momentum relation for a massive particle.

How do we quantise this field theory? The four-dimensional analogue of the Fourier expansion of the field ϕ takes the form

$$\hat{\phi}(\boldsymbol{x}, t) = \int_{-\infty}^{\infty} \frac{\mathrm{d}^3\boldsymbol{k}}{(2\pi)^3 2\omega} [\hat{a}(k)\mathrm{e}^{-\mathrm{i}k\cdot x} + \hat{a}^\dagger(k)\mathrm{e}^{\mathrm{i}k\cdot x}] \tag{4.130}$$

with a similar expansion for the 'conjugate momentum' $\hat{\pi} = \dot{\hat{\phi}}$:

$$\hat{\pi}(\boldsymbol{x}, t) = \int_{-\infty}^{\infty} \frac{\mathrm{d}^3\boldsymbol{k}}{(2\pi)^3 2\omega} (-\mathrm{i}\omega)[\hat{a}(k)\mathrm{e}^{-\mathrm{i}k\cdot x} - \hat{a}^\dagger(k)\mathrm{e}^{\mathrm{i}k\cdot x}]. \tag{4.131}$$

The $1/2\omega$ normalisation factors now look better motivated, since the volume element $\mathrm{d}^3\boldsymbol{k}/2\omega$ is an invariant (see (E.26)). The Hamiltonian is found to be

$$\hat{H}_{\mathrm{KG}} = \int \mathrm{d}^3\boldsymbol{x}\tfrac{1}{2}[\hat{\pi}^2 + \boldsymbol{\nabla}\hat{\phi}\cdot\boldsymbol{\nabla}\hat{\phi} + m^2\hat{\phi}^2] \tag{4.132}$$

and this can be expressed in terms of the $\hat{a}$'s and $\hat{a}^\dagger$'s using the expansions for $\hat{\phi}$ and $\hat{\pi}$ and the commutator

$$[\hat{a}(k), \hat{a}^\dagger(k')] = (2\pi)^3\cdot 2\omega\delta^3(\boldsymbol{k} - \boldsymbol{k}') \tag{4.133}$$

with all others vanishing. The result is, as expected,

$$\hat{H}_{\mathrm{KG}} = \frac{1}{2}\int \frac{\mathrm{d}^3\boldsymbol{k}}{(2\pi)^3 2\omega}[\hat{a}^\dagger(k)\hat{a}(k) + \hat{a}(k)\hat{a}^\dagger(k)]\omega \tag{4.134}$$

and, using the commutation relations and dropping the resulting infinite zero-point energy as usual, we arrive at

$$\hat{H}_{\mathrm{KG}} = \int \frac{\mathrm{d}^3\boldsymbol{k}}{(2\pi)^3 2\omega}\hat{a}^\dagger(k)\hat{a}(k)\omega. \tag{4.135}$$

This supports the physical interpretation of the mode operators $\hat{a}^\dagger$ and $\hat{a}$ as creation and destruction operators for quanta of the field $\hat{\phi}$ as before, except that now the energy–momentum relation for these particles is the relativistic one.

Since $\hat{\phi}$ is real ($\hat{\phi} = \hat{\phi}^\dagger$) and has no spin degrees of freedom, it is called a real scalar field. Only field quanta of one type enter—those created by $\hat{a}^\dagger(k)$ and destroyed by $\hat{a}(k)$. Thus $\hat{\phi}$ would correspond physically to a case where there was a unique particle state of a given mass m—for example, the π^0 field. It frequently happens, however, that for a given mass we find two distinct states, which are interpreted physically as 'particle' and 'antiparticle' states—for example (π^+, π^-), $(\mathrm{K}^0, \bar{\mathrm{K}}^0)$ and so on. To describe this situation, we clearly need to introduce two real scalar fields with the same mass, since we need two distinct types of mode operators. Consider, for example, the Lagrangian for two such free fields $\hat{\phi}_1$ and $\hat{\phi}_2$ of the same mass m:

$$\hat{\mathscr{L}} = \tfrac{1}{2}\partial_\mu\hat{\phi}_1\partial^\mu\hat{\phi}_1 - \tfrac{1}{2}m^2\hat{\phi}_1^2 + \tfrac{1}{2}\partial_\mu\hat{\phi}_2\partial^\mu\hat{\phi}_2 - \tfrac{1}{2}m^2\hat{\phi}_2^2. \tag{4.136}$$

We shall now see how this is appropriate to a 'particle–antiparticle' situation.

In general, 'particle' and 'antiparticle' are distinguished by having opposite values of one or more conserved additive quantum numbers. Since these quantum numbers are conserved, the operators corresponding to them commute with the Hamiltonian and are constant in time (in the Heisenberg formulation—see equation (4.50)); such operators are called *symmetry operators* and will be increasingly important in later chapters. For the present we consider the simplest case in which 'particle' and 'antiparticle' are distinguished by having opposite eigenvalues of just one symmetry operator. This situation is already realised in the simple Lagrangian of (4.136). The symmetry involved is just this: $\hat{\mathscr{L}}$

of (4.136) is left unchanged (is *invariant*) if $\hat{\phi}_1$ and $\hat{\phi}_2$ are replaced by $\hat{\phi}'_1$ and $\hat{\phi}'_2$, where

$$\begin{aligned}\hat{\phi}'_1 &= (\cos\alpha)\hat{\phi}_1 - (\sin\alpha)\hat{\phi}_2\\ \hat{\phi}'_2 &= (\sin\alpha)\hat{\phi}_1 + (\cos\alpha)\hat{\phi}_2.\end{aligned} \tag{4.137}$$

This is like a rotation of coordinates about the z axis of ordinary space, but of course it mixes *field* degrees of freedom, not simple spatial coordinates. The symmetry transformation (4.137) is sometimes called an 'O(2) transformation', referring to the two-dimensional rotation group O(2). We can easily check the invariance of $\hat{\mathcal{L}}$, i.e.

$$\hat{\mathcal{L}}(\hat{\phi}'_1, \hat{\phi}'_2) = \hat{\mathcal{L}}(\hat{\phi}_1, \hat{\phi}_2); \tag{4.138}$$

see problem 4.8.

Now let us see what is the conservation law associated with this symmetry. It is simpler (and sufficient) to consider an infinitesimal rotation ε for which

$$\begin{aligned}\hat{\phi}'_1 &= \hat{\phi}_1 - \varepsilon\hat{\phi}_2\\ \hat{\phi}'_2 &= \varepsilon\hat{\phi}_1 + \hat{\phi}_2\end{aligned} \tag{4.139}$$

so that we can define changes $\delta\hat{\phi}_i$ by

$$\begin{aligned}\delta\hat{\phi}_1 &\equiv \hat{\phi}'_1 - \hat{\phi}_1 = -\,\varepsilon\hat{\phi}_2\\ \delta\hat{\phi}_2 &\equiv \hat{\phi}'_2 - \hat{\phi}_2 = \varepsilon\hat{\phi}_1.\end{aligned} \tag{4.140}$$

Under this transformation $\hat{\mathcal{L}}$ is invariant, and so $\delta\hat{\mathcal{L}} = 0$. But $\hat{\mathcal{L}}$ is an explicit function of $\hat{\phi}_1$, $\hat{\phi}_2$, $\partial_\mu\hat{\phi}_1$ and $\partial_\mu\hat{\phi}_2$. Thus we can write

$$\begin{aligned}0 = \delta\hat{\mathcal{L}} = {}& \frac{\partial\hat{\mathcal{L}}}{\partial(\partial_\mu\hat{\phi}_1)}\delta(\partial_\mu\hat{\phi}_1) + \frac{\partial\hat{\mathcal{L}}}{\partial(\partial_\mu\hat{\phi}_2)}\delta(\partial_\mu\hat{\phi}_2)\\ &+ \frac{\partial\hat{\mathcal{L}}}{\partial\hat{\phi}_1}\delta\hat{\phi}_1 + \frac{\partial\hat{\mathcal{L}}}{\partial\hat{\phi}_2}\delta\hat{\phi}_2.\end{aligned} \tag{4.141}$$

This is a bit like the manipulations leading up to the derivation of the Euler–Lagrange equations, but now the changes $\delta\hat{\phi}_i (i = 1, 2)$ have nothing to do with space–time trajectories—they mix up the two fields. However, we can *use* the equations of motion for $\hat{\phi}_1$ and $\hat{\phi}_2$ to rewrite $\delta\hat{\mathcal{L}}$ as

$$\begin{aligned}0 = {}& \frac{\partial\hat{\mathcal{L}}}{\partial(\partial_\mu\hat{\phi}_1)}\delta(\partial_\mu\hat{\phi}_1) + \frac{\partial\hat{\mathcal{L}}}{\partial(\partial_\mu\hat{\phi}_2)}\delta(\partial_\mu\hat{\phi}_2)\\ &+ \left[\partial_\mu\left(\frac{\partial\hat{\mathcal{L}}}{\partial(\partial_\mu\hat{\phi}_1)}\right)\right]\delta\hat{\phi}_1 + \left[\partial_\mu\left(\frac{\partial\hat{\mathcal{L}}}{\partial(\partial_\mu\hat{\phi}_2)}\right)\right]\delta\hat{\phi}_2.\end{aligned} \tag{4.142}$$

Since $\delta(\partial_\mu\hat{\phi}_i) = \partial_\mu(\delta\hat{\phi}_i)$, this is just the total divergence

$$0 = \partial_\mu\left[\frac{\partial\hat{\mathscr{L}}}{\partial(\partial_\mu\hat{\phi}_1)}\delta\hat{\phi}_1 + \frac{\partial\hat{\mathscr{L}}}{\partial(\partial_\mu\hat{\phi}_2)}\delta\hat{\phi}_2\right]. \tag{4.143}$$

These formal steps are actually perfectly general, and will apply whenever a certain Lagrangian depending on two fields $\hat{\phi}_1$ and $\hat{\phi}_2$ is invariant under $\hat{\phi}_i \to \hat{\phi}_i + \delta\hat{\phi}_i$. In the present case, with $\delta\hat{\phi}_i$ given by (4.140), we have

$$\begin{aligned} 0 &= \partial_\mu\left[-\frac{\partial\hat{\mathscr{L}}}{\partial(\partial_\mu\hat{\phi}_1)}\varepsilon\hat{\phi}_2 + \frac{\partial\hat{\mathscr{L}}}{\partial(\partial_\mu\hat{\phi}_2)}\varepsilon\hat{\phi}_1\right] \\ &= \varepsilon\,\partial_\mu[(\partial^\mu\hat{\phi}_2)\hat{\phi}_1 - (\partial^\mu\hat{\phi}_1)\hat{\phi}_2] \end{aligned} \tag{4.144}$$

where the free field Lagrangian has been used in the second step. Since ε is arbitrary, we have proved that the 4-vector operator

$$\hat{N}^\mu = \hat{\phi}_1\partial^\mu\hat{\phi}_2 - \hat{\phi}_2\partial^\mu\hat{\phi}_1 \tag{4.145}$$

is conserved:

$$\partial_\mu\hat{N}^\mu = 0. \tag{4.146}$$

Such conserved 4-vector operators are called *symmetry currents*, often denoted generically by $\hat{J}^\mu$. There is a general theorem (due to Noether (1918) in the classical field case) that if a Lagrangian is invariant under a continuous transformation, then there will be an associated symmetry current. We shall consider Noether's theorem again in §8.2.

What does all this have to do with symmetry operators? Written out in full, (4.146) is

$$\partial\hat{N}^0/\partial t + \boldsymbol{\nabla}\cdot\hat{\boldsymbol{N}} = 0. \tag{4.147}$$

Integrating this equation over all space, we obtain

$$\frac{\mathrm{d}}{\mathrm{d}t}\int_{V\to\infty}\hat{N}^0\mathrm{d}^3\boldsymbol{x} + \int_{S\to\infty}\hat{\boldsymbol{N}}\cdot\mathrm{d}\boldsymbol{S} = 0 \tag{4.148}$$

where we have used the divergence theorem in the second term. Normally the fields die off sufficiently fast at infinity that the surface integral vanishes, and we can therefore deduce that the quantity $\hat{N}$ is constant in time, where

$$\hat{N} = \int\hat{N}^0\mathrm{d}^3\boldsymbol{x}; \tag{4.149}$$

that is, the volume integral of the $\mu = 0$ component of a symmetry current is a symmetry operator.

In order to see how $\hat{N}$ serves to distinguish 'particle' from 'antiparticle' in the simple example we are considering, it turns out to be convenient to regard $\hat{\phi}_1$ and $\hat{\phi}_2$ as components of a single *complex* field

$$\hat{\phi} = \frac{1}{\sqrt{2}}(\hat{\phi}_1 - \mathrm{i}\hat{\phi}_2)$$
$$\hat{\phi}^\dagger = \frac{1}{\sqrt{2}}(\hat{\phi}_1 + \mathrm{i}\hat{\phi}_2). \tag{4.150}$$

The plane wave expansions of the form (4.103) for $\hat{\phi}_1$ and $\hat{\phi}_2$ imply that $\hat{\phi}$ has the expansion

$$\hat{\phi} = \int \frac{\mathrm{d}^3\boldsymbol{k}}{(2\pi)^3\, 2\omega}[\hat{a}(k)\mathrm{e}^{-\mathrm{i}k\cdot x} + \hat{b}^\dagger(k)\mathrm{e}^{\mathrm{i}k\cdot x}] \tag{4.151}$$

where

$$\hat{a}(k) = \frac{1}{\sqrt{2}}(\hat{a}_1 - \mathrm{i}\hat{a}_2)$$
$$\hat{b}^\dagger(k) = \frac{1}{\sqrt{2}}(\hat{a}_1^\dagger - \mathrm{i}\hat{a}_2^\dagger). \tag{4.152}$$

The operators $\hat{a}$, $\hat{a}^\dagger$, $\hat{b}$, $\hat{b}^\dagger$ obey the commutation relations

$$[\hat{a}(k), \hat{a}^\dagger(k')] = (2\pi)^3\, 2\omega\delta^3(\boldsymbol{k} - \boldsymbol{k}')$$
$$[\hat{b}(k), \hat{b}^\dagger(k')] = (2\pi)^3\, 2\omega\delta^3(\boldsymbol{k} - \boldsymbol{k}') \tag{4.153}$$

with all others vanishing; this follows from the commutation relations

$$[\hat{a}_i(k), \hat{a}_j^\dagger(k')] = \delta_{ij}(2\pi)^3\, 2\omega\delta(\boldsymbol{k} - \boldsymbol{k}'), \text{ etc} \tag{4.154}$$

for the $\hat{a}_i$ operators. Note that now two distinct mode operators, $\hat{a}$ and $\hat{b}$, are appearing in the expansion (4.151) of the complex field.

In terms of $\hat{\phi}$ the Lagrangian of (4.136) becomes

$$\hat{\mathscr{L}} = \partial_\mu\hat{\phi}^\dagger\partial^\mu\hat{\phi} - m^2\hat{\phi}^\dagger\hat{\phi} \tag{4.155}$$

and the Hamiltonian is (dropping zero-point energies)

$$\hat{H} = \int \frac{\mathrm{d}^3\boldsymbol{k}}{(2\pi)^3\, 2\omega}[\hat{a}^\dagger(k)\hat{a}(k) + \hat{b}^\dagger(k)\hat{b}(k)]\omega. \tag{4.156}$$

The O(2) transformation (4.137) becomes a simple phase change

$$\hat{\phi}' = \mathrm{e}^{-\mathrm{i}\alpha}\hat{\phi} \tag{4.157}$$

which (see comment (iii) of §2.6) is called a U(1) rotation. The symmetry current $\hat{N}^\mu$ is

$$\hat{N}^\mu = \mathrm{i}(\hat{\phi}^\dagger\partial^\mu\hat{\phi} - \hat{\phi}\partial^\mu\hat{\phi}^\dagger) \tag{4.158}$$

and the symmetry operator $\hat{N}$ is (see problem 4.9)

$$\hat{N} = \int \frac{\mathrm{d}^3\boldsymbol{k}}{(2\pi)^3\, 2\omega}[\hat{a}^\dagger(k)\hat{a}(k) - \hat{b}^\dagger(k)\hat{b}(k)]. \tag{4.159}$$

We note that the Hamiltonian involves the *sum* of the number operators for 'a' quanta and for 'b' quanta, whereas $\hat{N}$ involves the *difference* of these number operators. Put differently, $\hat{N}$ counts $+1$ for each particle of type 'a' and -1 for each of type 'b'. This strongly suggests the interpretation that the 'b's' are the antiparticles of the 'a's': $\hat{N}$ is the conserved symmetry operator whose eigenvalues serve to distinguish them. For a general state, the eigenvalue of $\hat{N}$ is the number of 'a's' minus the number of 'anti-a's' and it is a constant of the motion, as is the total energy, which is the sum of the 'a' and 'anti-a' energies.

We have here the simplest form of the particle–antiparticle distinction: only one additive conserved quantity is involved. A more complicated example would be the (K^+, K^-) pair, which have opposite values of strangeness and of electromagnetic charge. Of course, in our simple Lagrangian (4.155) the electromagnetic interaction is absent, and so no electric charge can be defined (we shall remedy this below); the complex field $\hat{\phi}$ would be suitable (in respect of strangeness) for describing the (K^0, $\bar{K}^0$) pair.

The symmetry operator $\hat{N}$ has a number of further important properties. First of all, we have shown above that $d\hat{N}/dt = 0$ from the general (Noether) argument, but we ought also to check that

$$[\hat{N}, \hat{H}] = 0 \tag{4.160}$$

as is required for consistency, and expected for a symmetry. This is indeed true (problem 4.9(a)). We can also show that

$$\begin{aligned} [\hat{N}, \hat{\phi}] &= -\hat{\phi} \\ [\hat{N}, \hat{\phi}^\dagger] &= \hat{\phi}^\dagger \end{aligned} \tag{4.161}$$

and, by expansion of the exponential (problem 4.9(b)), that

$$\hat{U}(\alpha)\hat{\phi}\hat{U}^{-1}(\alpha) = e^{-i\alpha}\hat{\phi} = \hat{\phi}' \tag{4.162}$$

with

$$\hat{U}(\alpha) = e^{i\alpha\hat{N}}. \tag{4.163}$$

This shows that the unitary operator $\hat{U}(\alpha)$ effects finite U(1) rotations.

Consider now a state $|N\rangle$ which is an eigenstate of $\hat{N}$ with eigenvalue N. What is the eigenvalue of $\hat{N}$ for the state $\hat{\phi}|N\rangle$? It is easy to show, using (4.161), that

$$\hat{N}\hat{\phi}|N\rangle = (N-1)\hat{\phi}|N\rangle \tag{4.164}$$

so $\hat{\phi}$ lowers the $\hat{N}$ eigenvalue by 1. This is consistent with our interpretation that the $\hat{\phi}$ field destroys particles 'a' via the $\hat{a}$ piece in (4.151). (This '$\hat{\phi}$ destroys particles' convention is the reason for choosing $\hat{\phi} = (\hat{\phi}_1 - i\hat{\phi}_2)/\sqrt{2}$ in (4.150), which in turn led to the minus sign in the relation (4.161) and to the eigenvalue $N-1$ above.) That $\hat{\phi}$ lowers the $\hat{N}$ eigenvalue by 1 is also consistent with the interpretation that the same field $\hat{\phi}$ creates an antiparticle via the $\hat{b}^\dagger$ piece in (4.151). In the same way, by considering $\hat{\phi}^\dagger|N\rangle$, one easily verifies that $\hat{\phi}^\dagger$ increases N by 1, by creating a particle via $\hat{a}^\dagger$ or destroying an antiparticle via $\hat{b}$. The vacuum state (no particles and no antiparticles present) is defined by

$$\hat{a}(k)|0\rangle = \hat{b}(k)|0\rangle = 0 \quad \text{for all } k.$$

4.5 Fermionic fields and the spin–statistics connection

We must also consider the Lagrangian approach for the Dirac equation. The required unquantised Lagrangian is (problem 4.10)

$$\mathcal{L}_\mathrm{D} = \mathrm{i}\psi^\dagger\dot{\psi} + \mathrm{i}\psi^\dagger\boldsymbol{\alpha}\cdot\boldsymbol{\nabla}\psi - m\psi^\dagger\beta\psi. \tag{4.165}$$

The relativistic invariance of this is more evident in γ matrix notation:

$$\mathcal{L}_\mathrm{D} = \bar{\psi}(\mathrm{i}\gamma^\mu\partial_\mu - m)\psi. \tag{4.166}$$

We can now attempt to quantise ψ, $\bar{\psi}$ and make a mode expansion in terms of plane wave solutions in a similar fashion to that for the complex scalar field $\hat{\phi}$. We obtain (see problem 3.5 for the definition of u and v)

$$\hat{\psi} = \int \frac{\mathrm{d}^3\boldsymbol{k}}{(2\pi)^3\, 2\omega} \sum_{s=1,2} [\hat{c}_s(k)u(k,s)\mathrm{e}^{-ik\cdot x} + \hat{d}_s^\dagger(k)v(k,s)\mathrm{e}^{ik\cdot x}] \tag{4.167}$$

and we wish to interpret $\hat{c}_s^\dagger(k)$ as the creation operator for a Dirac particle of spin s and momentum k. Presumably we must define the vacuum by

$$\hat{c}_s(k)|0\rangle = 0 \quad \text{for all } k, \text{ and } s = 1, 2 \tag{4.168}$$

and a two-fermion state by

$$|k_1, s_1; k_2, s_2\rangle = \hat{c}_{s_1}^\dagger(k_1)\hat{c}_{s_2}^\dagger(k_2)|0\rangle. \tag{4.169}$$

But it is here that there must be a difference from the boson case. We require a state containing two identical fermions to be antisymmetric under exchange of state labels $k_1 \leftrightarrow k_2$, $s_1 \leftrightarrow s_2$, and to be forbidden by Pauli if the two sets of quantum numbers are the same.

The solution to this dilemma is simple but radical: for fermions, commutation relations are replaced by *anticommutation* relations! The anticommutator of two operators $\hat{A}$ and $\hat{B}$ is written

$$\{\hat{A}, \hat{B}\} \equiv \hat{A}\hat{B} + \hat{B}\hat{A}. \tag{4.170}$$

If two different $\hat{c}^\dagger$'s anticommute, then

$$\hat{c}_{s_1}^\dagger(k_1)\hat{c}_{s_2}^\dagger(k_2) + \hat{c}_{s_2}^\dagger(k_2)\hat{c}_{s_1}^\dagger(k_1) = 0 \tag{4.171}$$

so that we have the desired antisymmetry

$$|k_1, s_1; k_2, s_2\rangle = -|k_2, s_2; k_1, s_1\rangle. \tag{4.172}$$

In general we postulate

$$\begin{aligned}\{\hat{c}_{s_1}(k_1), \hat{c}^\dagger_{s_2}(k_2)\} &= (2\pi)^3\, 2\omega\delta^3(\boldsymbol{k}_1 - \boldsymbol{k}_2)\delta_{s_1 s_2} \\ \{\hat{c}_{s_1}(k_1), \hat{c}_{s_2}(k_2)\} &= \{\hat{c}^\dagger_{s_1}(k_1), \hat{c}^\dagger_{s_2}(k_2)\} = 0\end{aligned} \tag{4.173}$$

and similarly for the $\hat{d}$'s and $\hat{d}^\dagger$'s. The factor in front of the δ function depends on the convention for normalising Dirac wavefunctions.

One may well wonder *why* things have to be this way—'bosons commute, fermions anticommute'. To gain further insight, we turn again to a consideration of symmetries and the question of particle and antiparticle for the Dirac field.

The Dirac field $\hat{\psi}$ is a complex field, as is reflected in the two distinct mode operators in the expansion (4.167); as in the scalar field case, there is only one mass parameter, and we expect the quanta to be interpretable as particle and antiparticle. The symmetry operator which distinguishes them is found by analogy with the complex scalar field case. We note that $\hat{\mathscr{L}}_{\mathrm{D}}$ (the quantised version of (4.166)) is invariant under the U(1) transformation

$$\hat{\psi} \to \hat{\psi}' = \mathrm{e}^{-\mathrm{i}\alpha}\hat{\psi} \tag{4.174}$$

which is

$$\hat{\psi} \to \hat{\psi}' = \hat{\psi} - \mathrm{i}\varepsilon\hat{\psi} \tag{4.175}$$

in infinitesimal form. The corresponding (Noether) symmetry current can be calculated as

$$\hat{N}^\mu = \bar{\hat{\psi}}\gamma^\mu\hat{\psi} \tag{4.176}$$

and the associated symmetry operator is

$$\hat{N} = \int \hat{\psi}^\dagger\hat{\psi}\, \mathrm{d}^3\boldsymbol{x}. \tag{4.177}$$

$\hat{N}$ is clearly some kind of number operator for the fermion case. Inserting the plane wave expansion (4.167), we obtain after some effort

$$\hat{N} = \int \frac{\mathrm{d}^3\boldsymbol{k}}{(2\pi)^3\, 2\omega} \sum_{s=1,2} [\hat{c}^\dagger_s(k)\hat{c}_s(k) + \hat{d}_s(k)\hat{d}^\dagger_s(k)]. \tag{4.178}$$

Similarly the Dirac Hamiltonian may be shown to have the form

$$\hat{H} = \int \frac{\mathrm{d}^3\boldsymbol{k}}{(2\pi)^3\, 2\omega} \sum_{s=1,2} [\hat{c}^\dagger_s(k)\hat{c}_s(k) - \hat{d}_s(k)\hat{d}^\dagger_s(k)]\omega. \tag{4.179}$$

We have *not* so far assumed either commutation or anticommutation relations for the mode operators $\hat{c}$, $\hat{c}^\dagger$, $\hat{d}$ and $\hat{d}^\dagger$, but have only used properties of the Dirac spinors to reduce $\hat{N}$ and $\hat{H}$ to these forms. Suppose first that we assume commutation relations, so that the last terms in (4.178) and (4.179) are rewritten as $\hat{d}_s^\dagger(k)\hat{d}_s(k)$. Dropping the zero-point energy as usual, we see that $\hat{H}$ will then contain the *difference* of two number operators for 'c' and 'd' particles, and is therefore not positive-definite as we require for a sensible theory. Moreover, we suspect that, as in the $\hat{\phi}$ case, the 'd's' ought to be the antiparticles of the 'c's', carrying opposite $\hat{N}$ value: but $\hat{N}$ is just proportional (with this assumption) to the *sum* of the 'c' and 'd' number operators, counting +1 for each type, which does not fit this interpretation. On the other hand, if anticommutation relations are assumed, these problems disappear: dropping the usual infinite terms, we obtain

$$\hat{N} = \int \frac{\mathrm{d}^3\boldsymbol{k}}{(2\pi)^3\, 2\omega} \sum_{s=1,2} [\hat{c}_s^\dagger(k)\hat{c}_s(k) - \hat{d}_s^\dagger(k)\hat{d}_s(k)] \tag{4.180}$$

$$\hat{H} = \int \frac{\mathrm{d}^3\boldsymbol{k}}{(2\pi)^3\, 2\omega} \sum_{s=1,2} [\hat{c}_s^\dagger(k)\hat{c}_s(k) + \hat{d}_s^\dagger(k)\hat{d}_s(k)]\omega \tag{4.181}$$

which are satisfactory and allow us to interpret the 'd' quanta as the antiparticles of the 'c' quanta. Similar difficulties would have occurred in the complex scalar field case if we had assumed anticommutation relations for the boson operators. It is in this way that quantum field theory enforces the connection between spin and statistics.

4.6 A preview of QED

We have so far considered only non-interacting Klein–Gordon and Dirac fields. But from our discussion in Chapter 2, we have a strong indication of how to introduce electromagnetic interactions into our theory. The 'gauge principle' in quantum mechanics consisted in elevating a global (space–time-independent) U(1) phase invariance into a local (space–time-dependent) U(1) invariance—the compensating fields being then identified with the electromagnetic ones. In quantum field theory, exactly the same principle exists, and leads to the form of the electromagnetic interactions. Indeed, in the field theory formalism we are able to exhibit explicitly the symmetry current, and symmetry operator, associated with the U(1) invariance—and identify them precisely with the electromagnetic current and charge.

We have seen that for both the complex scalar and the Dirac fields the free Lagrangian is invariant under U(1) transformations (see (4.157) and (4.174)) which we now specifically recognise as being *global*. Let us therefore promote these global invariances into local ones in the way learned in Chapter 2. In the case of the Dirac Lagrangian

$$\mathcal{L}_{\mathrm{D}} = \bar{\hat{\psi}}(\mathrm{i}\gamma^{\mu}\partial_{\mu} - m)\hat{\psi} \tag{4.166}$$

we know that it will be invariant under the *local* U(1) phase transformation

$$\hat{\psi}(\boldsymbol{x}, t) \rightarrow \hat{\psi}'(\boldsymbol{x}, t) = \mathrm{e}^{-\mathrm{i}e\hat{\chi}(\boldsymbol{x},t)}\hat{\psi}(\boldsymbol{x}, t) \tag{4.182}$$

provided we make the replacement

$$\partial^{\mu} \rightarrow \hat{D}^{\mu} = \partial^{\mu} + \mathrm{i}e\hat{A}^{\mu} \tag{4.183}$$

and demand that the (quantised) 4-vector potential transform as

$$\hat{A}^{\mu} \rightarrow \hat{A}'^{\mu} = \hat{A}^{\mu} + \partial^{\mu}\hat{\chi}. \tag{4.184}$$

In this case the Lagrangian has gained an interaction term

$$\hat{\mathcal{L}}_{\mathrm{D}} \rightarrow \hat{\mathcal{L}}_{\mathrm{D}} + \hat{\mathcal{L}}_{\mathrm{int}} \tag{4.185}$$

where

$$\hat{\mathcal{L}}_{\mathrm{int}} = -e\bar{\hat{\psi}}\gamma^{\mu}\hat{\psi}\hat{A}_{\mu}. \tag{4.186}$$

Since $\hat{\mathcal{L}} = \hat{T} - \hat{V}$, we can interpret this $\hat{\mathcal{L}}_{\mathrm{int}}$ as an operator potential

$$\hat{V}_{\mathrm{D}} = e\hat{\psi}^{\dagger}\hat{\psi}\hat{A}_0 - e\hat{\psi}^{\dagger}\boldsymbol{\alpha}\hat{\psi}\cdot\hat{\boldsymbol{A}} \tag{4.187}$$

which is the field theory analogue of (3.83). It has the expected form '$\rho A_0 - \boldsymbol{j}\cdot\boldsymbol{A}$' if we identify the electromagnetic charge density operator with $e\hat{\psi}^{\dagger}\hat{\psi}$ (the charge times the density operator) and the electromagnetic current density operator with $e\hat{\psi}^{\dagger}\boldsymbol{\alpha}\hat{\psi}$. The electromagnetic 4-vector current operator $\hat{j}^{\mu}_{\mathrm{em}}$ is thus identified as

$$\hat{j}^{\mu}_{\mathrm{em}} = e\bar{\hat{\psi}}\gamma^{\mu}\hat{\psi}. \tag{4.188}$$

We now note that $\hat{j}^{\mu}_{\mathrm{em}}$ is just e times the symmetry current $\hat{N}^{\mu}$ of the previous section (see equation (4.176)). Conservation of $\hat{j}^{\mu}_{\mathrm{em}}$ would follow from *global* U(1) invariance alone (i.e. $\hat{\chi}$ a constant in equation (4.182)); but many Lagrangians, including interactions, could be constructed obeying this global U(1) invariance. The force of the *local* U(1) invariance requirement is that it has specified a unique form of the interaction (i.e. $\hat{\mathcal{L}}_{\mathrm{int}}$ of equation (4.186)). Indeed, this is just $-\hat{j}^{\mu}_{\mathrm{em}}\hat{A}_{\mu}$, so that in this type of theory the current $\hat{j}^{\mu}_{\mathrm{em}}$ is not only a symmetry current, but also determines the precise way in which the vector potential $\hat{A}^{\mu}$ couples to the matter field $\hat{\psi}$.

To complete the Lagrangian for QED, we must include a piece to give life to the Maxwell field $\hat{A}^{\mu}$. We shall determine the form of this term by requiring that the Euler–Lagrange equation for $\hat{A}^{\mu}$ must have the Maxwell form (cf (2.19))

$$\Box\hat{A}^{\mu} - \partial^{\mu}(\partial_{\nu}\hat{A}^{\nu}) = \hat{j}^{\mu}_{\mathrm{em}}. \tag{4.189}$$

The equation for $\hat{A}^\mu$ is

$$\partial_\nu\left(\frac{\partial\hat{\mathscr{L}}}{\partial(\partial_\nu\hat{A}_\mu)}\right) - \frac{\partial\hat{\mathscr{L}}}{\partial\hat{A}_\mu} = 0 \tag{4.190}$$

and it is clear from (4.186) that

$$-\partial\hat{\mathscr{L}}_{\text{int}}/\partial\hat{A}_\mu = \hat{j}^{\,\mu}_{\text{em}} \tag{4.191}$$

so we just need a 'kinetic' term for the $\hat{A}^\mu$ field such that the first term in (4.190) yields (minus) the left-hand side of (4.189). The answer is (see problem 4.11)

$$\hat{\mathscr{L}}_{\text{em}} = -\tfrac{1}{4}\hat{F}_{\mu\nu}\hat{F}^{\mu\nu} \tag{4.192}$$

where

$$\hat{F}_{\mu\nu} = \partial_\mu\hat{A}_\nu - \partial_\nu\hat{A}_\mu \tag{4.193}$$

is the quantised form of the field strength tensor introduced in Chapter 2. $\hat{F}_{\mu\nu}$ is clearly invariant under gauge transformations (4.184).

In the same way the global U(1) invariance (4.157) of the complex scalar field may be generalised to a local invariance incorporating electromagnetism. We have

$$\hat{\mathscr{L}}_{\text{KG}} \to \hat{\mathscr{L}}_{\text{KG}} + \hat{\mathscr{L}}_{\text{int}} \tag{4.194}$$

where

$$\hat{\mathscr{L}}_{\text{KG}} = \partial_\mu\hat{\phi}^\dagger\partial^\mu\hat{\phi} - m^2\hat{\phi}^\dagger\hat{\phi} \tag{4.155}$$

and (under $\partial_\mu \to \hat{D}_\mu$)

$$\hat{\mathscr{L}}_{\text{int}} = -\mathrm{i}e(\hat{\phi}^\dagger\partial^\mu\hat{\phi} - (\partial^\mu\hat{\phi}^\dagger)\hat{\phi})\hat{A}_\mu + e^2\hat{A}^\mu\hat{A}_\mu\hat{\phi}^\dagger\hat{\phi} \tag{4.195}$$

which is the field theory analogue of (3.80).

The electromagnetic current is

$$\hat{j}^{\,\mu}_{\text{em}} = -\partial\hat{\mathscr{L}}_{\text{int}}/\partial\hat{A}_\mu \tag{4.191}$$

as before, which from (4.195) is

$$\hat{j}^{\,\mu}_{\text{em}} = \mathrm{i}e(\hat{\phi}^\dagger\partial^\mu\hat{\phi} - (\partial^\mu\hat{\phi}^\dagger)\hat{\phi}) - 2e^2\hat{A}^\mu\hat{\phi}^\dagger\hat{\phi}. \tag{4.196}$$

We note that for the boson case the electromagnetic current is *not* just e times the (number) current appropriate to the global phase invariance. This has its origin in the fact that the boson current involves a derivative, and so the gauge covariant boson current must develop a term involving $\hat{A}^\mu$ itself, as is evident in (4.196), and as we also saw in the wavefunction case (cf equation (2.29)). However, the conserved electromagnetic charge (the space integral of the $\mu = 0$ component of (4.196)) may be identified with e times the boson number operator in the gauge $\hat{A}^0 = 0$ (which is a possible gauge choice). The full scalar QED

Lagrangian is completed by the inclusion of $\hat{\mathscr{L}}_{\text{em}}$ as before.

The extra interaction Lagrangian $\hat{\mathscr{L}}_{\text{int}}$ gives rise to an interaction Hamiltonian $\hat{H}_{\text{int}}$ that can be constructed following the rule of §4.2.5. In the fermion case $\hat{H}_{\text{int}}$ is just

$$\hat{H}_{\text{int}} = \int \hat{j}^{\mu}_{\text{em}} \hat{A}_{\mu} \mathrm{d}^3 x \tag{4.197}$$

with

$$\hat{j}^{\mu}_{\text{em}} = e \hat{\bar{\psi}} \gamma^{\mu} \hat{\psi}. \tag{4.188}$$

The form of this interaction can be interpreted very simply in terms of the particle aspect of the quantum fields. We note that it involves the product of three fields, $\hat{\psi}$, $\hat{\bar{\psi}}$ and $\hat{A}^{\mu}$, all at the same point, and it is the generalisation of the '$\hat{\phi}^3$' type of interaction considered in §4.3. The boson case also involves a '$\hat{\phi}^4$' type of term, $e^2 \hat{A}^{\mu} \hat{A}_{\mu} \hat{\phi}^{\dagger} \hat{\phi}$. Consider the simpler fermion case, for definiteness. The field $\hat{\psi}$ destroys electrons or creates positrons, while $\hat{\bar{\psi}}$ creates electrons and destroys positrons, and the interaction is guaranteed to conserve charge and fermion number. The electromagnetic field $\hat{A}^{\mu}$ creates and destroys photons. Thus this single interaction term incorporates many processes, some of which are shown in figure 4.6. More precisely, the operator $\hat{H}_{\text{int}}$ has non-zero matrix elements between the types of states in figure 4.6: for example,

$$\langle \mathrm{e}^- | \hat{H}_{\text{int}} | \gamma \mathrm{e}^- \rangle \neq 0. \tag{4.198}$$

Note that in these diagrams 'time' is understood to be flowing from left to right across the page (initial → final), not vertically. This is the usual convention which we shall follow henceforth. Such amplitudes (which we will consider in more detail in later chapters) are, of course, all of first order in the interaction term. In fact, the corresponding physical

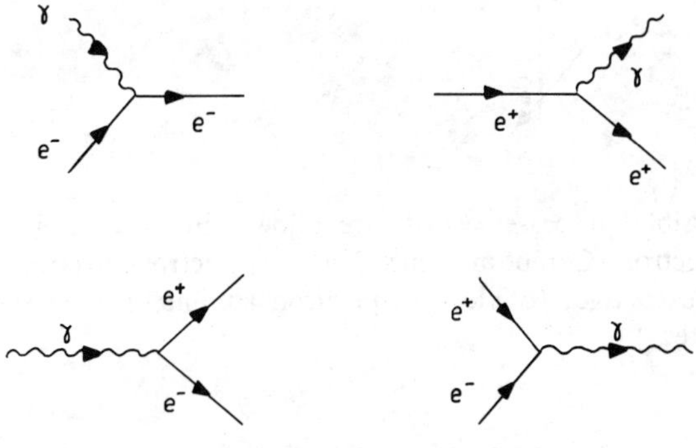

Figure 4.6 Possible basic '*vertices*' associated with the interaction density $e\hat{\bar{\psi}}\gamma^{\mu}\hat{\psi}\hat{A}_{\mu}$; these cannot occur as physical processes due to energy–momentum constraints.

processes cannot occur while conserving energy and momentum between initial and final states (see problems 1.1(a) and 1.1(b) from Chapter 1). However, in *second*-order quantum mechanical perturbation theory, we must include contributions of the form, symbolically,

$$\sum_n \langle \mathrm{f}|\hat{H}_{\mathrm{int}}|n\rangle \frac{1}{E_{\mathrm{i}} - E_n} \langle n|\hat{H}_{\mathrm{int}}|\mathrm{i}\rangle \tag{4.199}$$

in which $E_n \neq E_{\mathrm{i}}$, $E_n \neq E_{\mathrm{f}}$, although $E_{\mathrm{i}} = E_{\mathrm{f}}$ (see equation (A.84)). Thus we are allowed to 'stick' the processes of figure 4.6 together to form second-order amplitudes of the form shown in figure 4.7. Figure 4.7(*a*) contributes to electron Compton scattering, figure 4.7(*b*) is a photon exchange process in e^-e^- scattering, and figure 4.7(*c*) is an e^+e^- annihilation via a single intermediate photon state.

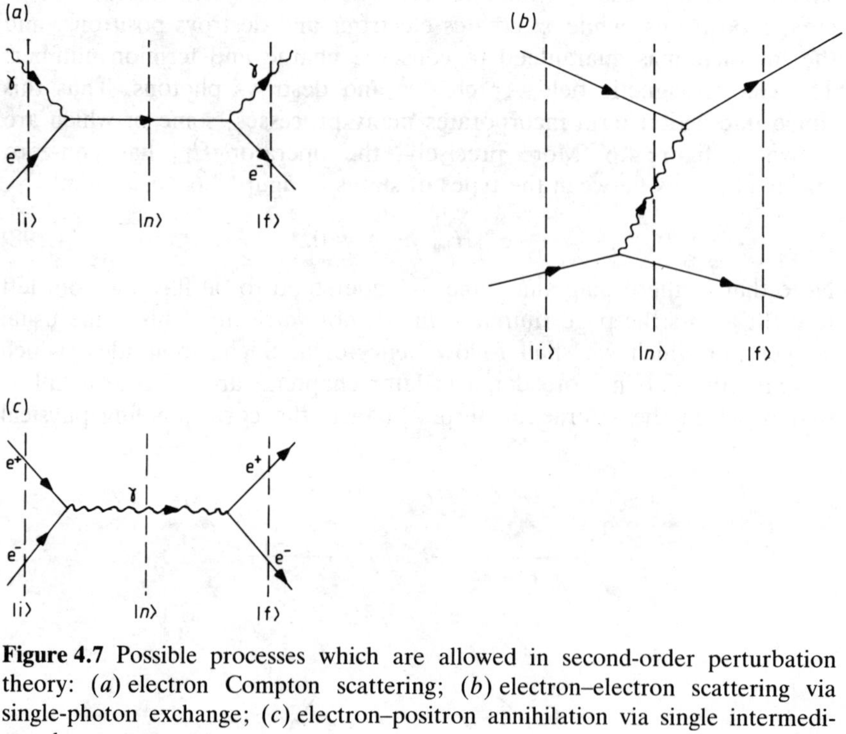

Figure 4.7 Possible processes which are allowed in second-order perturbation theory: (*a*) electron Compton scattering; (*b*) electron–electron scattering via single-photon exchange; (*c*) electron–positron annihilation via single intermediate photon state.

The precise way in which these physically intuitive pictures correspond (as indeed they do) to definite mathematical formulae for the corresponding amplitudes will emerge in Chapters 5 and 6. The reader may now begin to appreciate why we were reluctant to follow the 'field theory' way. Yet the basic structure of these interactions is really quite simple: we could have imagined far more complicated possibilities, involving simultaneous emission of many photons, for example. Expressions for the basic particle *vertices*, like those shown in figure 4.6, come when we take appropriate matrix elements of $\hat{H}_{\text{int}}$.

Summary

Discrete systems
classical coupled oscillators;
quantum mechanical coupled oscillators.

Continuous systems
classical field mechanics;
quantum field mechanics.

Photons as elementary quanta of excitation of the electromagnetic field

Lagrangians and the action principle.

Quantum mechanical harmonic oscillator.

Lagrange–Hamilton formulation of classical and quantum field mechanics.

Particles as excitation quanta and the vacuum as the ground state of a field system.

Connection to wavefunction quantum mechanics.

Interactions in quantum field theory.

Quantum field theory for the Klein–Gordon field.

Symmetry and Noether's theorem.

Particles and antiparticles.

Fermionic fields and the spin–statistics relation.

A preview of QED and Feynman diagrams.

Problems

4.1 Verify equation (4.35).

4.2 Consider one-dimensional motion under gravity so that $V(x) = -mgx$ in (4.38). Evaluate S of (4.37) for $t_1 = 0$, $t_2 = t_0$, for three possible trajectories:

(a) $x(t) = at$,

(b) $x(t) = \frac{1}{2}gt^2$ (the Newtonian result),

(c) $x(t) = bt^3$,

where the constants a and b are to be chosen so that all trajectories end at the same point $x(t_0)$.

4.3 (a) Use (4.48) and (4.54) to verify that

$$\hat{p} = m\dot{\hat{q}}$$

is consistent with the Heisenberg equation of motion for $\hat{A} = \hat{q}$.

(b) By similar methods verify that

$$\dot{\hat{p}} = -m\omega^2\hat{q}.$$

4.4 (a) Rewrite the Hamiltonian $\hat{H}$ of (4.52) in terms of the operators $\hat{a}$ and $\hat{a}^\dagger$.

(b) Evaluate the commutator between $\hat{a}$ and $\hat{a}^\dagger$ and use this result together with your expression for $\hat{H}$ from part (a) to verify equation (4.61).

(c) Verify that for $|n\rangle$ given by equation (4.67) the normalisation condition

$$\langle n|n\rangle = 1$$

is satisfied.

(d) Verify (4.69) directly using the commutation relation (4.60).

4.5 Treating ψ and ψ^* as independent classical fields, show that the Lagrangian density

$$\mathscr{L} = \mathrm{i}\psi^*\dot{\psi} - (1/2m)\nabla\psi^*\cdot\nabla\psi$$

gives the Schrödinger equation for ψ and ψ^* correctly.

4.6 Verify that the commutation relations for $\hat{a}(k)$ and $\hat{a}^\dagger(k')$ (equations (4.104)) are consistent with the fundamental commutation relations between $\hat{\phi}$ and $\hat{\pi}$ (equation (4.95)).

4.7 Insert the plane wave expansions for the operators $\hat{\phi}$ and $\hat{\pi}$ into the equation for $\hat{H}$, (4.98), and verify equation (4.106). [Hint: note that ω is defined to be always positive, so that (4.100) should strictly be written $\omega = |k|$.]

4.8 Verify that the Lagrangian $\hat{\mathscr{L}}$ of (4.136) is invariant (i.e. $\hat{\mathscr{L}}(\hat{\phi}_1, \hat{\phi}_2) = \hat{\mathscr{L}}(\hat{\phi}_1', \hat{\phi}_2')$) under the transformation (4.137) of the fields $(\hat{\phi}_1, \hat{\phi}_2) \to (\hat{\phi}_1', \hat{\phi}_2')$.

4.9 (a) Verify that, for $\hat{N}^\mu$ given by (4.158), the corresponding $\hat{N}$ of (4.149) reduces to the form (4.159); and that, with $\hat{H}$ given by (4.156),

$$[\hat{N}, \hat{H}] = 0.$$

(b) Verify equation (4.162).

4.10 Verify that varying $\psi^\dagger$ in the action principle with Lagrangian (4.165) gives the Dirac equation.

4.11 Verify that if $\hat{\mathscr{L}} = -\frac{1}{4}\hat{F}_{\mu\nu}\hat{F}^{\mu\nu} - \hat{j}^{\mu}_{\text{em}}\hat{A}_\mu$, where $\hat{F}_{\mu\nu} = \partial_\mu\hat{A}_\nu - \partial_\nu\hat{A}_\mu$, the Euler–Lagrange equations for $\hat{A}_\mu$ yield the Maxwell form

$$\Box\hat{A}^\mu - \partial^\mu(\partial_\nu\hat{A}^\nu) = \hat{j}^{\mu}_{\text{em}}.$$

[Hint: it is helpful to use the antisymmetry of $\hat{F}_{\mu\nu}$ to rewrite the '$F \cdot F$' term as $-\frac{1}{2}\hat{F}_{\mu\nu}\partial^\mu\hat{A}^\nu$.]

PART II

THE BEST THEORY WE HAVE: QUANTUM ELECTRODYNAMICS

5

THE ELECTROMAGNETIC INTERACTIONS OF SPIN-0 PARTICLES: FORCES AND PARTICLE EXCHANGE PROCESSES

5.1 First-order perturbation theory

5.1.1 The matrix element (1)

Consider the scattering of a spin-0 particle of charge e (for example, a π^+) in an electromagnetic field described by the (classical) potential A^μ. The appropriate potential for use in the KG equation has been given in §3.6:

$$\hat{V}_{\mathrm{KG}} = \mathrm{i}e(\partial_\mu A^\mu + A^\mu \partial_\mu) - e^2 A^2. \tag{5.1}$$

As we shall see in more detail as we go along, the parameter characterising each order of perturbation theory based on this potential is found to be $e^2/4\pi$. In natural units (see Appendices B and C) this has the value

$$\alpha = e^2/4\pi \approx \tfrac{1}{137} \tag{5.2}$$

for the elementary charge e. α is called the fine structure constant. The smallness of α is the reason why a perturbation approach has been very successful for QED.

To first order in α we can neglect the e^2A^2 term, and the perturbing potential is then

$$\hat{V} = \mathrm{i}e(\partial_\mu A^\mu + A^\mu \partial_\mu). \tag{5.3}$$

For a scattering process we shall assume† the same formula for the transition amplitude as in NRQM time-dependent perturbation theory (see Appendix A):

$$\mathscr{A}_{\mathrm{fi}}^{(1)} = -\mathrm{i}\int \mathrm{d}^4x\, \phi_{\mathrm{f}}^* \hat{V} \phi_{\mathrm{i}} \tag{5.4}$$

where ϕ_{i} and ϕ_{f} are the initial and final state free particle solutions. The latter are (recall (3.11))

$$\phi_{\mathrm{i}} = N_{\mathrm{i}} \mathrm{e}^{-\mathrm{i}p_{\mathrm{i}}\cdot x} \tag{5.5}$$

$$\phi_{\mathrm{f}} = N_{\mathrm{f}} \mathrm{e}^{-\mathrm{i}p_{\mathrm{f}}\cdot x} \tag{5.6}$$

†Justification may be found in Chapter 9 of Bjorken and Drell (1964).

and we shall fix the normalisation factors later. Inserting the expression for $\hat{V}$, and doing some integration by parts (problem 5.1), we obtain

$$\mathscr{A}_{\mathrm{fi}}^{(1)} = -\mathrm{i}\int \mathrm{d}^4x\{\mathrm{i}e[\phi_{\mathrm{f}}^*(\partial_\mu\phi_{\mathrm{i}}) - (\partial_\mu\phi_{\mathrm{f}}^*)\phi_{\mathrm{i}}]\}A^\mu. \tag{5.7}$$

This is very reminiscent of the probability current expression (3.20). Indeed, we can write this as

$$\mathscr{A}_{\mathrm{fi}}^{(1)} = -\mathrm{i}\int \mathrm{d}^4x j_{\mathrm{fi}}^\mu(x)A_\mu(x) \tag{5.8}$$

where

$$j_{\mathrm{fi}}^\mu(x) = \mathrm{i}e(\phi_{\mathrm{f}}^*\partial^\mu\phi_{\mathrm{i}} - (\partial^\mu\phi_{\mathrm{f}}^*)\phi_{\mathrm{i}}) \tag{5.9}$$

can be regarded as an electromagnetic 'transition current', analogous to the simple probability current for a single state. In the following section we shall see the exact meaning of this idea, using quantum field theory. Meanwhile, we insert the plane wave free particle solutions for ϕ_{i} and ϕ_{f} and find that

$$\mathscr{A}_{\mathrm{fi}}^{(1)} = -\mathrm{i}N_{\mathrm{i}}N_{\mathrm{f}}\int \mathrm{d}^4x e(p_{\mathrm{i}} + p_{\mathrm{f}})_\mu \mathrm{e}^{-\mathrm{i}(p_{\mathrm{i}} - p_{\mathrm{f}})\cdot x}A^\mu(x). \tag{5.10}$$

What shall we take for A^μ? The simplest case of all is that in which A^μ is a free field, i.e. the solution of the Maxwell equations in which the current source j^μ is set to zero. The resulting form of A^μ is a sufficiently important topic to merit a new subsection.

5.1.2 The form of A^μ for a free electromagnetic field

When j^μ is set equal to zero, A^μ satisfies the equation

$$\partial_\mu F^{\mu\nu} = \Box A^\nu - \partial^\nu(\partial_\mu A^\mu) = 0. \tag{5.11}$$

As we have seen in §2.3, these equations are left unchanged if we perform the gauge transformation

$$A^\mu \to A'^\mu = A^\mu - \partial^\mu\chi. \tag{5.12}$$

We can use this freedom to *choose* the A^μ with which we work to satisfy the condition

$$\partial_\mu A^\mu = 0. \tag{5.13}$$

This is called the *Lorentz condition*. The process of choosing a particular condition on A^μ so as to define it (ultimately) uniquely is called 'choosing a gauge'; actually, the condition (5.13) does not yet define A^μ uniquely, as we shall see shortly. The Lorentz condition is a very convenient one, since it decouples the different components of A^μ in Maxwell's equations (5.11)—in a covariant way, moreover, leaving the very simple equation

$$\Box A^{\mu} = 0. \tag{5.14}$$

This has plane wave solutions of the form

$$A^{\mu} = N\varepsilon^{\mu} \mathrm{e}^{-\mathrm{i}k \cdot x} \tag{5.15}$$

with $k^2 = 0$ (i.e. $k_0^2 = \boldsymbol{k}^2$), where N is a normalisation factor and ε^{μ} is a *polarisation vector* for the wave. The gauge condition (5.13) now reduces to a condition on ε^{μ}:

$$k \cdot \varepsilon = 0. \tag{5.16}$$

However, we have not yet exhausted all the gauge freedom. We are still free to make another shift in the potential

$$A^{\mu} \to A^{\mu} - \partial^{\mu} \tilde{\chi} \tag{5.17}$$

provided $\tilde{\chi}$ satisfies the massless KG equation

$$\Box \tilde{\chi} = 0. \tag{5.18}$$

This condition on $\tilde{\chi}$ ensures that, even after the further shift, the resulting potential still satisfies $\partial_{\mu} A^{\mu} = 0$. For our plane wave solutions, this residual gauge freedom corresponds to changing ε^{μ} by a multiple of k^{μ}:

$$\varepsilon^{\mu} \to \varepsilon^{\mu} + \beta k^{\mu} \equiv \varepsilon'^{\mu} \tag{5.19}$$

which still satisfies $\varepsilon' \cdot k = 0$ *since* $k^2 = 0$ *for these free field solutions*.

This freedom has important consequences. Consider a solution with

$$k^{\mu} = (k^0, \boldsymbol{k}), \qquad k_0^2 = \boldsymbol{k}^2 \tag{5.20}$$

and polarisation vector

$$\varepsilon^{\mu} = (\varepsilon^0, \boldsymbol{\varepsilon}) \tag{5.21}$$

satisfying the Lorentz condition

$$k \cdot \varepsilon = 0. \tag{5.16}$$

Gauge invariance now implies that we can add multiples of k^{μ} to ε^{μ} and still have a satisfactory polarisation vector.

It is therefore clear that we can arrange for the time component of ε^{μ} to vanish so that the Lorentz condition reduces to the 3-vector condition

$$\boldsymbol{k} \cdot \boldsymbol{\varepsilon} = 0. \tag{5.22}$$

This means that there are only two independent polarisation vectors, both transverse to $\boldsymbol{k}$, the propagation direction. For a wave travelling in the z direction ($k^{\mu} = (k^0, 0, 0, k^0)$) these may be chosen to be

$$\boldsymbol{\varepsilon}_{(1)} = (1, 0, 0) \tag{5.23}$$

$$\boldsymbol{\varepsilon}_{(2)} = (0, 1, 0). \tag{5.24}$$

Such a choice corresponds to *linear polarisation* of the associated $\boldsymbol{E}$ and $\boldsymbol{B}$ fields—which can of course be easily calculated from (2.9) and (2.10), given

$$A^{\mu}_{(i)} = N(0, \boldsymbol{\varepsilon}_{(i)})\mathrm{e}^{-\mathrm{i}k\cdot x}, \quad i = 1, 2. \tag{5.25}$$

A commonly used alternative choice is

$$\boldsymbol{\varepsilon}(\lambda = +1) = -\frac{1}{\sqrt{2}}(1, \mathrm{i}, 0) \tag{5.26}$$

$$\boldsymbol{\varepsilon}(\lambda = -1) = \frac{1}{\sqrt{2}}(1, -\mathrm{i}, 0) \tag{5.27}$$

(linear combinations of (5.23) and (5.24)), which correspond to circularly polarised radiation. The phase convention in (5.26) and (5.27) is the standard one in quantum mechanics for states of definite spin projection ('helicity') $\lambda = \pm 1$ along the direction of motion (the z axis here). We may easily check that

$$\boldsymbol{\varepsilon}^*(\lambda)\cdot\boldsymbol{\varepsilon}(\lambda') = \delta_{\lambda\lambda'} \tag{5.28a}$$

or, in terms of the corresponding 4-vectors $\varepsilon^{\mu} = (0, \boldsymbol{\varepsilon})$,

$$\varepsilon^*(\lambda)\cdot\varepsilon(\lambda') = -\delta_{\lambda\lambda'}. \tag{5.28b}$$

We have therefore arrived at the result, familiar in classical electromagnetic theory, that the free electromagnetic fields are purely *transverse*. Though they are described in this formalism by a vector potential with apparently four independent components $(V, \boldsymbol{A})$, the condition (5.13) reduces this number by one, and the further gauge freedom exploited in (5.17)–(5.19) reduces it by one more.

The above discussion has been entirely in terms of the (unquantised) classical field A^{μ}. What happens when we quantise A^{μ} so that we deal with the quantum field $\hat{A}^{\mu}$? In view of the discussion in the preceding chapter, especially that at the end of §4.2, it would seem plausible that the A^{μ} of (5.15) represents the *wavefunction of a photon*, which has 4-momentum k and polarisation vector ε^{μ}. This is certainly consistent with the constraint which we found, $k^2 = 0$, which means simply that the photon is massless. The two independent ε^{μ} vectors are now interpreted as the two independent polarisation—or helicity—wavefunctions for the photon, which therefore has only two independent polarisation states. A very important point is that the reduction to only *two* independent states can be traced back to the fact that the free photon is massless (cf the remark after equation (5.19)); we expect that massive 'analogues' of the photon—such as the Z^0 particle—will have *three* independent polarisation states, and this will be verified in §12.2.

We shall give a quantum field treatment in §5.2. For the moment we continue with the evaluation of the matrix element (5.10), making the above interpretation of A^{μ} in terms of the photon idea.

5.1.3 The matrix element (2)

For an incoming photon with momentum k and polarisation state λ, we therefore take

$$A^\mu = N_k \varepsilon^\mu(\lambda) \mathrm{e}^{-\mathrm{i}k\cdot x} \tag{5.29}$$

as the appropriate wavefunction in (5.10); for an outgoing photon we take

$$A^\mu = N_k \varepsilon^{\mu *}(\lambda) \mathrm{e}^{\mathrm{i}k\cdot x}. \tag{5.30}$$

Inserting (5.29) into (5.10), we obtain finally

$$\begin{aligned} \mathscr{A}_{\mathrm{fi}}^{(1)} &= -\mathrm{i}eN_{\mathrm{i}}N_{\mathrm{f}}N_k(p_{\mathrm{i}} + p_{\mathrm{f}})\cdot\varepsilon \int \mathrm{d}^4x \mathrm{e}^{-\mathrm{i}(p_{\mathrm{i}}-p_{\mathrm{f}})\cdot x} \mathrm{e}^{-\mathrm{i}k\cdot x} \\ &= -\mathrm{i}eN_{\mathrm{i}}N_{\mathrm{f}}N_k(p_{\mathrm{i}} + p_{\mathrm{f}})\cdot\varepsilon(2\pi)^4\delta^4(p_{\mathrm{f}} - p_{\mathrm{i}} - k). \end{aligned} \tag{5.31}$$

The energy part of the 4-momentum-conserving δ function forces $E_{\mathrm{f}} = E_{\mathrm{i}} + k^0$, indicating that energy k^0 has been absorbed. This corresponds to the time variation $\sim \mathrm{e}^{-\mathrm{i}k^0 t}$ of (5.29), as discussed in Appendix A.1. We therefore interpret the amplitude (5.31) as describing a process in which a photon of 4-momentum k is absorbed by a spinless particle of charge e and 4-momentum p_{i}, which then goes into the final state with momentum p_{f}, as represented by figure 5.1.

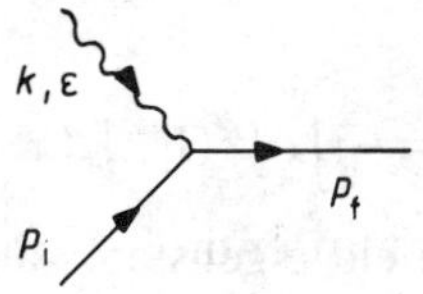

Figure 5.1 One-photon absorption.

Consider the constraints imposed by the energy–momentum δ function in our result:

$$E_{\mathrm{f}} = E_{\mathrm{i}} + k^0 \tag{5.32}$$

$$\boldsymbol{p}_{\mathrm{f}} = \boldsymbol{p}_{\mathrm{i}} + \boldsymbol{k}. \tag{5.33}$$

These are all free particles, so we also have

$$(k^0)^2 - \boldsymbol{k}^2 = 0, \qquad E_{\mathrm{f}}^2 - \boldsymbol{p}_{\mathrm{f}}^2 = E_{\mathrm{i}}^2 - \boldsymbol{p}_{\mathrm{i}}^2 = m^2. \tag{5.34}$$

Conditions (5.32) and (5.33) cannot be satisfied consistently with (5.34). Thus the process of figure 5.1 cannot occur as a real transition, as we remarked at the end of §4.6. However, exactly such a process does occur if the final particle is an excited state of the initial one (as in the excitation of an atomic or nuclear level, for instance); in this case the energy–momentum conditions can be satisfied. Correspondingly, the emission process of figure 5.2 is possible if an initial excited state is de-excited—as in spontaneous emission. The theory of spontaneous emission was first given by Dirac in the paper (Dirac 1927) which laid the foundations of the quantum field concept.

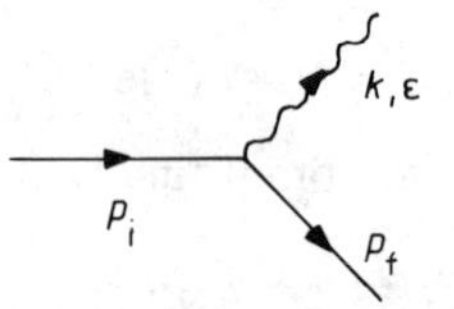

Figure 5.2 One-photon emission.

In the above we have been freely using the language of the quantised electromagnetic field ('photons')—let us now see how these results indeed follow from the field theory formalism of Chapter 4.

5.2 First-order quantum field theory

We show in Appendix A, equation (A.82), that the quantum field theory analogue of (5.4) is

$$\mathscr{A}_{\mathrm{fi}}^{(1)} = -\mathrm{i}\int \mathrm{d}^4 x \langle \mathrm{f}|\hat{\mathscr{H}}_{\mathrm{int}}^{\mathrm{I}}(\boldsymbol{x}, t)|\mathrm{i}\rangle \tag{5.35}$$

where $|\mathrm{i}\rangle$ and $|\mathrm{f}\rangle$ are free field eigenstates and the superscript I means 'in the interaction picture' (see Appendix A.3), implying that the field operators in $\hat{\mathscr{H}}_{\mathrm{int}}^{\mathrm{I}}(\boldsymbol{x}, t)$ have ordinary free field expansions of the type (4.151). What is $\hat{\mathscr{H}}_{\mathrm{int}}^{\mathrm{I}}(\boldsymbol{x}, t)$ in the present case? Normally, an interaction Hamiltonian density will be simply the negative of the corresponding interaction Lagrangian density, as can be seen as follows. Considering for simplicity just a single scalar field $\hat{\phi}$, $\hat{\mathscr{H}}$ is related to $\hat{\mathscr{L}}$ by $\hat{\mathscr{H}} = \hat{\pi}\dot{\hat{\phi}} - \hat{\mathscr{L}} = \hat{\pi}\dot{\hat{\phi}} - \hat{\mathscr{L}}_0 - \hat{\mathscr{L}}_{\mathrm{int}}$. Provided that $\hat{\mathscr{L}}_{\mathrm{int}}$ contains no time derivatives, $\hat{\pi} = \partial\hat{\mathscr{L}}/\partial\dot{\hat{\phi}}$ will remain unchanged by the interaction, and we can therefore write $\hat{\mathscr{H}} = \hat{\mathscr{H}}_0 + \hat{\mathscr{H}}_{\mathrm{int}}$, where $\hat{\mathscr{H}}_{\mathrm{int}} = -\hat{\mathscr{L}}_{\mathrm{int}}$. This situation occurs, for example, in the electromagnetic interactions of spin-$\frac{1}{2}$ particles: $\hat{\mathscr{L}}_{\mathrm{int}}$ of (4.186) involves no derivatives, and the $\hat{\mathscr{H}}_{\mathrm{int}}$ of (4.197) is indeed the negative of (4.186). However, for charged spin-0 particles, $\hat{\mathscr{L}}_{\mathrm{int}}$ of (4.195), *does* involve derivatives, and one finds that $\hat{\mathscr{H}}_{\mathrm{int}} = -\hat{\mathscr{L}}_{\mathrm{int}} + e^2(\hat{A}^0)^2\hat{\phi}^\dagger\hat{\phi}$. The additional (non-covariant) contribution appears to pose a problem. In fact, a careful treatment involving the precise definitions of the 'interaction picture', and of time-ordered products of the derivatives of field operators (Nishijima 1969, §5.5), shows that all such non-covariant additions are cancelled in physical matrix elements.

For $\hat{\mathscr{H}}^{\rm I}_{\rm int}$ in (5.35) we therefore adopt (the negative of) equation (4.195) – but, as in the step from (5.1) to (5.3), we discard the e^2 term to first order and use

$$\hat{\mathscr{H}}^{\rm I}_{\rm int} = {\rm i}e(\hat{\phi}^\dagger \partial^\mu \hat{\phi} - (\partial^\mu \hat{\phi}^\dagger)\hat{\phi})\hat{A}_\mu. \tag{5.36}$$

Equation (5.36) can be written as $\hat{j}^{\,\mu}_{\rm em}\,\hat{A}_\mu$, where

$$\hat{j}^{\,\mu}_{\rm em} = {\rm i}e(\hat{\phi}^\dagger \partial^\mu \hat{\phi} - (\partial^\mu \hat{\phi}^\dagger)\hat{\phi}) \tag{5.37}$$

is the appropriate generalisation of (5.9).

The expansions for the fields $\hat{\phi}$ and $\hat{A}^\mu$ are (cf (4.151))

$$\hat{\phi}(x) = \int \frac{{\rm d}^3\boldsymbol{p}}{(2\pi)^3\, 2E}[\hat{a}(p){\rm e}^{-{\rm i}p\cdot x} + \hat{b}^\dagger(p){\rm e}^{{\rm i}p\cdot x}] \tag{5.38}$$

and

$$\hat{A}^\mu(x) = \sum_\lambda \int \frac{{\rm d}^3\boldsymbol{k}'}{(2\pi)^3\, 2\omega'}[\varepsilon^\mu(k',\lambda)\hat{\alpha}(k',\lambda){\rm e}^{-{\rm i}k'\cdot x} + \varepsilon^{*\mu}(k',\lambda)\hat{\alpha}^\dagger(k',\lambda){\rm e}^{{\rm i}k'\cdot x}]. \tag{5.39}$$

Comparing (5.38) and (5.39), we see that indeed the forms (5.29) and (5.30) appear in (5.39), in a way analogous to the plane wave expansion functions in (5.38). We interpret $\hat{\alpha}(k',\lambda)$ as the operator which destroys a photon of the indicated momentum and polarisation, and $\hat{\alpha}^\dagger(k',\lambda)$ as the corresponding creation operator.

It is clear from the form of $\hat{j}^{\,\mu}_{\rm em}\hat{A}_\mu$ with (5.38) and (5.39) inserted, and from the preliminary discussion in §4.6, that the interaction will produce exactly such emission and absorption processes as shown in figures 5.1 and 5.2. Consider an absorption process, with an initial state $|\boldsymbol{p}_{\rm i}; k, \lambda\rangle$ of a pion and a photon, with the indicated momenta and polarisation, and a final state $|p_{\rm f}\rangle$ of one pion. We want to evaluate the matrix element of $\hat{\mathscr{H}}^{\rm I}_{\rm int}$ (discarding the e^2 term) between these states. We write the initial state as

$$|p_{\rm i}; k, \lambda\rangle = \hat{a}^\dagger(p_{\rm i})\hat{\alpha}^\dagger(k,\lambda)|0\rangle \tag{5.40}$$

where we note that the symbol $|0\rangle$ for the vacuum state must now be interpreted—by a slight extension—to mean the no-photon *and* no-pion state. More formally, $|0\rangle = |0\rangle_\gamma |0\rangle_\pi$.

The final state is just

$$|p_{\rm f}\rangle = \hat{a}^\dagger(p_{\rm f})|0\rangle \tag{5.41}$$

appropriate to single-photon *absorption*. The required matrix amplitude is then

$$-\mathrm{i}\langle 0|\hat{a}(p_\mathrm{f})\left\{\int \mathrm{d}^4x\,\hat{j}^\mu_{\mathrm{em}}(x)\cdot\hat{A}_\mu(x)\right\}\hat{a}^\dagger(p_\mathrm{i})\hat{\alpha}^\dagger(k,\lambda)|0\rangle. \tag{5.42}$$

Since all the matter field operators commute with the quite independent photon field operators $\hat{\alpha}$, $\hat{\alpha}^\dagger$, this separates into a product of two distinct pieces, one for the photon

$$\langle 0|\hat{A}_\mu(x)|k,\lambda\rangle = \langle 0|\hat{A}_\mu(x)\hat{\alpha}^\dagger(k,\lambda)|0\rangle \tag{5.43}$$

and the other for the charged particle

$$\langle p_\mathrm{f}|\hat{j}^\mu_{\mathrm{em}}(x)|p_\mathrm{i}\rangle = \langle 0|\hat{a}(p_\mathrm{f})\hat{j}^\mu_{\mathrm{em}}(x)\hat{a}^\dagger(p_\mathrm{i})|0\rangle. \tag{5.44}$$

The photon part (5.43) is just a 'one-particle to vacuum' matrix element of the photon field $\hat{A}_\mu$, which we expect to be the photon wavefunction, as in §4.2. Using

$$[\hat{\alpha}(k,\lambda),\hat{\alpha}^\dagger(k',\lambda')] = 2\omega(2\pi)^3\delta^3(\boldsymbol{k}-\boldsymbol{k}')\delta_{\lambda\lambda'} \tag{5.45}$$

(other commutators vanishing) and the vacuum condition

$$\hat{\alpha}(k,\lambda)|0\rangle = 0 \quad \text{for all } k,\lambda \tag{5.46}$$

we easily find that this does reduce to

$$N_k\varepsilon_\mu \mathrm{e}^{-\mathrm{i}k\cdot x} \tag{5.47}$$

as expected (cf (4.112)), just as in (5.29). The normalisation factor N_k will be attended to later.

The part (5.44) involving the scalar fields is more complicated since products of *four* operators (in various combinations) will appear in it—two from $\hat{j}^\mu_{\mathrm{em}}$ and one each from $|p_\mathrm{i}\rangle$ and $\langle p_\mathrm{f}|$. For example, we shall encounter things like

$$\langle 0|\hat{a}(p_\mathrm{f})\hat{a}^\dagger(p_1)\hat{a}(p_2)\hat{a}^\dagger(p_\mathrm{i})|0\rangle. \tag{5.48}$$

To evaluate such a thing, the straightforward strategy is to use the commutation relations repeatedly, so as to manouevre the $\hat{a}$'s to the far right and the $\hat{a}^\dagger$'s to the far left, and then utilise the vacuum conditions

$$\hat{a}(p)|0\rangle = 0, \qquad \langle 0|\hat{a}^\dagger(p) = 0. \tag{5.49}$$

Thus we can write (5.48) as

$$\langle 0|[\hat{a}^\dagger(p_1)\hat{a}(p_\mathrm{f}) + (2\pi)^3\,2\omega_\mathrm{f}\delta^3(\boldsymbol{p}_1-\boldsymbol{p}_\mathrm{f})] \times [\hat{a}^\dagger(p_\mathrm{i})\hat{a}(p_2) + (2\pi)^3\,2\omega_\mathrm{i}\delta^3(\boldsymbol{p}_2-\boldsymbol{p}_\mathrm{i})]|0\rangle \tag{5.50}$$

so that actually only the single term

$$(2\pi)^6\delta^3(\boldsymbol{p}_1-\boldsymbol{p}_\mathrm{f})\delta^3(\boldsymbol{p}_2-\boldsymbol{p}_\mathrm{i})2\omega_\mathrm{f}\cdot 2\omega_\mathrm{i} \tag{5.51}$$

survives.

However, the methodical application of this procedure is cumbersome, especially in more complicated cases (imagine a higher-order process!). A more powerful one is available. The matrix elements we need here are all of the form

$$\langle 0|\hat{A}\hat{B}\hat{C}\hat{D}|0\rangle \tag{5.52}$$

where $\hat{A}$, $\hat{B}$, $\hat{C}$, $\hat{D}$ are $\hat{a}$, $\hat{a}^\dagger$, $\hat{b}$ or $\hat{b}^\dagger$—or more generally, any linear combination of these, for instance field operators like $\hat{\phi}$. Let $\hat{A}$ have the form $\hat{a} + \hat{a}^\dagger$, say. Then using the vacuum conditions (5.49), we can write

$$\langle 0|\hat{A}\hat{B}\hat{C}\hat{D}|0\rangle = \langle 0|[\hat{a}, \hat{B}\hat{C}\hat{D}]|0\rangle. \tag{5.53}$$

Now it is an algebraic identity that

$$[\hat{a}, \hat{B}\hat{C}\hat{D}] = [\hat{a}, \hat{B}]\hat{C}\hat{D} + \hat{B}[\hat{a}, \hat{C}]\hat{D} + \hat{B}\hat{C}[\hat{a}, \hat{D}]. \tag{5.54}$$

Hence

$$\begin{aligned}\langle 0|\hat{A}\hat{B}\hat{C}\hat{D}|0\rangle &= \langle 0|[\hat{a}, \hat{B}]\hat{C}\hat{D}|0\rangle + \langle 0|\hat{B}[\hat{a}, \hat{C}]\hat{D}|0\rangle \\ &\quad + \langle 0|\hat{B}\hat{C}[\hat{a}, \hat{D}]|0\rangle.\end{aligned} \tag{5.55}$$

Now all the commutators in this equation—if non-vanishing—are just ordinary numbers, like $(2\pi)^3 2\omega\delta^3(\boldsymbol{p} - \boldsymbol{p}')$: that is, they do not contain operators. Thus we can rewrite this equation as

$$\begin{aligned}\langle 0|\hat{A}\hat{B}\hat{C}\hat{D}|0\rangle &= [\hat{a}, \hat{B}]\langle 0|\hat{C}\hat{D}|0\rangle + [\hat{a}, \hat{C}]\langle 0|\hat{B}\hat{D}|0\rangle \\ &\quad + [\hat{a}, \hat{D}]\langle 0|\hat{B}\hat{C}|0\rangle.\end{aligned} \tag{5.56}$$

We can rewrite this in a more suggestive form by noting that

$$[\hat{a}, \hat{B}] = \langle 0|[\hat{a}, \hat{B}]|0\rangle = \langle 0|\hat{a}\hat{B}|0\rangle = \langle 0|\hat{A}\hat{B}|0\rangle. \tag{5.57}$$

Thus the vacuum expectation value of a product of four operators is just the sum of the products of the possible pairwise 'contractions' (the name given to the vacuum expectation value of the product of two fields):

$$\begin{aligned}\langle 0|\hat{A}\hat{B}\hat{C}\hat{D}|0\rangle &= \langle 0|\hat{A}\hat{B}|0\rangle\langle 0|\hat{C}\hat{D}|0\rangle + \langle 0|\hat{A}\hat{C}|0\rangle\langle 0|\hat{B}\hat{D}|0\rangle \\ &\quad + \langle 0|\hat{A}\hat{D}|0\rangle\langle 0|\hat{B}\hat{C}|0\rangle.\end{aligned} \tag{5.58}$$

The general version of this result is known as Wick's theorem (Wick 1950) and is indispensable for a general discussion of quantum field perturbation theory.

In our case, each of $\hat{A}$, $\hat{B}$, $\hat{C}$, $\hat{D}$ is an $\hat{a}$, $\hat{b}$, $\hat{a}^\dagger$ or $\hat{b}^\dagger$. Inspection shows that any product with an odd number of creation or annihilation operators has vanishing expectation value. This vastly reduces the number of surviving terms. It is a good exercise to check (problem 5.2) that the value of (5.44) is

$$\langle p_{\rm f}|\hat{j}^\mu_{\rm em}(x)|p_{\rm i}\rangle = eN_{\rm f}N_{\rm i}(p_{\rm i} + p_{\rm f})^\mu {\rm e}^{-{\rm i}(p_{\rm i}-p_{\rm f})\cdot x} \tag{5.59}$$

exactly as in (5.10). Thus our *wave mechanical transition current is indeed the matrix element of the field theoretic electromagnetic current operator*:

$$j^{\mu}_{\text{fi}}(x) = \langle p_{\text{f}}|\hat{j}^{\mu}_{\text{em}}(x)|p_{\text{i}}\rangle. \tag{5.60}$$

Combining all these results, we have therefore connected the 'wavefunction' amplitude and the 'field theory' amplitude via

$$\begin{aligned}\mathscr{A}^{(1)}_{\text{fi}} &= -\mathrm{i}\int \mathrm{d}^4x j^{\mu}_{\text{fi}}(x)A_{\mu}(x)\\ &= -\mathrm{i}\int \mathrm{d}^4x \langle p_{\text{f}}|\hat{j}^{\mu}_{\text{em}}(x)|p_{\text{i}}\rangle\langle 0|\hat{A}_{\mu}(x)|k,\lambda\rangle.\end{aligned} \tag{5.61}$$

We can now confidently interpret the amplitude in the diagrammatic form, figure 5.1, and give precise rules for associating different parts of the diagram with different parts of the amplitude, so as to arrive at the complete expression (5.31). These rules are (so far):

(i) There is a normalisation factor 'N' for each line in the diagram.
(ii) At the vertex where the photon is absorbed or emitted, the factor is $-\mathrm{i}e(p_{\text{i}} + p_{\text{f}})^{\mu}$ for charge e.
(iii) For an incoming photon use polarisation vector ε_{μ}; for an outgoing one use ε^*_{μ}.
(iv) Multiply these factors together and then multiply the result by an overall 4-momentum-conserving δ function factor ($(2\pi)^4\delta^4(p_{\text{f}} - p_{\text{i}} - k)$ in this case).

5.3 Electromagnetic scattering of charged spinless bosons

We next consider the electromagnetic interaction of one particle with charge e_{a} with another of charge e_{b}, shown symbolically in figure 5.3. The particles 'a' and 'b' are taken to be *distinguishable*.

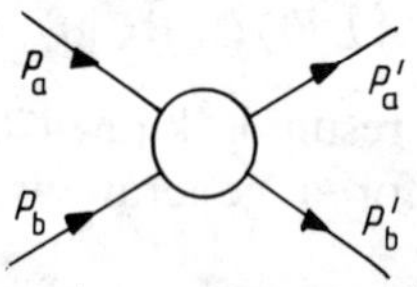

Figure 5.3 Scattering process a + b → a′ + b′.

We shall treat this process as an extension of that considered in §5.1: we regard it as the scattering of one of the particles (say 'a') in the field produced by the other. We shall see at the end that the answer is symmetrical with respect to 'a' and 'b'. The lowest-order amplitude is therefore

$$\mathscr{A} = -\mathrm{i}\int \mathrm{d}^4x\, j^{\mu}_{\mathrm{a'a}}(x) A_{\mu}(x) \tag{5.62}$$

as before, except that $A_{\mu}(x)$ is now the field caused by the motion of 'b'. It seems very plausible that $A_{\mu}(x)$ should be given by the solution of the Maxwell equations (2.19), with j_{μ} this time not zero, but rather the transition current due to the motion $\mathrm{b} \to \mathrm{b}'$:

$$\Box A^{\mu} - \partial^{\mu}(\partial_{\nu} A^{\nu}) = j^{\mu}_{\mathrm{b'b}} \tag{5.63}$$

where (cf (5.59))

$$j^{\mu}_{\mathrm{b'b}}(x) = e_{\mathrm{b}} N_{\mathrm{b}} N_{\mathrm{b'}}(p'_{\mathrm{b}} + p_{\mathrm{b}})^{\mu} \mathrm{e}^{-\mathrm{i}(p_{\mathrm{b}} - p'_{\mathrm{b}})\cdot x}. \tag{5.64}$$

Equation (5.63) will be much easier to solve if, once again, we can decouple the components of A^{μ} via the Lorentz condition (5.13). But A^{μ} is now no longer a free field, so perhaps this condition is now inconsistent with the Maxwell equations (5.63)? Actually, it works: from (5.63) we require

$$\partial_{\mu}\, j^{\mu}_{\mathrm{b'b}} = 0 \quad \text{if } \partial_{\mu} A^{\mu} = 0 \tag{5.65}$$

and from (5.64) we find

$$\partial_{\mu}\, j^{\mu}_{\mathrm{b'b}} = \mathrm{i} e_{\mathrm{b}} N_{\mathrm{b}} N_{\mathrm{b'}}(p'^2_{\mathrm{b}} - p^2_{\mathrm{b}}) \mathrm{e}^{-\mathrm{i}(p_{\mathrm{b}} - p'_{\mathrm{b}})\cdot x} = 0 \tag{5.66}$$

since $p^2_{\mathrm{b}} = p'^2_{\mathrm{b}} = M^2$. Thus we have to solve

$$\Box A^{\mu} = e_{\mathrm{b}} N_{\mathrm{b}} N'_{\mathrm{b}}(p'_{\mathrm{b}} + p_{\mathrm{b}})^{\mu} \mathrm{e}^{\mathrm{i}q\cdot x} \tag{5.67}$$

for A^{μ}, where $q^{\mu} = (p'_{\mathrm{b}} - p_{\mathrm{b}})$. Noting that

$$\Box \mathrm{e}^{\mathrm{i}q\cdot x} = -q^2 \mathrm{e}^{\mathrm{i}q\cdot x} \tag{5.68}$$

we obtain, by inspection,

$$A^{\mu} = -(q^2)^{-1} e_{\mathrm{b}} N_{\mathrm{b}} N'_{\mathrm{b}}(p'_{\mathrm{b}} + p_{\mathrm{b}})^{\mu} \mathrm{e}^{\mathrm{i}q\cdot x}. \tag{5.69}$$

Inserting this into (5.62), we find that the matrix element is

$$\mathscr{A} = -\mathrm{i}\int j^{\mu}_{\mathrm{a'a}}(x)\cdot \frac{-1}{q^2} \cdot j_{\mu\mathrm{b'b}}(x)\,\mathrm{d}^4x \tag{5.70}$$

$$= -\mathrm{i} e_{\mathrm{a}} e_{\mathrm{b}} N_{\mathrm{a}} N_{\mathrm{b}} N'_{\mathrm{a}} N'_{\mathrm{b}} (p_{\mathrm{a}} + p'_{\mathrm{a}})^{\mu} (p_{\mathrm{b}} + p'_{\mathrm{b}})_{\mu} \cdot -\frac{1}{q^2} \cdot \int \mathrm{d}^4x\, \mathrm{e}^{\mathrm{i}(p'_{\mathrm{a}} + p'_{\mathrm{b}} - p_{\mathrm{a}} - p_{\mathrm{b}})\cdot x} \tag{5.71}$$

$$= -\mathrm{i} e_{\mathrm{a}} e_{\mathrm{b}} N_{\mathrm{a}} N_{\mathrm{b}} N'_{\mathrm{a}} N'_{\mathrm{b}} (p_{\mathrm{a}} + p'_{\mathrm{a}})^{\mu} \cdot -\frac{g_{\mu\nu}}{q^2} \cdot (p_{\mathrm{b}} + p'_{\mathrm{b}})^{\nu}$$

$$\times (2\pi)^4 \delta^4(p'_{\mathrm{a}} + p'_{\mathrm{b}} - p_{\mathrm{a}} - p_{\mathrm{b}}) \tag{5.72}$$

$$\equiv -\mathrm{i}(2\pi)^4 \delta^4(p'_{\mathrm{a}} + p'_{\mathrm{b}} - p_{\mathrm{a}} - p_{\mathrm{b}}) N_{\mathrm{a}} N'_{\mathrm{a}} N_{\mathrm{b}} N'_{\mathrm{b}} F. \tag{5.73}$$

As promised, the expression is symmetrical in $\mathrm{a} \leftrightarrow \mathrm{b}$, bearing in mind that the energy–momentum-conserving δ function implies that

$$q = p'_b - p_b = p_a - p'_a.$$

Can we get a diagrammatic interpretation of our result? We have written it in a way which suggests what the diagram should be. We already know that the $(p_a + p'_a)^\mu$ and $(p_b + p'_b)^\nu$ bits are associated with emission or absorption vertices; and we know that in this process there are no free photons in the initial or final states. Therefore the photon must be emitted from one particle and absorbed by the other, as in figure 5.4 (note that there are two ways in which this can happen). In that case, the $-g_{\mu\nu}/q^2$ part must somehow correspond to the *sum* of the two processes shown in figure 5.4! We shall pursue this in more detail in §5.6. For the moment, it is high time that we calculated a cross section.

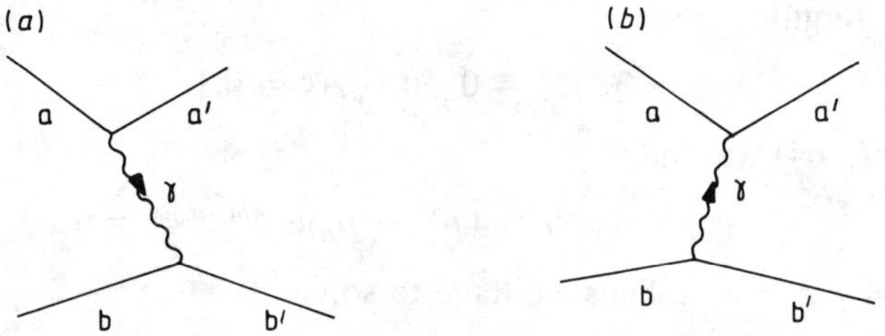

Figure 5.4 The two possible one-photon exchange processes: in (*a*) particle 'a' emits the photon and particle 'b' absorbs it, while in (*b*) particle 'b' emits and particle 'a' absorbs.

5.4 Cross section for two-particle scattering

Consider the general two-body process $1 + 2 \rightarrow 3 + 4$, where now the numerals label the momenta of the particles. We must first decide on the normalisation of our wavefunctions. We have defined plane wave solutions

$$\phi_i = N_i e^{-ip_i \cdot x} \tag{5.74}$$

so that the corresponding probability density for a positive-energy KG particle is

$$\rho_i = i\left[\phi_i^*\left(\frac{\partial \phi_i}{\partial t}\right) - \left(\frac{\partial \phi_i^*}{\partial t}\right)\phi_i\right] = 2N_i^2 E_i. \tag{5.75}$$

Instead of choosing to normalise our wavefunctions to one particle in a box of volume V, as in NRQM, we shall normalise instead to $2E$ *particles in a volume* V. This is in accord with the fact that ρ_i is the time component of a 4-vector and is therefore called a 'covariant normalisation' condition. Either choice is perfectly viable so long as one uses the appropriate flux and phase space factors.

With the condition

$$\int_V \rho_i \mathrm{d}^3 x = 2E_i \tag{5.76}$$

we obtain the result

$$N_i = V^{-1/2}. \tag{5.77}$$

The three steps to derive the cross section may be summarised as follows.

(1) The transition rate per unit volume is defined by

$$P_{fi} = |\mathscr{A}_{fi}|^2/VT \tag{5.78}$$

where T is the time of interaction. Either by cavalierly 'squaring the δ function' or rigorously using wavepackets (as discussed, for example, in Chapter 3 of Taylor (1972)), one derives

$$P_{fi} = (2\pi)^4 \delta^4(p_3 + p_4 - p_1 - p_2)(N_1 N_2 N_3 N_4)^2 |F|^2 \tag{5.79}$$

where (cf 5.73)

$$\mathscr{A}_{fi} = -i(2\pi)^4 \delta^4(p_3 + p_4 - p_1 - p_2) N_1 N_2 N_3 N_4 \cdot F. \tag{5.80}$$

(2) In order to obtain a quantity which may be compared from experiment to experiment, we must remove the dependence of the transition rate on the incident flux of particles and the number of target particles per unit volume.

(a) The flux of beam particles incident on a stationary target is just the number of particles per unit area which can reach the target in unit time. Thus the 'active' volume is just $|\boldsymbol{v}|$, the velocity of the beam particles, and we have normalised to $2E_1$ particles in a volume V, so the flux factor is

$$|\boldsymbol{v}|\frac{2E_1}{V}. \tag{5.81}$$

(b) The number of target particles per unit volume is just $2E_2/V$ (actually $2m_2/V$ for particle 2 at rest).

To obtain a normalisation-independent quantity, we therefore divide P_{fi} by the flux factor and by the number of target particles per unit volume, i.e. by the factor

$$|\boldsymbol{v}|\frac{2E_1 \cdot 2E_2}{V^2}. \tag{5.82}$$

(3) For a physical scattering cross section we must sum over the

available two-particle final states. For one particle in a volume V this amounts to integrating over the usual two-body phase space factor

$$\frac{V}{(2\pi)^3}\mathrm{d}^3p_3\frac{V}{(2\pi)^3}\mathrm{d}^3p_4. \tag{5.83}$$

For our normalisation of $2E$ particles in volume V the available phase space per particle is thus

$$\frac{V}{(2\pi)^3}\frac{\mathrm{d}^3p_3}{2E_3}\frac{V}{(2\pi)^3}\frac{\mathrm{d}^3p_4}{2E_4}. \tag{5.84}$$

Putting all this together, the cross section is given by

$$\mathrm{d}\sigma = P_{\mathrm{fi}}\frac{V^2}{2E_1\cdot 2E_2|\boldsymbol{v}|}\frac{V}{(2\pi)^3}\frac{\mathrm{d}^3p_3}{2E_3}\frac{V}{(2\pi)^3}\frac{\mathrm{d}^3p_4}{2E_4}. \tag{5.85}$$

Using our expressions for P_{fi} and N_{i}, the final result is

$$\mathrm{d}\sigma = \frac{|F|^2}{2E_1\cdot 2E_2|\boldsymbol{v}|}(2\pi)^4\delta^4(p_3+p_4-p_1-p_2)\frac{\mathrm{d}^3p_3}{2E_3(2\pi)^3}\frac{\mathrm{d}^3p_4}{2E_4(2\pi)^3}. \tag{5.86}$$

Note that:

(i) the factors involving the normalisation volume V have cancelled;

(ii) we can write the flux factor for collinear collisions in invariant form using the relation (easily verified in a particular frame (see problem 5.4))

$$E_1E_2|\boldsymbol{v}| = [(p_1\cdot p_2)^2 - m_1^2m_2^2]^{1/2}; \tag{5.87}$$

(iii) the four-dimensional δ function together with the phase space (including the $(1/2E)$ factors) is sometimes called Lorentz invariant phase space:

$$\mathrm{dLips}(s; p_3, p_4) \tag{5.88}$$

where the Mandelstam s variable is

$$s = (p_1 + p_2)^2 \tag{5.89}$$

(see e.g. Perkins 1987).

5.5 Explicit evaluation of the a + b → a′ + b′ cross section

We identify $p_{\mathrm{a}} = p_1$, $p'_{\mathrm{a}} = p_3$, $p_{\mathrm{b}} = p_2$ and $p'_{\mathrm{b}} = p_4$ in §5.3, and apply the results of §5.4. The differential cross section is

$$\boxed{\mathrm{d}\sigma = \frac{|F|^2}{4[(p_1\cdot p_2)^2 - m_1^2m_2^2]^{1/2}}\mathrm{dLips}(s; p_3, p_4)} \tag{5.90}$$

where the invariant amplitude F for the electromagnetic interaction is

$$F = (e_a e_b/q^2)(p_1 + p_3)\cdot(p_2 + p_4) \tag{5.91}$$

and the two-particle Lorentz invariant phase space is

$$\boxed{\begin{aligned} \mathrm{dLips}(s; p_3, p_4) &= (2\pi)^4 \delta^4(p_3 + p_4 - p_1 - p_2) \\ &\quad \times \frac{1}{(2\pi)^3}\frac{\mathrm{d}^3 \boldsymbol{p}_3}{2E_3}\frac{1}{(2\pi)^3}\frac{\mathrm{d}^3 \boldsymbol{p}_4}{2E_4}. \end{aligned}} \tag{5.92}$$

To gain familiarity with these factors, we shall evaluate these expressions in the centre-of-momentum (CM) frame defined by

$$\boldsymbol{p}_1 + \boldsymbol{p}_2 = \boldsymbol{p}_3 + \boldsymbol{p}_4 = 0. \tag{5.93}$$

The scattering is described by a CM scattering angle θ_{CM}, as in figure 5.5, and the four 4-momenta are given by

$$\begin{aligned} p_1^\mu &= (E_1, \boldsymbol{p}), \qquad & p_3^\mu &= (E_1, \boldsymbol{p}') \\ p_2^\mu &= (E_2, -\boldsymbol{p}), \qquad & p_4^\mu &= (E_2, -\boldsymbol{p}') \end{aligned} \tag{5.94}$$

where

$$|\boldsymbol{p}| = |\boldsymbol{p}'| = p \tag{5.95}$$

and

$$W = E_1 + E_2 = E_3 + E_4 = (p^2 + m_1^2)^{1/2} + (p^2 + m_2^2)^{1/2} \tag{5.96}$$

is the CM energy.

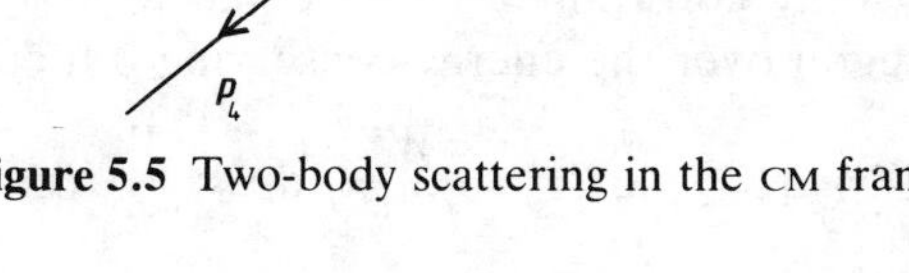

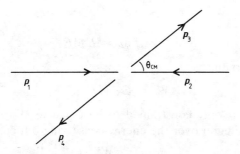

Figure 5.5 Two-body scattering in the CM frame.

5.5.1 Evaluation of two-body phase space in the CM frame

Before we specialise to the CM frame, it is convenient to simplify our expression for dLips:

$$\mathrm{dLips}(s; p_3, p_4) = \frac{1}{(4\pi)^2}\delta^4(p_3 + p_4 - p_1 - p_2)\frac{\mathrm{d}^3 \boldsymbol{p}_3}{E_3}\frac{\mathrm{d}^3 \boldsymbol{p}_4}{E_4}. \tag{5.97}$$

Using the 3-momentum δ function, we can eliminate the integral over $\mathrm{d}^3\boldsymbol{p}_4$:

$$\int \frac{\mathrm{d}^3 p_4}{E_4} \delta^4(p_3 + p_4 - p_1 - p_2) = \frac{1}{E_4} \delta(E_3 + E_4 - E_1 - E_2). \quad (5.98)$$

On the right-hand side $\boldsymbol{p}_4$ and E_4 are no longer independent variables but are determined by the conditions

$$\boldsymbol{p}_4 = \boldsymbol{p}_1 + \boldsymbol{p}_2 - \boldsymbol{p}_3, \qquad E_4 = (|\boldsymbol{p}_4|^2 + m_2^2)^{1/2}. \quad (5.99)$$

Next, convert $\mathrm{d}^3\boldsymbol{p}_3$ to angular variables

$$\mathrm{d}^3 p_3 = \mathrm{d}\Omega p_3^2 \mathrm{d}p_3 \quad (5.100)$$

where p_3 now stands for the magnitude of the 3-momentum. The energy and momentum are related by

$$E_3^2 = p_3^2 + m_1^2 \quad (5.101)$$

so that

$$E_3 \mathrm{d}E_3 = p_3 \mathrm{d}p_3. \quad (5.102)$$

With all these changes we arrive at the result (*valid in any frame*)

$$\mathrm{dLips}(s; p_3, p_4) = \frac{1}{(4\pi)^2} \mathrm{d}\Omega \frac{p_3 \mathrm{d}E_3}{E_4} \delta(E_3 + E_4 - E_1 - E_2). \quad (5.103)$$

Now *specialise to the* CM *frame* for which

$$E_3^2 = p^2 + m_1^2 \quad (5.104)$$

$$E_4^2 = p^2 + m_2^2 \quad (5.105)$$

and

$$E_3 \mathrm{d}E_3 = p\mathrm{d}p = E_4 \mathrm{d}E_4. \quad (5.106)$$

Introduce the variable

$$W' = E_3 + E_4 \quad (5.107)$$

(since $E_3 + E_4$ is only constrained to be equal to $W = E_1 + E_2$ *after* performing the integral over the energy-conserving δ function). Then

$$\mathrm{d}W' = \mathrm{d}E_3 + \mathrm{d}E_4 = \frac{W'}{E_3 E_4} p\mathrm{d}p = \frac{W'}{E_4} \mathrm{d}E_3 \quad (5.108)$$

where we have used equation (5.106) in each of the last two steps. Thus the factor

$$p_3 \frac{\mathrm{d}E_3}{E_4} \delta(E_3 + E_4 - E_1 - E_2)$$

becomes

$$\frac{p}{W'}\mathrm{d}W'\delta(W' - W) = \frac{p}{W} \tag{5.109}$$

and we arrive at the important result

$$\boxed{\mathrm{dLips}(s; p_3, p_4) = \frac{1}{(4\pi)^2}\frac{p}{W}\mathrm{d}\Omega} \tag{5.110}$$

for two-body phase space in the CM frame.

5.5.2 *Flux factor in the* CM *frame*

Define

$$f^2 = (p_1 \cdot p_2)^2 - m_1^2 m_2^2. \tag{5.111}$$

Using the CM result

$$p_1 \cdot p_2 = E_1 E_2 + p^2 \tag{5.112}$$

a straightforward calculation shows that

$$f = pW. \tag{5.113}$$

We can now write down the *CM differential cross section*

$$\mathrm{d}\sigma = \frac{1}{4f}|F|^2 \frac{1}{(4\pi)^2}\frac{p}{W}\mathrm{d}\Omega. \tag{5.114}$$

The result is

$$\boxed{\left.\frac{\mathrm{d}\sigma}{\mathrm{d}\Omega}\right|_{\mathrm{CM}} = \frac{1}{(8\pi W)^2}|F|^2.} \tag{5.115}$$

Before we evaluate $|F|^2$ let us convert this to invariant form as an exercise in changing variables. The 4-momentum transfer squared is defined as the Mandelstam t variable:

$$t = q^2 = (p_1 - p_3)^2 = (p_4 - p_2)^2. \tag{5.116}$$

In the CM frame this reduces to the relation

$$t = -2p^2(1 - \cos\theta_{\mathrm{CM}}) \tag{5.117}$$

and hence

$$\mathrm{d}t = 2p^2\mathrm{d}\,(\cos\theta_{\mathrm{CM}}). \tag{5.118}$$

For spinless particles there is cylindrical symmetry about the beam axis

$$\mathrm{d}\Omega_{\mathrm{CM}} = 2\pi\mathrm{d}(\cos\theta_{\mathrm{CM}}) \tag{5.119}$$

and so we arrive at the relation

$$\frac{\mathrm{d}}{\mathrm{d}t} = \frac{\pi}{p^2} \frac{\mathrm{d}}{\mathrm{d}\Omega_{\mathrm{CM}}}. \tag{5.120}$$

Thus the *two-body differential cross section in invariant form* is easily shown to be

$$\frac{\mathrm{d}\sigma}{\mathrm{d}t} = \frac{1}{64\pi} \frac{1}{(p_1 \cdot p_2)^2 - m_1^2 m_2^2} |F|^2 \tag{5.121}$$

or, in terms of s,

$$\boxed{\frac{\mathrm{d}\sigma}{\mathrm{d}t} = \frac{1}{16\pi} \frac{1}{[s - (m_1 + m_2)^2][s - (m_1 - m_2)^2]} |F|^2.} \tag{5.122}$$

These expressions are valid for any unpolarised $2 \rightarrow 2$ elastic scattering reaction. For a + b scattering it remains only to calculate $|F|^2$:

$$|F|^2 = \left(\frac{e_{\mathrm{a}} e_{\mathrm{b}}}{q^2}\right)^2 [(p_1 + p_3) \cdot (p_2 + p_4)]^2. \tag{5.123}$$

For elastic scattering the definitions

$$s = (p_1 + p_2)^2 = (p_3 + p_4)^2 \tag{5.124}$$

$$u = (p_1 - p_4)^2 = (p_2 - p_3)^2 \tag{5.125}$$

lead to the relations

$$2p_1 \cdot p_2 = 2p_3 \cdot p_4 = s - m_1^2 - m_2^2 \tag{5.126}$$

$$2p_1 \cdot p_4 = 2p_2 \cdot p_3 = m_1^2 + m_2^2 - u. \tag{5.127}$$

Finally we derive the result for $e_{\mathrm{a}} = e_{\mathrm{b}} = e$:

$$|F|^2 = (4\pi\alpha/t)^2 (s - u)^2 \tag{5.128}$$

where we have explicitly set

$$\alpha = e^2/4\pi \simeq \tfrac{1}{137} \tag{5.129}$$

the fine structure constant in natural units (see Appendices B and C).

It is straightforward to evaluate these invariant expressions in any desired frame. As an example, we consider the case in which 'a' is a particle of charge eZ_{a} and mass m, and 'b' is a particle of charge eZ_{b} and mass M, and we take $m \ll M$. This will then simulate either the electromagnetic scattering of a light particle by a heavy nucleus, say, or a 'spinless' electron by a 'spinless' proton (the complications of spin will occupy us in Chapter 6). We shall work in the laboratory frame

$$p_{\mathrm{b}} = (M, \mathbf{0}) \tag{5.130}$$

in which 'b' is initially at rest. The light particle 4-vectors are

$$p_{\mathrm{a}} = (\omega_{\mathrm{a}}, \boldsymbol{k}) \simeq (|\boldsymbol{k}|, \boldsymbol{k}) \tag{5.131}$$

$$p'_{\mathrm{a}} = (\omega'_{\mathrm{a}}, \boldsymbol{k}') \simeq (|\boldsymbol{k}'|, \boldsymbol{k}') \tag{5.132}$$

neglecting m. Then

$$\begin{aligned} t = q^2 = (p_{\mathrm{a}} - p'_{\mathrm{a}})^2 &= -2|\boldsymbol{k}|\cdot|\boldsymbol{k}'|\cdot(1 - \cos\theta) \\ &= -4|\boldsymbol{k}|\cdot|\boldsymbol{k}'|\cdot\sin^2\theta/2. \end{aligned} \tag{5.133}$$

If we write $p'_{\mathrm{b}} = p_{\mathrm{b}} + (p_{\mathrm{a}} - p'_{\mathrm{a}})$ and square it, we obtain $2p_{\mathrm{b}}\cdot q + q^2 = 0$, from which follows

$$|\boldsymbol{k}|/|\boldsymbol{k}'| = 1 + (2|\boldsymbol{k}|/M)\sin^2\theta/2. \tag{5.134}$$

If we now make the *further* approximation

$$|\boldsymbol{k}| \ll M \tag{5.135}$$

we can set $|\boldsymbol{k}| \simeq |\boldsymbol{k}'|$ and write

$$t \simeq -4|\boldsymbol{k}|^2\sin^2\theta/2 \tag{5.136}$$

$$\mathrm{d}t \simeq 2|\boldsymbol{k}|^2\mathrm{d}(\cos\theta). \tag{5.137}$$

The flux factor in (5.121) is

$$(p_{\mathrm{a}}\cdot p_{\mathrm{b}})^2 - m^2M^2 \simeq M^2\omega_{\mathrm{a}}^2 \simeq M^2|\boldsymbol{k}|^2 \tag{5.138}$$

in the approximation of neglecting m. We also have

$$s = (p_{\mathrm{a}} + p_{\mathrm{b}})^2 = (\omega_{\mathrm{a}} + M)^2 - \boldsymbol{k}^2 \simeq 2|\boldsymbol{k}|M \tag{5.139}$$

and

$$u = (p_{\mathrm{a}} - p'_{\mathrm{b}})^2 \simeq -2|\boldsymbol{k}|M \tag{5.140}$$

so that

$$(s - u)^2 \simeq 16|\boldsymbol{k}|^2M^2. \tag{5.141}$$

Putting the pieces together yields finally

$$\frac{\mathrm{d}\sigma}{\mathrm{d}\Omega} = \frac{\alpha^2 Z_{\mathrm{a}}^2 Z_{\mathrm{b}}^2}{4|\boldsymbol{k}|^2\sin^4\theta/2} \tag{5.142}$$

where $\mathrm{d}\Omega = 2\pi\mathrm{d}(\cos\theta)$. This is recognisable as the *Rutherford scattering cross section*.

We shall meet a similar laboratory frame cross section again in Chapter 6, §6.8, where we shall need to evaluate it without making the low-energy approximation (5.135). This is quite a tricky calculation and is described in Appendix E.

5.6 Scattering as an exchange process: forces in quantum field theory, and the Feynman graph for the matrix element

Of course, one does not have to wade through five chapters of this book in order to derive the Rutherford cross section. Nevertheless, it is satisfying that all this machinery has produced it, when required. Naturally, all sorts of more sophisticated corrections can eventually be calculated with this now quite powerful formalism. But we want first to consider how the familiar Rutherford formula, which we all learn initially as having to do with the *Coulomb potential,* emerges in the field theory language: we will see that in field theory a *force* arises from the *exchange of quanta of the field potentials*.

It will be useful to remind ourselves, very briefly, how the Rutherford formula for Coulomb scattering is obtained in ordinary non-relativistic quantum mechanics. One simply treats the static Coulomb potential $e_a e_b/r$ as a perturbation on the free particle Schrödinger equation, and obtains for the lowest-order amplitude (cf (A.20))

$$\mathscr{A}_{\mathrm{fi}}^{\mathrm{R}} = -\mathrm{i}\int \mathrm{d}^3\boldsymbol{x}\,\mathrm{d}t\,\mathrm{e}^{\mathrm{i}E_{\mathrm{f}}t - \mathrm{i}\boldsymbol{k}'\cdot\boldsymbol{x}}(e_a e_b/r)\mathrm{e}^{-\mathrm{i}E_{\mathrm{i}}t + \mathrm{i}\boldsymbol{k}\cdot\boldsymbol{x}} \tag{5.143}$$

$$= -\mathrm{i}\,2\pi\cdot\delta(E_{\mathrm{f}} - E_{\mathrm{i}})e_a e_b\int \mathrm{e}^{\mathrm{i}(\boldsymbol{k}-\boldsymbol{k}')\cdot\boldsymbol{x}}(1/r)\mathrm{d}^3\boldsymbol{x} \tag{5.144}$$

where $r = |\boldsymbol{x}|$. The Fourier transform here is apparently ill-defined for large r—the volume element being $r^2\mathrm{d}r\,\mathrm{d}\Omega$. A useful device helps us over the difficulty: we introduce a convergence factor $\mathrm{e}^{-\mu r}$ into the integral, which now becomes

$$\int \mathrm{e}^{\mathrm{i}\boldsymbol{q}\cdot\boldsymbol{x}}\cdot(1/r)\cdot\mathrm{e}^{-\mu r}r^2\mathrm{d}r\,\mathrm{d}\Omega \tag{5.145}$$

where $\boldsymbol{q} = \boldsymbol{k} - \boldsymbol{k}'$. Problem 5.5 evaluates this integral, with the result that the amplitude (5.144) becomes

$$\mathscr{A}_{\mathrm{fi}}^{\mathrm{R}} = -\mathrm{i}\,2\pi\cdot\delta(E_{\mathrm{f}} - E_{\mathrm{i}})e_a e_b\cdot 4\pi/(\boldsymbol{q}^2 + \mu^2). \tag{5.146}$$

Remarkably enough, we can now safely take the limit $\mu \to 0$ and obtain

$$\mathscr{A}_{\mathrm{fi}}^{\mathrm{R}} = -\mathrm{i}\,2\pi\cdot\delta(E_{\mathrm{f}} - E_{\mathrm{i}})e_a e_b\cdot 4\pi/\boldsymbol{q}^2. \tag{5.147}$$

Noting that

$$\boldsymbol{q}^2 \simeq 4\boldsymbol{k}^2\sin^2\theta/2 \tag{5.148}$$

we can see the tell-tale $\sin^{-4}\theta/2$ emerging from the square of $\mathscr{A}_{\mathrm{fi}}^{\mathrm{R}}$, which enters into the transition probability and hence the cross section. This is all we need—we shall not go through all the flux and phase space factors for this 'one particle in a fixed potential' problem (see any book on quantum mechanics).

We return to the calculation described in §5.3, only this time we shall reinterpret it using (a little) field theory. We note first of all that the amplitude (5.73) is of *second* order in e, since e_a and e_b are both proportional to e. This suggests that we try to regard it, in fact, as a second-order perturbation theory amplitude, in terms of the perturbation $\hat{\mathcal{H}}_{int}$.

The formalism we shall use for this is briefly described in Appendix A.3. From equations (A.80)–(A.82) we know that the first-order expression (5.35) is equivalent to

$$\mathcal{A}_{fi}^{(1)} = -i\langle f|\hat{H}_{int}|i\rangle 2\pi\delta(E_f - E_i) \tag{5.149}$$

where (from (A.78))

$$\langle f|\hat{H}_{int}|i\rangle = \int d^3\boldsymbol{x}\langle f|\hat{\mathcal{H}}_{int}(\boldsymbol{x}, 0)|i\rangle \tag{5.150}$$

and $\hat{\mathcal{H}}_{int}(\boldsymbol{x}, 0)$ is the interaction Hamiltonian in the Schrödinger picture. $\hat{\mathcal{H}}_{int}(\boldsymbol{x}, 0)$ is given by an expression of the form (5.36), except that the fields on the right-hand side are now to be interpreted as Schrödinger picture fields, having expansions of the form (5.38) but with $t = 0$; for example,

$$\hat{\phi}(\boldsymbol{x}, 0) = \int \frac{d^3\boldsymbol{p}}{(2\pi)^3\, 2E}[\hat{a}(p)e^{i\boldsymbol{p}\cdot\boldsymbol{x}} + \hat{b}^\dagger(p)e^{-i\boldsymbol{p}\cdot\boldsymbol{x}}]. \tag{5.151}$$

Since $\hat{\mathcal{H}}_{int}(\boldsymbol{x}, 0)$ is time-independent, we can take over from NRQM the familiar formalism of transition theory for time-independent potentials, arriving at the second-order amplitude (cf (A.84))

$$\mathcal{A}_{fi}^{(2)} = -i\sum_{n\neq i}\langle f|\hat{H}_{int}|n\rangle\frac{1}{E_i - E_n}\langle n|\hat{H}_{int}|i\rangle 2\pi\delta(E_f - E_i). \tag{5.152}$$

For the a + b → a′ + b′ process of figure 5.3, equation (5.152) becomes

$$\mathcal{A}_{fi}^{(2)} = -i\sum_{\substack{n \\ E_n\neq E_a+E_b}}\left\{\int d^3\boldsymbol{x}\langle p_a' p_b'|\hat{\mathcal{H}}_{int}(\boldsymbol{x}, 0)|n\rangle\right\}\frac{1}{E_a + E_b - E_n}$$

$$\times\left\{\int d^3\boldsymbol{x}'\langle n|\hat{\mathcal{H}}_{int}(\boldsymbol{x}', 0)|p_a p_b\rangle\right\}2\pi\delta(E_f - E_i). \tag{5.153}$$

We must stress an important feature of this equation. The integrals are over three-dimensional space only, not over time. This means that when we insert plane wave expansions like (5.151) into the matrix elements and integrate, we shall get momentum conservation δ functions, but *no energy conservation δ function*, as far as the 'intermediate states' $|n\rangle$ are concerned. Indeed, this is always the case in this form of perturbation theory: there is an explicit δ function in (5.153) which guarantees overall energy conservation $E_a + E_b = E_a' + E_b'$, but if $E_a + E_b$ (or $E_a' + E_b'$) were also to equal E_n, the denominator of (5.153) would vanish and the expression would not make sense. This asymmetry in the treatment of energy and momentum seems to imply that we run the risk of violating Lorentz invariance. However, we shall see that the formalism, which is called 'non-covariant perturbation theory', does in the end produce a fully Lorentz invariant answer.

We proceed with the evaluation of (5.153). To save ourselves operator algebra, and because it is more physical, we shall freely use the pictorial representation of the matrix elements, already introduced. We first determine what are the 'intermediate' states $|n\rangle$. $\hat{\mathcal{H}}_{int}$ induces single-photon emission or absorption processes. Thus the only possibility for the states $|n\rangle$ is that they consist of two particles and one photon—which is created by the 'first' operator (on the right) in (5.153) and destroyed by the second. However, this photon could be created from either a or b, and absorbed by either a′ or b′. If it is absorbed by the same particle as emitted it, we shall be evaluating amplitudes corresponding to the pictures shown in figure 5.6. These are interesting processes, to which we shall return in due course (§5.9)—however, it seems intuitively clear that they do *not* correspond to any interaction *between* the particles. Thus we are led to the two processes already introduced in figure 5.4, but pictured again in figures 5.7(a) and 5.7(b). The reason for drawing them again is to make explicit what is the nature of the state $|n\rangle$ in the two cases. Consider figure 5.7(a). In this one, the second matrix element in (5.153) is

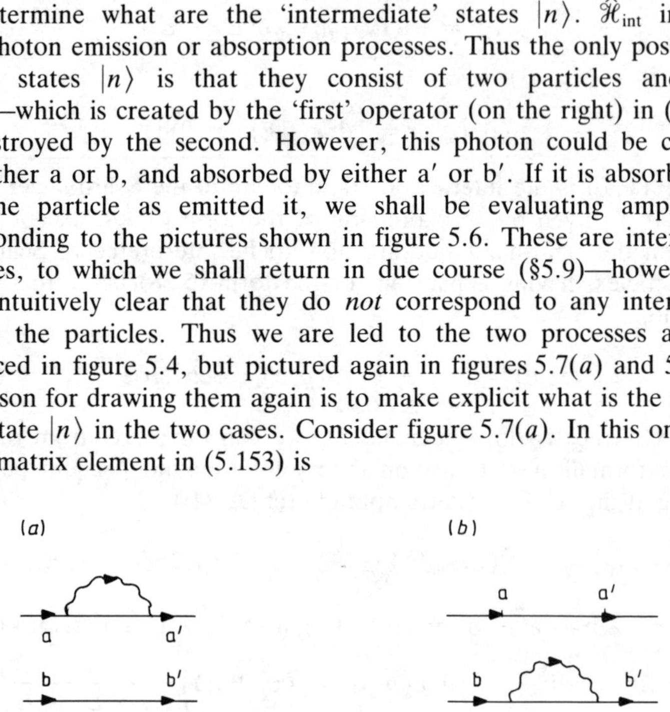

Figure 5.6 Processes in which a single photon is emitted and then absorbed by the same particle.

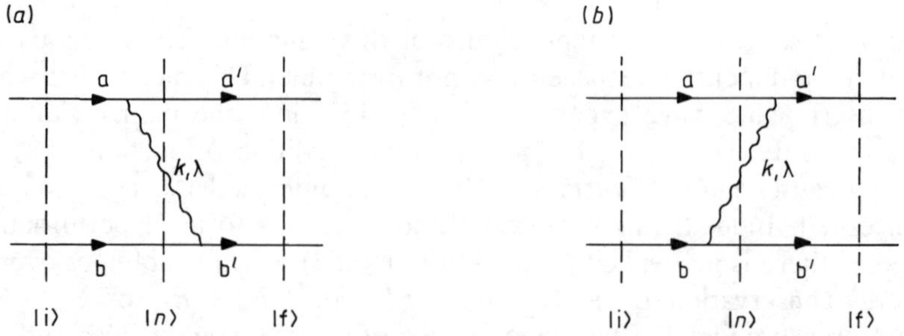

Figure 5.7 The two one-photon exchange processes.

$$\int d^3x' \langle p'_a, p_b; k, \lambda | \hat{\mathscr{H}}_{int}(x', 0) | p_a, p_b \rangle; \tag{5.154}$$

there are two particles and an emitted photon in the state $\langle n|$. Similarly, the first matrix element is

$$\int d^3x \langle p'_a, p'_b | \hat{\mathscr{H}}_{int}(x, 0) | p'_a, p_b; k, \lambda \rangle. \tag{5.155}$$

The intermediate state $|n\rangle$ is therefore

$$|n\rangle = |p'_a, p_b; k, \lambda\rangle \tag{5.156}$$

and the summation over n is reduced to summing over all possible k and λ. The energy E_n in (5.153) is thus

$$E_n = E'_a + E_b + E_k. \tag{5.157}$$

It is clear, in fact, that the 'b' in (5.154) and the 'a' in (5.155) are unaffected by the $\hat{\mathscr{H}}_{int}$ operators. We are really dealing with

$$\int d^3x' \langle p'_a; k, \lambda | \hat{\mathscr{H}}_{int}(x', 0) | p_a \rangle, \qquad \int d^3x \langle p'_b | \hat{\mathscr{H}}_{int}(x, 0) | p_b; k, \lambda \rangle. \tag{5.158}$$

But these are almost exactly the same as the matrix elements considered in §§5.2 and 5.1: the difference is simply that the time coordinate is set equal to zero, and the integrals are over three-dimensional space only. The net result of this is that all '$(2\pi)^4 \delta^4(\ )$' factors in expressions such as (5.31) are replaced by '$(2\pi)^3 \delta^3(\ )$' factors expressing momentum—but not energy—conservation. Thus for figure 5.7(*a*) we obtain the amplitude

$$\begin{aligned}\mathscr{A}_{fi}^{(2)} = &-i\sum_{k,\lambda} e_b N_b N'_b N_k (p_b + p'_b)^\mu \varepsilon_\mu(k, \lambda)(2\pi)^3 \delta^3(\boldsymbol{p}'_b - \boldsymbol{p}_b - \boldsymbol{k}) \\ &\times \frac{1}{E_a + E_b - (E'_a + E_b + E_k)} \cdot e_a N_a N'_a N_k (p_a + p'_a)^\nu \\ &\times \varepsilon^*_\nu(k, \lambda)(2\pi)^3 \delta^3(\boldsymbol{p}_a - \boldsymbol{p}'_a - \boldsymbol{k})\, 2\pi\delta(E'_a + E'_b - E_a - E_b).\end{aligned} \tag{5.159}$$

The 'sum' on k in (5.159) means $\int d^3\boldsymbol{k}/(2\pi)^3 2E_k$, anticipating that all volume factors will as usual cancel (and therefore not putting them in). This integral can be done using one δ^3 function in (5.159), leaving the result

$$\begin{aligned}-i(2\pi)^4 \delta^4(p'_a + p'_b - p_a - p_b) N_a N_b N'_a N'_b e_a e_b (p_b + p'_b)^\mu (p_a + p'_a)^\nu \\ \times \left\{ \sum_\lambda \frac{\varepsilon^*_\nu(k, \lambda)\varepsilon_\mu(k, \lambda) N_k^2}{2E_k(E_a - E'_a - E_k)} \right\}.\end{aligned} \tag{5.160}$$

We are clearly getting quite near to (5.72).

There is also figure 5.7(*b*) to consider. This, however, differs from figure 5.7(*a*) only in that for figure 5.7(*b*) we evidently have

$$E_n = E'_{\mathrm{b}} + E_{\mathrm{a}} + E_k \tag{5.161}$$

and $\varepsilon \leftrightarrow \varepsilon^*$in the bracket { } of (5.160). In fact, we shall shortly see that $\Sigma_\lambda \varepsilon^*_\nu \varepsilon_\mu$ is real, so that the only material change is that the denominator of (5.160) is replaced by $E'_{\mathrm{a}} - E_{\mathrm{a}} - E_k$, where overall energy conservation has been used. We now note

$$\frac{1}{2E_k}\left\{\frac{1}{E_{\mathrm{a}} - E'_{\mathrm{a}} - E_k} - \frac{1}{E_{\mathrm{a}} - E'_{\mathrm{a}} + E_k}\right\} = \frac{1}{(E_{\mathrm{a}} - E'_{\mathrm{a}})^2 - E_k^2} \tag{5.162}$$

where, from the δ functions in (5.159),

$$E_k^2 = \boldsymbol{k}^2 = (\boldsymbol{p}_{\mathrm{a}} - \boldsymbol{p}'_{\mathrm{a}})^2. \tag{5.163}$$

Thus the right-hand side of (5.162) is just

$$1/q^2 \tag{5.164}$$

where $q^2 = (p_{\mathrm{a}} - p'_{\mathrm{a}})^2$! This of course is exactly the $1/q^2$ factor appearing in (5.72).

This is a remarkable result, since it shows that, as promised (see also Appendix A.3), the final answer, which is the sum of the *two* separate and individually *non*-covariant amplitudes, is actually a *single invariant* expression (anticipating the far from obvious conclusion—see below—that the polarisation sum in (5.160) yields the $g_{\mu\nu}$ factor in (5.72)), which can be represented by the *single* 'Feynman graph' shown in figure 5.8. Note that in this graph—the *Feynman graph* for the one-photon exchange process—the photon line is drawn vertically, implying the inclusion of the *two* differently sequenced processes of figure 5.4.

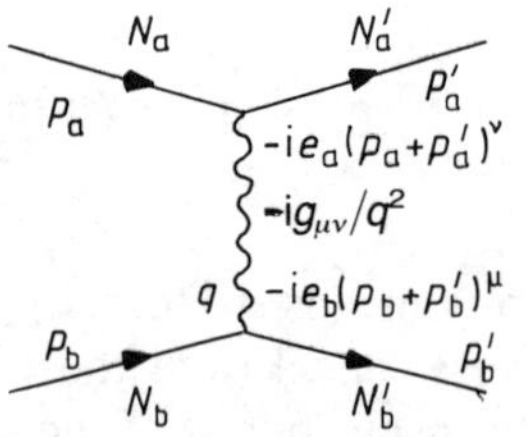

Figure 5.8 The single one-photon exchange Feynman graph.

With this step, we have made a fundamental reinterpretation. In figures 5.7(*a*) and 5.7(*b*) the photon has zero mass—in the sense that $E_k^2 - \boldsymbol{k}^2 = 0$—but energy is not conserved at the vertices. This is, as repeatedly emphasised above, perfectly normal in second-order quantum mechanical perturbation theory. One calls the states $|n\rangle$ 'virtual', as well as 'intermediate', for this reason. However, this is an intrinsically non-covariant statement, since *energy* is singled out. By contrast, we can interpret figure 5.8 quite differently. We can interpret $q = p_{\mathrm{a}} - p'_{\mathrm{a}}$ as the *4-momentum* of the photon, assuming covariant 4-momentum conservation at each vertex. But then the mass of the photon is q^2, which is not zero! Here also we call the photon 'virtual'. The great advantages of figure 5.8 are that it is a single expression and it is an invariant. There is no conservation of rest mass in relativity, but there is of 4-momentum: thus—by allowing the intermediate photon to go 'off mass shell' (i.e. $q^2 \neq 0$)'—we can remain fully covariant. It is the amplitude of figure 5.8 that 'covariant' field theoretic perturbation theory would have given us.

We are nearly through. The remaining job is to justify the association of the $g_{\mu\nu}$ factor with the polarisation sum in (5.160). The essential question is: exactly what polarisations λ have to be considered? If the photons in figure 5.7 were free quanta, we know from §5.1.2 what the answer would be—namely, λ would run over just the two transverse polarisation states. In the present case, however, the photons are interacting with the charged particle sources, and the story is more complicated. We are trying to derive the Lorentz-invariant form of the matrix element, (5.72). In this case, we would want to work in the Lorentz gauge $\partial_\mu \hat{A}^\mu = 0$. Indeed, we imposed just this condition in (5.65), in the course of our first (non field-theoretic) derivation of (5.72). In field theory, however, it can be shown (Aitchison 1982 §4.1, Mandl and Shaw, 1984, §5.2) that the constant $\partial_\mu \hat{A}^\mu = 0$ conflicts with the field commutation relations, and thus it cannot be imposed as an operator condition. Instead, a condition must be imposed on the allowed photon *states*. The result of this turns out to be that the sum over λ in (5.160) is actually to be taken over all *four* possible polarisation states, defined as follows. We take $\boldsymbol{k}$ along the z-direction, so that two transverse polarisation vectors may be chosen to be

$$\varepsilon_\mu(\lambda = 1) = (0, 1, 0, 0), \qquad \varepsilon_\mu(\lambda = 2) = (0, 0, 1, 0) \tag{5.165}$$

(cf (5.23), (5.24)). The other two ε's are

$$\varepsilon_\mu(\lambda = 0) = (1, 0, 0, 0) \quad \text{'timelike polarisation'} \tag{5.166}$$

and

$$\varepsilon_\mu(\lambda = 3) = (0, 0, 0, 1) \quad \text{'longitudinal polarisation'.} \tag{5.167}$$

But in performing the 'sum' over these four λ's there is an unexpected subtlety: the contribution from the $\lambda = 0$ case must be taken with a minus sign! Some idea of the reason for this can be obtained by noting the result (problem 5.6)

$$\varepsilon_\mu(\lambda = 1)\varepsilon_\nu(\lambda = 1) + \varepsilon_\mu(\lambda = 2)\,\varepsilon_\nu(\lambda = 2) + \varepsilon_\mu(\lambda = 3)\varepsilon_\nu(\lambda = 3) - \varepsilon_\mu(\lambda = 0)\varepsilon_\nu(\lambda = 0) = -g_{\mu\nu}. \quad (5.168)$$

This equation shows us that if, indeed, we interpret

$$'\sum_\lambda \varepsilon_\mu(\lambda)\ \varepsilon_\nu(\lambda)'$$

to mean the left-hand side of (5.168), rather than the *sum* of all four terms, the result will be the covariant expression $-g_{\mu\nu}$: and this of course is precisely the term we were seeking.

It is clear that we have fully reconstructed the whole of (5.72). Furthermore, since we know all the other parts of figure 5.8 from §5.2, we can deduce the rule for the *internal photon* line, namely:

(v) for an internal photon of 4-momentum q, a factor $-\mathrm{i}g_{\mu\nu}/q^2$. This rule, and the final expression (5.72) for the amplitude are relevant to the *single* figure 5.8, in which, by a re-interpretation, the photon's 4-momentum q is now such that $q^2 \neq 0$.

We can gain some insight into the difference between such a 'virtual' photon and a 'real' one by the following considerations. We have arrived at an expression of the form

$$j_{\mathrm{b}}^{\mu}\cdot -g_{\mu\nu}/q^2 \cdot j_{\mathrm{a}}^{\nu} \quad (5.169)$$

with $q^2 \neq 0$, i.e.

$$q = (q^0, 0, 0, |\boldsymbol{q}|) \quad (5.170)$$

with $q^0 \neq |\boldsymbol{q}|$. The current conservation relation then reads

$$q^0 j^0 - |\boldsymbol{q}|\cdot j^3 = 0 \quad (5.171)$$

i.e.

$$j^3 = q^0 j^0/|\boldsymbol{q}|. \quad (5.172)$$

Expression (5.169) can be written as

$$(j_{\mathrm{b}}^1 j_{\mathrm{a}}^1 + j_{\mathrm{b}}^2 j_{\mathrm{a}}^2)/q^2 + (j_{\mathrm{b}}^3 j_{\mathrm{a}}^3 - j_{\mathrm{b}}^0 j_{\mathrm{a}}^0)/q^2 \quad (5.173)$$

$$= (j_{\mathrm{b}}^1 j_{\mathrm{a}}^1 + j_{\mathrm{b}}^2 j_{\mathrm{a}}^2)/q^2 + j_{\mathrm{b}}^0 j_{\mathrm{a}}^0/\boldsymbol{q}^2 \quad (5.174)$$

using (5.172). The first term may be interpreted as being due to the exchange of a transversely polarised photon (only the 1, 2 components enter, perpendicular to $\boldsymbol{q}$). For *real* photons $q^2 \to 0$, so that this term will completely dominate the second. The latter, however, must obviously be included when $q^2 \neq 0$. We note that it depends on the 3-momentum squared, $\boldsymbol{q}^2$, rather than the 4-momentum squared q^2, and that it involves the charge densities j_{b}^0 and j_{a}^0. Referring back to the start of this section, we can interpret it as the *instantaneous Coulomb interaction* between these charge densities, since

$$\int d^4x\, e^{iq\cdot x}\, \delta(t)/r = \int d^3x\, e^{i\boldsymbol{q}\cdot\boldsymbol{x}}/r = 4\pi/\boldsymbol{q}^2. \qquad (5.175)$$

Thus, in summary, the single covariant amplitude (5.72), corresponding to the graph of figure 5.8, includes contributions from the exchange of transversely polarised photons *and* from the familiar Coulomb potential. This is the true relativistic extension of the static Coulomb result of (5.147). We shall, however, have more to say about rule (v) in Chapter 13.

The factors associated with rules (i), (ii) and (v) have been attached to the vertices and internal photon line in figure 5.8. Implementation of rule (iv) then reconstructs the amplitude (5.72).

5.7 More examples: electromagnetic $\pi^+\pi^+$ and $\pi^+\pi^-$ scattering

It is clear that we can construct the amplitudes for many more processes involving photons and charged spinless particles by 'sticking together', in various ways, the different vertices contained in $\hat{\mathcal{H}}_{\text{int}}$. We first consider the modification of §5.3 which is necessary when the particles are *indistinguishable*. Rather than talk about general particles 'a', 'b', we shall now call them simply π's, though it must be remembered that in actual $\pi\pi$ scattering the strong interaction will generally dominate the electromagnetic one which we are calculating here.

The process is

$$\pi^+(p_1) + \pi^+(p_2) \to \pi^+(p_3) + \pi^+(p_4). \qquad (5.176)$$

We have identical particles in the initial and final states, and we must use appropriately symmetrised states. In terms of the scattering amplitude, the full amplitude must be symmetric under exchange of the initial or final π^+'s. In graphical terms we have, as before, the diagram of figure 5.9, but to this we must add the diagram of figure 5.10 in which

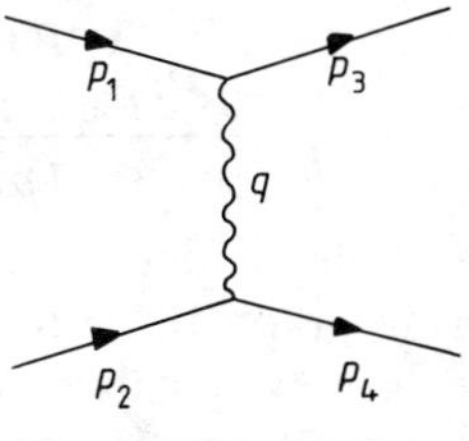

Figure 5.9 Feynman diagram for $\pi^+\pi^+$ scattering, in the one-photon exchange approximation; direct amplitude.

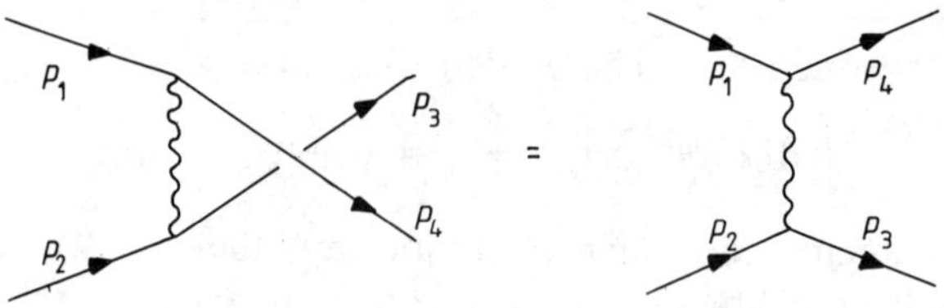

Figure 5.10 Exchange amplitude for $\pi^+\pi^+$ scattering.

p_3 and p_4 are interchanged. Using the rules already given, we can write down the result of the complete amplitude

$$\mathscr{A} = \mathscr{A}_{\text{direct}} + \mathscr{A}_{\text{exchange}}$$

and obtain

$$\begin{aligned}\mathscr{A}_{\pi^+\pi^+} &= -\mathrm{i}(2\pi)^4\,\delta^4(p_3 + p_4 - p_1 - p_2)N_1N_3N_2N_4 \\ &\quad\times\left(\frac{-e^2(p_1 + p_3)_\mu(p_2 + p_4)^\mu}{(p_2 - p_4)^2} + \frac{-e^2(p_1 + p_4)_\mu(p_2 + p_3)^\mu}{(p_2 - p_3)^2}\right)\end{aligned} \tag{5.177}$$

$$\equiv -\mathrm{i}(2\pi)^4\delta^4(p_3 + p_4 - p_1 - p_2)N_1N_2N_3N_4F_{\pi^+\pi^+}(p_1, p_2; p_3, p_4). \tag{5.178}$$

We have defined the invariant (Feynman) amplitude F for $\pi^+\pi^+$ scattering by removing all the 'standard' normalisation and momentum conservation factors, as in (5.73).

Now consider $\pi^+\pi^-$ elastic scattering. Using the antiparticle hypothesis, we have the correspondences shown in figure 5.11. In terms of the Feynman amplitudes this corresponds to the relation

$$F_{\pi^+\pi^-}(p_{\mathrm{a}}, p_{\mathrm{b}}; p_{\mathrm{c}}, p_{\mathrm{d}}) \equiv F_{\pi^+\pi^+}(p_{\mathrm{a}}, -p_{\mathrm{d}}; p_{\mathrm{c}}, -p_{\mathrm{b}}) \tag{5.179}$$

and we can immediately write down the answer. (Alternatively, one can derive this in detail using negative-energy solutions, as we did for the energy conservation factors of the earlier examples, or the appropriate bits of $\hat{\mathscr{H}}_{\text{int}}$.) We obtain

$$\begin{aligned}F_{\pi^+\pi^-}(p_{\mathrm{a}}, p_{\mathrm{b}}; p_{\mathrm{c}}, p_{\mathrm{d}}) = \Bigg(&\frac{-\,e^2(p_{\mathrm{a}} + p_{\mathrm{c}})_\mu(-\,p_{\mathrm{d}} - p_{\mathrm{b}})^\mu}{(-p_{\mathrm{d}} + p_{\mathrm{b}})^2} \\ &+ \frac{-\,e^2(p_{\mathrm{a}} - p_{\mathrm{b}})_\mu(-\,p_{\mathrm{d}} + p_{\mathrm{c}})^\mu}{(p_{\mathrm{a}} + p_{\mathrm{b}})^2}\Bigg)\end{aligned} \tag{5.180}$$

corresponding to the two graphs of figures 5.12 and 5.13. Figure 5.12 is just like the a + b scattering graph of §5.6, with $e_{\mathrm{a}} = e$ and $e_{\mathrm{b}} = -e$: the sign of the amplitude is now negative with respect to the corresponding $\pi^+\pi^+$ scattering, as we expect, since like charges repel and unlike charges attract. Figure 5.13 is a new process: particle–antiparticle annihilation to a virtual photon state, with consequent rematerialisation. The effect of symmetrisation in the $\pi^+\pi^+$ process shows up as an annihilation graph in the 'crossed ' $\pi^+\pi^-$ channel.

Figure 5.11 Feynman reinterpretation of $\pi^+\pi^-$ scattering.

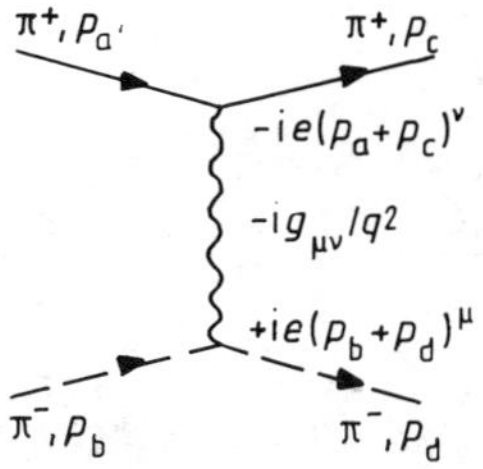

Figure 5.12 One-photon exchange amplitude in $\pi^+\pi^-$ scattering.

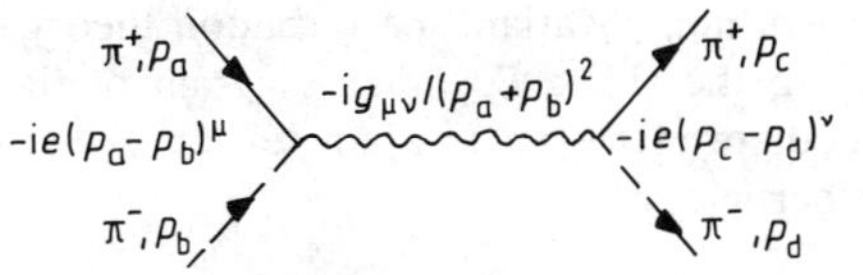

Figure 5.13 One-photon annihilation amplitude in $\pi^+\pi^-$ scattering.

5.8 Pion Compton scattering: the pion propagator, and gauge invariance

5.8.1 The pion propagator

We have discussed exclusively, so far, cases in which the exchanged—or, in diagram terms, 'internal'—particle is a photon. However, we can stick the vertices together in one more way, shown in figure 5.14. This is a contribution to the process

$$\gamma(k) + \pi^+(p) \to \gamma(k') + \pi^+(p') \tag{5.181}$$

which is called π^+ Compton scattering. We know how to write down factors for all the parts of figure 5.14 except for one: we do not yet know what is the factor for an internal or *virtual* pion. We could get this by following the route which led to the virtual photon factor, $-\mathrm{i}g_{\mu\nu}/q^2$, namely by considering the sum of the *two different processes* which are actually both contained in figure 5.14, and which are shown in figures 5.15(*a*) and 5.15(*b*): see problem 5.7.

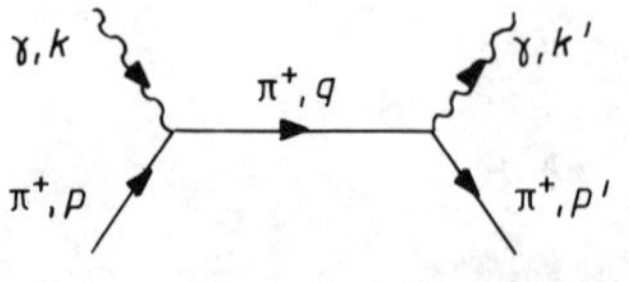

Figure 5.14 Feynman diagram for pion Compton scattering.

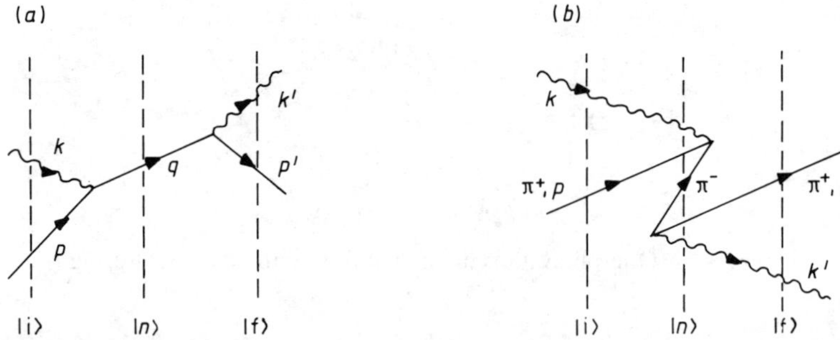

Figure 5.15 The two 'non-covariant' perturbation theory diagrams (cf figure 5.7) which correspond to the single Feynman diagram of figure 5.14 (cf figure 5.8). In (*a*) the intermediate state $|n\rangle$ is a single π^+; in (*b*) it contains five quanta (including an antiparticle, π^-).

However, we shall adopt a different tactic, relying on analogy with the photon case. Let us remember where the vital $1/q^2$ factor came from in (5.70)–(5.73). We wrote down the 'interacting' Maxwell equation (in Lorentz gauge)

$$\Box A^{\mu} = j^{\mu} \tag{5.182}$$

and, since $j^{\mu} \sim \mathrm{e}^{-\mathrm{i}q\cdot x}$ for this case (lowest-order perturbation), we were able to invert (5.182) and obtain

$$A^{\mu} = -(g^{\mu\nu}/q^2)\cdot j_{\nu}. \tag{5.183}$$

For a two-particle process the amplitude had the form

$$j^{\mu}_{\mathrm{b}}(-g_{\mu\nu}/q^2)j^{\nu}_{\mathrm{a}} \tag{5.184}$$

and we identified $-\mathrm{i}g_{\mu\nu}/q^2$ as the virtual photon factor. Let us follow the same steps for interacting pions. In this case the corresponding equation is

$$(\Box + m^2)\phi = -V\phi = -j \quad \text{(say)} \tag{5.185}$$

and for a plane wave 'current' $j \sim e^{-iq\cdot x}$ we can invert (5.185) by

$$\phi = \frac{1}{q^2 - m^2}\, j. \tag{5.186}$$

Comparing (5.183) with (5.186), we conjecture the rule

(vi) a factor $i/(q^2 - m^2)$ for an internal spin-0 particle of 4-momentum q.

These factors for internal particles are called *propagators*, an important term. Our conjecture (vi) is indeed the correct answer for the pion propagator. For a proper derivation the reader is strongly recommended to consult Chapter 6 of Bjorken and Drell (1964), which gives a discussion of non-relativistic and relativistic propagator theory. It is explicitly shown there that the propagator is just the momentum space Green function of the appropriate free particle wave equation, and that the Feynman prescription for negative- and positive-energy solutions corresponds to the so-called $+i\varepsilon$ prescription for handling the poles in the momentum space propagator.

5.8.2 Pion exchange and the Yukawa potential

Before proceeding with the $\gamma\pi$ scattering calculation, we discuss a very simple—but fundamental—interpretation of the form just given for the pion propagator. It is essentially the part of an amplitude that corresponds to an internal—i.e. *exchanged*—pion. Now we have learned that an exchanged photon is the field theoretic way of describing an electromagnetic force between the particles which exchanged it. Further, there was a close correspondence between the nature of the force law ($1/r$ potential) and the form of the photon propagator: compare (5.143)–(5.147) and note that, since for the static potential $E_i = E_f$, $q^2 = (E_f - E_i)^2 - (\boldsymbol{k} - \boldsymbol{k}')^2 = -\boldsymbol{q}^2$. It is natural therefore to ask: suppose two particles interacted by the exchange of a pion; what static potential would this correspond to?

The answer is found from (5.146). The expression $(\boldsymbol{q}^2 + \mu^2)^{-1}$ is precisely (up to a factor $-i$) the propagator for a spin-0 particle of mass μ, for the static case $q^2 = -\boldsymbol{q}^2$. Equation (5.146) shows that it corresponds to a potential of *Yukawa form*

$$V \propto e^{-\mu r}/r. \tag{5.187}$$

In (5.187) the range of the potential is of order μ^{-1} (i.e. $\hbar/\mu c$ in full units): beyond this distance V becomes rapidly negligible. But μ was the mass of the exchanged quantum: thus *the exchange of a quantum of* *mass μ gives rise to a force of range μ^{-1}*. This fundamental relationship was used by Yukawa (1935) (see also Wick (1938)) to estimate the mass of the then hypothetical particle responsible for the short-range (~1–2 fm) nucleon–nucleon potential. For a range of 1 fm, μ~200 MeV—very similar to the actual pion mass (and indeed it is true that pion exchange does give a good description of the longest-range part of the nuclear potential). As $\mu \to 0$ we recover once again the Coulomb potential, which is said to be of 'infinite range', having no exponentially decaying factor.

5.8.3 Pion Compton scattering and gauge invariance

Returning to $\gamma\pi$ scattering, we now have all the ingredients assembled to evaluate the contribution of the Feynman diagram of figure 5.14 (which includes both of figures 5.15(*a*) and 5.15(*b*)). Putting together the factors for the external photons and the internal pion gives an expression for the invariant amplitude (defined as in (5.73) by separating off the normalisation and δ factors),

$$F_{\gamma\pi}^{(\mathrm{d})} = e^2\varepsilon'^*\cdot(q + p')[1/(q^2 - m^2)]\varepsilon\cdot(q + p) \tag{5.188}$$

where the 4-momentum of the virtual pion is

$$q = k + p = k' + p'. \tag{5.189}$$

If we impose the Lorentz conditions

$$\varepsilon\cdot k = \varepsilon'\cdot k' \tag{5.190}$$

and define

$$s = (k + p)^2 = (k' + p')^2 \tag{5.191}$$

equation (5.188) may be written

$$F_{\gamma\pi}^{(\mathrm{d})} = 4e^2(\varepsilon^*\cdot p')(\varepsilon\cdot p)/(s - m^2). \tag{5.192}$$

The superscript (d), which stands for 'direct', indicates that this is not the whole story. Imagine the graph of figure 5.14 as describing the 'crossed' process $\pi^+\pi^- \to \gamma\gamma$. For two identical bosons in the final state we must symmetrise the two-photon wavefunction. This leads us to consider the diagram of figure 5.16, *which must also contribute to the $\gamma\pi \to \gamma\pi$ process.* For Compton scattering, the corresponding diagram may be redrawn as in figure 5.17. Following the same arguments as before, we associate the following invariant amplitude to this diagram:

$$F_{\gamma\pi}^{(\mathrm{e})} = e^2\varepsilon'^*\cdot(p + q')[1/(q'^2 - m^2)]\varepsilon\cdot(q' + p') \tag{5.193}$$

where

$$q' = p - k' = p' - k. \tag{5.194}$$

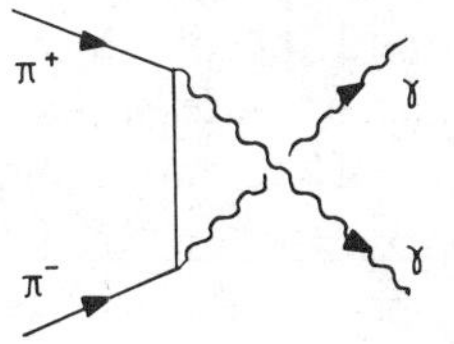

Figure 5.16 Exchange contribution to $\pi^+\pi^- \to \gamma\gamma$.

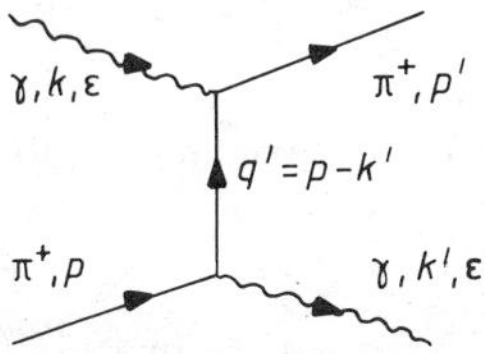

Figure 5.17 Exchange contribution to $\gamma\pi^+ \to \gamma\pi^+$.

With the Lorentz condition, as before, and defining the Mandelstam variable

$$u = q'^2 = (p - k')^2 = (k - p')^2 \tag{5.195}$$

equation (5.193) reduces to

$$F_{\gamma\pi}^{(e)} = 4e^2(\varepsilon^* \cdot p)(\varepsilon \cdot p')/(u - m^2). \tag{5.196}$$

This is sometimes referred to as the exchange (or u channel) contribution, in contrast to the direct (or s channel) term $F_{\gamma\pi}^{(d)}$.

At this point it would seem that we now have all possible order e^2 contributions to pion Compton scattering. In fact, this is not so. The long-forgotten term

$$-e^2 A^2 \tag{5.197}$$

in the KG potential $\hat{V}_{\mathrm{KG}}$ of equation (5.1) *will* contribute to this process. In field theory terms, (5.197) corresponds to the part of (4.195) which was neglected when we wrote (5.36)—that is, to the missing term

$$-e^2 \hat{A}^2 \hat{\phi}^\dagger \hat{\phi} \tag{5.198}$$

in $\hat{\mathscr{H}}_{\mathrm{int}}$. The term (5.198) will give rise to a *new* vertex, in which four quanta—two photons and two mesons—interact at a point, as shown in figure 5.18 for the part of (5.198) referring to the process (5.181). This is called a 'contact' term, since no intermediate state is involved. The amplitude for this diagram can be calculated from first principles by evaluating the matrix element

$$-e^2 \langle p'; k', \varepsilon' | \hat{A}^2 \hat{\phi}^\dagger \hat{\phi} | p; k, \varepsilon \rangle \tag{5.199}$$

using the field expansions and the operator algebra. We shall, instead, proceed rather indirectly, by a route which will emphasise the importance of the *gauge invariance* of the entire amplitude.

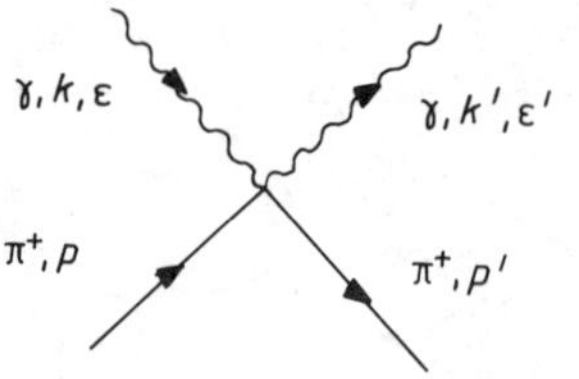

Figure 5.18. Four-point contact term for $\gamma\pi^+ \to \gamma\pi^+$.

Consider a process, shown in figure 5.19, involving a photon of momentum k^μ, whose polarisation state is described by the vector ε^μ. The amplitude for this process must be linear in the photon polarisation vector and thus we may write

$$\mathscr{A} = \varepsilon^\mu T_\mu \tag{5.200}$$

where T_μ depends on the particular process under consideration. With the Lorentz choice for ε^μ we have

$$k \cdot \varepsilon = 0. \tag{5.201}$$

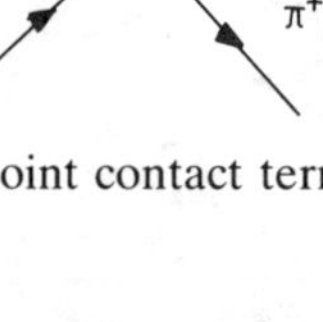

Figure 5.19 General one-photon process.

But, as we have seen in §5.1, gauge invariance for real photons allows us to replace ε^μ by ε'^μ where

$$\varepsilon'^\mu = \varepsilon^\mu + \beta k^\mu \tag{5.202}$$

without affecting any physical quantities, since this transformation corresponds only to a change of gauge (within the Lorentz gauge). For our amplitude $\mathscr{A}$, the requirement of gauge invariance reduces to the condition

$$\boxed{k^\mu T_\mu = 0 \quad \text{gauge invariance.}} \tag{5.203}$$

Let us apply this test of gauge invariance to the sum of (5.192) and (5.196). We have

$$F_{\gamma\pi}^{(\mathrm{d})} + F_{\gamma\pi}^{(\mathrm{e})} = 4e^2\varepsilon'^{\mu *}\varepsilon^{\nu}\left(\frac{p'_\mu p_\nu}{s - m^2} + \frac{p_\mu p'_\nu}{u - m^2}\right) \tag{5.204}$$

$$\equiv \varepsilon'^{\mu *}\varepsilon^{\nu} T_{\mu\nu}. \tag{5.205}$$

Now replace ε by k and use

$$2p\cdot k = s - m^2 \tag{5.206}$$

$$2p'\cdot k = -(u - m^2) \tag{5.207}$$

to obtain

$$\varepsilon'^{\mu *}k^{\nu}T_{\mu\nu} = 2e^2\varepsilon'^{\mu *}(p' - p)_\mu \tag{5.208}$$

$$= 2e^2\varepsilon'^{*}\cdot k \tag{5.209}$$

which is *not* zero. If we replace ε' by k' in this, we are left with

$$2e^2 k'\cdot k \tag{5.210}$$

which is still not zero. We therefore deduce that the two amplitudes so far considered cannot constitute the complete amplitude, which has to satisfy the gauge invariance requirement (5.203) for each photon.

At this point we have to recall that we are doing a perturbation theory calculation with e^2 (or, more precisely, the dimensionless quantity $\alpha = e^2/4\pi$) as the expansion parameter. Condition (5.203) of course refers to the complete amplitude, whereas $F_{\gamma\pi}^{(\mathrm{d})} + F_{\gamma\pi}^{(\mathrm{e})}$ is a contribution of order e^2. It would, however, seem unreasonable to expect higher-order terms in the perturbation expansion to restore the gauge invariance of the second-order terms, since in principle e^2 is an arbitrary parameter. Rather, we expect (5.203) to be true *order by order* in perturbation theory. We must therefore suppose that in addition to $F_{\gamma\pi}^{(\mathrm{d})} + F_{\gamma\pi}^{(\mathrm{e})}$ we need an amplitude

$$F_{\gamma\pi}^{(\mathrm{c})} = -\,2e^2\varepsilon'^{*}\cdot\varepsilon. \tag{5.211}$$

The complete amplitude

$$F_{\gamma\pi}^{(\mathrm{d})} + F_{\gamma\pi}^{(\mathrm{e})} + F_{\gamma\pi}^{(\mathrm{c})} \tag{5.212}$$

then satisfies the gauge invariance condition, at this order of e^2. Thus, even if we had omitted the e^2A^2 term in our interaction potential, consistency with gauge invariance would have demanded its existence. The condition (5.203) provides an extremely useful check in practical calculations. The term (5.211) is, of course, the result obtained by evaluating the corresponding part of (5.197) or (5.198). The factor 2 arises from the fact that each $\hat{A}$ in the $\hat{A}^2$ term can be associated with either the emission or the absorption of a photon.

The total invariant amplitude (to order e^2) for pion Compton scattering is then

$$F_{\gamma\pi} = e^2 \varepsilon'^{\mu *} \varepsilon^{\nu} \left(\frac{4p'_{\mu} p_{\nu}}{s - m^2} + \frac{4p_{\mu} p'_{\nu}}{u - m^2} - 2g_{\mu\nu} \right) \tag{5.213}$$

and the full amplitude to be used in calculating the $\gamma\pi$ Compton cross section is obtained by multiplying this by the usual normalisation and δ function factors.

5.9 Trees and loops: higher-order calculations

The sample calculations which we have presented so far all have one very important feature in common: the intermediate states $|n\rangle$ are all—when interpreted in the (invariant) Feynman graph form—*single-particle* states, e.g. a single γ or a single π. For such states, as we saw explicitly in equations (5.159), the momentum is uniquely determined by the momentum conservation laws at the vertices of the diagram, enforced by the δ functions. This means that the integral $\int d^3\boldsymbol{k}$ over the intermediate state momenta can be done using a δ function, as we saw, leaving no residual integral. Consider, however, the discarded diagram of figure 5.6, the interesting piece of which we draw again in figure 5.20.

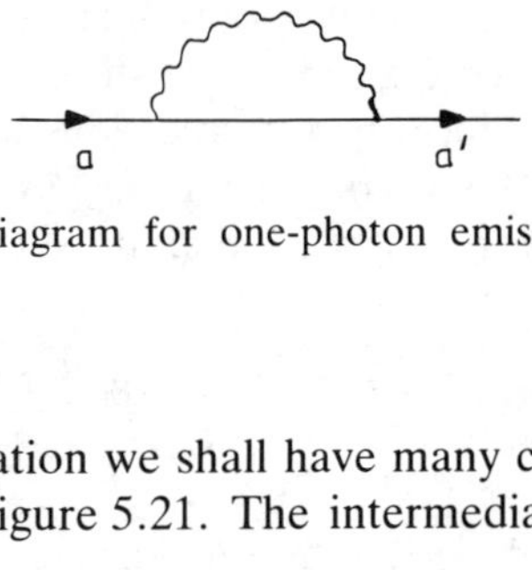

Figure 5.20 Feynman diagram for one-photon emission and reabsorption process.

In 'non-covariant' notation we shall have many contributions to this; one of them is shown in figure 5.21. The intermediate state summation will now involve

$$\iint d^3\boldsymbol{p}' d^3\boldsymbol{k}\, \delta^3(\boldsymbol{p}_1 - \boldsymbol{p}' - \boldsymbol{k}) \delta^3(\boldsymbol{p}' + \boldsymbol{k} - \boldsymbol{p}_2) \ldots \tag{5.214}$$

among other factors. One of the integrals in (5.214) can be done at once, leaving, say,

$$\delta^3(\boldsymbol{p}_2 - \boldsymbol{p}_1) \int d^3\boldsymbol{k} \ldots . \tag{5.215}$$

There is clearly no way in which the momentum of the photon can be determined—the photon is emitted and then absorbed by the same particle, and the overall conservation of momentum has appeared correctly in (5.215).

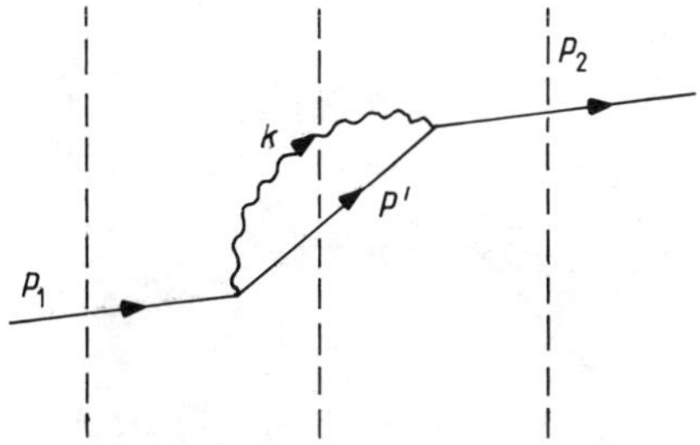

Figure 5.21 One 'non-covariant' contribution to the process of figure 5.20.

Of course, this has been a 'non-covariant' argument. It is very plausible, though, that the appropriate generalisation holds for the covariant Feynman graphs. That is, we expect to have to integrate over those 4-momenta of intermediate state particles which are not determined by the energy–momentum conservation conditions at each vertex. An example is given in figure 5.22: this is a higher-order process, contributing in $O(\alpha^2)$ to the amplitude for $1 + 2 \rightarrow 3 + 4$. It is clear that the 4-momentum k is undetermined, and the correct procedure is

(vii) integrate over any undetermined 4-momentum k with factor $\int d^4k/(2\pi)^4$.

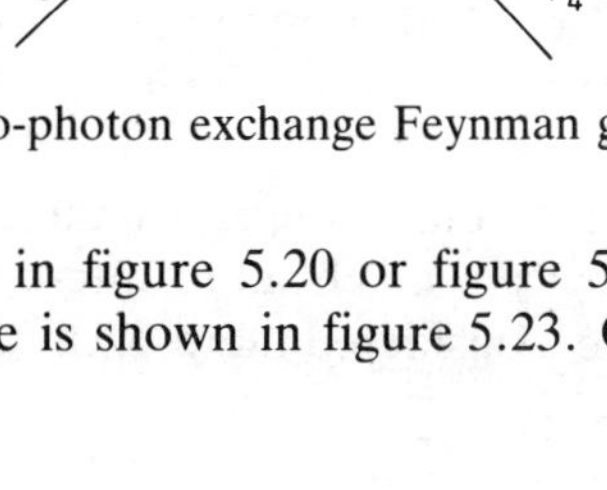

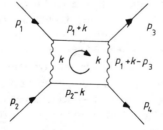

Figure 5.22 Two-photon exchange Feynman graph.

Graphs such as those in figure 5.20 or figure 5.22 are called one-loop graphs: a two-loop one is shown in figure 5.23. One has to integrate, in general, over the number of independent loop momenta, as indicated in figure 5.22 for the one-loop graph.

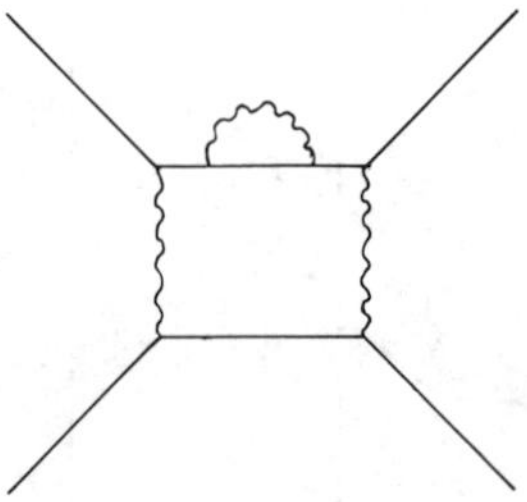

Figure 5.23 Two-loop Feynman graph.

The interested reader might care to write down the invariant amplitude for figure 5.20, for example. A nasty surprise is in store for those who do, and who then try to evaluate the integral: it is infinite (an example of such a divergent integral is given in problem 6.13 of the following chapter). The same is true of figure 5.22—and, unfortunately, of most loop diagrams. This is extremely annoying: figure 5.22 should represent a *small* correction ($O(\alpha^2)$) to the lowest-order amplitude of figure 5.9, yet in fact it is infinite! Only after a long struggle with such infinities has it been understood how to obtain physically sensible results from such a perturbation expansion, i.e. to ensure that contributions from graphs that are higher-order in α are indeed smaller than the lowest-order contribution. The precise programme for manipulating and 'taming' these infinities is known as 'renormalisation' of the theory. 'Renormalisation' because all the infinities are miraculously swept up into formal expressions for the quantities like physical mass and charge of the particles: having done this, these (formally ill-defined) expressions are *replaced* by their finite physical values. We postpone further discussion till Chapter 6. We note, however, the useful distinction between 'tree' diagrams (ones with no loops) and 'loop' diagrams (ones with one or more loops). For theories in which the physical coupling constant is small, such as weak and electromagnetic interactions, the tree diagram calculations are usually a good approximation to compare with experiment. Moreover, at the tree level, all the subtleties of renormalisation never enter, and so for many purposes it is sufficient to know how to calculate just these tree diagrams. This is what this book aims to explain.

We now turn to a derivation of the Feynman rules for electrons and photons, which is physically a much more interesting example than the spin-0 example we have considered here. All the necessary concepts have been developed already: the only new thing is the (considerable) complication of spin.

Summary

First-order perturbation theory in wave mechanical formalism: absorption of a single photon by a spin-0 particle.

Justification of these amplitudes using quantum field theory formalism.

Derivation of the amplitude, and associated Feynman diagram, for electromagnetic scattering of two charged spin-0 particles, in lowest order using heuristic argument.

Explicit cross section calculation involving two-body phase space.

Rederivation of the two-body amplitude in quantum field theory.

Interpretation in terms of photon exchange.

The covariant Feynman graph and 'off-mass-shell' particles.

The exchange of transverse photons, and the instantaneous Coulomb interaction, both contained in the Feynman graph.

Identical particle scattering; 'crossed' channels.

The propagator for a spin-0 particle.

Pion exchange and the Yukawa potential.

Pion Compton scattering, and illustration of how gauge invariance can determine the form of interactions.

A few words about higher-order calculations and the distinction between tree diagrams and loop diagrams.

Problems

5.1 Consider a matrix element of the form

$$M = \int \mathrm{d}^3 x \int \mathrm{d}t \, \mathrm{e}^{+\mathrm{i}p_\mathrm{f}\cdot x} \partial_\mu A^\mu \mathrm{e}^{-\mathrm{i}p_\mathrm{i}\cdot x}.$$

Assuming the integration is over all space–time and that

$$A^0 \to 0 \quad \text{as } t \to \pm\infty$$

and

$$|\boldsymbol{A}| \to 0 \quad \text{as } |\boldsymbol{x}| \to \infty$$

use integration by parts to show

(a) $$\int \mathrm{d}t\, \mathrm{e}^{+\mathrm{i}p_f\cdot x}\partial_0 A^0 \mathrm{e}^{-\mathrm{i}p_i\cdot x} = (-\,\mathrm{i}p_{f0})\int \mathrm{d}t\, \mathrm{e}^{+\mathrm{i}p_f\cdot x} A^0 \mathrm{e}^{-\mathrm{i}p_i\cdot x}$$

(b) $$\int \mathrm{d}^3 x\, \mathrm{e}^{+\mathrm{i}p_f\cdot x}\boldsymbol{\nabla}\cdot\boldsymbol{A}\,\mathrm{e}^{-\mathrm{i}p_i\cdot x} = +\,\mathrm{i}\boldsymbol{p}_f\cdot\left(\int \mathrm{d}^3 x\, \mathrm{e}^{+\mathrm{i}p_f\cdot x}\boldsymbol{A}\,\mathrm{e}^{-\mathrm{i}p_i\cdot x}\right).$$

Hence show that

$$\int \mathrm{d}^3 x \int \mathrm{d}t\, \mathrm{e}^{+\mathrm{i}p_f\cdot x}(\partial_\mu A^\mu + A^\mu \partial_\mu)\mathrm{e}^{-\mathrm{i}p_i\cdot x}$$
$$= -\mathrm{i}(p_f + p_i)_\mu \int \mathrm{d}^3 x \int \mathrm{d}t\, \mathrm{e}^{+\mathrm{i}p_f\cdot x} A^\mu \mathrm{e}^{-\mathrm{i}p_i\cdot x}.$$

5.2 Verify equation (5.59).

5.3 The amplitude for lowest-order electromagnetic scattering of particles 'a' and 'b' was calculated in §5.3. Using the Feynman interpretation of (antiparticle) ā states as negative 4-momentum (particle) a states, calculate the corresponding matrix element for āb scattering. Associate the factors in your expression with the corresponding Feynman diagram.

5.4 (a) In both the CM and laboratory frames, verify that the incident flux factor satisfies the equation

$$4E_1E_2|\boldsymbol{v}| = 4[(p_1\cdot p_2)^2 - m_1^2 m_2^2]^{1/2}.$$

(b) Introducing the variable $s = (p_1 + p_2)^2$, show that

$$4[(p_1\cdot p_2)^2 - m_1^2 m_2^2] = [s - (m_1 + m_2)^2][s - (m_1 - m_2)^2].$$

(c) Evaluate the differential cross section for the elastic āb electromagnetic scattering considered in problem 5.3. Express your result in terms of the CM variables E_1, E_2, p and θ_{CM}.

5.5 Evaluate the integral in expression (5.145), as follows. Since $\boldsymbol{q}$ is a fixed vector as far as the integration is concerned, let us take it to be the z axis of the $\boldsymbol{x}$ coordinate system. Then $\exp(\mathrm{i}\boldsymbol{q}\cdot\boldsymbol{x}) = \exp(\mathrm{i}|\boldsymbol{q}|r\cos\theta)$, where θ is the angle between $\boldsymbol{x}$ and $\boldsymbol{q}$. The integrand is now independent of the azimuthal angle ϕ, where $\mathrm{d}\Omega = \sin\theta\,\mathrm{d}\theta\,\mathrm{d}\phi$, so the ϕ integral contributes just 2π. The θ integral can also be done, using $\sin\theta\,\mathrm{d}\theta = -\mathrm{d}(\cos\theta)$. Finally, do the r integral and obtain the result

$$\int \mathrm{e}^{\mathrm{i}\boldsymbol{q}\cdot\boldsymbol{x}}\frac{1}{r}\mathrm{e}^{-\mu r}r^2\,\mathrm{d}r\,\mathrm{d}\Omega = 4\pi/(\boldsymbol{q}^2 + \mu^2).$$

5.6 Verify the equation (5.168).

5.7 Consider the two processes shown in figure 5.15, but for simplicity suppose that the 'photon' has spin 0, and that consequently all '$\varepsilon\cdot(p_i + p_f)$' factors at the vertices are set equal to unity. Write down the amplitudes for these two processes, and show that, for them, the analogue of equation (5.162) is

$$\frac{1}{2E_n}\left\{\frac{1}{E_k + E_p - E_n} - \frac{1}{E_k + E_p + E_n}\right\} = \frac{1}{(E_k + E_p)^2 - E_n^2}$$

where $E_n^2 = m^2 + \boldsymbol{p}_n^2 = m^2 + (\boldsymbol{p} + \boldsymbol{k})^2$. Hence show that the factor $1/q^2$ for the photon is here replaced by $1/(q^2 - m^2)$, where q is the 4-momentum carried by the off-shell internal pion in the *single* Feynman graph of figure 5.14, i.e. $q = (k + p)$.

6

THE ELECTROMAGNETIC INTERACTIONS OF SPIN-$\frac{1}{2}$ PARTICLES

6.1 First-order perturbation theory

We develop perturbation theory in the same way as we did for the KG case. The appropriate potential has been given in §3.6:

$$\hat{V}_{\mathrm{D}} = eA^0\mathbf{1} - e\boldsymbol{\alpha}\cdot\boldsymbol{A} = e\begin{pmatrix} A^0 & -\boldsymbol{\sigma}\cdot\boldsymbol{A} \\ -\boldsymbol{\sigma}\cdot\boldsymbol{A} & A^0 \end{pmatrix} \tag{6.1}$$

for a particle of charge e (>0). This potential is a 4×4 matrix, and to obtain an amplitude in the form of a single complex number, we must use $\psi_{\mathrm{f}}^{\dagger}$ instead of ψ_{f}^{*} in the matrix element. The first-order amplitude (figure 6.1) is therefore

$$\mathscr{A} = -\mathrm{i}\int \mathrm{d}^4x\, \psi_{\mathrm{f}}^{\dagger}(k', s')\hat{V}_{\mathrm{D}}\psi_{\mathrm{i}}(k, s) \quad \text{(first order)} \tag{6.2}$$

where s and s' label the spin components. The spin labels are necessary since the spin configuration may be changed by the interaction. In terms of A^{μ}, the matrix product is

$$\psi_{\mathrm{f}}^{\dagger}\hat{V}_{\mathrm{D}}\psi_{\mathrm{i}} = e\psi_{\mathrm{f}}^{\dagger}(A^0\mathbf{1} - \boldsymbol{\alpha}\cdot\boldsymbol{A})\psi_{\mathrm{i}} \tag{6.3}$$

and the Lorentz transformation character (e.g. scalar, 4-vector, etc) of this matrix element is unclear. In order to display the Lorentz properties of the interaction, it is convenient to use 'γ matrices' instead of the original $\boldsymbol{\alpha}$ and β matrices. These are defined as follows:

$$\begin{aligned} &\gamma^0 = \beta, && (\gamma^0)^2 = \mathbf{1} \\ &\gamma^i = \beta\alpha_i, && (\gamma^i)^2 = -\mathbf{1}, \qquad i = 1, 2, 3. \end{aligned} \tag{6.4}$$

The Dirac equation may then be written in γ matrix notation by multiplying from the left by β. We find the elegant result (see problem 3.4)

$$(\mathrm{i}\gamma^{\mu}\partial_{\mu} - m)\psi = 0 \tag{6.5}$$

where

$$\gamma^{\mu}\partial_{\mu} = \gamma^0\partial/\partial t + \boldsymbol{\gamma}\cdot\boldsymbol{\nabla} \tag{6.6}$$

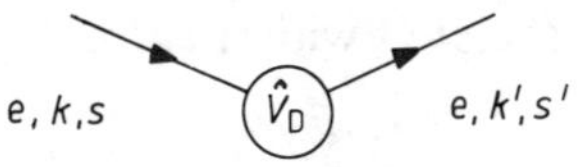

Figure 6.1 First-order scattering in potential $\hat{V}_D$.

since $\partial_\mu = (\partial_0, \boldsymbol{\nabla})$. The anticommutation relations of the $\boldsymbol{\alpha}$'s and β also find compact expression in terms of γ's:

$$\{\gamma^\mu, \gamma^\nu\} = 2g^{\mu\nu}\mathbf{1}. \tag{6.7}$$

Since the product of γ^μ with a 4-vector occurs so frequently, a special notation is used:

$$\gamma^\mu a_\mu = \gamma^0 a_0 - \boldsymbol{\gamma}\cdot\boldsymbol{a} \equiv \not{a}. \tag{6.8}$$

This is pronounced 'a-slash'. Thus the Dirac equation becomes even 'simpler' (i.e. more condensed):

$$(\mathrm{i}\not{\partial} - m)\psi = 0. \tag{6.9}$$

Finally we introduce the quantity 'ψ-bar' defined (problem 3.4(b)) by

$$\bar{\psi} \equiv \psi^\dagger\gamma^0 \tag{6.10}$$

in order to make evident the Lorentz transformation properties of matrix products involving $\psi^\dagger$ and ψ. For example, consider the Dirac probability current

$$\rho = \psi^\dagger\psi \tag{6.11}$$

$$\boldsymbol{j} = \psi^\dagger\boldsymbol{\alpha}\psi \tag{6.12}$$

which, as we saw in Chapter 3, satisfies a conservation law of the form

$$\partial\rho/\partial t + \boldsymbol{\nabla}\cdot\boldsymbol{j} = 0. \tag{6.13}$$

Since we want this conservation law to be covariant—i.e. true in all frames—and we know that

$$\partial^\mu = (\partial/\partial t, -\boldsymbol{\nabla}) \tag{6.14}$$

transforms as a 4-vector, it follows that $\psi^\dagger\psi$ is not an invariant, but rather the '0' component of a 4-vector. The advantage of γ matrix notation is that it compactly indicates the Lorentz properties of products of ψ's and γ matrices—by a 4-dimensional generalisation of the notation $\chi^\dagger\boldsymbol{\sigma}\chi$ for Pauli spinors (see Appendix A). In γ matrix notation the probability density is

$$\rho = \bar{\psi}\gamma^0\psi \tag{6.15}$$

and the current is

$$\boldsymbol{j} = \bar{\psi}\boldsymbol{\gamma}\psi. \tag{6.16}$$

These are now naturally associated with a current 4-vector

$$j^\mu = \bar{\psi}\gamma^\mu\psi \tag{6.17}$$

and the conservation law becomes

$$\partial_\mu j^\mu = 0 \tag{6.18}$$

which clearly shows its covariance. One can in fact show formally (Bjorken and Drell 1964, chapter 2; Aitchison 1972, §11.2) that it is the combination $\bar{\psi}\psi$ which is Lorentz invariant rather than $\psi^\dagger\psi$, and that $\bar{\psi}\gamma^\mu\psi$ is a 4-vector.

Returning to our interaction matrix element, we can now rewrite this in 4-vector notation as

$$\psi_{\mathrm{f}}^\dagger \hat{V}_{\mathrm{D}}\psi_{\mathrm{i}} = e\psi_{\mathrm{f}}^\dagger(A^0\mathbf{1} - \boldsymbol{\alpha}\cdot\boldsymbol{A})\psi_{\mathrm{i}} \tag{6.19}$$

$$= e\bar{\psi}_{\mathrm{f}}\gamma^\mu A_\mu\psi_{\mathrm{i}} \tag{6.20}$$

$$= e\,\bar{\psi}_{\mathrm{f}}\not{A}\psi_{\mathrm{i}}. \tag{6.21}$$

Thus we may write the transition amplitude in the form of a current interacting with A^μ:

$$\mathscr{A} = -\mathrm{i}\int \mathrm{d}^4x\,\bar{\psi}_{\mathrm{f}}(e\gamma^\mu A_\mu)\psi_{\mathrm{i}} \tag{6.22}$$

$$= -\mathrm{i}\int \mathrm{d}^4x\, j_{\mathrm{fi}}^\mu A_\mu \tag{6.23}$$

where we have defined a fermion electromagnetic transition current

$$j_{\mathrm{fi}}^\mu = e\bar{\psi}_{\mathrm{f}}\gamma^\mu\psi_{\mathrm{i}} \tag{6.24}$$

exactly analogous to the boson one introduced in §5.1. As in the latter case, we can easily check that, for the simple fermion plane wave states we are considering,

$$j_{\mathrm{fi}}^\mu = \langle k', s'|\hat{j}_{\mathrm{em}}^\mu|k, s\rangle \tag{6.25}$$

where the fermion current field operator $\hat{j}_{\mathrm{em}}^\mu$ is (cf §(4.6))

$$\hat{j}_{\mathrm{em}}^\mu = e\bar{\hat{\psi}}\gamma^\mu\hat{\psi} \tag{6.26}$$

for a particle of charge e. This is all we shall need of the field theory for the moment. We can now insert plane wave solutions, as usual, to calculate this lowest-order amplitude. Before we do this, let us look carefully at the normalisation of the Dirac plane waves and spinors.

6.2 Normalisation of Dirac spinors

We begin by considering a plane wave positive-energy state

$$\psi^{1,2} = N\omega^{1,2}\mathrm{e}^{-\mathrm{i}p\cdot x}. \tag{6.27}$$

How shall we normalise ψ? For KG particles we chose to normalise to $2E$ particles in a box of volume V:

$$\int_V \rho \, \mathrm{d}^3x = 2E \tag{6.28}$$

where ρ is the probability density. For spin-$\frac{1}{2}$ Dirac particles we shall follow the same convention. With the explicit form for $\omega^{1,2}$

$$\omega^{1,2} = \begin{pmatrix} \phi^{1,2} \\ \dfrac{\boldsymbol{\sigma}\cdot\boldsymbol{p}}{E+m}\phi^{1,2} \end{pmatrix} \tag{6.29}$$

and with

$$\rho = \psi^\dagger \psi \tag{6.30}$$

we find (see problem 6.1)

$$N = V^{-1/2}(E+m)^{1/2}. \tag{6.31}$$

It is convenient to define positive-energy spinors with the factor $(E+m)^{1/2}$ incorporated; namely, to define the spinors $u(p, s)$ by

$$u(p, s) = (E+m)^{1/2} \begin{pmatrix} \phi^s \\ \dfrac{\boldsymbol{\sigma}\cdot\boldsymbol{p}}{E+m}\phi^s \end{pmatrix}, \quad s = 1, 2. \tag{6.32}$$

Similarly, we define negative-energy spinors $v(p, s)$ from $\omega^{3,4}$ by

$$v(p, s) = (E+m)^{1/2} \begin{pmatrix} \dfrac{\boldsymbol{\sigma}\cdot\boldsymbol{p}}{E+m}\chi^s \\ \chi^s \end{pmatrix} \quad s = 1, 2. \tag{6.33}$$

It is vital to realise that we have set up these spinors $v(p, s)$ so that E and $\boldsymbol{p}$, and $s = 1$ and 2 in the equation above, correspond to the *physical positron energy* (i.e. positive) and *momentum*, and *spin projections* $\uparrow$ or $\downarrow$.

One may verify (problem 3.5) that these spinors satisfy

$$u^\dagger u = v^\dagger v = 2E \tag{6.34}$$

and that the momentum space Dirac equations for u and v read (problem 6.1)

$$(\not{p} - m)u = 0 \tag{6.35a}$$

$$(\not{p} + m)v = 0. \tag{6.35b}$$

As for $\bar{\psi}$, it is useful to define the quantities

$$\bar{u} \equiv u^\dagger \gamma^0 \tag{6.36a}$$

$$\bar{v} \equiv v^\dagger \gamma^0. \tag{6.36b}$$

It is then straightforward to verify (problem 6.1) that

$$\bar{u}u = 2m \tag{6.37a}$$

$$\bar{v}v = -2m \tag{6.37b}$$

and that $\bar{u}$ and $\bar{v}$ satisfy the following conjugate Dirac equations:

$$\bar{u}(\not{p} - m) = 0 \tag{6.38a}$$

$$\bar{v}(\not{p} + m) = 0. \tag{6.38b}$$

A word of warning about conventions is in order here. We have used the so-called covariant normalisation and have defined spinors u and v accordingly. Our definitions of u and v differ by $(1/2m)^{1/2}$ from those defined in Bjorken and Drell (1964) and Aitchison (1972). These authors, moreover, use a non-covariant normalisation for ψ with corresponding changes in the phase space and flux factors. The final results for cross sections are of course independent of convention.

With our conventions we have (since the e^- is the 'particle')

$$\psi(e^-) = V^{-1/2}u(p, s)e^{-ip\cdot x} \tag{6.39}$$

and

$$\psi(e^+) = V^{-1/2}v(p, s)e^{+ip\cdot x} \tag{6.40}$$

for positive- and negative-energy plane wave solutions, and we can identify $V^{-1/2}$ as the usual plane wave box normalisation factor as in the KG case. For example, for positive-energy electrons, the electromagnetic transition current

$$j^\mu(e^-) = (-e)\bar{\psi}_f\gamma^\mu\psi_i \tag{6.41}$$

is just

$$j^\mu(e^-) = (-e)\mathcal{N}_f\mathcal{N}_i\bar{u}_f\gamma^\mu u_i e^{+i(k'-k)\cdot x} \tag{6.42}$$

where

$$\bar{u}_f \equiv \bar{u}(k', s'), \qquad u_i \equiv u(k, s) \tag{6.43}$$

and the normalisation factors for fermions have been denoted by a script $\mathcal{N}$:

$$\mathcal{N}_f = \mathcal{N}_i = V^{-1/2}$$

(with our conventions these factors $\mathcal{N}$ are identical to those for bosons, N; this is not so in all conventions).

6.3 π^+e^- elastic scattering

We are at last in a position to calculate a cross section that can be measured. We consider elastic π^+e^- scattering: the notation for 4-

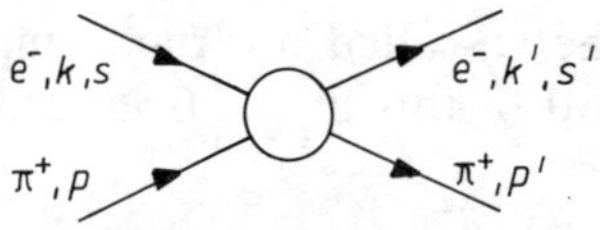

Figure 6.2 $e^-\pi^+$ scattering amplitude.

momenta and spins is defined in figure 6.2. We attack the problem in two stages as we did for 'a + b' scattering. We know the amplitude for electrons scattering in a potential A^μ:

$$\mathscr{A} = -\mathrm{i}\int \mathrm{d}^4x\, j_\mu(\mathrm{e}^-)A^\mu \tag{6.44}$$

where†

$$j_\mu(\mathrm{e}^-) = (-e)\mathcal{N}\mathcal{N}'\bar{u}(k', s')\gamma_\mu u(k, s)\mathrm{e}^{+\mathrm{i}(k'-k)\cdot x} \tag{6.45}$$

is the appropriate electromagnetic transition current for the electrons. For A^μ we need the potential produced by the motion of charged pions: from our previous 'a + b' calculations this is given by (cf (5.69))

$$A^\mu = -q^{-2}j^\mu(\pi^+) \tag{6.46}$$

where (cf (5.59))

$$j^\mu(\pi^+) = eNN'(p + p')^\mu \mathrm{e}^{+\mathrm{i}(p'-p)\cdot x} \tag{6.47}$$

is the electromagnetic transition current for the pions and, as usual, the 4-momentum transfer q has been defined by

$$q = p' - p = k - k'. \tag{6.48}$$

The entire lowest-order $e^-\pi^+$ scattering amplitude may therefore be written down

$$\mathscr{A}_{\mathrm{e}\pi} = -\mathrm{i}\int \mathrm{d}^4x\, j_\mu(\mathrm{e}^-)\left(-\frac{1}{q^2}\right)j^\mu(\pi^+). \tag{6.49}$$

Putting everything together, we arrive at

$$\begin{aligned}\mathscr{A}_{\mathrm{e}\pi} = -\mathrm{i}\mathcal{N}\mathcal{N}'NN'\int \mathrm{d}^4x[\bar{u}(k', s')\gamma_\mu u(k, s)\mathrm{e}^{+\mathrm{i}(k'-k)\cdot x}]\\ \times (+e^2/q^2)[(p + p')^\mu \mathrm{e}^{+\mathrm{i}(p'-p)\cdot x}]\end{aligned} \tag{6.50}$$

and the integral over d^4x may be performed to produce the expected 4-momentum-conserving δ function

$$\begin{aligned}\mathscr{A}_{\mathrm{e}\pi} = -\mathrm{i}\mathcal{N}\mathcal{N}'NN'(2\pi)^4\,\delta^4(k' + p' - k - p)\\ \times (-e)\bar{u}(k', s')\gamma_\mu u(k, s)(-g^{\mu\nu}/q^2)(+e)(p + p')_\nu.\end{aligned} \tag{6.51}$$

†γ_μ is defined by $\gamma_\mu = g_{\mu\nu}\gamma^\nu$: see (A.64) of Appendix A.2. Thus $\gamma_\mu A^\mu = g_{\mu\nu}\gamma^\nu A^\mu = \gamma^\nu g_{\mu\nu}A^\mu = \gamma^\nu A_\nu$.

It is useful to define the so-called invariant amplitude for eπ scattering by dropping the standard factors at the front of this expression; namely, we define

$$F_{ss'} \equiv (-e)\bar{u}(k', s')\gamma_\mu u(k, s)(-g^{\mu\nu}/q^2)(+e)(p + p')_\nu \tag{6.52}$$

where we have only explicitly specified the electron spin labels s and s' for the amplitude F.

As in our boson example, it is helpful to represent the various factors in the amplitude by associating them with elements of a Feynman diagram, as shown in figure 6.3. The association of factors to the pieces of the diagram is exactly as before for the pion and photon lines. The new features are the appearance of initial and final electron spinors u and $\bar{u}'$ respectively, and the matrix vertex factor $ie\gamma^\mu$. These give the following rules:

(viii) For each fermion entering the graph include the spinor $u(k, s)$, and for each fermion leaving the graph include the spinor $\bar{u}(k', s')$, together with the corresponding normalisation factors.

(ix) The fermion–photon–fermion vertex is $-iQ\gamma^\mu$ for charge Q.

The ordering of the spinors is important: it is implied to be

$$\bar{u}'\gamma_\mu u \equiv \bar{u}(k', s')\gamma_\mu u(k, s) \tag{6.53}$$

so that for a specific choice of the index μ we obtain a *number*. These four numbers will then transform as a 4-vector under Lorentz transformations and 'contract' with the 4-vector $(p + p')^\mu$ coming from the other vertex, giving a Lorentz invariant amplitude.

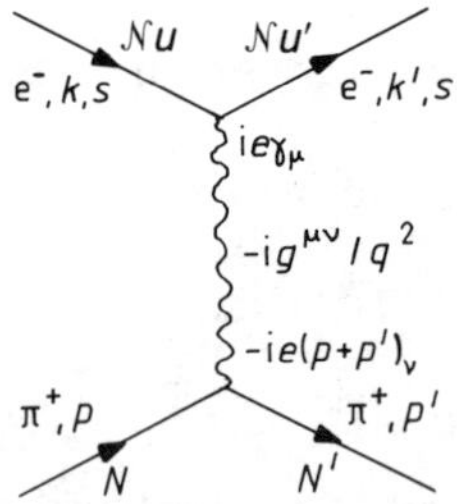

Figure 6.3 Feynman diagram for $e^-\pi^+$ scattering, in the one-photon exchange approximation.

The steps to the cross sections are now exactly as for the spin-0 case. The cross section for the scattering of an electron in spin state s to one spin state s' is (cf (5.86))

$$d\sigma_{ss'} = \frac{1}{4E\omega|\boldsymbol{v}|}|F_{ss'}|^2(2\pi)^4\,\delta^4(k' + p' - k - p) \times \frac{1}{(2\pi)^3}\frac{d^3k'}{2\omega'}\frac{1}{(2\pi)^3}\frac{d^3p'}{2E'} \tag{6.54}$$

where we have defined

$$\begin{aligned} k^\mu &= (\omega, \boldsymbol{k}), \qquad & k'^\mu &= (\omega', \boldsymbol{k}') \\ p^\mu &= (E, \boldsymbol{p}), \qquad & p'^\mu &= (E', \boldsymbol{p}'). \end{aligned} \tag{6.55}$$

The matrix product in $F_{ss'}$ may be evaluated using the explicit form of the γ matrices and spinors, i.e. using

$$u(k, s) = (\omega + m)^{1/2}\begin{pmatrix} \phi^s \\ \dfrac{\boldsymbol{\sigma}\cdot\boldsymbol{k}}{\omega + m}\phi^s \end{pmatrix} \tag{6.56}$$

and so on. In practice, however, this is both tedious and, moreover, not related directly to what experimentalists usually measure. The simplest experiments (and certainly all electron–pion scattering ones to date) use unpolarised initial electrons: this can be regarded as an average of equal amounts of spin-$\uparrow$ and spin-$\downarrow$ electrons. It is important to note that this is an *incoherent* average, in the sense that we average the *cross section*, rather than the amplitude. Furthermore, most experiments usually measure only the direction and energy of the scattered electron and are not sensitive to its polarisation state $\uparrow$ or $\downarrow$. Thus what we really wish to calculate is the *unpolarised cross section* defined by

$$d\bar{\sigma} \equiv \tfrac{1}{2}(d\sigma_{\uparrow\uparrow} + d\sigma_{\uparrow\downarrow} + d\sigma_{\downarrow\uparrow} + d\sigma_{\downarrow\downarrow}) \tag{6.57}$$

$$= \tfrac{1}{2}\sum_s\sum_{s'} d\sigma_{ss'}\,; \tag{6.58}$$

i.e. an average over the initial spin polarisations ($2 = 2s + 1$) and a sum over the final spin states arising from each initial spin state.

It is of course perfectly possible to work out all these cross sections explicitly and then add them up. Fortunately, we can avoid this by using a powerful labour-saving device due to Feynman. We discuss these so-called 'trace' techniques in the next section.

6.4 Trace techniques for spin summations

We begin by listing some properties of the trace of a matrix

$$\mathrm{Tr}\,\mathbf{A} = \sum_i A_{ii}; \tag{6.59}$$

i.e. the sum of its diagonal elements. This definition implies an important property for the trace of a matrix product:

$$\mathrm{Tr}(\mathbf{AB}) = \sum_i (\mathbf{AB})_{ii} = \sum_{i,j} A_{ij}B_{ji}. \tag{6.60}$$

Note that we have not used the summation convention here. Once we have added the labels to make explicit the sum over matrix elements, the ordering of matrix elements is immaterial—the indices will keep track of this automatically. Thus we can rewrite Tr(**AB**) as

$$\mathrm{Tr}(\mathbf{AB}) = \sum_{i,j} B_{ji}A_{ij} \tag{6.61}$$

and hence prove the desired property

$$\mathrm{Tr}(\mathbf{AB}) = \mathrm{Tr}(\mathbf{BA}). \tag{6.62}$$

Similarly, it is easy to show

$$\mathrm{Tr}(\mathbf{ABC}) = \mathrm{Tr}(\mathbf{CAB}) \tag{6.63}$$

and so on.

Consider the expression $|F_{ss'}|^2$. We have (cf (6.52))

$$F_{ss'} = (e^2/q^2)\bar{u}(k', s')\gamma_\mu u(k, s)(p + p')^\mu \tag{6.64}$$

and so

$$\begin{aligned} |F_{ss'}|^2 = {} & (e^2/q^2)^2[\bar{u}(k', s')\gamma_\mu u(k, s)(p + p')^\mu] \\ & \times [\bar{u}(k', s')\gamma_\nu u(k, s)(p + p')^\nu]^*. \end{aligned} \tag{6.65}$$

Now $F_{ss'}$ is just a complex number (a 1×1 matrix) and taking its complex conjugate is equivalent to taking the Hermitian conjugate

$$\begin{aligned} [\bar{u}(k', s')\gamma_\nu u(k, s)(p + p')^\nu]^\dagger &= [u^\dagger(k, s)\gamma_\nu^\dagger\gamma_0^\dagger u(k', s')](p + p')^\nu \\ &= \bar{u}(k, s)\gamma_\nu u(k', s')(p + p')^\nu \end{aligned} \tag{6.66}$$

since (see problem 6.2)

$$\bar{\gamma}_\nu \equiv \gamma_0\gamma_\nu^\dagger\gamma_0 = \gamma_\nu. \tag{6.67}$$

Thus for the spin sums we are required to evaluate the quantity

$$\begin{aligned} \tfrac{1}{2}\sum_s\sum_{s'}|F_{ss'}|^2 &= \left(\frac{e^2}{q^2}\right)^2 \tfrac{1}{2}\sum_{s,\, s'}\bar{u}(k', s')\gamma_\mu u(k, s)\bar{u}(k, s)\gamma_\nu u(k', s') \\ &\quad \times (p + p')^\mu(p + p')^\nu \end{aligned} \tag{6.68}$$

$$\equiv \left(\frac{e^2}{q^2}\right)^2 L_{\mu\nu}T^{\mu\nu}; \tag{6.69}$$

i.e. the summation factorises into the product of a 'lepton tensor' $L_{\mu\nu}$

involving all the Dirac algebra

$$L_{\mu\nu} = \tfrac{1}{2}\sum_{s}\sum_{s'}\bar{u}(k', s')\gamma_\mu u(k, s)\bar{u}(k, s)\gamma_\nu u(k', s') \tag{6.70}$$

and a 'hadron tensor' $T^{\mu\nu}$ for the pion vertex ,

$$T^{\mu\nu} = (p + p')^\mu (p + p')^\nu. \tag{6.71}$$

This separation occurs because of our 'one-photon exchange' approximation, illustrated again in figure 6.4.

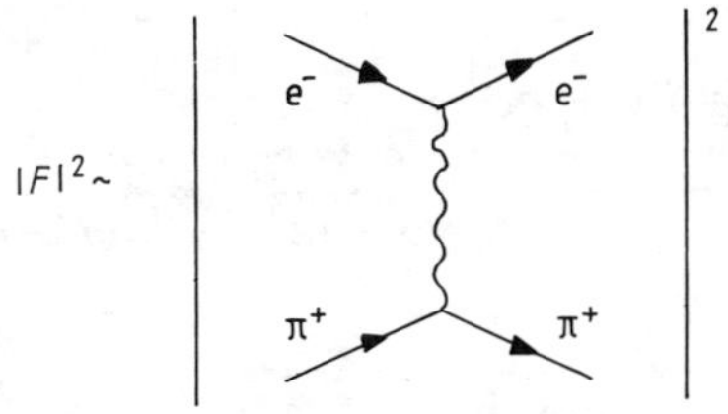

Figure 6.4 One-photon exchange approximation for $|F|^2$ in $e^-\pi^+$ scattering.

All the hard work is in evaluating the lepton tensor. We now write this including the *matrix indices* with a summation convention implied for these:

$$L_{\mu\nu} = \tfrac{1}{2}\sum_{s}\sum_{s'}\bar{u}_\alpha(k', s')(\gamma_\mu)_{\alpha\beta}u_\beta(k, s)\bar{u}_\gamma(k, s)(\gamma_\nu)_{\gamma\delta}u_\delta(k', s'). \tag{6.72}$$

Since all quantities are now matrix elements (i.e. numbers), we can look at the factors in isolation. Consider the terms depending on s,

$$\sum_{s} u_\beta(k, s)\bar{u}_\gamma(k, s). \tag{6.73}$$

These products, 16 in all corresponding to possible β and γ combinations, cannot depend on s since the sum is over $s = \uparrow$ and $\downarrow$. We can arrange these 16 quantities in a 4×4 matrix

$$\begin{pmatrix} u_1 \\ u_2 \\ u_3 \\ u_4 \end{pmatrix} (\bar{u}_1, \bar{u}_2, \bar{u}_3, \bar{u}_4) = \begin{pmatrix} u_1\bar{u}_1 & u_1\bar{u}_2 & \cdots \\ u_2\bar{u}_1 & u_2\bar{u}_2 & \cdots \\ \vdots & \vdots & \vdots \end{pmatrix} \tag{6.74}$$

and use our explicit spinors $u(k, s)$ to calculate this matrix:

$$u(k,s) = (\omega + m)^{1/2} \begin{pmatrix} \phi^s \\ \dfrac{\boldsymbol{\sigma}\cdot\boldsymbol{k}}{\omega + m}\phi^s \end{pmatrix}. \tag{6.75}$$

Notice that the sum over polarisations for the two-component spinors

$$\phi^1\phi^{1\dagger} + \phi^2\phi^{2\dagger} = \begin{pmatrix} 1 & 0 \\ 0 & 1 \end{pmatrix} \tag{6.76}$$

gives just the 2×2 unit matrix. We therefore arrive at the result (problem 6.3)

$$u(k, \uparrow)\bar{u}(k, \uparrow) + u(k, \downarrow)\bar{u}(k, \downarrow)$$

$$= (\omega + m)\begin{pmatrix} \mathbf{1} & -\dfrac{\boldsymbol{\sigma}\cdot\boldsymbol{k}}{\omega + m} \\ \dfrac{\boldsymbol{\sigma}\cdot\boldsymbol{k}}{\omega + m} & -\dfrac{\omega - m}{\omega + m}\mathbf{1} \end{pmatrix} \tag{6.77}$$

$$= (\not{k} + m) \tag{6.78}$$

where $\not{k}$ is the matrix $\omega\gamma^0 - \boldsymbol{k}\cdot\boldsymbol{\gamma}$ and we are adopting the accepted convention that the 4×4 unit matrix multiplying m is not shown explicitly. Thus we can replace the sum over initial spin states by the corresponding matrix element of the matrix $(\not{k} + m)$:

$$L_{\mu\nu} = \tfrac{1}{2}\sum_{s'}\bar{u}_\alpha(k', s')(\gamma_\mu)_{\alpha\beta}(\not{k} + m)_{\beta\gamma}(\gamma_\nu)_{\gamma\delta}u_\delta(k', s'). \tag{6.79}$$

Since we can reorder matrix elements as we wish, we can use the same trick to perform the second spin sum

$$\sum_{s'} u_\delta(k', s')\bar{u}_\alpha(k', s') = (\not{k}' + m)_{\delta\alpha}. \tag{6.80}$$

Thus $L_{\mu\nu}$ takes the form of a matrix product, summed over diagonal elements, i.e. a *trace*

$$L_{\mu\nu} = \tfrac{1}{2}(\not{k}' + m)_{\delta\alpha}(\gamma_\mu)_{\alpha\beta}(\not{k} + m)_{\beta\gamma}(\gamma_\nu)_{\gamma\delta} \tag{6.31}$$

$$= \tfrac{1}{2}\mathrm{Tr}[(\not{k}' + m)\gamma_\mu(\not{k} + m)\gamma_\nu]. \tag{6.82}$$

How does all this help? It helps because the special property of the Dirac γ matrices

$$\{\gamma^\mu, \gamma^\nu\} = 2g^{\mu\nu}\mathbf{1} \tag{6.83}$$

allows simple evaluation of traces of products of such matrices. Here we just list the 'trace theorems' that we shall use to evaluate $L_{\mu\nu}$: more complete statements of trace theorems and γ matrix algebra, together with proofs of these theorems, are given in Appendix D.

We need the following results:

(a) $\mathrm{Tr}\,\mathbf{1} = 4$ (6.84a)

(b) Tr (odd number of γ's) = 0 (6.84b)

(c) $\mathrm{Tr}\,(\not{a}\not{b}) = 4(a\cdot b)$ (6.84c)

(d) $\mathrm{Tr}\,(\not{a}\not{b}\not{c}\not{d}) = 4[(a\cdot b)(c\cdot d) + (a\cdot d)(b\cdot c) - (a\cdot c)(b\cdot d)].$ (6.84d)

Then

$$\mathrm{Tr}[(\not{k}' + m)\gamma_\mu(\not{k} + m)\gamma_\nu] = \mathrm{Tr}(\not{k}'\gamma_\mu\not{k}\gamma_\nu) + m\,\mathrm{Tr}(\gamma_\mu\not{k}\gamma_\nu) + m\,\mathrm{Tr}(\not{k}'\gamma_\mu\gamma_\nu) + m^2\,\mathrm{Tr}(\gamma_\mu\gamma_\nu). \tag{6.85}$$

The terms linear in m are zero by theorem (b), and using (c) in the form

$$\mathrm{Tr}(\gamma_\mu\gamma_\nu)a^\mu b^\nu = 4g_{\mu\nu}a^\mu b^\nu = 4a\cdot b \tag{6.86}$$

and (d) in a similar form, we obtain (problem 6.4)

$$\mathrm{Tr}[(\not{k}' + m)\gamma_\mu(\not{k} + m)\gamma_\nu] = 4[k'_\mu k_\nu + k'_\nu k_\mu - (k\cdot k')g_{\mu\nu}] + 4m^2 g_{\mu\nu}.$$

Since

$$q^2 = (k - k')^2 = 2m^2 - 2k\cdot k' \tag{6.87}$$

the final result for the *lepton tensor* (6.82) may be written

$$\boxed{L_{\mu\nu} = 2[k'_\mu k_\nu + k'_\nu k_\mu + (q^2/2)g_{\mu\nu}].} \tag{6.88}$$

The tensor $L_{\mu\nu}$ is important since it appears not only in elastic scattering calculations but also in the calculation of inelastic electron scattering cross sections and, in fact, in most quark parton calculations! We must now perform the contraction with $T^{\mu\nu}$ to obtain the cross section.

6.5 Cross sections for e$^-\pi^+$ and e$^+\pi^+$ scattering

For the unpolarised e$^-\pi^+$ cross section we need to evaluate (cf (6.69))

$$\tfrac{1}{2}\sum_s\sum_{s'}|F_{ss'}|^2 = \left(\frac{4\pi\alpha}{q^2}\right)^2 L_{\mu\nu}T^{\mu\nu}. \tag{6.89}$$

Using the form for $L_{\mu\nu}$ together with

$$T^{\mu\nu} = (p + p')^\mu(p + p')^\nu \tag{6.90}$$

we find (problem 6.5)

$$L_{\mu\nu}T^{\mu\nu} = 8[2(p\cdot k)(p\cdot k') + (q^2/2)M^2] \tag{6.91}$$

since $k'\cdot p' = k\cdot p$ and $k\cdot p' = k'\cdot p$ from momentum conservation, and $p^2 = p'^2 = M^2$ (we are using m for the electron mass and M for the pion mass). After the usual 'routine struggle' one can now obtain the differential cross section, expressed either in invariant form or in the form appropriate to a particular frame. We give the result for the 'laboratory frame' which we define (unrealistically for actual eπ scattering) by the condition $p^\mu = (M, \mathbf{0})$:

$$\frac{\mathrm{d}\bar{\sigma}}{\mathrm{d}\Omega} = \frac{\alpha^2}{4k^2 \sin^4(\theta/2)} \cos^2(\theta/2)\frac{k'}{k}. \tag{6.92}$$

In this formula we have neglected the electron mass in the kinematics so that

$$k \equiv |\boldsymbol{k}| = \omega \tag{6.93a}$$

$$k' \equiv |\boldsymbol{k}'| = \omega' \tag{6.93b}$$

and

$$q^2 = -4kk' \sin^2(\theta/2) \tag{6.94}$$

where θ is the electron scattering angle in this frame, as shown in figure 6.5. The evaluation of the phase space integral requires some care and this is detailed in Appendix E. We shall denote the cross section (6.92) by

$$\left(\frac{\mathrm{d}\sigma}{\mathrm{d}\Omega}\right)_{\mathrm{ns}} \quad \text{'no-structure' cross section.} \tag{6.95}$$

It describes essentially the 'kinematics' of a relativistic electron scattering from a target which recoils. This 'no-structure' cross section also occurs in the cross section for the scattering of electrons by protons or muons: the appellation 'no-structure' will be made clearer in the discussion of form factors.

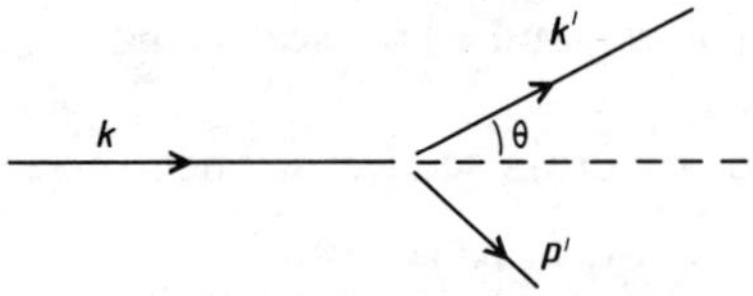

Figure 6.5 Two-body scattering in the 'laboratory' frame.

Consider now $e^+\pi^+$ scattering. In the boson case the 'wavefunction' current j^μ_{fi} for π^+ was of the form $e(p_\mathrm{f} + p_\mathrm{i})^\mu$, so that reversing the signs of the p's gave the current correctly for a particle of charge $-e$. In the fermion case the wavefunction current for e^- is $-e\bar{u}_\mathrm{f}\gamma^\mu u_\mathrm{i}$, which does not change sign when $k_\mathrm{i} \to -k_\mathrm{i}$, $k_\mathrm{f} \to -k_\mathrm{f}$: all that happens is that the

u's become v's. This is why, as we pointed out in §3.5, a minus sign has to be put in by hand in the wavefunction approach for fermions. However, the correct result is automatic (it had better be!) if we use the field theoretic current operator $-e\hat{\bar{\psi}}\gamma^\mu\hat{\psi}$. As we saw in §§4.5 and 4.6, the spatial integral of the $\mu = 0$ part of this—the charge operator—is

$$\hat{\rho}_{\text{em}} = -e\int \frac{d^3k}{(2\pi)^3}\frac{1}{2E}\sum_s [\hat{c}_s^\dagger(k)\hat{c}_s(k) - \hat{d}_s^\dagger(k)\hat{d}_s(k)] \tag{6.96}$$

from which we see that the fermion and antifermion parts enter indeed with opposite signs. The same is true for $\hat{\boldsymbol{j}}_{\text{em}}$. Thus the amplitude for a positron of 4-momentum k and spin s to scatter to the state k', s' under a potential $\hat{V}_{\text{D}}$ is (figure 6.6)

$$\underset{\nearrow\atop\text{'outgoing'}}{\mathcal{N}\bar{v}(k, s)}\underset{\uparrow\atop\text{extra minus}}{(-\mathrm{i}e\gamma^\mu)}\underset{\nwarrow\atop\text{'incoming'}}{v(k', s')}. \tag{6.97}$$

Rules (viii) and (ix) are now extended in the obvious way for antifermions.

Figure 6.6 Feynman reinterpretation of first-order e$^+$ scattering process.

Using the result

$$\sum_s v(k, s)\bar{v}(k, s) = (\not{k} - m) \tag{6.98}$$

one obtains a trace in exactly the same way as before, and in this approximation (lowest order in α) the cross sections for $e^-\pi^+$ and $e^+\pi^+$ scattering are identical.

6.6 The pion form factor and invariance arguments

Our treatment of pion–electron scattering has so far ignored the fact that the pion is a hadron and may participate in strong interactions. In reality we expect there to be 'virtual' strong interaction effects at the pion–photon vertex which will modify our simple perturbative picture. For example, we might expect contributions from graphs such as those shown in figures 6.7(a) and 6.7(b).

Since the coupling constant for strong interactions is not small, we cannot, in any sensible approximation, ignore such effects. What can we

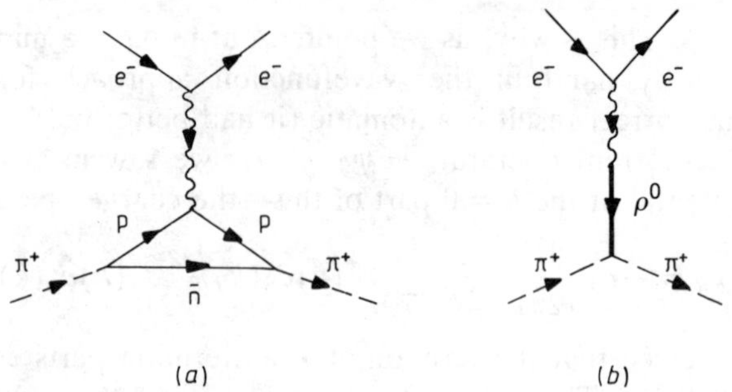

Figure 6.7 Virtual processes contributing at the $\pi\pi\gamma$ vertex: (*a*) virtual nucleon loop; (*b*) virtual ρ meson.

do? Fortunately, although our present knowledge of the strong interaction is not enough to calculate all such effects, the kinematic form of all these contributions is severely restricted by two important symmetry principles—Lorentz invariance and electromagnetic current conservation. The use of invariance arguments to place restrictions on the form of interactions is an extremely general and important tool in the absence of a complete theory. We consider the application of these arguments to the coupling of a pion to a virtual photon. In figure 6.8 all the strong interaction effects are indicated by a 'blob' at the vertex, summarising our ignorance of these processes.

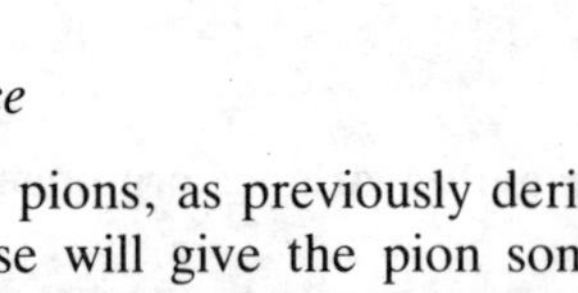

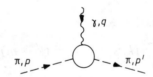

Figure 6.8 General $\pi\pi\gamma$ vertex.

6.6.1 *Lorentz invariance*

Our Feynman rules for pions, as previously derived, neglected all strong interaction effects: these will give the pion some non-trivial 'structure' and a finite size. It is nevertheless useful to consider the hypothetical case of a 'point' pion—a pion with no strong interaction effects. For such a 'point' pion we know that the electromagnetic transition current may be written

$$j_\mu(\pi) = eNN'(p + p')_\mu e^{+i(p'-p)\cdot x} \tag{6.99}$$

which is the matrix element, between free single-pion states, of the electromagnetic current operator (see (5.59)). In particular, we may identify

$$\langle \pi^+(p')|\hat{j}^{\mu}_{\text{em}}(0)|\pi^+(p)\rangle = eNN'(p + p')^{\mu} \tag{6.100}$$

since all the x dependence is carried by the exponential factor. This is the form for a point pion. Strong interaction effects are expected to modify this form of the pion electromagnetic current but they cannot destroy its 4-vector character. To construct the general form for a real pion therefore, we must first enumerate the independent momentum 4-vectors we have at our disposal to parametrise the 4-vector nature of the current. These are just

$$p,\ p' \text{ and } q \tag{6.101}$$

subject to the condition

$$p' = p + q. \tag{6.102}$$

There are two independent combinations; these we can choose to be the linear combinations

$$(p' + p)_{\mu} \tag{6.103}$$

and

$$(p' - p)_{\mu} = q_{\mu}. \tag{6.104}$$

Both of these 4-vectors can, in general, parametrise the 4-vector nature of the electromagnetic current of a real pion. Moreover, they can be multiplied by an unknown scalar function of the available Lorentz scalar products for this process. Since

$$p^2 = p'^2 = M^2 \tag{6.105}$$

and

$$q^2 = 2M^2 - 2p{\cdot}p' \tag{6.106}$$

there is only one independent scalar in the problem, which we may take to be q^2, the 4-momentum transfer to the vertex. Thus, from Lorentz invariance, we are led to write the electromagnetic current of a pion in the form

$$\langle \pi^+, p'|\hat{j}^{\mu}_{\text{em}}(0)|\pi^+, p\rangle = eNN'[F(q^2)(p' + p)^{\mu} + G(q^2)q^{\mu}]. \tag{6.107}$$

The functions F and G are called 'form factors'. This is as far as Lorentz invariance can take us. To identify the pion form factor, we must consider our second symmetry principle.

6.6.2 Current conservation and gauge transformations

As we have discussed in §5.1, the gauge transformation

$$A^{\mu} \to A^{\mu} - \partial^{\mu}\chi \tag{6.108}$$

leaves Maxwell's equations invariant. We are therefore free to choose a gauge for A^μ, and it is often convenient to impose the Lorentz condition

$$\partial_\mu A^\mu = 0. \tag{6.109}$$

In this gauge the Maxwell equations are

$$\Box A^\mu = j^\mu \tag{6.110}$$

and the gauge condition is the same as the familiar charge conservation condition

$$\partial_\mu j^\mu = 0. \tag{6.111}$$

For our plane wave pion current (5.59) the derivative brings down a factor q_μ

$$-\mathrm{i}\partial_\mu j^\mu(\pi) = q_\mu j^\mu(\pi) \tag{6.112}$$

and the current conservation condition is equivalent to the condition

$$q_\mu\langle\pi^+(p')|\hat{j}^\mu_{\text{em}}(0)|\pi^+(p)\rangle = 0 \tag{6.113}$$

on the matrix element of the electromagnetic current operator $\hat{j}^\mu_{\text{em}}(0)$.

In the case of a 'point' pion this is clearly satisfied since

$$q\cdot(p' + p) = 0. \tag{6.114}$$

In the general case we obtain the condition

$$q_\mu[F(q^2)(p' + p)^\mu + G(q^2)q^\mu] = 0. \tag{6.115}$$

The first term vanishes as before, but $q^2 \neq 0$ in general, and we therefore conclude that current conservation implies that

$$G(q^2) = 0. \tag{6.116}$$

In other words, all the virtual strong interaction effects at the $\pi^+\pi^+\gamma$ vertex are described by one scalar function of the virtual photon's squared 4-momentum:

$$\underset{\text{'point pion'}}{e(p' + p)^\mu} \rightarrow \underset{\text{'real pion'}}{eF(q^2)(p' + p)^\mu}. \tag{6.117}$$

$F(q^2)$ is the electromagnetic form factor of the pion.

The electric charge is defined to be the coupling at zero momentum transfer, so the form factor is normalised by the condition

$$F(0) = 1. \tag{6.118}$$

Experiment can measure this form factor at non-zero values of q^2: for $q^2 \leqslant 0$, i.e. 'spacelike', in electron–pion scattering, and for $q^2 \geqslant 4M^2$, i.e. 'timelike', in the 'crossed reaction' $e^+e^- \to \pi^+\pi^-$. It is a challenge for any theory of the pion to explain or predict the results of such measurements. Physically, we expect the form factor to fall as $|q^2|$ increases, since, roughly speaking, it represents the probability for a pion to remain a pion. As $|q^2|$ increases, the probability of inelastic processes which involve the creation of extra particles becomes greater, and the probability for elastic scattering is correspondingly reduced.

6.7 Electron–muon elastic scattering

We now consider $e^-\mu^-$ elastic scattering: our notation is indicated in figure 6.9. In the lowest order of perturbation theory—the one-photon exchange approximation—we can draw the relevant Feynman graph for this process. This is shown in figure 6.10. All the elements for the graph have been met before and so we can immediately write down the invariant amplitude $F_{sr;s'r'}$, which now depends on four spin labels:

$$\boxed{F_{sr;s'r'} = (-e)\bar{u}(k', s')\gamma_\mu u(k, s)(-g^{\mu\nu}/q^2)(-e)\bar{u}(p', r')\gamma_\nu u(p, r).} \tag{6.119}$$

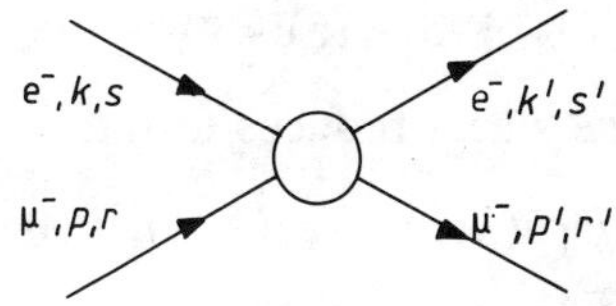

Figure 6.9 $e^-\mu^-$ scattering amplitude.

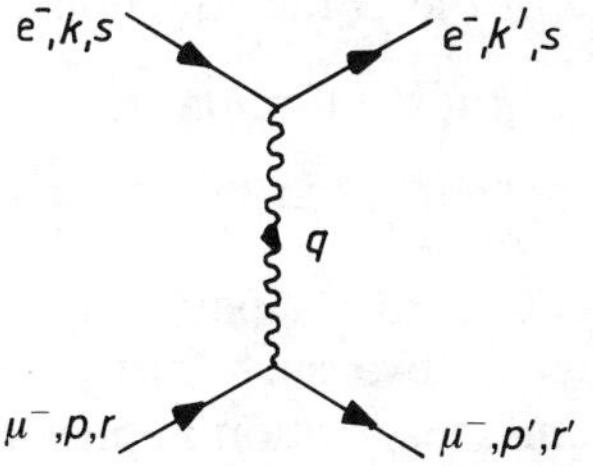

Figure 6.10 One-photon exchange amplitude in $e^-\mu^-$ scattering.

Although experiments with polarised leptons are not uncommon, we shall only be concerned with the unpolarised cross section

$$\mathrm{d}\sigma \sim \tfrac{1}{4}\sum_{s,r}\sum_{s',r'}|F_{sr;s'r'}|^2. \tag{6.120}$$

We perform the same manipulations as in our $e\pi$ example, and the cross section reduces to a factorised form involving two traces:

$$\tfrac{1}{4}\sum_{s,r}\sum_{s',r'}|F_{sr;s'r'}|^2 = \left(\frac{e^2}{q^2}\right)^2 \{\tfrac{1}{2}\mathrm{Tr}[(\not{k}' + m)\gamma_\mu(\not{k} + m)\gamma_\nu]\}$$
$$\times \{\tfrac{1}{2}\mathrm{Tr}[(\not{p}' + M)\gamma^\mu(\not{p} + M)\gamma^\nu]\} \tag{6.121}$$

$$\equiv \left(\frac{e^2}{q^2}\right)^2 L_{\mu\nu}M^{\mu\nu} \tag{6.122}$$

where $L_{\mu\nu}$ is the 'electron tensor' calculated before

$$L_{\mu\nu} = 2[k'_\mu k_\nu + k'_\nu k_\mu + (q^2/2)g_{\mu\nu}] \tag{6.123}$$

but now $M^{\mu\nu}$ is the appropriate tensor for the muon coupling, with the same structure as $L_{\mu\nu}$:

$$M^{\mu\nu} = 2[p'^\mu p^\nu + p'^\nu p^\mu + (q^2/2)g^{\mu\nu}]. \tag{6.124}$$

To evaluate the cross section, we must perform the 'contraction' $L_{\mu\nu}M^{\mu\nu}$. A useful trick to simplify this calculation is to use current conservation for the electron tensor $L_{\mu\nu}$. For the electron transition current, the electromagnetic current conservation condition is (cf (6.111))

$$\partial^\mu j_\mu(\mathrm{e}) = 0 \tag{6.125}$$

which for our plane wave states reduces to the condition

$$q^\mu[\bar{u}(k', s')\gamma_\mu u(k, s)] = 0. \tag{6.126}$$

This is just

$$\bar{u}(k', s')(\not{k} - \not{k}')u(k, s) \tag{6.127}$$

which vanishes by virtue of the Dirac equations

$$\not{k}u(k, s) = mu(k, s) \tag{6.128}$$

$$\bar{u}(k', s')\not{k}' = m\bar{u}(k', s'). \tag{6.129}$$

This is obviously independent of the particular spin projections s and s'. Since $L_{\mu\nu}$ is the product of two such currents, summed and averaged over polarisations, current conservation implies the conditions

$$q^\mu L_{\mu\nu} = q^\nu L_{\mu\nu} = 0 \tag{6.130}$$

which can be explicitly checked using our result for $L_{\mu\nu}$. The usefulness of this result is that in the contraction $L_{\mu\nu}M^{\mu\nu}$ we can replace p' in $M^{\mu\nu}$ by $(p + q)$ and then drop all the terms involving q's, i.e.

$$L_{\mu\nu}M^{\mu\nu} = L_{\mu\nu}M^{\mu\nu}_{\text{eff}} \tag{6.131}$$

where

$$M^{\mu\nu}_{\text{eff}} = 2[2p^\mu p^\nu + (q^2/2)g^{\mu\nu}]. \tag{6.132}$$

The calculation of the cross section is now straightforward. In the 'laboratory' system, defined (unrealistically) by the initial muon at rest

$$p^\mu = (M, 0, 0, 0) \tag{6.133}$$

with M now the muon mass, the result is (problem 6.6(a))

$$\frac{\mathrm{d}\sigma}{\mathrm{d}\Omega} = \left(\frac{\mathrm{d}\sigma}{\mathrm{d}\Omega}\right)_{\text{ns}}\left(1 - \frac{q^2\tan^2\theta/2)}{2M^2}\right). \tag{6.134}$$

Notice the following points:

(a) The 'no-structure' cross section for $e\pi$ scattering now appears modified by an additional $\tan^2(\theta/2)$ constribution. This is due to the spin-$\frac{1}{2}$ nature of the muon which gives rise to scattering from both the charge *and* the magnetic moment.

(b) In the kinematics the electron mass has been neglected, which is usually a good approximation at high energies. We should add a word of explanation for the 'laboratory' cross sections we have calculated, with both pions and muons unrealistically initially at rest. These forms of the cross section, $(\mathrm{d}\sigma/\mathrm{d}\Omega)_{\text{ns}}$, and of the cross section for the scattering of two Dirac point particles, will be of great value in our discussion of the quark parton model in the next chapter.

(c) The crossed version of this process, namely $e^+e^- \to \mu^+\mu^-$, is a very important monitoring reaction for electron–positron colliding beam machines. It is also basic to a discussion of the predictions of the quark parton model for $e^+e^- \to$ hadrons. An instructive calculation similar to the one above leads to the result (see problem 6.7)

$$\frac{\mathrm{d}\sigma}{\mathrm{d}\Omega} = \frac{\alpha^2}{4q^2}(1 + \cos^2\theta) \tag{6.135}$$

where all variables are defined in the e^+e^- CM frame, and all masses have been neglected.

6.8 Electron–proton elastic scattering and nucleon form factors

In the one-photon exchange approximation, the Feynman diagram for elastic ep scattering may be drawn as in figure 6.11, where the 'blob' at the ppγ vertex signifies the expected modification of the point coupling due to strong interactions. The structure of the proton vertex can be analysed using symmetry principles in the same way as for the pion vertex. The presence of Dirac spinors and γ matrices makes this a somewhat involved procedure: problem 6.8 is an example of the type of complication that arises. Full details of such an analysis can be found in Bernstein (1968), for example. In order to generalise more easily to inelastic electron scattering, we proceed in a different way, although again we rely on Lorentz invariance and current conservation. Before we embark on this analysis let us look at what we expect to find.

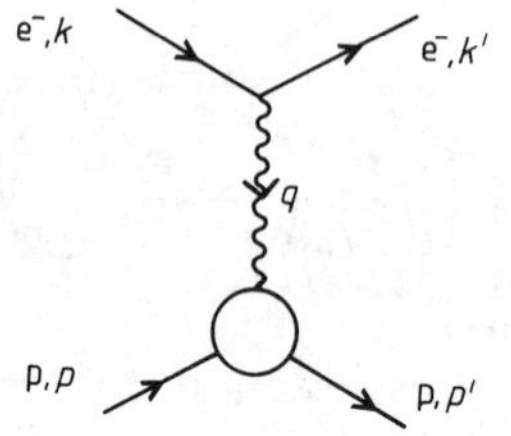

Figure 6.11 One-photon exchange amplitude in e$^-$p scattering, including hadronic corrections at the ppγ vertex.

Consider the example of electron scattering from a 'point' proton. This is just the same as our calculation of eμ scattering apart from the change of sign of the charge. The cross section factorises into two tensors, $L_{\mu\nu}$ (for the electron) and $T^{\mu\nu}$ (for the 'point' proton). The cross section is the same as the eμ case (except that M is now the proton mass)

$$\frac{\mathrm{d}\sigma}{\mathrm{d}\Omega} = \left(\frac{\mathrm{d}\sigma}{\mathrm{d}\Omega}\right)_{\mathrm{ns}}\left(1 - \frac{q^2 \tan^2(\theta/2)}{2M^2}\right) \qquad (6.136)$$

and the two terms in the brackets may be regarded as corresponding to electric and magnetic scattering. Since a real proton is not a point particle, the virtual strong interaction effects will modify both the charge and the magnetic moment distribution. Hence we may expect that two form factors will be needed to describe the deviation from pointlike behaviour. This is in fact the case, as we now show.

For a point proton we have (cf (6.100))

$$\langle \mathrm{p}; p', s'|\hat{j}^{\mu}_{\mathrm{em}}(0)|\mathrm{p}; p, s\rangle = (+e)\mathcal{N}\mathcal{N}'\bar{u}(p', s')\gamma^{\mu}u(p, s) \quad (6.137)$$

for the matrix element of the electromagnetic current. The tensor $T^{\mu\nu}$ is essentially two such factors summed and averaged over polarisations. In the case of a real proton we can deduce the form of $T^{\mu\nu}$ by symmetry arguments.

6.8.1 Lorentz invariance

$T^{\mu\nu}$ must retain its tensor character: this must be made up using the available 4-vectors and tensors at our disposal. For the spin-averaged case we have only

$$p,\ q \text{ and } g_{\mu\nu} \tag{6.138}$$

since $p' = p + q$. The antisymmetric tensor $\varepsilon_{\mu\nu\alpha\beta}$ (see Appendix D) must actually be ruled out using parity invariance: the tensor $T^{\mu\nu}$ is not a pseudotensor since $\hat{j}^{\mu}_{\text{em}}$ is a vector. It is helpful to remember that $\varepsilon_{\mu\nu\alpha\beta}$ is the generalisation of ε_{ijk} in three dimensions, and that the vector product of two 3-vectors—a pseudovector—may be written

$$(\boldsymbol{a} \times \boldsymbol{b})_i = \varepsilon_{ijk} a_j b_k. \tag{6.139}$$

6.8.2 *Current conservation*

For a real proton, current conservation gives the condition (cf (6.113))

$$q_\mu \langle \text{p};\ p',\ s' | \hat{j}^{\mu}_{\text{em}}(0) | \text{p};\ p,\ s \rangle = 0 \tag{6.140}$$

which translates to the conditions (cf (6.130))

$$q_\mu T^{\mu\nu} = q_\nu T^{\mu\nu} = 0 \tag{6.141}$$

on the tensor $T^{\mu\nu}$.

There are only two possible tensors we can make that satisfy both these requirements. One involves p and is constructed to be orthogonal to q. We introduce a vector

$$\tilde{p}_\mu = p_\mu + \alpha q_\mu \tag{6.142}$$

and require

$$q \cdot \tilde{p} = 0. \tag{6.143}$$

Hence we find

$$\tilde{p}_\mu = p_\mu - (p \cdot q / q^2) q_\mu \tag{6.144}$$

and thus the tensor

$$\tilde{p}^\mu \tilde{p}^\nu = [p^\mu - (p \cdot q/q^2) q^\mu][p^\nu - (p \cdot q/q^2) q^\nu] \tag{6.145}$$

satisfies all our requirements. The second tensor must involve $g^{\mu\nu}$ and may be chosen to be

$$-g^{\mu\nu} + q^\mu q^\nu / q^2 \tag{6.146}$$

which again satisfies our conditions. Thus from invariance arguments alone, the tensor $T^{\mu\nu}$ for the proton vertex may be parametrised by these two tensors, each multiplied by an unknown function of q^2. If we define

$$\begin{aligned} T^{\mu\nu} = {} & 4A(q^2)[p^\mu - (p \cdot q/q^2) q^\mu][p^\nu - (p \cdot q/q^2) q^\nu] \\ & + 2M^2 B(q^2)(-g^{\mu\nu} + q^\mu q^\nu / q^2) \end{aligned} \tag{6.147}$$

the cross section in the laboratory frame is

$$\frac{d\sigma}{d\Omega} = \left(\frac{d\sigma}{d\Omega}\right)_{ns}[A + B\tan^2(\theta/2)]. \tag{6.148}$$

The functions A and B may be related to the 'charge' and 'magnetic' form factors of the proton. The Dirac 'charge' and Pauli 'anomalous magnetic moment' form factors are defined by

$$\langle \mathrm{p}; p', s'|\hat{j}^{\mu}_{em}(0)|\mathrm{p}; p, s\rangle = (+e)\mathcal{N}\mathcal{N}'\bar{u}(p', s') \times \left[\gamma^{\mu}\mathcal{F}_1(q^2) + \frac{i\kappa\mathcal{F}_2(q^2)}{2M}\sigma^{\mu\nu}q_{\nu}\right]u(p,s) \tag{6.149}$$

with the normalisation

$$\mathcal{F}_1(0) = 1 \tag{6.150a}$$

$$\mathcal{F}_2(0) = 1 \tag{6.150b}$$

and the magnetic moment of the proton is not one (nuclear) magneton, as for an electron or muon (neglecting higher-order corrections), but rather $\mu_p = 1 + \kappa$ with $\kappa = 1.79$. Problem 6.8 shows that the $\bar{u}\gamma^{\mu}u$ piece in (6.149) can be rewritten in terms of $\bar{u}(p + p')^{\mu}u/2M$ and $\bar{u}i\sigma^{\mu\nu}q_{\nu}u/2M$. The first of these is analogous to the interaction of a charged spin-0 particle. As regards the second, we note that $\sigma^{\mu\nu}$ is just

$$\sigma^{\mu\nu} = \tfrac{1}{2}i[\gamma^{\mu}, \gamma^{\nu}] \tag{6.151}$$

which reduces to the Pauli spin matrices for the spacelike components

$$\sigma^{ij} = \begin{pmatrix}\sigma^k & 0\\ 0 & \sigma^k\end{pmatrix} \tag{6.152}$$

with our representation of γ matrices (σ^{ij} is a 4×4 matrix, σ^k is 2×2, and i, j and k are in cyclic order). The second term in this 'Gordon decomposition' of $\bar{u}\gamma^{\mu}u$ thus corresponds to an interaction via the spin magnetic moment—with, in fact, $g = 2$. Thus the addition of the κ term in (6.149) corresponds to an 'anomalous' magnetic moment piece. In terms of $\mathcal{F}_1$ and $\mathcal{F}_2$ one can show that

$$A = \mathcal{F}_1^2 + \tau\kappa^2\mathcal{F}_2^2 \tag{6.153}$$

$$B = 2\tau(\mathcal{F}_1 + \kappa\mathcal{F}_2)^2 \tag{6.154}$$

where

$$\tau = -q^2/4M^2. \tag{6.155}$$

The cross section formula for electron–proton elastic scattering is called the 'Rosenbluth' cross section. The pointlike cross section (6.136) is recovered from (6.148) by setting $\mathcal{F}_1 = 1$ and $\kappa = 0$ in (6.153) and (6.154).

Before we leave elastic scattering it is helpful to look in some more detail at the kinematics. It will be sufficient to consider the 'pointlike' case, which we shall call $e^-\mu^+$, for definiteness. Energy and momentum conservation at the μ^+ vertex gives the condition

$$p + q = p' \tag{6.156}$$

with the mass-shell conditions (M is the μ^+ mass)

$$p^2 = p'^2 = M^2. \tag{6.157}$$

Hence for elastic scattering we have the relation

$$2p\cdot q = -q^2. \tag{6.158}$$

It is conventional to relate these invariants to the corresponding laboratory frame ($p^\mu = (M, \mathbf{0})$) expressions. Neglecting the electron mass so that

$$k \equiv |\boldsymbol{k}| = \omega \tag{6.159a}$$

$$k' \equiv |\boldsymbol{k}'| = \omega' \tag{6.159b}$$

we have

$$q^2 = -2kk'(1 - \cos\theta) = -4kk'\sin^2(\theta/2) \tag{6.160}$$

and

$$p\cdot q = M(k - k') = M\nu \tag{6.161}$$

where ν is the energy transfer q^0 in this frame. To avoid unnecessary minus signs, it is convenient to define

$$Q^2 = -q^2 = 4kk'\sin^2(\theta/2) \tag{6.162}$$

and the elastic scattering relation between $p\cdot q$ and q^2 reads

$$\nu = Q^2/2M \tag{6.163}$$

or

$$\frac{k'}{k} = \frac{1}{1 + (2k/M)\sin^2(\theta/2)}. \tag{6.164}$$

Remembering, therefore, that for elastic scattering k' and θ are not independent variables, we can perform a change of variables (see Appendix E) in the laboratory frame

$$\mathrm{d}\Omega = 2\pi\mathrm{d}(\cos\theta) = (\pi/k'^2)\mathrm{d}Q^2 \tag{6.165}$$

and write the differential cross section for $e^-\mu^+$ scattering as

$$\frac{\mathrm{d}\sigma}{\mathrm{d}Q^2} = \frac{\pi\alpha^2}{4k^2\sin^4(\theta/2)}\frac{1}{kk'}[\cos^2(\theta/2) + 2\tau\sin^2(\theta/2)]. \tag{6.166}$$

For elastic scattering ν is not independent of Q^2 but we may formally write this as a double-differential cross section by inserting the δ function to ensure this condition is satisfied:

$$\frac{d^2\sigma}{dQ^2 d\nu} = \frac{\pi\alpha^2}{4k^2\sin^4(\theta/2)}\frac{1}{kk'}\left[\cos^2(\theta/2) + \left(\frac{Q^2}{2M^2}\right)\sin^2(\theta/2)\right]\delta\left(\nu - \frac{Q^2}{2M}\right). \tag{6.167}$$

This is the cross section for the scattering of an electron from a pointlike fermion target of charge e and mass M.

It is illuminating to plot out the physically allowed regions of Q^2 and ν (figure 6.12). Elastic ep scattering corresponds to the line $Q^2 = 2M\nu$. Resonance production ep $\rightarrow$ eN* with $p'^2 = M'^2$ corresponds to lines parallel to the elastic line, shifted to the right by $M'^2 - M^2$ since

$$2M\nu = Q^2 + M'^2 - M^2. \tag{6.168}$$

Experiments with real photons, $Q^2 = 0$, correspond to exploring along the ν axis. In the next chapter we switch our attention to so-called deep inelastic electron scattering—the region of large Q^2 and large ν.

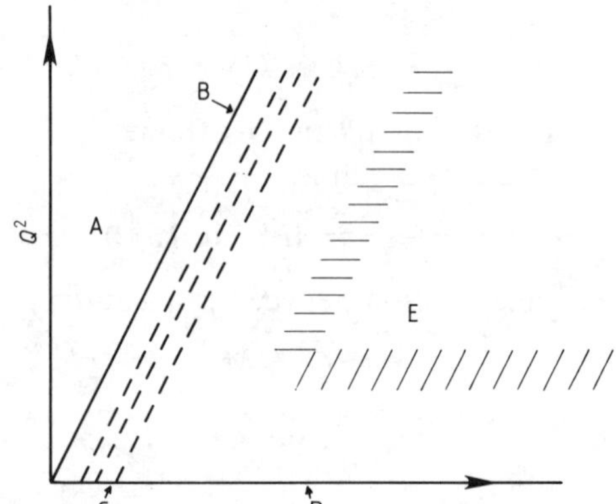

Figure 6.12 Physical regions in the Q^2, ν variables: A, kinematically forbidden region; B, line of elastic scattering ($Q^2 = 2M\nu$); C, lines of resonance electroproduction; D, photoproduction; E, deep inelastic region (Q^2 and ν large).

6.9 Electron Compton scattering: the electron propagator

6.9.1 The electron propagator

Just as in the case of the pion Compton scattering amplitudes discussed in §5.8, we may also consider joining our basic electron–photon processes together in such a way that an electron appears as an *internal* particle. The two such lowest-order Feynman graphs that we can draw are shown in figures 6.13(*a*) and 6.13(*b*) (cf. figures 5.14 and 5.17). These are plainly the $O(e^2)$ contributions to the electron Compton scattering process

$$\gamma(k) + e^-(p) \rightarrow \gamma(k') + e^-(p'). \tag{6.169}$$

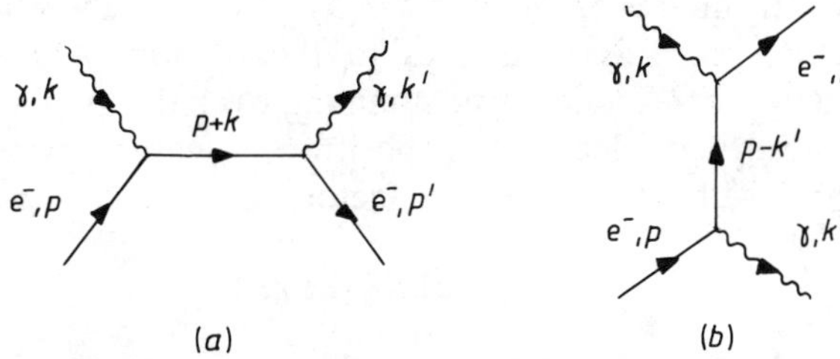

Figure 6.13 Order e^2 contributions to electron Compton scattering.

Once again, we know all the factors in figures 6.13 except one—the amplitude for the internal electron, i.e. the *electron propagator* in the terminology of §5.8.1. We can find its form in either of two ways, as discussed for the pion in §5.8.2. First, we can repeat the simple argument used in §5.8.1, that the form of the propagator is just the inverse of the appropriate wave operator, as evaluated when acting on plane wave solutions. For spin-$\frac{1}{2}$ particles, the free Dirac equation

$$(\mathrm{i}\not\partial - m)\psi = 0 \tag{6.170}$$

after the substitution $\partial^\mu \rightarrow \partial^\mu - \mathrm{i}eA^\mu$ gives the equation

$$(\mathrm{i}\not\partial - m)\psi = -e\not A\psi \tag{6.171}$$

as the analogue of (5.185). We therefore expect the propagator to be proportional to $(\not q - m)^{-1}$, leading to the rule:

(x) a factor $\mathrm{i}(\not q - m)^{-1}$ for an internal spin-$\frac{1}{2}$ particle of 4-momentum q.

Using the result

$$(\not q - m)(\not q + m) = q^2 - m^2 \tag{6.172}$$

the inverse operator $(\not q - m)^{-1}$ may be written as

$$(\not q + m)/(q^2 - m^2). \tag{6.173}$$

In this form the denominator has the basic form now familiar from both the photon and pion cases. The spin-$\frac{1}{2}$ propagator has the extra 4×4 matrix factor in the numerator.

The origin of the factor may be understood if we consider the other possible way we have learned to find propagators (cf §5.6 for the photon one and problem 5.7 for the pion one)—that is, by explicitly calculating the sum of the two different non-covariant graphs which correspond to just one of either figure 6.13(a) or 6.13(b). For example, we could consider figures 6.14(a) and 6.14(b) which add to produce figure 6.13(b). In these processes, an electron or positron is emitted and then absorbed in the intermediate states indicated in the figures, just as the photon was in figures 5.7(a) and 5.7(b). We shall therefore have to include a factor, precisely analogous to the photon polarisation sum $\Sigma_\lambda \varepsilon_\nu^* \varepsilon_\mu$ of (5.160), corresponding to summing over the spin states of the emitted and absorbed electron or positron wavefunctions. Thus for figure 6.14(a) we will get a numerator factor

$$\sum_s u(q_1, s)\bar{u}(q_1, s) \tag{6.174}$$

while for figure 6.14(b) the u's will be replaced by v's. We leave it as a not entirely trivial problem for the reader (problem 6.9) to show, using (6.80) and (6.98), that the sum of figure 6.14(a) and 6.14(b) does correctly reconstruct the electron propagator $(\not{q} - m)^{-1}$ in the form (6.173).

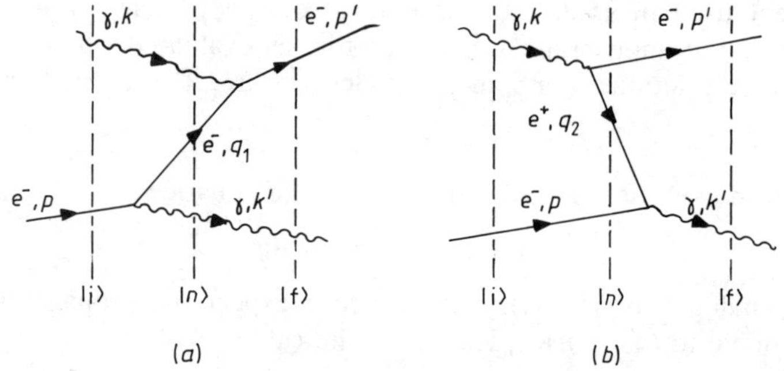

Figure 6.14 The two non-covariant amplitudes which add together to make the covariant amplitude of figure 6.13(b).

6.9.2 *Electron Compton scattering*

The invariant amplitude is

$$F_{\gamma e} = F_{\gamma e}^{(d)} + F_{\gamma e}^{(e)} \tag{6.175}$$

where

$$F_{\gamma e}^{(d)} = e^2 \varepsilon_\nu^*(k', \lambda')\varepsilon_\mu(k, \lambda)\bar{u}(p', s')\gamma^\nu \frac{(\not{p} + \not{k} + m)}{(p + k)^2 - m^2}\gamma^\mu u(p,s) \tag{6.176}$$

and

$$F_{\gamma e}^{(e)} = e^2 \varepsilon_\nu^*(k', \lambda')\varepsilon_\mu(k, \lambda)\bar{u}(p', s')\gamma^\mu \frac{(\not{p} - \not{k}' + m)}{(p - k')^2 - m^2}\gamma^\nu u(p,s) \tag{6.177}$$

represent the 'direct' and 'exchange' contributions of figures 6.13(*a*) and 6.13(*b*) respectively. There is, of course, no 'contact' interaction like that of figure 5.18 in this case, and we can check (problem 6.10) that the replacement of ε_ν^* by k'_ν, or of ε_μ by k_μ, reduces the *sum* of (6.176) and (6.177)—that is the complete amplitude—to zero, thus verifying gauge invariance to this order in α (cf (5.203)).

The calculation of the cross section is of considerable interest, since it is required when considering lowest-order QCD corrections to the parton model for deep inelastic scattering of leptons from nucleons (see the following chapter and Chapter 9). We must average $|F_{\gamma e}|^2$ over initial electron and photon spins, and sum over final ones. Consider first the direct amplitude $F_{\gamma e}^{(d)}$. For this contribution we must evaluate

$$\frac{e^4}{4(s - m^2)^2} \cdot \sum_{\lambda, \lambda', s, s'} \varepsilon_\nu^{*\prime} \varepsilon_\mu \varepsilon_\rho^* \varepsilon'_\sigma \bar{u}' \gamma^\nu (\not{p} + \not{k} + m)\gamma^\mu u \bar{u} \gamma^\rho (\not{p} + \not{k} + m)\gamma^\sigma u' \tag{6.178}$$

where we have shortened the notation in an obvious way and introduced the invariant variable $s = (p + k)^2$. Using (5.170), the term (6.178) becomes

$$\frac{e^4}{4(s - m^2)^2} \cdot \sum_{s,s'} \bar{u}' \gamma^\nu (\not{p} + \not{k} + m)\gamma^\mu u \bar{u} \gamma_\mu (\not{p} + \not{k} + m)\gamma_\nu u' \tag{6.179}$$

$$= \frac{e^4}{4(s - m^2)^2} \cdot \mathrm{Tr}[\gamma_\nu(\not{p}' + m)\gamma^\nu(\not{p} + \not{k} + m)\gamma^\mu(\not{p} + m)\gamma_\mu(\not{p} + \not{k} + m)] \tag{6.180}$$

where in the second step we have moved the γ_ν to the front of the trace, using (6.62). Expression (6.180) involves the trace of eight γ matrices, which is beyond the power of the machinery given so far. However, it simplifies greatly if we neglect the electron mass—that is, if we are interested in the high-energy limit, as we shall be in parton model applications. In that case, (6.180) becomes

$$\frac{e^4}{4s^2} \mathrm{Tr}[\gamma_\nu \not{p}' \gamma^\nu (\not{p} + \not{k}) \gamma^\mu \not{p} \gamma_\mu (\not{p} + \not{k})] \tag{6.181}$$

which we can simplify using the result (D.3) to

$$\frac{e^4}{s^2}\mathrm{Tr}[\not{p}'(\not{p} + \not{k})\not{p}(\not{p} + \not{k})] \tag{6.182}$$

$$= \frac{e^4}{s^2}\mathrm{Tr}[\not{p}'\not{k}\not{p}\not{k}] \quad \text{using } \not{p}^2 = p^2 = 0 \tag{6.183}$$

$$= \frac{4e^4}{s^2}\cdot 2(p'\cdot k)(p\cdot k) \quad \text{using (6.84d) and } k^2 = 0 \tag{6.184}$$

$$= -2e^4 u/s \tag{6.185}$$

where $u = (p - k')^2$. Problem 6.11 finishes the calculation, with the result that the spin-averaged squared amplitude is

$$\frac{1}{4}\sum_{s,s',\lambda,\lambda'} |F_{\gamma e}|^2 = -2e^4\left(\frac{u}{s} + \frac{s}{u}\right). \tag{6.186}$$

The cross section in the CMS is then (cf (5.115))

$$\frac{\mathrm{d}\sigma}{\mathrm{d}(\cos\theta)} = \frac{2\pi\cdot 2e^4}{64\pi^2 s}\left(\frac{-u}{s} - \frac{s}{u}\right) = \frac{\pi\alpha^2}{s}\left(\frac{-u}{s} - \frac{s}{u}\right). \tag{6.187}$$

For parton model calculations, what is actually required is the analogous quantity calculated for the case in which the initial photon is virtual (see §9.4.2). However, the discussion of §5.6 shows that we may still use the polarisation sum (5.170). A difference will arise in passing from (6.183) to (6.184) where we must remember that $k^2 \neq 0$. Since k^2 will be spacelike, we put $k^2 = -Q^2$ (cf (6.162)) and find (problem 6.12) that the spin-averaged squared amplitude for the virtual Compton process

$$\gamma^*(k^2 = -Q^2) + \mathrm{e}^- \to \gamma + \mathrm{e}^- \tag{6.188}$$

is given by

$$-2e^4\left(\frac{u}{s} + \frac{s}{u} - \frac{2Q^2 t}{su}\right). \tag{6.189}$$

6.10 Higher-order calculations: renormalisation and tests of QED

In the same way that the spin-0 Feynman rules generalised for use in high-order 'loop' calculations, so too do the spin-$\frac{1}{2}$ rules developed in this chapter. As in the spin-0 case (§5.9), divergent integrals are encountered as soon as one tries to go beyond the lowest-order (tree) diagrams. Our whole approach has been based on perturbation theory, and something has to be done about these infinities if perturbation theory is to make any sense. Without going into too many technical details, we shall now try to give a simple qualitative discussion of what can be done, by the technique of *renormalisation*, to tame these infinities which arise in perturbation theory. It is as well to emphasise at the outset that our discussion will be conducted entirely within the framework of perturbation theory; it could be that one day someone will come along and explain how to get finite answers from theories that are not renormalisable in perturbation theory. For the moment no such alternative scheme of calculation exists, and we therefore accept perturbative renormalisability† as a necessary requirement for a sensible physical theory.

This last point is actually very non-trivial, since it turns out that this requirement is remarkably restrictive: it severely limits the class of possible theories. The power of this limitation was used to great effect in the creation of the electroweak theory (see Chapters 13 and 14), principally by Weinberg (see e.g. Weinberg 1980); the proof of the theory's renormalisability was given by 't Hooft (1971b). This is one good reason for learning something about renormalisation theory—another, of course, is the remarkable precision with which the predictions of *renormalised* QED agree with experiments (see below).

Consider then $e^-\mu^-$ scattering, for example, as calculated in perturbation theory in powers of α. The lowest-order amplitude, corresponding to figure 6.15, was calculated in §6.7. Let us examine the contribution of one of the graphs which enters in order α^2, the one shown in figure 6.16, in which the 'bubble' in the photon propagator is formed by an electron–positron pair being created and subsequently annihilated. The amplitude for the graph can (almost) be calculated from the rules already given (including the one in §5.9 for loops). The result is that the amplitude for figure 6.16 is the same as that of figure 6.15 except that the photon propagator in the latter is replaced by a more complicated expression

$$-\frac{ig^{\mu\nu}}{q^2} \to (-1)\left(\frac{-i}{q^2}\right)\int\frac{d^4k}{(2\pi)^4}\mathrm{Tr}\left(ie\gamma^\mu\frac{i}{\not{k}-m}ie\gamma^\nu\frac{i}{\not{k}-\not{q}-m}\right)\left(\frac{-i}{q^2}\right) \tag{6.190}$$

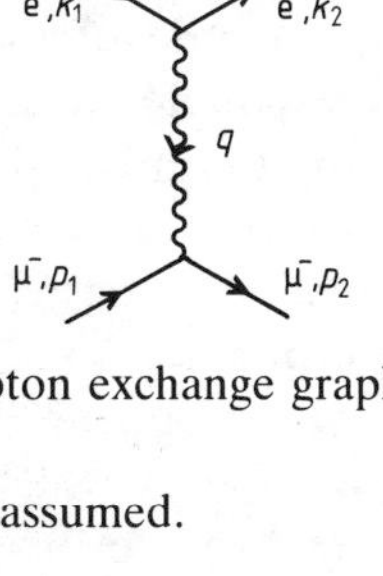

Figure 6.15 One-photon exchange graph for $e^-\mu^- \to e^-\mu^-$.

†Henceforth the epithet will be assumed.

where k is the 'loop momentum' involved in the bubble, and m is the electron mass. This expression certainly has recognisable features: the two factors of $-i/q^2$ at each end correspond to the photon pieces, while the $ie\gamma$ factors correspond to the coupling of the photons to the internal fermions, whose propagators appear more or less as expected. The overall minus sign at the front occurs whenever there is a closed fermion loop: this can only be proved using quantum field theory, which, of course, also explains the particular ordering ('follow round the loop') of the factors in (6.190) (see Bjorken and Drell 1964, §8.2).

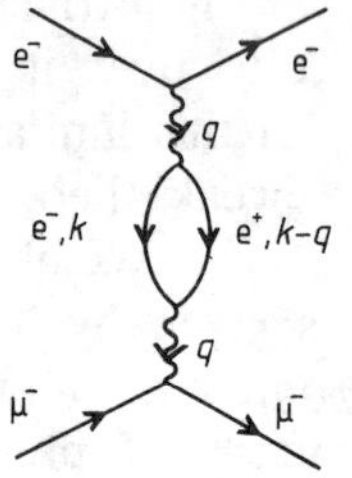

Figure 6.16 Higher-order (vacuum polarisation) correction for $e^-\mu^- \to e^-\mu^-$.

The integral over k in (6.190) extends up to infinity, and it is obvious that it does diverge, just by naively counting powers in the numerator and denominator. This divergence is not a feature peculiar to the particular contribution shown in figure 6.16; there are divergences in most diagrams involving loops. To carry out the renormalisation procedure (see below), one requires a consistent way of identifying the precise nature of the divergences and of separating off a meaningful finite contribution. The most popular method, in the context of gauge theories, is that called 'dimensional regularisation' ('t Hooft and Veltman 1972), which proved an invaluable technical tool in the proof of renormalisability of non-Abelian gauge theories ('t Hooft 1971a,b) and also in performing actual higher-order calculations in those theories. Our purpose here, however, is not at all to equip the reader to do such calculations, but rather to discuss in a quite elementary way the physics of renormalisation, with only a little of the mathematics. For this, the 'cut-off' method is perfectly adequate. In this approach, the large-k end of a divergent integration—for example, that in (6.190)—is cut off at a finite value Λ (which we must eventually allow to tend to infinity). The divergence will then be identified in terms of Λ. An example involving spinless particles is given in problem 6.13.

In the case of (6.190) it would appear that the integral grows as Λ^2 (by counting powers) for large $k \sim \Lambda$. Such an integral would be called 'quadratically divergent'. In fact, however, it can be shown that the real rate of growth is only logarithmic; explicit evaluation yields the result (Bjorken and Drell 1964, §8.2) that, whereas (disregarding inessential factors) figure 6.15 gives the amplitude

$$e^2\bar{u}(k')\gamma_\mu u(k)(-g^{\mu\nu}/q^2)\bar{u}(p')\gamma_\nu u(p) \tag{6.191}$$

the sum of figures 6.15 and 6.16 is given by

$$e^2\left[1 - \frac{\alpha}{3\pi}\ln\left(\frac{\Lambda^2}{m^2}\right) + X(q^2)\right]\bar{u}(k')\gamma_\mu u(k)\frac{-g^{\mu\nu}}{q^2}\bar{u}(p')\gamma_\nu u(p) \tag{6.192}$$

where X is a *finite* function of q^2 which tends to zero as $q^2 \to 0$.

The interesting thing about the expression (6.192) is that it has the same general form as the lowest-order expression (6.191). In particular, it can still be written in the fundamental form

$$(\text{current}) \times (\text{photon propagator}) \times (\text{current}) \tag{6.193}$$

except that the lowest-order expression for the electromagnetic current (cf equation (6.42)—we here omit normalisation factors) which is appropriate to (6.191), namely

$$\langle k'|\hat{j}^{\mu}_{\text{em}}(0)|k\rangle = -e\bar{u}(k')\gamma^\mu u(k) \tag{6.194}$$

(and similarly for $p \to p'$), is replaced in (6.192) by

$$\langle k'|\hat{j}^{\mu}_{\text{em}}(0)|k\rangle = -e\left[1 - \frac{\alpha}{3\pi}\ln\left(\frac{\Lambda^2}{m^2}\right) + X(q^2)\right]^{1/2}\bar{u}(k')\gamma^\mu u(k) \tag{6.195}$$

if, as seems reasonable, we associate the correction equally with each vertex in figure 6.16.

The fact that this higher-order contribution can be regarded as effectively changing typical current matrix elements from (6.194) to (6.195) is crucial to the next step. Let us ask: what do we mean by 'charge'? How can we define it operationally? The charge on a particle can be measured by observing the motion of the particle in an electromagnetic field—which may be an external field or the field of another particle. This is a procedure which has, in principle, nothing to do with quantum mechanics as such: charge has a perfectly good classical meaning. This suggests that a sensible and convenient definition of charge could be obtained by considering the long-wavelength ($q^2 \to 0$) limit of our quantum expressions for the currents. Actually, this is precisely how we did define charge in our earlier discussions (§§6.6 and 6.8) of electromagnetic form factors; indeed, equation (6.195) has just the form of equation (6.149), with the Dirac 'charge' form factor $\mathscr{F}_1(q^2)$ replaced by an explicit expression, which is associated with the particular higher-order process of figure 6.16. Since, as we have said, $X(q^2)$ in (6.195) tends to zero as $q^2 \to 0$, we see that, due to the correction represented by figure 6.16, the charge of the particles must be regarded as being not e, the parameter describing the lowest-order processes, but rather e', where

$$e' = e\left[1 - \frac{\alpha}{3\pi}\ln\left(\frac{\Lambda^2}{m^2}\right)\right]^{1/2}. \tag{6.196}$$

Thus at least part of the effect of figure 6.16 is to change the charge from e to e', according to (6.196). In particular, the part that has this effect is the *divergent* part; the rest, involving $X(q^2)$, is *finite*—it can provisionally be regarded as a contribution to the form factor $\mathscr{F}_1$, and we shall come back to it in a little while.

The idea that the effective interaction strength of a theory can be changed from its 'free space' value by the presence of interactions is a familiar one in solid state physics. The charge of an electron in a solid gets changed into an 'effective' one by reason of its interactions with the other electrons in the solid, and with the lattice ions. In QED the situation, as illustrated by (6.196), is different in two respects. The first is that the correction is infinite as $\Lambda \to \infty$. Let us pause for a moment to think what sort of Λ values are actually required before the correction factor in (6.196) departs much from unity. Clearly the answer is that nothing will happen for $\Lambda \ll m \exp(1/2\alpha) \sim 10^{26}$ GeV! This is a fantastically high energy. Put differently, it corresponds to a length scale $l_\Lambda = \hbar/\Lambda c \sim 10^{-27}$ fm. Do we really believe that the theory of QED will stand as it is, without modification, at such vast energies and infinitesimal distances? Most people think not: for example, a measure of the likely distance at which quantum gravity effects must be included is given by the Planck length $l_P = (\hbar G_N/c^3)^{1/2}$ where G_N is Newton's gravitational constant. The value of l_P is $\sim 10^{-20}$ fm, far greater than the above value of l_Λ. Thus surely some new physics will enter before the presence of the Λ in (6.196) becomes really serious. However, this reassuring state of affairs is no accident: it is precisely due to the fact that the correction in (6.196) is logarithmically dependent on Λ^2, and not—for example—exponentially. These considerations give us an important insight into the physical content of renormalisability: namely, a renormalisable theory is one which is *insensitive to ignorance about its behaviour at very short distances*.

From this point of view we can now understand in a different way the importance of renormalisability. What we are saying is that physics at our scales of energy or distance is not going to be sensitive to the behaviour of matter at exceedingly remote scales, if our theories are renormalisable. It would be very unpleasant if it were otherwise, for then we should have to wait until everything was known, down to the smallest conceivable distances, before we could get on with our mundane calculations at our own accessible scales. We can also understand how the 'taming' is likely to go. In other areas of physics one frequently manages very well to do calculations relevant to a certain energy and length scale, describing what goes on at small distances by a few parameters which are uncalculable as far as the physics of the original scale is concerned. For example, as we pointed out in Chapter 1, atomic physics is fairly well separated from nuclear physics, and all an atomic physicist usually needs to know about a nucleus are certain gross parameters such as mass, charge, spin, etc. It is the same in the case of QED: we might hope that all the short-distance troubles can be lumped together into a few effective parameters—like the electron's mass and charge—which are the ones we in fact deal with in our measurements. Perhaps there are 'frozen' ultra-short-distance degrees of freedom which determine these parameters—but that is a separate story.

Indeed, some such procedure is almost forced upon us by the second way in which the QED situation differs from that in our solid state analogy. In QED there is, apparently, no medium to provide the interactions which, for the electron in a solid, alter its effective charge. In relativistic quantum mechanics, however, the remarkable possibility exists of *quantum fluctuations in the vacuum*. The process we are considering, figure 6.16, is a typical example of such a vacuum fluctuation. The photon exchanged between the two leptons can create, from the vacuum, a virtual electron–positron pair. This pair has only a fleeting existence; it lives on borrowed time, in the sense that the energy necessary to call it into existence, $\Delta E \sim 2mc^2$, is only available for a time of order $\Delta t \sim \hbar/2mc^2$ (we are using the so-called 'energy–time uncertainty relation', which states that there is a spread ΔE in the energy of a quantum system given by $\Delta E \gtrsim \hbar/\Delta t$, where Δt is the time over which the system's energy is measured; over short times quite large energy fluctuations are possible). In relativistic quantum mechanics, therefore, we must beware of thinking of the vacuum, too naively, as being 'nothing'. Even when a particle is supposedly free and propagating in the vacuum, the parameters describing it (such as the charge) are changed, from the values appearing in the original one-particle Hamiltonian, to ones which include the effect of its interactions with the virtual particles produced from the vacuum. It is appropriate to mention at this point that further remarkable properties of the vacuum seem to be necessary in order to understand fundamental aspects of both the theory of weak interactions (namely, how gauge bosons can acquire mass) and QCD (the problem of confinement). These matters will be taken up in Chapters 13 and 15.

In the case of an electron propagating in a solid, we can always take it out of the solid and have a look at it in 'free space'. Thus its interaction-modified effective charge and its free space charge can each be separately defined experimentally. However, we cannot perform the same process to get 'outside' the fluctuating interactions present in the relativistic vacuum. There seems, in short, to be no sense in which the original one-particle e can be experimentally identified, free of all the vacuum fluctuation corrections; only the fully corrected charge is experimentally accessible. Equation (6.196) gives the form of one particular correction to e, the unobservable original charge. The suggestion is, then, that we forget about e and Λ, and identify e' with the measured charge. The miracle of renormalisability consists in this: after rewriting the theory in terms of e' and a few other parameters taken from experiment, *no dependence on* Λ *will remain*: all quantities of physical interest will be finite as $\Lambda \to \infty$.

Let us give some idea of how this comes about. First, we generalise (6.196) a little, writing

$$e_{\mathrm{R}} = e\left(1 + \sum_{n=1}^{\infty} \alpha^n c_n(\Lambda^2)\right) \tag{6.197}$$

to express the fact that there are, of course, an infinite set of possible higher-order corrections to e; e_{R} is the full renormalised charge. Going back now to figure 6.16, and putting in for definiteness the explicit expression (Bjorken and Drell 1964) for $X(q^2)$ which is valid for low q^2, the effect of including this process as a correction to figure 6.15 is to change

$$e^2 \bar{u}\gamma_\mu u(-g^{\mu\nu}/q^2)\bar{u}\gamma_\nu u \tag{6.198}$$

into

$$e^2\left[1 - \frac{\alpha}{3\pi}\ln\left(\frac{\Lambda^2}{m^2}\right) - \frac{\alpha}{15\pi}\frac{q^2}{m^2} + \mathrm{O}(\alpha^2)\right]\bar{u}\gamma_\mu u \frac{-g^{\mu\nu}}{q^2}\bar{u}\gamma_\nu u. \tag{6.199}$$

However, we have (cf (6.196))

$$e_{\mathrm{R}}^2 = e^2\left[1 - \frac{\alpha}{3\pi}\ln\left(\frac{\Lambda^2}{m^2}\right) + \mathrm{O}(\alpha^2)\right] \tag{6.200}$$

which can be inverted to yield

$$e^2 = e_{\mathrm{R}}^2\left[1 + \frac{\alpha_{\mathrm{R}}}{3\pi}\ln\left(\frac{\Lambda^2}{m^2}\right) + \mathrm{O}(\alpha_{\mathrm{R}}^2)\right]. \tag{6.201}$$

Inserting (6.201) into (6.199) we find that the latter becomes

$$e_{\mathrm{R}}^2\left(1 - \frac{\alpha_{\mathrm{R}}}{15\pi}\frac{q^2}{m^2} + \mathrm{O}(\alpha_{\mathrm{R}}^2)\right)\bar{u}\gamma_\mu u \frac{-g^{\mu\nu}}{q^2}\bar{u}\gamma_\nu u \tag{6.202}$$

which is finite to this order. This, of course, is an illustration which is so simple as to border on the trivial. Nevertheless, the essence of the matter is contained in it: explicit expressions for the coefficients $c_n(\Lambda^2)$ in (6.197) can be obtained by considering all graphs contributing at a given order of perturbation theory, and, by an inversion procedure of the above form, infinities in physical amplitudes can be systematically removed, order by order.

The particular type of process we have been considering (figure 6.16) has, as we have seen, the effect of altering the photon propagator, and thence, according to the interpretation offered, the value of the charge on the leptons. Such corrections are called *vacuum polarisation* processes because of the effect the pairs have, when split apart (polarised) in the field of a 'test' charge, in reducing the effective charge of the test particle. This will be amplified in §9.3, but at this point—after so much talk—we must point out a vital piece of physics contained in (6.202). The form given is valid for small $q^2 (q^2 \ll m^2)$ and therefore amounts—when evaluated in the non-relativistic limit and Fourier transformed (cf §5.6)—to a modification at short distances of Coulomb's law, which leads to a first-order shift in the energy levels of hydrogen-like atoms, as first calculated by Uehling (1935). In hydrogen the effect contributes about −27 MHz out of a total *Lamb shift* (the difference in energy between the $2^2P_{1/2}$ and $2^2S_{1/2}$ states) of some +1000 MHz. The effect is quite clearly observable in hydrogen, but is very much more pronounced in muonic helium where the heavier muon's orbit samples distances ~ 1/200 closer to the nucleus than does the electron's orbit. The early Lamb shift measurements were an enormous impetus to the development of quantum field theory (specifically QED). A qualitative discussion of the Lamb shift is given in Aitchison (1985).

Vacuum polarisation effects are not the only processes leading to infinities, of course. Another is shown in figure 6.17, which may be regarded as a correction to the lowest-order Compton diagram, figure 6.18. We note here the appearance of an emission and absorption process by the *same* particle, which we first came across in §5.9. It turns out that the amplitude for figure 6.17 also contains a part which diverges logarithmically as $\Lambda^2 \to \infty$. This infinity can, remarkably enough, be interpreted as being associated with a change in the mass of the electron, in the same way as the infinity in figure 6.16 was with the charge. In general, the measured mass m_R can be expressed as an infinite series

$$m_R = m\left(1 + \sum_{n=1}^{\infty} \alpha^n d_n(\Lambda^2)\right) \tag{6.203}$$

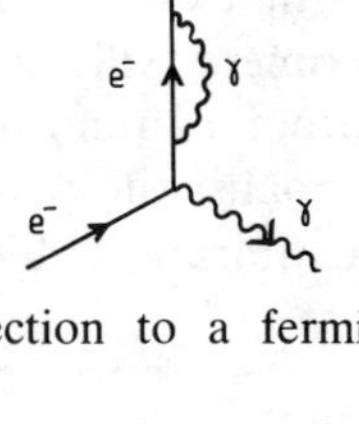

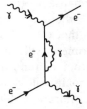

Figure 6.17 Higher-order correction to a fermion propagator in the process $\gamma e^- \to \gamma e^-$.

each term of which diverges as $\Lambda^2 \to \infty$ but is calculable in perturbation theory with Λ^2 held finite.

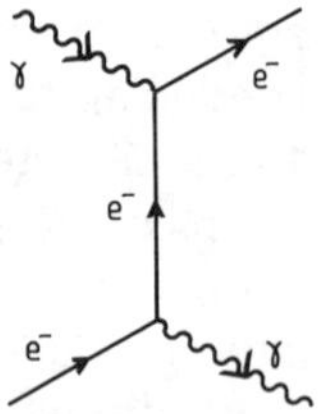

Figure 6.18 Lowest-order graph in $\gamma e^- \to \gamma e^-$.

Once again, in solid state physics it is natural to regard the mass of a particle as being affected by interactions with its surroundings—which are, in the QED case, virtual interactions of the sort affecting the electron propagator, as illustrated by figure 6.17; these interactions affect the ease with which the particle responds to an external force, and hence change its inertia. In QED the mass shifts represented by (6.203) are, however, infinite as $\Lambda^2 \to \infty$. By an inversion procedure similar to that outlined for charge, the 'bare' mass m appearing in amplitudes calculated according to perturbation theory can be eliminated, order by order, in favour of the finite quantity m_R; when this is done, many further divergences are cured.

Still further divergences can be identified, and associated with (infinite) changes in the normalisation of the wavefunctions. However, we need not be more specific about these, since we are now in a position to state what is meant by a *renormalisable theory*. It is one in which only a *finite* number of parameters e_R, m_R, . . . have to be taken from experiment; when, by the inversion procedure outlined above, all amplitudes derived in perturbation theory are re-expressed in terms of these physical parameters, *all* infinities are removed. Thus, in relativistic quantum mechanics, the process of replacing e, m, . . . by e_R, m_R, . . . is a 'renormalisation' which exorcises infinities from physical amplitudes.

We may alternatively say that only a finite number of *types* of divergence occur in a renormalisable theory; curing these removes all infinities, to all orders in perturbation theory. The identification of these types, and the demonstration that their effective incorporation into finite parameters taken from experiment will render an arbitrary graph finite, is what is involved in the proof of renormalisability. It is a tall order, and the proofs of the renormalisability of QED (see e.g. Bjorken and Drell 1964, Bogoliubov and Shirkov 1980) and of non-Abelian gauge theories†, both spontaneously broken ('t Hooft 1971a) and unbroken ('t Hooft 1971b), are landmark achievements.

By contrast, a non-renormalisable theory is one in which new types of divergence appear at each order of perturbation theory; to cure all of them by the same procedure would entail introducing an infinite number of constants to be taken from experiment. This makes such theories remarkably unpredictive. There is no discernible principle to tell us how big any of these constants are supposed to be; therefore none of them can be excluded *a priori*. Of course, we can *arrange* all sorts of things to work out right by suitable choice of these parameters, but we should have simply accommodated the data.

This concept of a *type* of divergence will be important when we come to discuss the question of renormalisability for weak interactions. Its physical meaning is not yet very clear from what we have said above. In fact, it appears much more understandably in the context of an alternative way of formulating the renormalisation process, which we now describe. While simple to interpret physically, the procedure we have tried to explain is not very pleasant to contemplate mathematically. There is an awkward 'convulsion' (Lautrup 1976) when the theory turns itself inside out, at the stage when the amplitudes calculated in terms of $e, m, \ldots$ are re-expressed in terms of $e_R, m_R, \ldots$. An alternative procedure would be to insist from the start that the parameters appearing in the perturbation expansion are already the physical ones $e_R, m_R, \ldots$. Considering, to begin with, just the question of the electron mass parameter, this latter procedure would amount to basing perturbation theory on the equation

$$(\mathrm{i}\not{\partial} - m_R + e\not{A} + \delta m)\psi = 0 \tag{6.204}$$

rather than on

$$(\mathrm{i}\not{\partial} - m + e\not{A})\psi = 0 \tag{6.205}$$

where in (6.204) we have introduced the mass shift $\delta m = m_R - m$; δm is to be regarded as an 'additional interaction', the Dirac equation for the physical free particle being now $(\mathrm{i}\not{\partial} - m_R)\psi = 0$. Similarly, we should have to rewrite all quantities appearing in (6.204) in terms of their renormalised equivalents and various corrections like δm.

Naturally, when we proceed to calculate amplitudes in this new perturbation theory we shall encounter the same old divergences that we met before. Now, however, there is an important difference, which is that we have to include the contributions of the new pieces, like δm, which have appeared via the rearrangement $e, m, \ldots \to e_R, m_R, \ldots$.

†For a review, see for example Ryder (1985) and Collins (1984).

These additional terms are, in fact, unknown *a priori* (just as the bare charge e and bare mass m themselves are). *A renormalisable theory is one in which these terms can be so chosen as to cancel all divergences order by order in the new perturbation theory based on* e_R. These additional terms are known as 'counter terms', and there is a finite number of them in a renormalisable theory.

This finite number corresponds exactly to the number of 'types' of divergence, but now we can see the physics of the situation. The only *available* counter terms are the ones which arise in the process of rewriting the original theory in terms of renormalised quantities plus extra bits (the counter terms). All the counter terms must correspond to masses, interactions, etc which are present in the original theory—which is, in fact, *the* theory we are trying to make sense of! We are not free to add in any old kind of counter term—if we did, we would be defining a new theory.

We can illustrate this important point by considering, for example, a process which is new to us, $\gamma\gamma$ scattering. For this, the lowest-order contribution is already a one-loop graph, figure 6.19, *because there is no* $\gamma\gamma \to \gamma\gamma$ *interaction in the original theory*. Classically, there is no interaction between photons, and this process—Delbruck scattering—is a purely *quantum* field effect, arising from the virtual e^+e^- pairs visible in the loop. (One day, high-intensity lasers may test the predicted cross section.) This graph had better be finite, after renormalisation of the electron charge and mass, because there is *no* counter term to adjust so as to absorb its infinity. Indeed, it is finite—due, ultimately, to gauge invariance (see e.g. Itzykson and Zuber 1980, §7.3.1). A non-renormalisable theory, on the other hand, is hardly a theory at all—it has constantly to be redefined as more and more infinities appear!

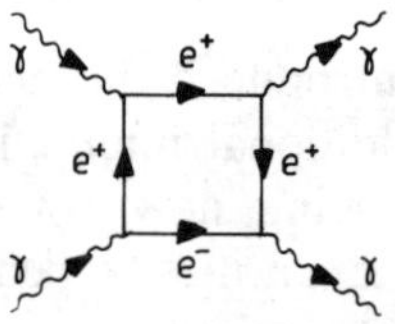

Figure 6.19 Lowest-order contribution to $\gamma\gamma \to \gamma\gamma$.

This concludes what we have to say about renormalisation in QED. It would be understandable if both the mathematically sophisticated and the more practically inclined reader were somewhat uneasy about our bland assertion that all the infinities can be manipulated and 'defined away'. Ultimately, our faith in the renormalisation procedure for handling these apparently infinite contributions is based on very pragmatic grounds—the astonishing accuracy to which renormalised QED has been tested.

The example of the magnetic moment of the electron will serve to make this point about accuracy. In natural units, the magnetic moment may be written

$$\mu = ge/4m$$

where e and m are the charge and mass of the electron respectively and g is the gyromagnetic ratio or g factor of the electron. The Dirac equation for the electron, treated in lowest-order perturbation theory, predicts $g = 2$ exactly. However, as in our discussion of hadronic form factors, we expect that this result will be modified by virtual electromagnetic interactions, such as the 'vertex correction' shown in figure 6.20. But in this case, there is an important difference from the previous case of form factors caused by virtual strong interactions: since the strength of electromagnetic interactions, as measured by the fine-structure constant α, is small, we ought to be able to calculate these corrections reliably in perturbation theory. In fact, defining the 'anomaly' by

$$g = 2(1 + a) \tag{6.206}$$

we find in renormalised QED that the contributions to a may be expressed as a power series in α/π:

$$a = A(\alpha/\pi) + B(\alpha/\pi)^2 + C(\alpha/\pi)^3 + \ldots \tag{6.207}$$

The diagram of figure 6.20 is a typical 'second-order' α/π contribution: figure 6.21 shows one 'fourth-order' $(\alpha/\pi)^2$ graph. The contributing graphs can equivalently be classified by the number of loops they contain, which is just one half of the 'order' in α: thus figure 6.21 is a two-loop contribution.

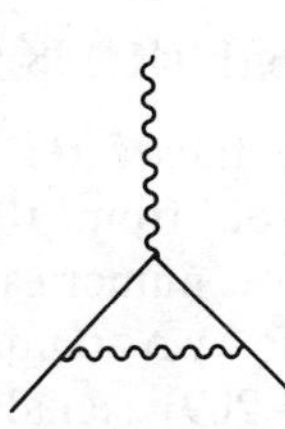

Figure 6.20 Vertex correction to electron–photon coupling: a second-order contribution to $g - 2$.

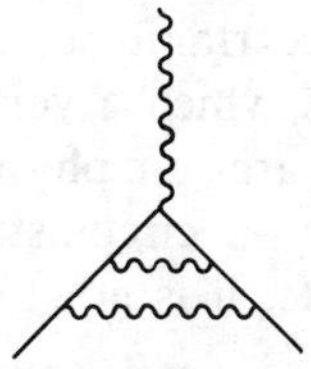

Figure 6.21 A fourth-order contribution to $g - 2$.

Since $\alpha/\pi \sim 2.3 \times 10^{-3}$, it is clear that the contributions of successive orders become very small quite quickly (the coefficients A, B, C, ... turn out to be of order 1). However, 'a' can be determined experimentally to a quite extraordinary degree of precision. The latest published value is (Van Dyck *et al* 1987)

$$a(\mathrm{e}^-)_{\mathrm{exp}} = 1159\,652\,188.4(4.3) \times 10^{-12} \tag{6.208}$$

for the electron and

$$a(\mathrm{e}^+)_{\mathrm{exp}} = 1159\,652\,187.9(4.3) \times 10^{-12} \tag{6.209}$$

for the positron, where the figure in parentheses is the experimental uncertainty. (Note that the equality between (6.208) and (6.209) provides the most stringent test of CPT invariance for a charged particle.) To confront these values with the theoretical prediction requires the calculation of up to eighth-order (four-loop) graphs. Some idea of the complexity of this undertaking can be gained from the following quotation from Kinoshita and Sapirstein (1984), in an article summarising the then current status of QED:

> The calculation of the 4 loop correction is of an extremely large scale; while the use of self energy instead of vertex graphs cuts down the number of contributing terms from 518 to 47 for the most important class of graphs, those without vacuum polarization loops, each of the integrands has up to 15,000 terms which must be integrated over 10 dimensional Feynman parameter space .... The present answer, which is still very preliminary, was obtained with 50 CDC-7600 equivalent days of computation. The error quoted is statistical .... Substantial progress will have to wait for massively parallel processors.

The theoretical value reported by Kinoshita and Sapirstein (1984), which includes their evaluation of the four-loop graphs, is

$$a_{\mathrm{th}} = 1159\,651\,941(128)(43) \times 10^{-12} \tag{6.210}$$

where the first figure in parentheses reflects the experimental error in the determination of α (taken from the Josephson effect) and the second represents the level of numerical accuracy. As these authors remark, it is possible that the two-standard-deviation discrepancy between (6.210) and (6.208), (6.209) signals a serious problem for QED; alternatively it may be that the extraction of α from the Josephson effect may not be reliable at this level. This epic confrontation between theory and experiment will certainly continue to be pursued; it is a classic example of the way in which a very high-precision measurement in a thoroughly 'low-energy' area of physics can have profound impact on the 'high-energy' frontier—a circumstance we may be increasingly dependent upon. At all events, it is not for nothing that we have called QED 'the best theory we have'.

Summary

First-order perturbation theory for the interaction of an electron with an electromagnetic field: vector potential and electromagnetic transition current.

Amplitude and Feynman diagram for lowest-order $e^-\pi^+$ scattering.

Trace techniques for spin summations.

Explicit two-body cross section calculations.

The notion of a form factor, illustrated by the pion form factor.

Application of Lorentz invariance and current conservation to restrict the parametrisation of the pion form factor.

$e^-\mu^-$ scattering in lowest order (two pointlike fermions).

e^-p scattering and nucleon form factors: parametrisation subject to Lorentz invariance and current conservation.

Kinematics of elastic and inelastic e^-p scattering.

Electron Compton scattering and fermion propagator.

Renormalisation,and accuracy of QED predictions.

Problems

6.1 (a) If the four-component spinors ψ are normalised in a box of volume V by the condition

$$\int_V \mathrm{d}^3x\psi^\dagger\psi = 2E$$

find the necessary normalisation factor N in the expression

$$\psi = N\begin{pmatrix} \phi \\ \dfrac{\boldsymbol{\sigma}\cdot\boldsymbol{p}}{E+m}\phi \end{pmatrix} \mathrm{e}^{-\mathrm{i}p\cdot x}$$

(b) For $E > 0$ the positive-energy spinor $u(p,s)$ is defined by

$$u(p,\, s) = (E + m)^{1/2}\begin{pmatrix} \phi^s \\ \dfrac{\boldsymbol{\sigma}\cdot\boldsymbol{p}}{E+m}\phi^s \end{pmatrix}.$$

Find the equation satisfied by $u(p, s)$ and write this in γ matrix notation. Show also that the conjugate spinor $\bar{u}(p, s)$ satisfies the equation

$$\bar{u}(p, s)(\not{p} - m) = 0.$$

(c) Find the analogous equations satisfied by the negative-energy spinors $v(p, s)$ and $\bar{v}(p,s)$ where

$$v(p, s) = (E + m)^{1/2}\begin{pmatrix} \dfrac{\boldsymbol{\sigma}\cdot\boldsymbol{p}}{E + m}\chi^s \\ \chi^s \end{pmatrix}.$$

6.2 (a) For arbitrary matrices **A** and **B** prove the result

$$(\mathbf{AB})^\dagger = \mathbf{B}^\dagger\mathbf{A}^\dagger$$

where $\mathbf{M}^\dagger$ denotes the Hermitian conjugate matrix defined by

$$(\mathbf{M}^\dagger)_{ij} = \mathbf{M}^*_{ji}.$$

(b) Show that the matrices

$$\alpha = \begin{pmatrix} \mathbf{0} & \boldsymbol{\sigma} \\ \boldsymbol{\sigma} & \mathbf{0} \end{pmatrix}, \qquad \beta = \begin{pmatrix} \mathbf{1} & \mathbf{0} \\ \mathbf{0} & \mathbf{1} \end{pmatrix} = \gamma^0$$

are Hermitian ($\mathbf{M} = \mathbf{M}^\dagger$). Show also that $\boldsymbol{\gamma}$ is anti-Hermitian ($\mathbf{M} = -\mathbf{M}^\dagger$).

(c) If the matrix $\bar{\Gamma}$ is defined by the operation

$$\bar{\Gamma} \equiv \gamma^0\Gamma^\dagger\gamma^0$$

for an arbitrary γ matrix Γ, show that: (i) $\overline{\gamma^\mu} = \gamma^\mu$, (ii) $\overline{(\mathrm{i}\gamma_5)} = \mathrm{i}\gamma_5$, where $\gamma_5 = \mathrm{i}\gamma^0\gamma^1\gamma^2\gamma^3$.

(d) Defining the matrices

$$\sigma^{\mu\nu} = \tfrac{1}{2}\mathrm{i}[\gamma^\mu, \gamma^\nu]$$

find $\overline{\sigma^{\mu\nu}}$.

6.3 (a) Use the γ matrix anticommutation relations

$$\{\gamma^\mu, \gamma^\nu\} = 2g^{\mu\nu}$$

to prove the result

$$\not{a}\not{b} = -\not{b}\not{a} + 2a\cdot b\mathbf{1}$$

where a and b are arbitrary 4-vectors.

(b) Hence prove the identity

$$(\not{k} + m)(\not{k} - m) = (\not{k} - m)(\not{k} + m) = 0$$

if $k^2 = m^2$. Use this result to show that the 4×4 matrix

$$\Lambda_+(k) = \not{k} + m$$

when applied to any four-component spinor ψ projects out only positive-energy solutions. Similarly, show that

$$\Lambda_-(k) = \not{k} - m$$

projects out negative-energy solutions.

(c) Use the explicit forms for the spinors $u(k, s)$ to evaluate the spin sum

$$\sum_s u(k, s)\bar{u}(k, s)$$

in the form of a 4×4 matrix, and compare this with the matrix $(\not{k} + m)$.

6.4 (a) For arbitrary matrices **A, B, C** and **D** prove the result

$$\mathrm{Tr}(\mathbf{ABCD}) = \mathrm{Tr}(\mathbf{DABC}) = \mathrm{Tr}(\mathbf{CDAB}) = \mathrm{Tr}(\mathbf{BCDA}).$$

(b) Use the γ matrix trace theorems (summarised in Appendix D) to evaluate the lepton tensor

$$L_{\mu\nu} = \tfrac{1}{2}\mathrm{Tr}[(\not{k}' + m)\gamma_\mu(\not{k} + m)\gamma_\nu].$$

Defining $q^2 = (k - k')^2$, with $k^2 = k'^2 = m^2$, show that your result may be written

$$L_{\mu\nu} = 2[k'_\mu k_\nu + k'_\nu k_\mu + (q^2/2)g_{\mu\nu}].$$

6.5 For unpolarised $e\pi$ scattering we need to evaluate the 'contraction' $T^{\mu\nu}L_{\mu\nu}$, where $L_{\mu\nu}$ is given in problem 6.4, and

$$T^{\mu\nu} = (p + p')^\mu(p + p')^\nu$$

where $p' = p + q$.

Using the explicit form for $L_{\mu\nu}$:

(a) show that

$$q^\mu L_{\mu\nu} = q^\nu L_{\mu\nu} = 0;$$

(b) use this last result to assist you in evaluating the contraction $T^{\mu\nu}L_{\mu\nu}$ and show that

$$T^{\mu\nu}L_{\mu\nu} = 8[2(k\cdot p)(k'\cdot p) + (q^2/2)M^2]$$

where M is the pion mass.

The remaining details of the evaluation of the laboratory cross section are given in Appendix E.

6.6 (a) Derive an expression for the spin-averaged differential cross section for lowest-order $e\mu$ scattering in the laboratory frame, defined by $p^\mu = (M, \mathbf{0})$, where M is now the muon mass, and show that it may be written in the form

$$\frac{d\sigma}{d\Omega} = \left(\frac{d\sigma}{d\Omega}\right)_{ns} [1 - (q^2/2M^2) \tan^2(\theta/2)]$$

where the 'no-structure' cross section is that of eπ scattering (Appendix E), and the electron mass has as usual been neglected.

(b) Neglecting *all* masses, evaluate $|\bar{F}|^2$ in terms of s, t and u, and use the result

$$\frac{d\sigma}{dt} = \frac{1}{16\pi s^2}|\bar{F}|^2$$

to show that the eμ cross section may be written in the form

$$\frac{d\sigma}{dt} = \frac{4\pi\alpha^2}{t^2}\frac{1}{2}\left(1 + \frac{u^2}{s^2}\right).$$

Show also that by introducing the variable y, defined in terms of laboratory variables by $y = (k - k')/k$, this reduces to the result

$$\frac{d\sigma}{dy} = \frac{4\pi\alpha^2}{t^2} s\tfrac{1}{2}[1 + (1 - y)^2].$$

6.7 Consider the process $e^+e^- \to \mu^+\mu^-$ in the CM frame.

(a) Draw the lowest-order Feynman diagram and write down the corresponding amplitude.

(b) Show that the spin-averaged squared matrix element has the form

$$|\bar{F}|^2 = \frac{(4\pi\alpha)^2}{Q^4} L(e)_{\mu\nu} L(\mu)^{\mu\nu}$$

where Q^2 is the square of the total CM energy.

(c) Evaluate the traces and the tensor contraction (neglecting lepton masses), and show that

$$|\bar{F}|^2 = (4\pi\alpha)^2(1 + \cos^2\theta)$$

where θ is the CM scattering angle.

(d) Hence show that the total elastic cross section is

$$\sigma = 4\pi\alpha^2/3Q^2.$$

6.8 Starting from the expression

$$\bar{u}(p')\mathrm{i}\frac{\sigma^{\mu\nu}}{2M} q_\nu u(p)$$

where $q = p' - p$ and $\sigma^{\mu\nu}$ is defined in problem 6.2, use the Dirac equation and properties of γ matrices to prove the 'Gordon decomposition' of the current

$$\bar{u}(p')\gamma^{\mu}u(p) = \bar{u}(p')\left(\frac{(p + p')^{\mu}}{2M} + \mathrm{i}\frac{\sigma^{\mu\nu}q_{\nu}}{2M}\right)u(p).$$

6.9 Omitting inessential factors, the amplitude for figure 6.14(*a*) is (cf §5.6)

$$\sum_{s} u(q_1, s)\bar{u}(q_1, s)/[E_k + E_p - (E_k + E_n + E_{k'})]\cdot 2E_n$$

where $q_1 = (E_n, \boldsymbol{p} - \boldsymbol{k}')$ and $E_n^2 = [m^2 + (\boldsymbol{p} - \boldsymbol{k}')^2]$. Similarly, the amplitude for figure 6.14(*b*) is, remembering the minus sign from §3.5.3,

$$-\sum_{s} v(q_2, s)\bar{v}(q_2, s)/[E_k + E_p - (E_p + E_{p'} + E_n)]\cdot 2E_n$$

where $q_2 = (E_n, \boldsymbol{k}' - \boldsymbol{p})$. Verify that the sum of these contributions is

$$(\not{q} + m)/(q^2 - m^2)$$

where $q = (E_p - E_{k'}, \boldsymbol{p} - \boldsymbol{k}')$.

6.10 Verify that when ε_{μ} is replaced by k_{μ}, or ε_{ν}^{*} by k'_{ν}, the amplitude $F_{\gamma\mathrm{e}}$ of equation (6.175) vanishes.

6.11 (a) The spin-averaged squared amplitude for lowest-order electron Compton scattering contains the interference term

$$\sum_{\lambda,\lambda',s,s'} F_{\gamma\mathrm{e}}^{(\mathrm{d})} F_{\gamma\mathrm{e}}^{(\mathrm{e})*}$$

where the *F*'s are given in (6.176) and (6.177). Obtain an expression analogous to (6.180) for this term, and prove that it is, in fact, zero. [Hint: use the relations (D.4) and (D.5).]

(b) Explain why the term

$$\sum_{\lambda,\lambda',s,s'} F_{\gamma\mathrm{e}}^{(\mathrm{e})*} F_{\gamma\mathrm{e}}^{(\mathrm{e})}$$

is given by (6.185) with *s* and *u* interchanged.

6.12 Recalculate the interference term of problem 6.11(a) for the case $k^2 = -Q^2$, and hence verify (6.189).

6.13 As a simple example of a divergent 'loop' integral, consider the d^4k integral in (6.190), replacing the fermion propagators by spin-0 propagators for simplicity. The essential piece is then

$$\int \frac{\mathrm{d}^4k}{(2\pi)^4}\,\frac{1}{(k^2 - m^2)}\,\frac{1}{[(k - q)^2 - m^2]}.$$

To compute this, proceed as follows:

(a) Verify the identity

$$\frac{1}{AB} = \int_0^1 \frac{\mathrm{d}\alpha}{[(1-\alpha)A + \alpha B]^2}.$$

(b) Applying the result to the d^4k integral, it becomes

$$\int_0^1 \mathrm{d}\alpha \int \frac{\mathrm{d}^4k}{(2\pi)^4} \frac{1}{\{(1-\alpha)(k^2-m^2) + \alpha[(k-q)^2 - m^2]\}^2}.$$

Show that this reduces to

$$\int_0^1 \mathrm{d}\alpha \int \frac{\mathrm{d}^4k}{(2\pi)^4} \frac{1}{[(k-\alpha q)^2 + \alpha q^2 - \alpha^2 q^2 - m^2]^2}.$$

Changing the integration variable to $k - \alpha q$, this is equal to

$$\int_0^1 \mathrm{d}\alpha \int \frac{\mathrm{d}^4k}{(2\pi)^4} \frac{1}{[k^2 - \Delta]^2}$$

where $\Delta = m^2 + \alpha^2 q^2 - \alpha q^2$.

(c) The d^4k integral means $\mathrm{d}k^0\,\mathrm{d}^3\boldsymbol{k}$, and $k^2 = (k^0)^2 - \boldsymbol{k}^2$. Introducing $k^0 = \mathrm{i}\kappa^0$ (a manoeuvre which, though correct, we shall not justify here), we have to evaluate

$$\mathrm{i}\int_0^1 \mathrm{d}\alpha \int \frac{\mathrm{d}\kappa^0\,\mathrm{d}^3\boldsymbol{k}}{(2\pi)^4} \frac{1}{[(\kappa^0)^2 + \boldsymbol{k}^2 + \Delta]^2}.$$

This is now an integral in four-dimensional Euclidean space (the d^4k one having been in Minkowski space), and we can use the four-dimensional analogue of spherical polars. Since only the square length $(\kappa^0)^2 + \boldsymbol{k}^2$ enters, we need not worry about any angle-type integrals, and the basic integral reduces to something proportional to

$$\int_0^1 \mathrm{d}\alpha \int \frac{\kappa^3\mathrm{d}\kappa}{[\kappa^2 + \Delta]^2}$$

where $\kappa = [(\kappa^0)^2 + \boldsymbol{k}^2]^{1/2}$, the four-dimensional length of $(\kappa^0, \boldsymbol{k})$. Changing variables to $u = \kappa^2$, the u integral is

$$\int_0^\infty \frac{u\,\mathrm{d}u}{(u+\Delta)^2}.$$

To investigate the divergence, we replace the upper limit by the '*cut-off*' Λ^2; show that

$$\int_0^{\Lambda^2} \frac{u\mathrm{d}u}{(u+\Delta)^2} = -1 + \frac{\Delta}{\Lambda^2 + \Delta} + \ln\left(\frac{(\Lambda^2 + \Delta)}{\Delta}\right)$$

and deduce that as $\Lambda^2 \to \infty$ the integral is *logarithmically divergent*.

Note that this degree of divergence can be guessed by arguing that, at large values of u, $(u+\Delta)^2 \simeq u^2$, so that the integral behaves as $\int \mathrm{d}u/u$, which is a logarithm. If we had one more power of u in the denominator, the integral would converge, going as $\int \mathrm{d}u/u^2$. If we had one fewer, the integral would go as $\int^{\Lambda^2} \mathrm{d}u \sim \Lambda^2$, which diverges 'quadratically'. In general, one estimates the degree of divergence of such a loop amplitude by going back to the $\mathrm{d}^4 k$ integral and 'counting powers' of k in numerator and denominator, the $\mathrm{d}^4 k$ counting as four powers. If the difference in power between numerator and denominator is $d > 0$, the divergence is expected to be as Λ^{2d}; and if $d = 0$, it is expected to be logarithmic—though sometimes these simple rules of thumb fail. If $d < 0$, the integral is convergent.

7

DEEP INELASTIC ELECTRON–NUCLEON SCATTERING AND THE QUARK PARTON MODEL

We have obtained the rules for doing calculations of simple processes in quantum electrodynamics for particles of spin 0 and spin $\frac{1}{2}$, and many explicit examples have been considered. We could now proceed directly to a discussion of the other types of interaction with which this book is concerned, namely weak and strong interactions. Instead, however, we choose to give an (admittedly brief) introduction to a topic of central importance in particle physics, the structure of hadrons as revealed by deep inelastic electron scattering experiments (the equally important neutrino scattering experiments will be discussed in §11.2). We do this partly because the necessary calculations involve straightforward, illustrative and eminently practical applications of the rules already obtained, but, more particularly, because it is from a comparison of these calculations with experiment that one obtains compelling evidence for the existence of the pointlike constituents of hadrons—'quarks'—the interactions of which, as we shall see, are believed to be described by gauge theories.

7.1 Inelastic ep scattering: kinematics and structure functions

At large momentum transfers there is very little elastic scattering: inelastic scattering, in which there is more than just the electron and proton in the final state, is much more probable. The simplest inelastic cross section to measure is the so-called 'inclusive' cross section, for which only the final electron is observed. This is therefore a sum over the cross sections for all the possible hadronic final states: no attempt is made to select any particular state from the hadronic debris created at the proton vertex. This process may be represented by the diagram of figure 7.1, assuming that the one-photon exchange amplitude dominates. The 'blob' at the proton vertex indicates our ignorance of the detailed structure: X indicates a sum over all possible hadronic final states. However, the assumption of one-photon exchange, which is known experimentally to be a very good approximation, means that, as in our previous examples (cf (6.69), (6.122)), the cross section must factorise into a leptonic tensor contracted with a tensor describing the hadron vertex

$$d\sigma \sim L_{\mu\nu} W^{\mu\nu}(q, p). \tag{7.1}$$

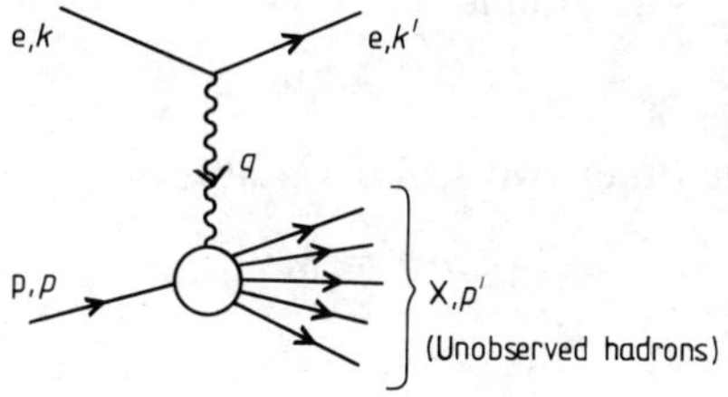

Figure 7.1 Inelastic electron–proton scattering, in one-photon exchange approximation.

The lepton vertex is well described by QED and takes the same form as before:

$$L_{\mu\nu} = 2[k'_\mu k_\nu + k'_\nu k_\mu + (q^2/2) g_{\mu\nu}]. \tag{6.88}$$

For the hadron tensor, on the other hand, we expect strong interactions to play an important role, and we must deduce its general structure by our powerful invariance arguments. We will only consider unpolarised scattering and therefore perform an average over the initial proton spins. The sum over final states, X, includes all possible quantum numbers for each hadronic state with total momentum p'. For an inclusive cross section, the final phase space involves only the scattered electron. Moreover, since we are not restricting the scattering process by picking out any specific state of X, the energy k' and the scattering angle θ of the final electron are now independent variables. In $W^{\mu\nu}(q, p)$ the sum over X includes the phase space for each hadronic state restricted by the usual 4-momentum-conserving δ function to ensure that each state in X has momentum p'. Including some conventional factors, we define $W^{\mu\nu}(q, p)$ by (see problem 7.1)

$$\begin{aligned} e^2 W^{\mu\nu}(q, p) = \frac{V}{4\pi M} \tfrac{1}{2} \sum_s \sum_X \langle \mathrm{p}; p, s | \hat{j}^\mu_{\mathrm{em}}(0) | \mathrm{X}; p' \rangle \langle \mathrm{X}; p' | \hat{j}^\nu_{\mathrm{em}}(0) | \mathrm{p}; p, s \rangle \\ \times (2\pi)^4 \delta^4(p + q - p'). \end{aligned} \tag{7.2}$$

How do we parametrise the tensor structure of $W^{\mu\nu}$? As usual, Lorentz invariance and current conservation come to our aid. There is one important difference compared with the elastic form factor case. For inclusive inelastic scattering there are now two independent scalar variables. The relation

$$p' = p + q \tag{7.3}$$

leads to

$$p'^2 = M^2 + 2p{\cdot}q + q^2 \tag{7.4}$$

where M is the proton mass. In this case, the invariant mass of the hadronic final state is a *variable*

$$p'^2 = W^2 \tag{7.5}$$

and is related to the other two scalar variables

$$p{\cdot}q = M\nu \tag{6.161}$$

and (cf (6.162))

$$q^2 = -Q^2 \tag{7.6}$$

by the condition (cf (6.168))

$$2M\nu = Q^2 + W^2 - M^2. \tag{7.7}$$

Our invariance arguments lead us to the same *tensor* structure as for *elastic* ep scattering, but now the functions $A(q^2)$, $B(q^2)$ are replaced by 'structure functions' which are functions of two variables, usually taken to be ν and Q^2. The conventional definition of the proton structure functions W_1 and W_2 is

$$\boxed{\begin{aligned} W^{\mu\nu}(q, p) = {} & (-g^{\mu\nu} + q^\mu q^\nu/q^2)W_1(Q^2, \nu) \\ & + [p^\mu - (p{\cdot}q/q^2)q^\mu][p^\nu - (p{\cdot}q/q^2)q^\nu]M^{-2}W_2(Q^2, \nu). \end{aligned}} \tag{7.8}$$

Inserting the usual flux factor together with the final electron phase space leads to the following expression for the inclusive differential cross section for inelastic electron–proton scattering (see problem 7.1):

$$\mathrm{d}\sigma = \left(\frac{4\pi\alpha}{q^2}\right)^2 \frac{1}{4[(k{\cdot}p)^2 - m^2M^2]^{1/2}} 4\pi M L_{\mu\nu} W^{\mu\nu} \frac{\mathrm{d}^3k'}{2\omega'(2\pi)^3}. \tag{7.9}$$

In terms of 'laboratory' variables, neglecting electron mass effects, this yields

$$\boxed{\frac{\mathrm{d}^2\sigma}{\mathrm{d}\Omega\,\mathrm{d}k'} = \frac{\alpha^2}{4k^2\sin^4(\theta/2)}[W_2\cos^2(\theta/2) + 2W_1\sin^2(\theta/2)].} \tag{7.10}$$

Remembering now that $\cos\theta$ and k' are independent variables for inelastic scattering, we can change variables from $\cos\theta$ and k' to Q^2 and ν, assuming azimuthal symmetry for the unpolarised cross section.

We have

$$Q^2 = 2kk'(1 - \cos\theta) \tag{7.11}$$

$$\nu = k - k' \tag{7.12}$$

so that

$$\mathrm{d}(\cos\theta)\,\mathrm{d}k' = \frac{1}{2kk'}\mathrm{d}Q^2\,\mathrm{d}\nu \tag{7.13}$$

and

$$\boxed{\frac{\mathrm{d}^2\sigma}{\mathrm{d}Q^2\,\mathrm{d}\nu} = \frac{\pi\alpha^2}{4k^2\sin^4(\theta/2)}\frac{1}{kk'}[W_2\cos^2(\theta/2) + 2W_1\sin^2(\theta/2)].} \tag{7.14}$$

Yet another choice of variables is sometimes used instead of these, namely the variables

$$x = Q^2/2M\nu \tag{7.15}$$

whose significance we shall see in the next section, and

$$y = \nu/k \tag{7.16}$$

which is the fractional energy transfer in the 'laboratory' frame. The Jacobian for the transformation from Q^2 and ν to x and y is (see problem 7.2)

$$\mathrm{d}Q^2\,\mathrm{d}\nu = 2Mk^2y\,\mathrm{d}x\,\mathrm{d}y. \tag{7.17}$$

We emphasise that the foregoing—in particular (7.8), (7.10) and (7.14)—is all completely general, given the initial one-photon approximation. The physics is all contained in the ν and Q^2 dependence of the two structure functions W_1 and W_2.

A priori, one might expect W_1 and W_2 to be complicated functions of ν and Q^2, reflecting the complexity of the inelastic scattering process. However, in 1969 Bjorken predicted that in the 'deep inelastic region'—large ν and Q^2, but Q^2/ν finite—there should be a very simple behaviour. He predicted that the structure functions should *scale*, i.e. become functions not of Q^2 and ν independently but only of their ratio Q^2/ν. It was the verification of approximate 'Bjorken scaling' that led to the development of the modern quark parton model. We therefore specialise our discussion of inelastic scattering to the deep inelastic region.

7.2 Bjorken scaling and the parton model

From considerations based on the quark model current algebra of Gell-Mann (for a review, see Adler and Dashen (1968)), Bjorken (1969) was led to propose the following 'scaling hypothesis': in the limit

$$\left.\begin{array}{r} Q^2 \to \infty \\ \nu \to \infty \end{array}\right\} \quad \text{with } x = Q^2/2M\nu \text{ fixed} \tag{7.18}$$

the structure functions scale as

$$MW_1(Q^2, \nu) \to F_1(x) \tag{7.19a}$$

$$\nu W_2(Q^2, \nu) \to F_2(x). \tag{7.19b}$$

We must emphasise that the *physical* content of Bjorken's hypothesis is that the functions $F_1(x)$ and $F_2(x)$ are *finite*†.

Early experimental support for these predictions (figure 7.2) led initially to an examination of the theoretical basis of Bjorken's argu-

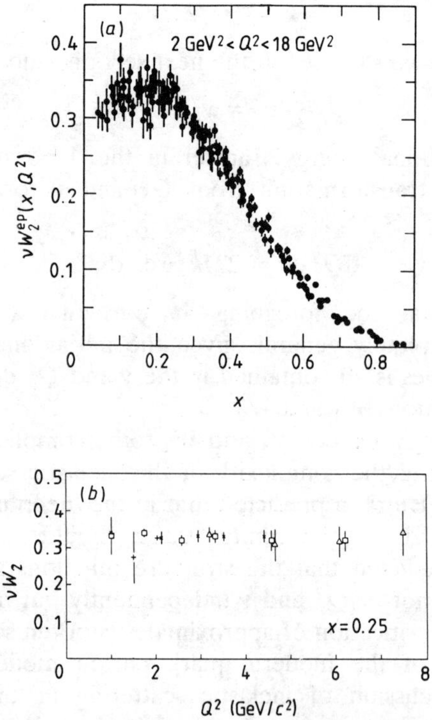

Figure 7.2 Bjorken scaling: the structure function νW_2 (a) plotted against x for different Q^2 values (Attwood 1980) and (b) plotted against Q^2 for the single x value, $x = 0.25$ (Friedman and Kendall 1972).

†It is always possible to write $W(Q^2, \nu) = f(x, Q^2)$, say, where $f(x, Q^2)$ will tend to some function $F(x)$ as $Q^2 \to \infty$ with x fixed. $F(x)$ may, however, be zero, finite or infinite. The physics lies in the hypothesis that, in this limit, a finite part remains.

ments, and to the formulation of the simple intuitive picture provided by the quark parton model. Closer scrutiny of figure 7.2(*a*) will encourage the (correct) suspicion that, in fact, there is a small but significant spread in the data for any given x value. Nevertheless, scaling is certainly the most immediate gross feature of the data, and an understanding of it is of fundamental importance. Later, in Chapter 9, we shall give an introduction to the way in which the *violations* of simple scaling behaviour can be predicted by QCD (compare figure 9.40 with figure 7.2(*b*)).

How can the scaling be understood? Feynman, when asked to explain Bjorken's arguments, gave an intuitive explanation in terms of elastic scattering from free pointlike constituents of the nucleon, which he dubbed 'partons'. The essence of the argument lies in the *kinematics of elastic scattering of electrons by pointlike partons*: we can therefore use the results of the previous chapters to derive the parton model results. At high Q^2 and ν it is intuitively reasonable (and in fact the basis for the light-cone and short-distance operator approach (Wilson 1969) to scaling) that the virtual photon is probing very short distance and time scales within the proton. In this situation, Feynman supposed that the photon interacts with small (pointlike) constituents within the proton, which carry only a certain fraction y of the proton's energy and momentum (figure 7.3). Over the short time scales involved in the transfer of a large amount of energy ν, the struck constituents can perhaps be treated as effectively free. We then have the idealised elastic scattering process shown in figure 7.4. It is the kinematics of the elastic scattering condition for the partons that leads directly to a relation between Q^2 and ν and hence to the observed scaling behaviour. The original discussion of the parton model took place in the infinite-momentum frame of the proton. While this has the merit that it eliminates the need for explicit statements about parton masses and so on, it also obscures the simple kinematic origin of scaling. For this reason, at the expense of some theoretical niceties, we prefer to perform a direct calculation of electron–parton scattering in close analogy with our previous examples.

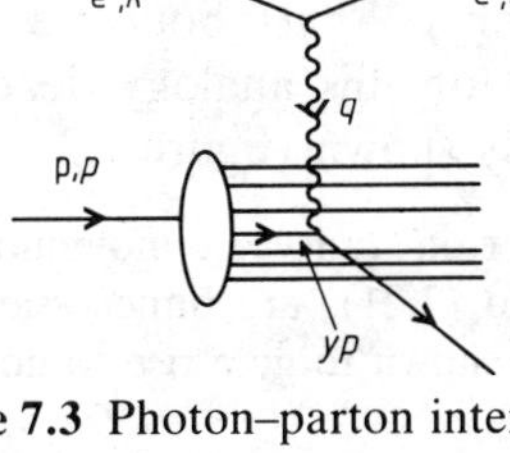

Figure 7.3 Photon–parton interaction.

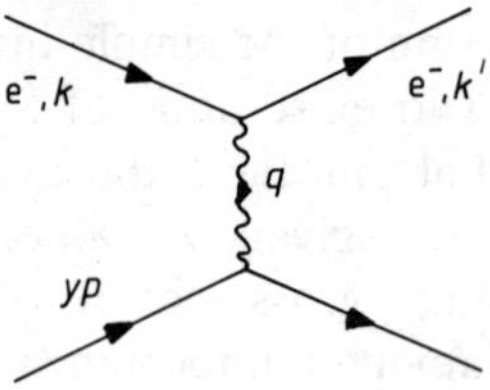

Figure 7.4 Elastic electron–parton scattering.

We first show that the fraction y is none other than Bjorken's variable x. For a parton of type i we write

$$p_i^\mu \approx yp^\mu \tag{7.20}$$

and, roughly speaking†, we can imagine that the partons have mass

$$m_i \approx yM. \tag{7.21}$$

Then, exactly as in (6.155) and (6.156), energy and momentum conservation at the parton vertex, together with the assumption that the struck parton remains on-shell (as indicated by the fact that in figure 7.4 the partons are free), imply that

$$(q + yp)^2 = m_i^2 \tag{7.22}$$

which, using (7.6), (6.161) and (7.21), gives

$$y = Q^2/2M\nu \equiv x. \tag{7.23}$$

Thus the fact that the nucleon structure functions do seem to depend (to good approximation) only on the variable x is interpreted physically as showing that the scattering is dominated by the 'quasi-free' electron–parton process shown in figure 7.4.

What sort of values of x do we expect? Consider an analogous situation—electron scattering from deuterium. Here the target (the deuteron) is undoubtedly composite, and its 'partons' are, to a first approximation, just the two nucleons. Since $m_N \simeq \frac{1}{2}m_D$, we expect to see the value $x \simeq \frac{1}{2}$ (cf (7.21)) favoured; $x = 1$ here would correspond to elastic scattering from the deuteron. A peak at $x \approx \frac{1}{2}$ is indeed observed (figure 7.5) in quasi-elastic e^-d scattering (the broadening of the peak is due to the fact that the constituent nucleons have some motion within the deuteron.) What about the nucleon itself? A simple three-quark model would, on this analogy, lead us to expect a peak at $x \simeq \frac{1}{3}$, but the data already shown (figure 7.2(a)) do not look much like

†Explicit statements about parton transverse momenta and masses, such as those made in equations (7.20) and (7.21), are unnecessary in a rigorous treatment, where such quantities can be shown to give rise to non-leading scaling behaviour (Sachrajda 1983).

that. Perhaps there is something else present too—which we shall uncover as our story proceeds.

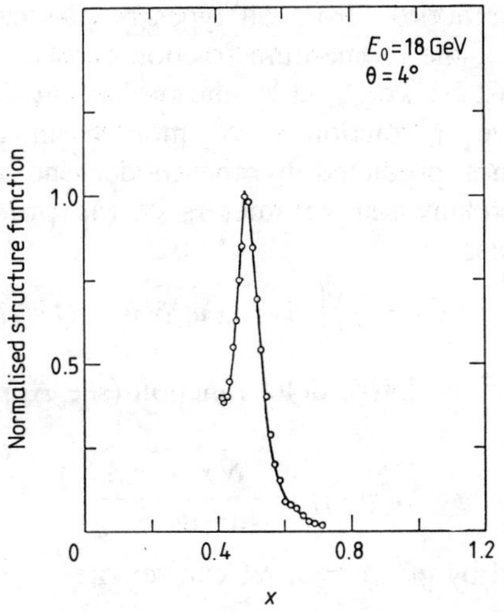

Figure 7.5 Structure function for quasi-elastic e^-d scattering, plotted against x (Attwood 1980).

Certainly it seems sensible to suppose that a nucleon contains *at least* some quarks (and also antiquarks) of the type introduced in the simple composite models of the nucleon (§1.2.3). If quarks are supposed to have spin $\frac{1}{2}$, then the scattering of an electron from a quark or antiquark—generically a *parton*—of type i, charge e_i (in units of e) is just given by the eμ scattering cross section (6.167), with obvious modifications:

$$\frac{d^2\sigma^i}{dQ^2\,d\nu} = \frac{\pi\alpha^2}{4k^2\sin^4(\theta/2)}\frac{1}{kk'}\left(e_i^2\cos^2(\theta/2) + e_i^2\frac{Q^2}{4m_i^2}2\sin^2(\theta/2)\right) \times \delta(\nu - Q^2/2m_i). \tag{7.24}$$

This is to be compared with the general inclusive inelastic cross section formula written in terms of W_1 and W_2:

$$\frac{d^2\sigma}{dQ^2\,d\nu} = \frac{\pi\alpha^2}{4k^2\sin^4(\theta/2)}\frac{1}{kk'}[W_2\cos^2(\theta/2) + W_1 2\sin^2(\theta/2)]. \tag{7.25}$$

Thus the contribution to W_1 and W_2 from one parton of type i is immediately seen to be

$$W_1^i = e_i^2\frac{Q^2}{4M^2x^2}\delta(\nu - Q^2/2Mx) \tag{7.26}$$

$$W_2^i = e_i^2\delta(\nu - Q^2/2Mx) \tag{7.27}$$

where we have set $m_i = xM$. At large ν and Q^2 it is assumed that the contributions from different partons add incoherently in cross section. Thus, to obtain the total contribution from all quark partons, we must sum over the contributions from all types of partons, i, and integrate over all values of x, the momentum fraction carried by the parton. The integral over x must be weighted by the probability $f_i(x)$ for the parton of type i to have a fraction x of momentum. These probability distributions are not predicted by the model and are, in this quark parton picture, fundamental parameters of the proton. The structure function W_2 becomes

$$W_2(\nu, Q^2) = \sum_i \int_0^1 \mathrm{d}x\, f_i(x) e_i^2 \delta(\nu - Q^2/2Mx). \tag{7.28}$$

Using the result for the Dirac delta function (see Appendix A, equation (A.118))

$$\delta(g(x)) = \frac{\delta(x - x_0)}{|\mathrm{d}g/\mathrm{d}x|_{x = x_0}} \tag{7.29}$$

where x_0 is defined by $g(x_0) = 0$, we can rewrite

$$\delta(\nu - Q^2/2Mx) = (x/\nu)\delta(x - Q^2/2M\nu) \tag{7.30}$$

under the x integral. Hence we obtain

$$\nu W_2(\nu, Q^2) = \sum_i e_i^2 x f_i(x) \equiv F_2(x) \tag{7.31}$$

which is the desired scaling behaviour. Similar manipulations lead to

$$MW_1(\nu, Q^2) = F_1(x) \tag{7.32}$$

where

$$2xF_1(x) = F_2(x). \tag{7.33}$$

This relation between F_1 and F_2 is called the Callan–Gross relation (see Callan and Gross 1969): it is a direct consequence of our assumption of spin-$\frac{1}{2}$ quarks. The physical origin of this relation is best discussed in terms of virtual photon total cross sections for transverse ($\lambda = \pm 1$) virtual photons and for a longitudinal/scalar ($\lambda = 0$) virtual photon contribution. The longitudinal/scalar photon is present because $q^2 \neq 0$ for a virtual photon (see §5.6). However, in the discussion of polarisation vectors a slight difference occurs for spacelike q^2. In a frame in which

$$q^\mu = (q^0, 0, 0, q^3) \tag{7.34}$$

the transverse polarisation vectors are as before

$$\varepsilon^\mu(\lambda = \pm 1) = \mp 2^{-1/2}(0, 1, \pm \mathrm{i}, 0) \tag{7.35}$$

with normalisation

$$\varepsilon^2(\lambda = \pm 1) = -1. \tag{7.36}$$

To construct the longitudinal/scalar polarisation vector, we must satisfy

$$q \cdot \varepsilon = 0 \tag{7.37}$$

and so are led to the result

$$\varepsilon^\mu(\lambda = 0) = (1/\sqrt{Q^2})(q^3, 0, 0, q^0) \tag{7.38}$$

with

$$\varepsilon^2(\lambda = 0) = +1. \tag{7.39}$$

The precise definition of a virtual photon cross section is obviously just a convention. It is usually taken to be

$$\sigma_\lambda(\gamma \mathrm{p} \to \mathrm{X}) = (4\pi^2\alpha/K)\varepsilon_\mu^*(\lambda)\varepsilon_\nu(\lambda)W^{\mu\nu} \tag{7.40}$$

by analogy with the total cross section for real photons of polarisation λ incident on an unpolarised proton target. Notice the presence of the factor $W^{\mu\nu}$ defined in (7.2). The factor K is the flux factor: for real photons, producing a final state of mass W, this is just

$$K = (W^2 - M^2)/2M. \tag{7.41}$$

In the so-called 'Hand convention', this same factor is used for virtual photons which produce a final state of mass W. With these definitions we find (see problem 7.3) that the transverse ($\lambda = \pm 1$) photon cross section

$$\sigma_\mathrm{T} = (4\pi^2\alpha/K)\tfrac{1}{2}\sum_{\lambda = \pm 1} \varepsilon_\mu^*(\lambda)\varepsilon_\nu(\lambda)W^{\mu\nu} \tag{7.42}$$

is given by

$$\sigma_\mathrm{T} = (4\pi^2\alpha/K)W_1 \tag{7.43}$$

and the longitudinal/scalar cross section

$$\sigma_\mathrm{S} = (4\pi^2\alpha/K)\varepsilon_\mu^*(\lambda = 0)\varepsilon_\nu(\lambda = 0)W^{\mu\nu} \tag{7.44}$$

by

$$\sigma_\mathrm{S} = (4\pi^2\alpha/K)[(1 + \nu^2/Q^2)W_2 - W_1]. \tag{7.45}$$

In fact these expressions give an intuitive explanation of the positivity properties of W_1 and W_2, namely

$$W_1 \geqslant 0 \tag{7.46}$$

$$(1 + \nu^2/Q^2)W_2 - W_1 \geqslant 0. \tag{7.47}$$

The combination in the $\lambda = 0$ cross section is sometimes denoted by W_L:

$$W_L = (1 + \nu^2/Q^2)W_2 - W_1. \tag{7.48}$$

The scaling limit of these expressions can be taken using

$$\nu W_2 \to F_2 \tag{7.49a}$$

$$MW_1 \to F_1 \tag{7.49b}$$

and $x = Q^2/2M\nu$ finite, as Q^2 and ν grow large. We find

$$\sigma_T \to \frac{4\pi^2\alpha}{MK} F_1(x) \tag{7.50}$$

and

$$\sigma_S \to (4\pi^2\alpha/MK)(1/2x)(F_2 - 2xF_1) \tag{7.51}$$

where we have negelected a term of order F_2/ν in the last expression. Thus the Callan–Gross relation corresponds to the result

$$\sigma_S/\sigma_T \to 0 \tag{7.52}$$

in terms of photon cross sections.

A parton calculation using pointlike spin-0 partons shows the opposite result, namely

$$\sigma_T/\sigma_S \to 0. \tag{7.53}$$

Both these results may be understood by considering the helicities of partons and photons in the so-called parton Breit or 'brick-wall' frame. The particular frame is the one in which the photon and parton are collinear and the 3-momentum of the parton is exactly reversed by the collision (see figure 7.6). In this frame, the photon transfers no energy, only 3-momentum. The vanishing of transverse photon cross sections for scalar partons is now obvious. The transverse photons bring in ± 1 units of the z component of angular momentum: spin-0 partons cannot absorb this. Thus only the scalar $\lambda = 0$ cross section is non-zero. For spin-$\frac{1}{2}$ partons the argument is slightly more complicated in that it depends on the helicity properties of the γ_μ coupling of the parton to the photon.

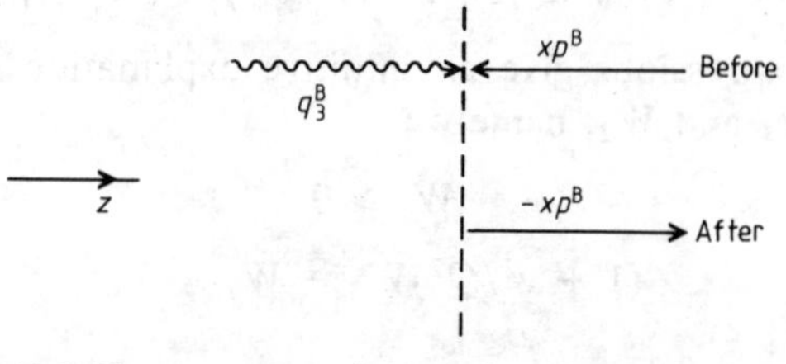

Figure 7.6 Photon–parton interaction in the Breit frame.

As we shall see in Chapter 10, for massless spin-$\frac{1}{2}$ particles the γ_μ coupling conserves helicity—i.e. the projection of spin along the direction of motion of the particle. Thus in the Breit frame, conservation of helicity necessitates a change in the z component of the parton's angular momentum by ± 1 unit, thereby requiring the absorption of a transverse photon (figure 7.7). The Lorentz transformation from the parton Breit frame to the 'laboratory' frame does not affect the ratio of transverse to longitudinal photons, if we neglect parton transverse momenta. These arguments therefore make clear the origin of the Callan–Gross relation.

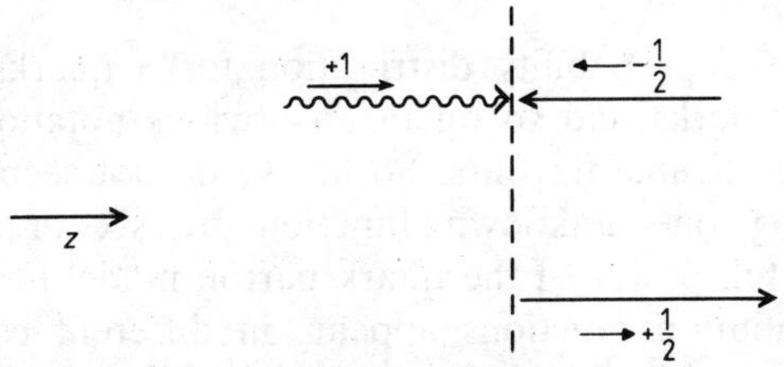

Figure 7.7 Angular momentum balance for absorption of photon by helicity-conserving spin-$\frac{1}{2}$ parton.

Experimentally, the Callan–Gross relation is reasonably well satisfied in that $R = \sigma_S/\sigma_T$ is small for most, if not all, of the deep inelastic regime (figure 7.8). This leads us to suppose that the *electrically charged partons coupling to photons have spin* $\frac{1}{2}$.

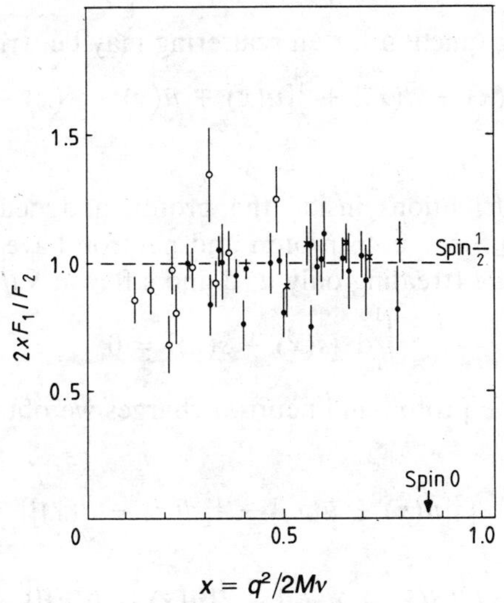

Figure 7.8 The ratio $2xF_1/F_2$: ○, $1.5 < q^2 < 4$ GeV^2; ●, $5 < q^2 < 11$ GeV^2; ×, $12 < q^2 < 16$ GeV^2. (From Perkins D 1987 *Introduction to High Energy Physics* 3rd edn, courtesy Addison-Wesley Publishing Company.)

7.3 The quark parton model

We now proceed a stage further with the idea that the 'partons' *are* quarks (and antiquarks). If we assume that the photon only couples to these objects, we can make more specific scaling predictions. The quantum numbers of the quarks have been given in Chapter 1. For a proton we have the result (cf. (7.31))

$$F_2^{\mathrm{ep}}(x) = x\{\tfrac{4}{9}[u(x) + \bar{u}(x)] + \tfrac{1}{9}[d(x) + \bar{d}(x) + s(x) + \bar{s}(x)] + \ldots\} \tag{7.54}$$

where $u(x)$ is the probability distribution for u quarks in the proton, $\bar{u}(x)$ for u antiquarks and so on in an obvious notation, and the dots indicate further possible flavours. So far we do not seem to have gained much, replacing one unknown function by six or more unknown functions. The full power of the quark parton model lies in the fact that the same distribution functions appear, in different combinations, for neutron targets, and in the analogous scaling functions for deep inelastic scattering with neutrino and antineutrino beams (see §11.2). For electron scattering from neutron targets we can use I-spin invariance (see e.g. Close 1979) to relate the distribution of u and d quarks in a neutron to the distributions in a proton. The result is

$$u^{\mathrm{p}}(x) = d^{\mathrm{n}}(x) \equiv u(x) \tag{7.55}$$

$$d^{\mathrm{p}}(x) = u^{\mathrm{n}}(x) \equiv d(x). \tag{7.56}$$

Hence the scaling function for en scattering may be written

$$F_2^{\mathrm{en}}(x) = x\{\tfrac{4}{9}[d(x) + \bar{d}(x)] + \tfrac{1}{9}[u(x) + \bar{u}(x) + s(x) + \bar{s}(x)] + \ldots\}. \tag{7.57}$$

The quark distributions inside the proton and neutron must satisfy some constraints. Since both proton and neutron have strangeness zero, we have a *sum rule* (treating only u, d and s flavours from now on)

$$\int_0^1 \mathrm{d}x[s(x) - \bar{s}(x)] = 0. \tag{7.58}$$

Similarly, from the proton and neutron charges we obtain two other sum rules:

$$\int_0^1 \mathrm{d}x\{\tfrac{2}{3}[u(x) - \bar{u}(x)] - \tfrac{1}{3}[d(x) - \bar{d}(x)]\} = 1 \tag{7.59a}$$

$$\int_0^1 \mathrm{d}x\{\tfrac{2}{3}[d(x) - \bar{d}(x)] - \tfrac{1}{3}[u(x) - \bar{u}(x)]\} = 0. \tag{7.59b}$$

These are equivalent to the sum rules

$$2 = \int_0^1 dx[u(x) - \bar{u}(x)] \tag{7.60a}$$

$$1 = \int_0^1 dx[d(x) - \bar{d}(x)] \tag{7.60b}$$

which are, of course, just the excess of u and d quarks over antiquarks inside the proton. Testing these sum rules requires neutrino data to separate the various structure functions, as we shall explain in Chapter 11.

One can gain some further insight if one is prepared to make a model. For example, one can introduce the idea of 'valence' quarks (those of the elementary constituent quark model) and 'sea' quarks ($q\bar{q}$ pairs created virtually). Then, in a proton, the u and d quark distributions would be parametrised by the sum of valence and sea contributions

$$u = u_V + q_S \tag{7.61a}$$

$$d = d_V + q_S \tag{7.61b}$$

while the antiquark and strange quark distributions are taken to be pure sea

$$\bar{u} = \bar{d} = s = \bar{s} = q_S \tag{7.62}$$

where we have assumed that the 'sea' is flavour-independent. Such a model replaces the six unknown functions now in play by three, and is consequently more predictive. The strangeness sum rule (7.58) is now satisfied automatically, while (7.60a) and (7.60b) are satisfied by the valence distributions alone:

$$\int_0^1 dx\, u_V(x) = 2 \tag{7.63a}$$

$$\int_0^1 dx\, d_V(x) = 1. \tag{7.63b}$$

It is interesting to consider the ratio F_2^{en}/F_2^{ep}. From (7.54), (7.57) and (7.61) we have

$$F_2^{en}/F_2^{ep} = \frac{4d_V + u_V + \Sigma}{4u_V + d_V + \Sigma} \tag{7.64}$$

where Σ is the total sea contribution. The data are shown in figure 7.9. Independent of any model, since all the probability functions must be positive-definite, we obtain immediately the bounds

$$\tfrac{1}{4} \leqslant F_2^{en}/F_2^{ep} \leqslant 4 \tag{7.65}$$

which are consistent with the data. However, we note that at small x the ratio appears to tend to 1, indicating—from (7.64)—that the sea

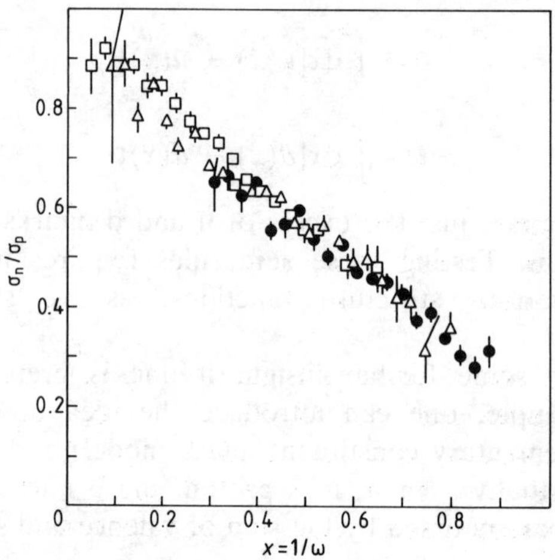

Figure 7.9 The differential cross section σ_n/σ_p versus x (Bodek *et al* 1974). ●, 15°, 19°, 26°, 34°; △, 18°, 26°, 34°, □, 6°, 10°. If, as the data indicate, σ_S/σ_T is the same for n and p, σ_n/σ_p may be interpreted as F_2^{en}/F_2^{ep}.

contribution dominates. This could then explain why the distribution shown in figure 7.2 differed so considerably from our simplest expectation: perhaps it is the sum of two contributions, one peaked roughly at $x \simeq \frac{1}{3}$ (valence) and the other coming in strongly at small x (sea). Such a picture may be at least qualitatively tested by considering the *difference* between the proton and neutron structure functions, from which the sea contribution may be assumed to cancel out. An example of such a distribution is shown in figure 7.10 which does roughly indicate the behaviour expected for the valence quarks. More quantitative separations between quark and antiquark contributions can be made with the help of the neutrino data (cf Chapter 11).

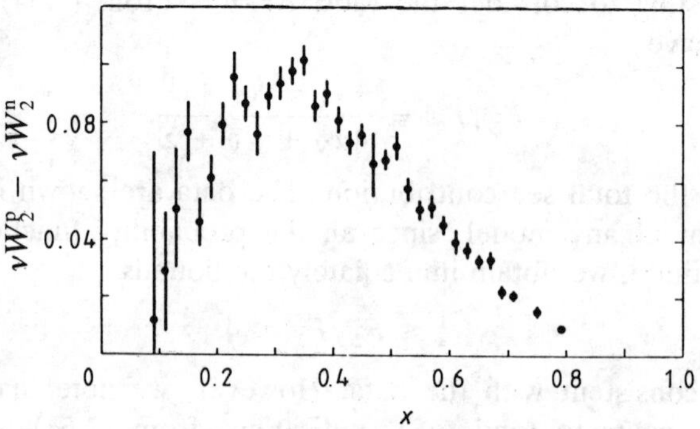

Figure 7.10 $\nu W_2^p - \nu W_2^n$ versus x (Bodek *et al* 1973).

An important sum rule emerges from the picture of $xf_i(x)$ as the fractional momentum carried by quark i. This is the *momentum sum rule*

$$\int_0^1 \mathrm{d}x\, x[u(x) + \bar{u}(x) + d(x) + \bar{d}(x) + s(x) + \bar{s}(x)] = 1 - \varepsilon \quad (7.66)$$

where ε is interpreted as the fraction of the proton momentum that is not carried by quarks. The integral in (7.66) is directly related to ν and $\bar{\nu}$ cross sections (§11.2), and its evaluation implies $\varepsilon \simeq \frac{1}{2}$ (the CHARM (1981) result was $1 - \varepsilon = 0.44 \pm 0.02$). This suggests that about half the total momentum is carried by uncharged objects. These remaining 'partons' will be identified with the gluons of QCD (see Chapter 9).

7.4 The Drell–Yan process

Much of the importance of the parton model lies outside its original domain of deep inelastic lepton scattering. In deep inelastic scattering it is possible to provide a more formal basis for the parton model in terms of light-cone and short-distance operator expansions (Frishman 1974). The advantage of the parton formulation lies in the fact that it suggests other processes for which a parton description may be relevant but for which formal operator arguments are not possible. The most important such process is the Drell–Yan process (Drell and Yan 1970)

$$\mathrm{p} + \mathrm{p} \rightarrow \mu^+\mu^- + \mathrm{X} \quad (7.67)$$

in which a $\mu^+\mu^-$ pair is produced in proton–proton collisions along with unobserved hadrons X, as shown in figure 7.11. The assumption of the parton model is that in the limit

$$s \rightarrow \infty, \quad \text{with } \tau = q^2/s \text{ finite} \quad (7.68)$$

the dominant process is that shown in figure 7.12: a quark and antiquark from different hadrons are assumed to annihilate to a virtual photon which then decays to a $\mu^+\mu^-$ pair (compare figures 7.3 and 7.4), the remaining quarks and antiquarks subsequently emerging as hadrons.

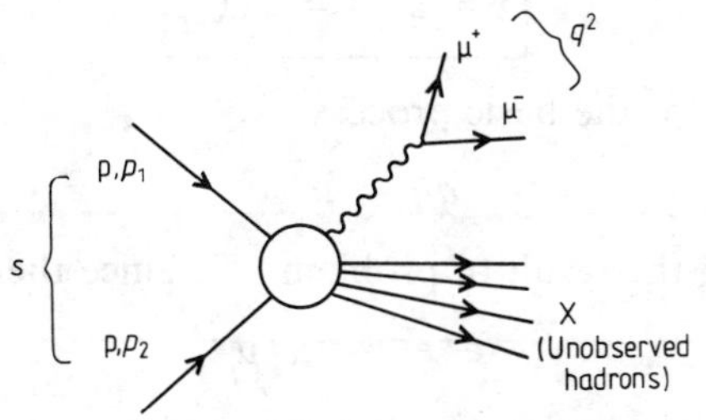

Figure 7.11 Drell–Yan process.

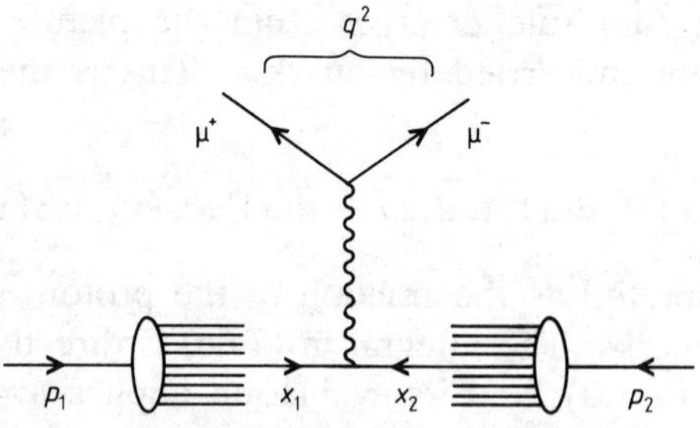

Figure 7.12 Parton model amplitude for Drell–Yan process..

Let us work in the CM system and neglect all masses. In this case we have

$$p_1^\mu = (P, 0, 0, P), \qquad p_2^\mu = (P, 0, 0, -P) \tag{7.69}$$

and

$$s = 4P^2. \tag{7.70}$$

Neglecting quark masses and transverse momenta, we have quark momenta

$$p_{q_1}^\mu = x_1(P, 0, 0, P) \tag{7.71a}$$

$$p_{q_2}^\mu = x_2(P, 0, 0, -P) \tag{7.71b}$$

and the photon momentum

$$q = p_{q_1} + p_{q_2} \tag{7.72}$$

has non-zero components

$$q^0 = (x_1 + x_2)P \tag{7.73a}$$

$$q^3 = (x_1 - x_2)P. \tag{7.73b}$$

Thus we find

$$q^2 = 4x_1x_2P^2 \tag{7.74}$$

and hence

$$\boxed{\tau = q^2/s = x_1x_2.} \tag{7.75}$$

The cross section for the basic process

$$q\bar{q} \rightarrow \mu^+\mu^- \tag{7.76}$$

is calculated using the result of problem 6.7. Since the QED process

$$e^+e^- \rightarrow \mu^+\mu^- \tag{7.77}$$

has the cross section (neglecting all masses)

$$\sigma(e^+e^- \rightarrow \mu^+\mu^-) = 4\pi\alpha^2/3q^2 \tag{7.78}$$

we expect the result for a quark of type a with charge e_a (in units of e) to be

$$\sigma(\mathrm{q}_a\bar{\mathrm{q}}_a \to \mu^+\mu^-) = (4\pi\alpha^2/3q^2)e_a^2. \tag{7.79}$$

To obtain the parton model prediction for proton–proton collisions, one merely multiplies this cross section by the probabilities for finding a quark of type a with momentum fraction x_1, and an antiquark of the same type with fraction x_2, namely

$$q_a(x_1)\mathrm{d}x_1\bar{q}_a(x_2)\mathrm{d}x_2. \tag{7.80}$$

There is of course another contribution for which the antiquark has fraction x_1 and the quark x_2:

$$\bar{q}_a(x_1)\mathrm{d}x_1 q_a(x_2)\mathrm{d}x_2. \tag{7.81}$$

Thus the Drell–Yan prediction is

$$\boxed{\begin{aligned}&\mathrm{d}^2\sigma(\mathrm{pp} \to \mu^+\mu^- + \mathrm{X})\\ &\quad = \frac{4\pi\alpha^2}{9q^2}\sum_a e_a^2[q_a(x_1)\bar{q}_a(x_2) + \bar{q}_a(x_1)q_a(x_2)]\mathrm{d}x_1\mathrm{d}x_2\end{aligned}} \tag{7.82}$$

where we have included a factor $\frac{1}{3}$ to account for the *colour* of the quarks (see Chapter 9): in order to make a colour singlet photon, one needs to match the colours of quark and antiquark. Equation (7.82) is the master formula. Its importance lies in the fact that the *same* quark distribution functions are measured in deep inelastic lepton scattering so one can make absolute predictions. For example, if the photon in figure 7.12 is replaced by a W(Z), one can predict W(Z) production cross sections, as we shall see in Chapter 14.

We would expect some 'scaling' property to hold for this cross section, following from the pointlike constituent cross section (7.79). One way to exhibit this is to use the variables q^2 and $x_\mathrm{F} = x_1 - x_2$ as discussed in problem 7.4. There it is shown that the dimensionless quantity

$$q^4\frac{\mathrm{d}^2\sigma}{\mathrm{d}q^2\,\mathrm{d}x_\mathrm{F}} \tag{7.83}$$

should be a function of x_F and the *ratio* $\tau = q^2/s$. The data bear out this prediction well—see figure 7.13.

Furthermore, the assumption that the lepton pair is produced via quark–antiquark annihilation to a virtual photon can be checked by observing the angular distribution of either lepton in the dilepton rest frame, relative to the incident proton beam direction. This distribution is expected to be the same as in $\mathrm{e}^+\mathrm{e}^- \to \mu^+\mu^-$, namely (cf (6.135))

$$\mathrm{d}\sigma/\mathrm{d}\Omega \propto (1 + \cos^2\theta) \tag{7.84}$$

as is indeed observed (figure 7.14). Note that figure 7.14 provides evidence that the quarks have spin $\frac{1}{2}$: if they are assumed to have spin 0, the angular distribution would be (see problem 7.5) proportional to $(1 - \cos^2\theta)$, and this is clearly ruled out.

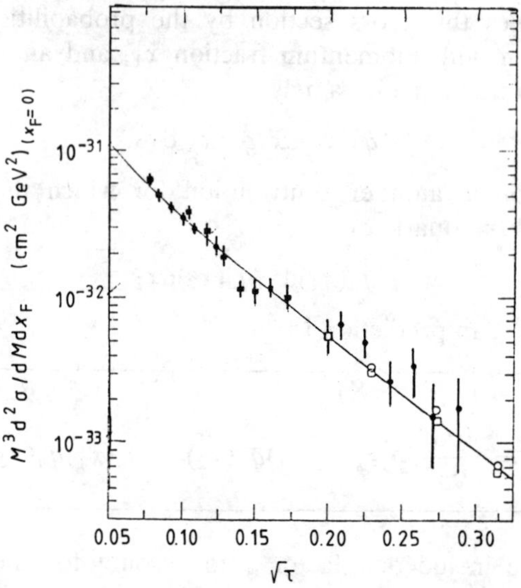

Figure 7.13 The dimensionless cross section $M^3 d^2\sigma/dM\, dx_F$ ($M = \sqrt{q^2}$) at $x_F = 0$ for pN scattering, plotted against $\sqrt{\tau} = M/\sqrt{s}$ (Scott 1985): ●, $\sqrt{s} = 62$ GeV; ■, 44; □, 27.4; ○, 23.8.

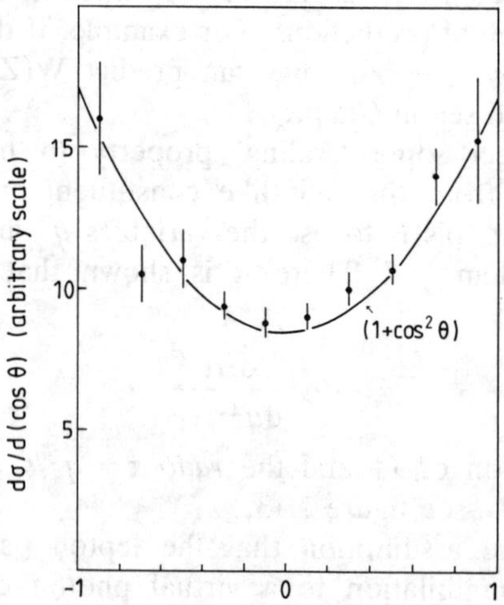

Figure 7.14 Angular distribution of muons, measured in the $\mu^+\mu^-$ rest frame, relative to the incident beam direction, in the Drell–Yan process (From Perkins D 1987 *Introduction to High Energy Physics* 3rd edn, courtesy Addison-Wesley Publishing Company.)

7.5 e^+e^- annihilation into hadrons

The last electromagnetic process we wish to consider is electron–positron annihilation into hadrons (figure 7.15):

$$e^+e^- \to X. \tag{7.85}$$

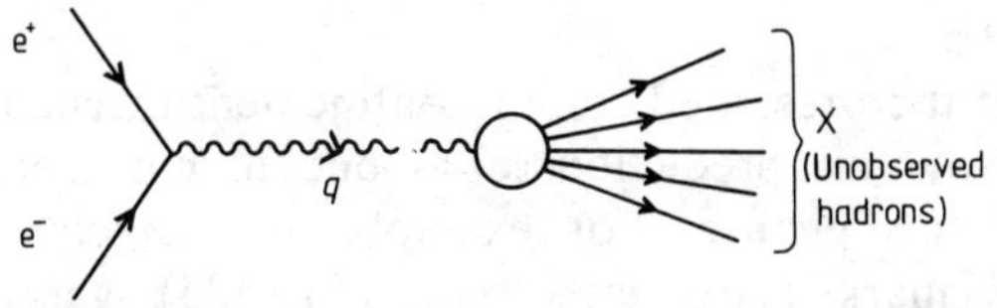

Figure 7.15 e^+e^- annihilation to hadrons, in one-photon approximation.

As usual, the dominance of the one-photon intermediate state is assumed. How does the virtual photon convert into hadrons? In terms of a parton picture, this process is similar to the Drell–Yan process

$$pp \to \mu^+\mu^- + X \tag{7.86}$$

and it is natural to imagine that at large q^2 the basic subprocess is quark–antiquark pair creation (figure 7.16). The total cross section for $q\bar{q}$ pair production is then (cf (7.79))

$$\sigma(e^+e^- \to q_a\bar{q}_a) = (4\pi\alpha^2/3q^2)e_a^2. \tag{7.87}$$

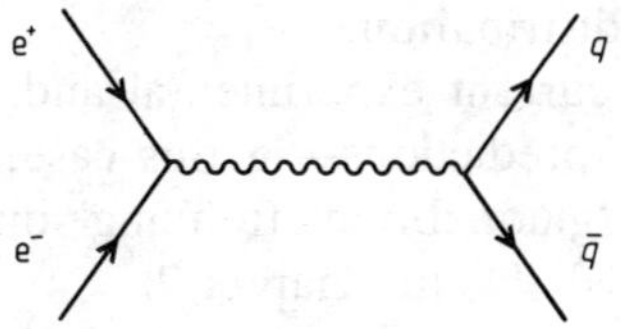

Figure 7.16 Parton model subprocess in $e^+e^- \to$ hadrons.

The produced quark and antiquark must then 'dress' themselves as ordinary, colourless hadrons (see Chapter 9), and it is assumed that this dressing process does not affect the cross section. Thus, in the quark parton model, we arrive at the result that at large q^2

$$\sigma(e^+e^- \to \text{hadrons}) = (4\pi\alpha^2/3q^2)\sum e_a^2. \tag{7.88}$$

This result is usually given in terms of the ratio

$$R = \frac{\sigma(e^+e^- \to \text{hadrons})}{(e^+e^- \to \mu^+\mu^-)} = \sum_a e_a^2. \tag{7.89}$$

For the light quarks u, d and s occurring in three colours, we therefore predict

$$R = 3[(\tfrac{2}{3})^2 + (-\tfrac{1}{3})^2 + (-\tfrac{1}{3})^2] = 2; \tag{7.90}$$

above the c threshold but below the b threshold we expect $R = \frac{10}{3}$, and above the b threshold $R = \frac{11}{3}$. These expectations are in reasonable accord with experiment, especially at energies well beyond the resonance region and the b threshold, as figure 7.17 shows. The important subject of QCD corrections to the parton model result (7.89) will be discussed in §9.5.

The fact that the presumed colour confinement mechanism (see §15.3) does not destroy this prediction leads one to consider more detailed predictions of this picture. For example, the angular distribution of massless spin-$\frac{1}{2}$ quarks is expected to be (cf (6.135) again)

$$\mathrm{d}\sigma/\mathrm{d}\Omega = (\alpha^2/4q^2)e_a^2(1 + \cos^2\theta) \tag{7.91}$$

just as for the $\mu^+\mu^-$ process. However, in this case there is an important difference: the quarks are not observed! Nevertheless a remarkable 'memory' of (7.91) is retained by the observed final state hadrons. Experimentally one observes events in which hadrons emerge from the interaction region in two relatively well collimated cones, or 'jets'—see figure 7.18 (already shown in §1.2.4 as figure 1.16). The distribution of events as a function of the (inferred) angle of the jet axis is shown in figure 7.19, and is in good agreement with (7.91). The interpretation is that the primary process is $e^+e^- \to q\bar{q}$, the quark and the antiquark then turning into hadrons as they separate and experience the very strong colour forces (Chapter 9), but without losing the memory of the original quark angular distribution.

Again, much of the current experimental and theoretical interest lies in corrections to these predictions—in this case, to three- and four-jet events attributable to 'gluon bremsstrahlung' diagrams. We provide a brief glimpse of these delights in Chapter 9.

Summary

Kinematics of inelastic electron scattering.

Definition of the nucleon structure functions.

Scaling and partons.

The parton model predictions for the structure functions.

Callan–Gross relation.

The quark parton model: sum rules.

The Drell–Yan process in the parton model, scaling.

$e^+e^- \to$ hadrons; jets.

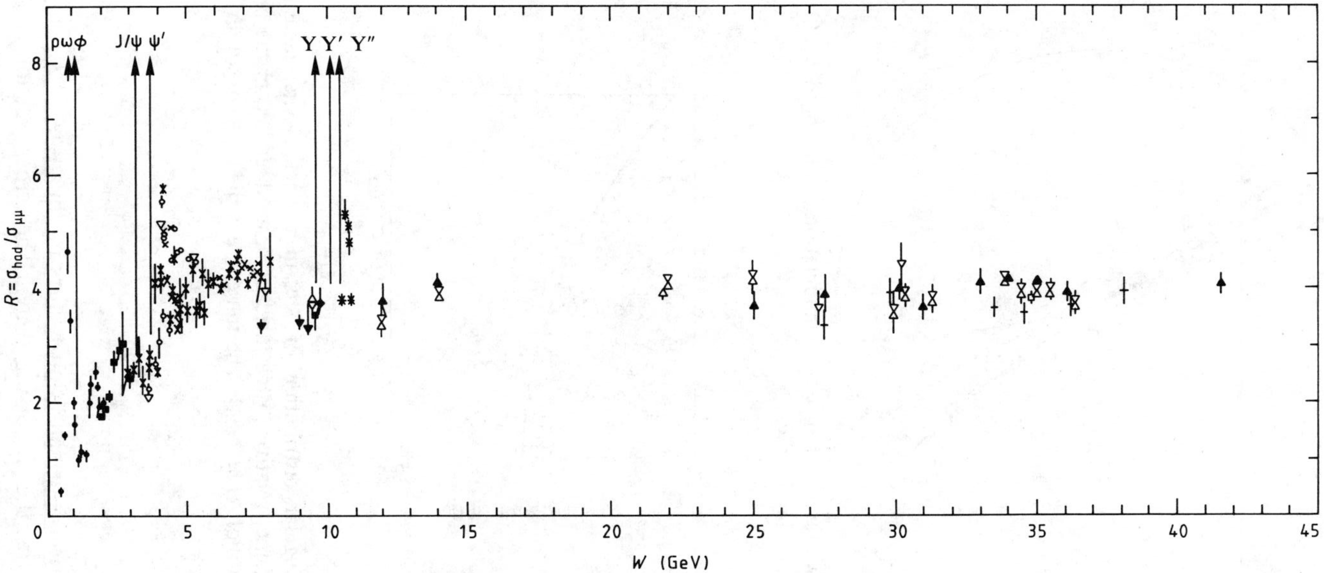

Figure 7.17 The ratio R (see equation (7.89)) (courtesy R Cashmore).

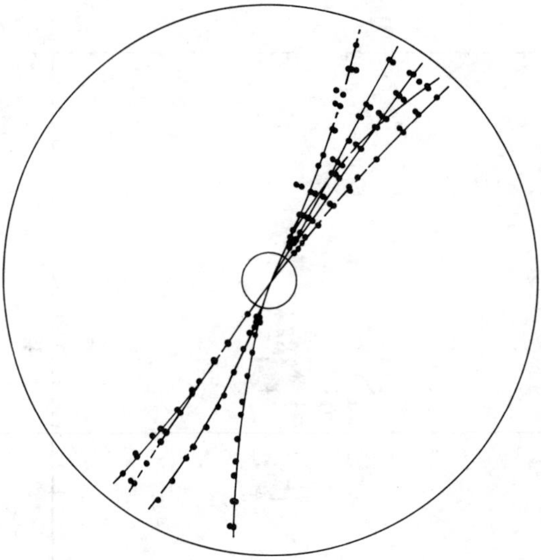

Figure 7.18 Two-jet event in e^+e^- annihilation, from the TASSO detector at the e^+e^- storage ring PETRA.

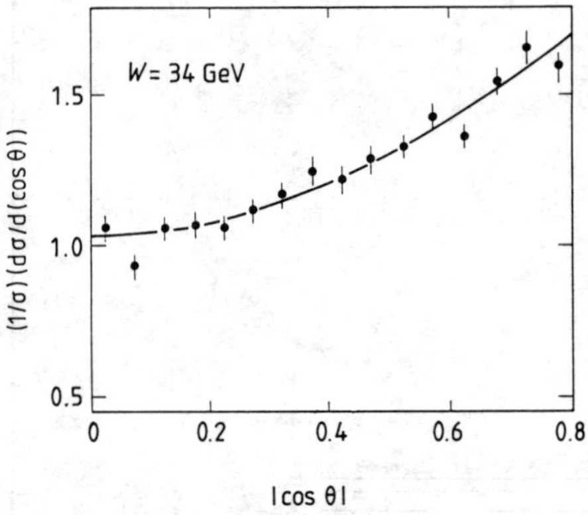

Figure 7.19 Angular distribution of jets in two-jet events, measured in the two-jet rest frame, relative to the incident beam direction, in the process $e^+e^- \rightarrow$ two jets (Althoff *et al* 1984). The full curve is the $(1 + \cos^2 \theta)$ distribution.

Problems

7.1 The various normalisation factors in equations (7.2) and (7.9) may be checked in the following way. The cross section for inclusive electron–proton scattering may be written (equation (7.9))

$$\mathrm{d}\sigma = \left(\frac{4\pi\alpha}{q^2}\right)^2 \frac{1}{4[(k\cdot p)^2 - m^2M^2]^{1/2}} 4\pi M L_{\mu\nu} W^{\mu\nu} \frac{\mathrm{d}^3k'}{2\omega'(2\pi)^3}$$

in the usual one-photon exchange approximation, and the tensor $W^{\mu\nu}$ is related to hadronic matrix elements of the electromagnetic current operator by equation (7.2):

$$e^2 W^{\mu\nu}(q, p) = \frac{V}{4\pi M}\tfrac{1}{2}\sum_s\sum_{\mathrm{X}}\langle \mathrm{p};\, p,\, s|\hat{j}^{\mu}_{\mathrm{em}}(0)|\mathrm{X};\, p'\rangle$$
$$\times\ \langle \mathrm{X};\, p'|\hat{j}^{\nu}_{\mathrm{em}}(0)|\mathrm{p};\, p,\, s\rangle(2\pi)^4\,\delta^4(p + q - p')$$

where the sum X is over all possible hadronic final states. If we consider the special case of elastic scattering, the sum over X is only over the final proton's degrees of freedom:

$$e^2 W^{\mu\nu}_{\mathrm{el}} = \frac{V}{4\pi M}\tfrac{1}{2}\sum_s\sum_{s'}\langle \mathrm{p};\, p,\, s|\hat{j}^{\mu}_{\mathrm{em}}(0)|\mathrm{p};\, p',\, s'\rangle\langle \mathrm{p};\, p',\, s'|\hat{j}^{\nu}_{\mathrm{em}}(0)|\mathrm{p};\, p,\, s\rangle$$
$$\times\ (2\pi)^4\,\delta^4(p + q - p')\frac{V}{(2\pi)^3}\frac{\mathrm{d}^3p'}{2E'}.$$

Now use equation (6.137) for the electromagnetic current matrix elements of a 'point' proton to show that the resulting cross section is identical to that for elastic eμ scattering.

7.2 (a) Perform the contraction $L_{\mu\nu}W^{\mu\nu}$ for inclusive inelastic electron–proton scattering (remember $q^{\mu}L_{\mu\nu} = q^{\nu}L_{\mu\nu} = 0$). Hence verify that the inclusive differential cross section in terms of 'laboratory' variables, and neglecting the electron mass, has the form

$$\frac{\mathrm{d}^2\sigma}{\mathrm{d}\Omega\,\mathrm{d}k'} = \frac{\alpha^2}{4k^2\sin^4(\theta/2)}[W_2\cos^2(\theta/2) + W_1 2\sin^2(\theta/2)].$$

(b) By calculating the Jacobian

$$J = \begin{vmatrix} \partial u/\partial x & \partial u/\partial y \\ \partial v/\partial x & \partial v/\partial y \end{vmatrix}$$

for a change of variables $(x, y) \rightarrow (u, v)$

$$\mathrm{d}u\,\mathrm{d}v = |J|\mathrm{d}x\,\mathrm{d}y$$

find expressions for $\mathrm{d}^2\sigma/\mathrm{d}Q^2\,\mathrm{d}\nu$ and $\mathrm{d}^2\sigma/\mathrm{d}x\,\mathrm{d}y$, where Q^2 and ν have their usual significance, and x is the scaling variable $Q^2/2M\nu$ and $y = \nu/k$.

7.3 Consider the description of inelastic electron–proton scattering in terms of virtual photon cross sections:

(a) In the 'laboratory' frame with

$$p^{\mu} = (M, 0, 0, 0)$$

and

$$q^\mu = (q^0, 0, 0, q^3)$$

evaluate the transverse spin sum

$$\tfrac{1}{2}\sum_{\lambda=\pm 1} \varepsilon_\mu(\lambda)\varepsilon_\nu^*(\lambda)W^{\mu\nu}.$$

Hence show that the 'Hand' cross section for transverse virtual photons is

$$\sigma_{\mathrm{T}} = (4\pi^2\alpha/K)W_1.$$

(b) Using the definition

$$\varepsilon_{\mathrm{S}}^\mu = (1/\sqrt{Q^2})(q^3, 0, 0, q^0)$$

and rewriting this in terms of the 'laboratory' 4-vectors p^μ and q^μ, evaluate the longitudinal/scalar virtual photon cross section. Hence show that

$$W_2 = \frac{K}{4\pi^2\alpha}\frac{Q^2}{Q^2 + \nu^2}(\sigma_{\mathrm{S}} + \sigma_{\mathrm{T}}).$$

7.4 (a) By the expedient of inserting a δ function, the differential cross section for Drell–Yan production of a lepton pair of mass $\sqrt{q^2}$ may be written as

$$\frac{\mathrm{d}\sigma}{\mathrm{d}q^2} = \int \mathrm{d}x_1\,\mathrm{d}x_2 \frac{\mathrm{d}^2\sigma}{\mathrm{d}x_1\,\mathrm{d}x_2}\delta(q^2 - sx_1x_2).$$

Show that this is equivalent to the form

$$\frac{\mathrm{d}\sigma}{\mathrm{d}q^2} = \frac{4\pi\alpha^2}{9q^4}\int \mathrm{d}x_1\,\mathrm{d}x_2\,x_1x_2\delta(x_1x_2 - \tau)$$

$$\times \sum_a e_a^2[q_a(x_1)\bar{q}_a(x_2) + \bar{q}_a(x_1)q_a(x_2)]$$

which, since $q^2 = s\tau$, exhibits a scaling law of the form

$$s^2\,\mathrm{d}\sigma/\mathrm{d}q^2 = F(\tau).$$

(b) Introduce the Feynman scaling variable

$$x_{\mathrm{F}} = x_1 - x_2$$

with

$$q^2 = sx_1x_2$$

and show that

$$\mathrm{d}q^2\,\mathrm{d}x_{\mathrm{F}} = (x_1 + x_2)s\,\mathrm{d}x_1\,\mathrm{d}x_2.$$

Hence show that the Drell–Yan formula can be rewritten as

$$\frac{d^2\sigma}{dq^2\,dx_F} = \frac{4\pi\alpha^2}{9q^4}\frac{\tau}{(x_F^2 + 4\tau)^{1/2}}\sum_a e_a^2[q_a(x_1)\bar{q}_a(x_2) + \bar{q}_a(x_1)q_a(x_2)].$$

7.5 Verify that if the quarks participating in the Drell–Yan subprocess $q\bar{q} \to \gamma \to \mu\bar{\mu}$ had spin 0, the CM angular distribution of the final $\mu^+\mu^-$ pair would be proportional to $(1 - \cos^2\theta)$.

PART III

NON-ABELIAN GAUGE THEORY AND QCD

. . . The difference between a neutron and a proton is then a purely arbitrary process. As usually conceived, however, this arbitrariness is subject to the following limitation: once one chooses what to call a proton, what a neutron, at one space–time point, one is then not free to make any choices at other space–time points.

It seems that this is not consistent with the localized field concept that underlies the usual physical theories. In the present paper we wish to explore the possibility of requiring all interactions to be invariant under *independent* rotations of the isotopic spin at all space–time points . . ..

Yang and Mills (1954)

8

NON-ABELIAN GAUGE THEORY

In the preceding chapters, a complete dynamical theory—QED—has been introduced, based on a remarkably simple basic principle: the invariance of the theory under local phase transformations on the wavefunctions (Chapter 2) or field operators (Chapter 4) of charged particles. Such transformations were characterised as Abelian in §2.6(iii), since the phase factors commuted. The remainder of this book is going to be concerned with the formulation and elementary application of theories which are built on a generalisation of this principle, in which the 'phase factors' involve matrix operators which do not in general commute with each other. Theories based on making such invariances local are called non-Abelian gauge theories. In this chapter we present an introduction to this type of theory, beginning with the simpler case of global (not local) non-Abelian phase invariance. The first such symmetry to be used in particle physics was the hadronic isospin 'SU(2) symmetry', proposed in the context of nuclear physics by Heisenberg (1932), and now seen as following from the near equality of the u and d quark masses, and the flavour independence of the QCD forces (see the following chapter) between them. This $SU(2)_f$ flavour symmetry was extended to $SU(3)_f$ by Gell-Mann (1961) and Ne'eman (1961)—an extension seen, in its turn, as reflecting the rough equality of the u, d and s quark masses, together with flavour independence of QCD.

After considering these global non-Abelian flavour symmetries, we will turn to the local symmetry case. As the quotation from the Yang and Mills (1954) paper shows, the whole modern development of non-Abelian gauge theories began with their attempt (see also Shaw 1955) to make hadronic isospin into a local symmetry. However, the beautiful formalism developed by these authors turned out *not* to describe interactions between hadrons. Instead, it describes the interactions between the *constituents* of the hadrons, namely quarks—and this in two respects. First, a local SU(2) symmetry (called 'weak isospin') governs the weak interactions of quarks (and leptons); and second, a local SU(3) symmetry (called '$SU(3)_c$') governs the strong interactions of quarks. It is important to realise that, despite the fact that each of these two local symmetries is based on the same group as one of the earlier global (flavour) symmetries, the physics involved is completely different. In the weak interaction case, since the group is an SU(2), it is natural to use 'isospin language' in talking about it; but we must always remember that it is *weak* isospin, which (as we shall see in Chapters 13 and 14) is an attribute of leptons as well as of quarks, and hence physically quite distinct from hadronic isospin. In the case of the strong quark interactions, the $SU(3)_c$ group refers to a new degree of freedom ('colour'), which is quite distinct from flavour u, d, s (see Chapter 9).

8.1 Non-Abelian global phase invariance: internal symmetry multiplets

The transformations initially considered in connection with the gauge principle in §2.5 were just global phase transformations on a single wavefunction

$$\psi' = \mathrm{e}^{\mathrm{i}\alpha}\psi. \tag{8.1}$$

The generalisation to non-Abelian invariances comes when we take the simple step—but one with many ramifications—of considering more than one wavefunction, or state, at a time. Quite generally in quantum mechanics, we know that whenever we have a set of states which are *degenerate* in energy (or mass) there is no unique way of specifying the states: any linear combination of some initially chosen set of states will do just as well, provided the normalisation conditions on the states are still satisfied. Consider, for example, the simplest case of just two such states—to be specific, the neutron and proton (figure 8.1). This single near coincidence of the masses was enough to suggest to Heisenberg (1932) that, as far as the strong nuclear forces were concerned (electromagnetism being negligible by comparison), the two states could be regarded as truly degenerate, so that any arbitrary linear combination of neutron and proton wavefunctions would be entirely equivalent, as far as this force was concerned, for a single 'neutron' or single 'proton' wavefunction. Thus redefinitions of neutron and proton wavefunctions could be allowed, of the form

$$\psi_\mathrm{p} \to \psi'_\mathrm{p} = \alpha\psi_\mathrm{p} + \beta\psi_\mathrm{n} \tag{8.2a}$$

$$\psi_\mathrm{n} \to \psi'_\mathrm{n} = \gamma\psi_\mathrm{p} + \delta\psi_\mathrm{n} \tag{8.2b}$$

for complex coefficients α, β, γ, δ. In particular, since ψ_p and ψ_n are degenerate, we have

$$H\psi_\mathrm{p} = E\psi_\mathrm{p}, \qquad H\psi_\mathrm{n} = E\psi_\mathrm{n} \tag{8.3}$$

from which it follows that

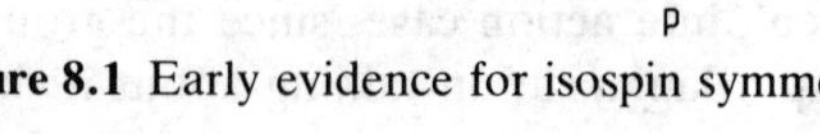

Figure 8.1 Early evidence for isospin symmetry.

$$H\psi'_{\text{p}} = H(\alpha\psi_{\text{p}} + \beta\psi_{\text{n}}) = \alpha H\psi_{\text{p}} + \beta H\psi_{\text{n}}$$
$$= E(\alpha\psi_{\text{p}} + \beta\psi_{\text{n}}) = E\psi'_{\text{p}} \qquad (8.4a)$$

and similarly

$$H\psi'_{\text{n}} = E\psi'_{\text{n}} \qquad (8.4b)$$

showing that the redefined wavefunctions still describe two states with the same energy degeneracy.

The two-fold degeneracy seen in figure 8.1 is suggestive of that found in spin-$\frac{1}{2}$ systems in the absence of any magnetic field; the $s_z = \pm\frac{1}{2}$ components are degenerate. The analogy can be brought out by introducing the *two-component nucleon isospinor*

$$\psi^{(1/2)} \equiv \begin{pmatrix}\psi_{\text{p}}\\ \psi_{\text{n}}\end{pmatrix} \equiv \psi_{\text{p}}\chi_{\text{p}}^{(1/2)} + \psi_{\text{n}}\chi_{\text{n}}^{(1/2)} \qquad (8.5)$$

where

$$\chi_{\text{p}}^{(1/2)} = \begin{pmatrix}1\\0\end{pmatrix}, \qquad \chi_{\text{n}}^{(1/2)} = \begin{pmatrix}0\\1\end{pmatrix}. \qquad (8.6)$$

In $\psi^{(1/2)}$, ψ_{p} is the amplitude for the nucleon to have 'isospin up' (cf (8.6)), and ψ_{n} that for it to have 'isospin down'. We shall see what the 'spin operators' are in a moment.

Equations (8.2a) and (8.2b) can be compactly written in terms of $\psi^{(1/2)}$ as

$$\psi^{(1/2)} \rightarrow \psi^{(1/2)\prime} = \mathbf{U}\psi^{(1/2)} \qquad (8.7)$$

where $\mathbf{U}$ is a complex 2×2 matrix. Heisenberg's proposal, then, was that the physics of strong interactions remained the same under the transformation (8.7): in other words, a symmetry was involved. We must emphasise that such a symmetry can *only* be exact in the *absence* of electromagnetic interactions: it is therefore an intrinsically approximate symmetry.

What are the restrictions on the 2×2 matrix $\mathbf{U}$ in the transformation? We have, explicitly,

$$\begin{pmatrix}\psi'_{\text{p}}\\ \psi'_{\text{n}}\end{pmatrix} = \mathbf{U}\begin{pmatrix}\psi_{\text{p}}\\ \psi_{\text{n}}\end{pmatrix}. \qquad (8.8)$$

To preserve the normalisation (probability), we require $\mathbf{U}$ to be unitary:

$$\mathbf{U}\mathbf{U}^{\dagger} = \mathbf{U}^{\dagger}\mathbf{U} = \mathbf{1}. \qquad (8.9)$$

Clearly this property is in no way restricted to the case of two states—the transformation coefficients for n degenerate states will form the entries of an $n \times n$ unitary matrix. A trivialisation is the case $n = 1$, for which, as we noted in §2.6, $\mathbf{U}$ reduces to a single phase factor, indicating how all the previous work is going to be contained as a special case of these more general transformations. Indeed, from elementary properties of determinants we have

$$\det \mathbf{U}\mathbf{U}^{\dagger} = \det \mathbf{U}\cdot\det \mathbf{U}^{\dagger} = \det \mathbf{U}\cdot\det \mathbf{U}^{*} = |\det \mathbf{U}|^2 = 1 \quad (8.10)$$

so that

$$\det \mathbf{U} = \exp(\mathrm{i}\theta) \quad (8.11)$$

where θ is a real number. We can separate off such an overall phase factor from the transformations mixing 'p' and 'n', because it corresponds to a rotation of the phase of both p and n wavefunctions by the *same* amount. Invariance of the theory under such a transformation corresponds physically to the conservation law of total number of p's and n's—as we saw in §4.3 in the field theory case.

The new physics will lie in the remaining transformations which satisfy

$$\det \mathbf{U} = +1. \quad (8.12)$$

Such a matrix is said to be a *special* unitary matrix—which simply means it has unit determinant. The set of all such matrices $\mathbf{U}$ constitutes the Lie group SU(2)—the group of special, unitary 2×2 matrices. Lie groups have the important property that their physical consequences may be found by considering 'infinitesimal' transformations, i.e. matrices $\mathbf{U}$ which differ only slightly from the 'no-change' situation corresponding to $\mathbf{U} = \mathbf{1}$. We may therefore write

$$\mathbf{U} = \mathbf{1} + \mathrm{i}\xi \quad (8.13)$$

where ξ is a 2×2 matrix whose entries are all first-order small quantities. The condition $\det \mathbf{U} = 1$ now reduces, on neglect of second-order terms $O(\xi^2)$, to the condition

$$\mathrm{Tr}\,\xi = 0. \quad (8.14)$$

The condition that $\mathbf{U}$ be unitary, i.e.

$$(\mathbf{1} + \mathrm{i}\xi)(\mathbf{1} - \mathrm{i}\xi^{\dagger}) = \mathbf{1} \quad (8.15)$$

similarly reduces (in first order) to the condition

$$\xi = \xi^{\dagger}. \quad (8.16)$$

Thus ξ is a 2×2 traceless Hermitian matrix. Such a matrix involves three independent parameters, as can be seen by counting the restrictions implied by $\mathrm{Tr}\,\xi = 0$ and $\xi = \xi^{\dagger}$, and can be written quite generally as

$$\xi = \boldsymbol{\varepsilon}\cdot\boldsymbol{\tau}/2 \quad (8.17)$$

where $\boldsymbol{\varepsilon}$ stands for the three quantities

$$\boldsymbol{\varepsilon} = (\varepsilon_1, \varepsilon_2, \varepsilon_3) \tag{8.18}$$

which are all first-order small, and the three matrices $\boldsymbol{\tau}$ are just the familiar Pauli matrices

$$\tau_1 = \begin{pmatrix} 0 & 1 \\ 1 & 0 \end{pmatrix}, \qquad \tau_2 = \begin{pmatrix} 0 & -\mathrm{i} \\ \mathrm{i} & 0 \end{pmatrix}, \qquad \tau_3 = \begin{pmatrix} 1 & 0 \\ 0 & -1 \end{pmatrix}. \tag{8.19}$$

The matrices $\tau/2$ obviously satisfy the same commutation relations as the angular momentum operators of a spin-$\frac{1}{2}$ particle, namely

$$[\tau_i/2, \tau_j/2] = \mathrm{i}\varepsilon_{ijk}\tau_k/2 \tag{8.20}$$

and the basic wavefunctions $\chi_{\mathrm{p}}^{(1/2)}$, $\chi_{\mathrm{n}}^{(1/2)}$ are eigenfunctions of total 'spin' and third-component of 'spin' according to

$$\tfrac{1}{4}\boldsymbol{\tau}^2\chi_{\mathrm{p}}^{(1/2)} = \tfrac{3}{4}\chi_{\mathrm{p}}^{(1/2)}, \qquad \tfrac{1}{2}\tau_3\chi_{\mathrm{p}}^{(1/2)} = \tfrac{1}{2}\chi_{\mathrm{p}}^{(1/2)} \tag{8.21a}$$

$$\tfrac{1}{4}\boldsymbol{\tau}^2\chi_{\mathrm{n}}^{(1/2)} = \tfrac{3}{4}\chi_{\mathrm{n}}^{(1/2)}, \qquad \tfrac{1}{2}\tau_3\chi_{\mathrm{n}}^{(1/2)} = -\tfrac{1}{2}\chi_{\mathrm{n}}^{(1/2)}. \tag{8.21b}$$

This two-fold degree of freedom in the 'charge space' of n and p is called *isospin*.

An infinitesimal SU(2) transformation of the p–n doublet is therefore specified by

$$\begin{pmatrix} \psi'_{\mathrm{p}} \\ \psi'_{\mathrm{n}} \end{pmatrix} = (\mathbf{1} + \mathrm{i}\boldsymbol{\varepsilon}\cdot\boldsymbol{\tau}/2)\begin{pmatrix} \psi_{\mathrm{p}} \\ \psi_{\mathrm{n}} \end{pmatrix}. \tag{8.22}$$

The form for a finite SU(2) transformation $\mathbf{U}$ may then be obtained from the infinitesimal form using the result

$$\mathrm{e}^A = \lim_{n\to\infty}(1 + A/n)^n \tag{8.23}$$

generalised to matrices. Let $\boldsymbol{\varepsilon} = \boldsymbol{\alpha}/n$ apply the infinitesimal transformation n times, and let n tend to infinity. We obtain

$$\mathbf{U} = \exp(\mathrm{i}\boldsymbol{\alpha}\cdot\boldsymbol{\tau}/2) \tag{8.24}$$

so that

$$\psi^{(1/2)\prime} \equiv \begin{pmatrix} \psi'_{\mathrm{p}} \\ \psi'_{\mathrm{n}} \end{pmatrix} = \exp(\mathrm{i}\boldsymbol{\alpha}\cdot\boldsymbol{\tau}/2)\begin{pmatrix} \psi_{\mathrm{p}} \\ \psi_{\mathrm{n}} \end{pmatrix} \equiv \exp(\mathrm{i}\boldsymbol{\alpha}\cdot\boldsymbol{\tau}/2)\psi^{(1/2)}. \tag{8.25}$$

In this form, it is clear that our 2×2 transformation is a generalisation of the global phase transformation of (8.1), except that:

(a) there are now *three* 'phase angles' $\boldsymbol{\alpha}$;
(b) there are non-commuting matrix operators (the $\boldsymbol{\tau}$'s) appearing in the exponent.

The last fact is the reason for the description 'non-Abelian' phase invariance. As the commutation relations for the $\boldsymbol{\tau}$ matrices show, SU(2) is a non-Abelian group in that two SU(2) transformations do not in general commute. By contrast, in the case of electric charge or particle number, successive transformations clearly commute: this corresponds to an Abelian phase invariance and, as noted in §2.6, to an Abelian U(1) group.

We must now ask what are the physical consequences of the assumption that the transformation (8.25) is an invariance, or symmetry, of the strong nuclear force. We expect there to be some associated conservation law or laws. For these single-particle states it is very easy to see what the conserved quantities are. Returning to (8.3) and (8.4), we may write them as

$$H\psi^{(1/2)} = E\psi^{(1/2)} \tag{8.26}$$

and

$$H\psi^{(1/2)\prime} = E\psi^{(1/2)\prime}. \tag{8.27}$$

But we also have

$$\psi^{(1/2)\prime} = \mathbf{U}\psi^{(1/2)}. \tag{8.28}$$

Thus

$$\begin{aligned} H\mathbf{U}\psi^{(1/2)} = H\psi^{(1/2)\prime} = E\psi^{(1/2)\prime} &= E\mathbf{U}\psi^{(1/2)} \\ &= \mathbf{U}E\psi^{(1/2)} = \mathbf{U}H\psi^{(1/2)}. \end{aligned} \tag{8.29}$$

However, $\psi^{(1/2)}$ is a completely arbitrary wavefunction in the two-dimensional isospin space, and hence for all states in this space we have

$$H\mathbf{U} - \mathbf{U}H = 0, \qquad \text{or} \qquad [H, \mathbf{U}] = 0. \tag{8.30}$$

In particular, for the infinitesimal $\mathbf{U}$ we have

$$[H, \boldsymbol{\varepsilon}\cdot\boldsymbol{\tau}] = 0 \tag{8.31}$$

and since each ε_i is arbitrary, it follows that

$$[H, \boldsymbol{\tau}] = 0. \tag{8.32}$$

But the $\boldsymbol{\tau}$'s are Hermitian and hence correspond to possible observables: (8.32) is the statement that their eigenvalues are constants of the motion (i.e. conserved quantities). Also (8.30) may be written as

$$\mathbf{U}H\mathbf{U}^{-1} = H \tag{8.33}$$

which is the statement that H is invariant under the transformation (8.25). Thus the existence of these constants of the motion is a consequence of the corresponding invariance of H.

Of course, since the τ_i's do not commute among themselves, we cannot give definite values to more than one of them at a time. The problem of finding a classification of the states which makes the maximum use of (8.32), given the commutation relations (8.20), is well known in the analogous angular momentum theory†. The answer is that the total squared 'spin'

$$(\mathbf{T}^{(1/2)})^2 \equiv (\tfrac{1}{2}\boldsymbol{\tau})^2 = \tfrac{1}{4}(\tau_1^2 + \tau_2^2 + \tau_3^2) \tag{8.34}$$

and one component of spin, say $T_3^{(1/2)} = \frac{1}{2}\tau_3$, can be given definite values simultaneously. The corresponding eigenfunctions are of course just the $\chi_\mathrm{p}^{(1/2)}$'s and $\chi_\mathrm{n}^{(1/2)}$'s (cf (8.21a) and (8.21b)). Thus the two degenerate energy levels are characterised by the values of $\boldsymbol{\tau}^2$ and τ_3.

So far all of this is no more than an elaborate formalism for single-particle states. The really interesting physics comes when we consider states of *several* nucleons from the isospin viewpoint. We introduce the total isospin operators

$$\mathbf{T} = \tfrac{1}{2}\boldsymbol{\tau}_1 + \tfrac{1}{2}\boldsymbol{\tau}_2 + \ldots + \tfrac{1}{2}\boldsymbol{\tau}_A \tag{8.35}$$

for the A nucleons in the nucleus. The Hamiltonian describing the system is presumed to be invariant under the transformation (8.33) for all the nucleons independently. It then follows that

$$[H, \mathbf{T}] = 0. \tag{8.36}$$

Thus the eigenvalues of the $\mathbf{T}$ operators are constants of the motion. Further, since the isospin operators for different nucleons commute with each other (they are quite independent), the commutation relations (8.20) for each of the individual $\boldsymbol{\tau}$'s imply

$$[T_i, T_j] = \mathrm{i}\varepsilon_{ijk} T_k. \tag{8.37}$$

Thus the energy levels of nuclei ought to be characterised—after allowance for electromagnetic effects, and correcting for the slight neutron–proton mass difference—by the eigenvalues of $\mathbf{T}^2$ and T_3, say, which—once again—form a maximal commuting set of operators. These eigenvalues should then be, to a good approximation, 'good quantum numbers' for nuclei, if the assumed isospin invariance is true.

What are the possible eigenvalues? Again this is answered by familiar angular momentum results, since the $\mathbf{T}$'s satisfy exactly the same commutation relations (8.37) as the angular momentum operators. The eigenvalues of $\mathbf{T}^2$ are of the form $T(T+1)$, where $T = 0, \frac{1}{2}, 1, \ldots$, and for a given T the eigenvalues of T_3 are $-T, -T+1, \ldots, T-1, T$; that is, there are $2T+1$ *degenerate states* for a given T. These states all have the same A value, and since T_3 counts $+\frac{1}{2}$ for every proton and $-\frac{1}{2}$ for every neutron, it is clear that successive values of T_3 correspond physically to changing one neutron into a proton or vice versa. Thus we

†See for example the books by Gasiorowicz (1966, 1974).

expect to see 'charge multiplets' of levels in neighbouring nuclear isobars. These are precisely the multiplets of which we have already introduced examples in Chapter 1: see figure 1.8, which we reproduce here as figure 8.2 for convenience. These level schemes (which have been adjusted for Coulomb energy differences and for $m_n - m_p$) provide clear evidence of $T = \frac{1}{2}$ (doublet), $T = 1$ (triplet) and $T = \frac{3}{2}$ (quartet) multiplets. It is important to note that states in the same T multiplet must have the same J^P quantum numbers (these are indicated on the levels): obviously the nuclear forces will depend on the space and spin degrees of freedom of the nucleons, and will only be the same between different nucleons if the space–spin part of the wavefunction is the same.

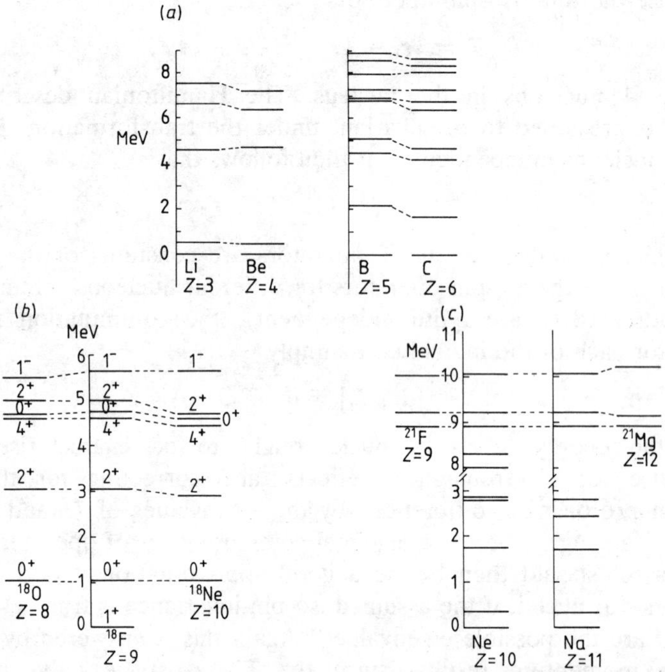

Figure 8.2 Energy levels (adjusted for Coulomb energy and neutron–proton mass differences) of nuclei of the same mass number but different charge, showing (*a*) 'mirror' doublets, (*b*) triplets and (*c*) doublets and quartets.

Thus the assumed invariance of the nuclear force has produced a much richer nuclear multiplet structure than the original n–p doublet. These higher-dimensional multiplets ($T = 1, \frac{3}{2}, \ldots$) are called 'irreducible representations' of SU(2). The commutation relations (8.37) are called the *Lie algebra* of SU(2), and the general group theoretical problem of understanding all *possible* multiplets for SU(2) is equivalent to the problem of finding matrices which satisfy these commutation relations. These are, in fact, precisely the $(2T+1) \times (2T+1)$ angular momentum matrices—generalisations of the $\boldsymbol{\tau}$'s, which themselves correspond to $T = \frac{1}{2}$. For example, the $T = 1$ matrices are 3×3 and can be compactly summarised by (problem 8.1)

$$(T_i^{(1)})_{jk} = -\mathrm{i}\varepsilon_{ijk} \tag{8.38}$$

where the numbers ε_{ijk} are deliberately chosen to be the *same* numbers that specify the algebra in (8.37); they are called the *structure constants* of the SU(2) group. In general, there will be $(2T+1) \times (2T+1)$ matrices $\mathbf{T}^{(T)}$ which satisfy (8.37), and correspondingly $(2T+1)$-dimensional wavefunctions $\psi^{(T)}$ analogous to the two-dimensional ($T = \frac{1}{2}$) case of (8.5). The generalisation of (8.25) to these higher-dimensional multiplets is then

$$\psi^{(T)\prime} = \exp(\mathrm{i}\boldsymbol{\alpha}\cdot\mathbf{T}^{(T)})\, \psi^{(T)}. \tag{8.39}$$

The matrices $\mathbf{T}^{(T)}$ are called the *generators* of infinitesimal SU(2) transformations (cf (8.22)).

Now let us look at isospin in elementary particle physics. The neutron and proton states themselves are actually only the ground states of a whole series of corresponding levels, as noted in Chapter 1 (see figure 1.10(a)). Another series of baryonic multiplets comes in *four* charge states, as shown in figure 1.10(b); and in the meson sector, the π's appear to be the lowest states of a sequence of triplets, shown in figure 1.11. Many other examples also exist. The most natural interpretation of these facts is that, just as in the nuclear case, the observed states are composites of more basic entities which carry different charges, while the forces between these entities are charge-independent. These entities are, of course, the quarks: the n contains (udd), the p (uud), and the Δ states (uuu, uud, udd, ddd). This u/d degree of freedom is what we now call the SU(2) isospin flavour symmetry at the quark level. Notice, however, that this simple constituent quark model predicts that there are no baryon or meson states with $T = 2$ or higher.

There are actually also *larger* hadronic multiplets, in which strange particles are grouped with non-strange ones. Here it was found by Gell-Mann (1961) and Ne'eman (1961) (see also Gell-Mann and Ne'eman 1964) that the correct generalisation of the SU(2) flavour symmetry was to SU(3). The u, d, s quarks now transform as the 'fundamental' three-dimensional representation of SU(3), while the nucleons belong to an *eight*-dimensional representation and the Δ's to a *ten*. The three quark flavours u, d, s fit into a three-dimensional generalisation of (8.5)

$$\psi = \begin{pmatrix} u \\ d \\ s \end{pmatrix} \tag{8.40}$$

and unitary transformations among these states are written as

$$\psi' = \mathbf{U}\psi \tag{8.41}$$

where $\mathbf{U}$ is a 3×3 unitary matrix of determinant 1. The set of such matrices forms the group SU(3). An infinitesimal SU(3) matrix has the form

$$\mathbf{U} = \mathbf{1} + \mathrm{i}\boldsymbol{\chi} \tag{8.42}$$

where $\boldsymbol{\chi}$ is a 3×3 traceless Hermitian matrix. Such a matrix involves *eight* independent parameters and can be written as

$$\boldsymbol{\chi} = \boldsymbol{\eta}\cdot\boldsymbol{\lambda}/2 \tag{8.43}$$

where $\boldsymbol{\eta} = (\eta_1, \ldots, \eta_8)$ and the $\boldsymbol{\lambda}$'s are eight matrices generalising the $\boldsymbol{\tau}$ matrices of (8.19). They are the generators of SU(3) in the three-dimensional representation, and their commutation relations define the algebra of SU(3):

$$[\tfrac{1}{2}\lambda_i, \tfrac{1}{2}\lambda_j] = \mathrm{i}f_{ijk}\cdot\tfrac{1}{2}\lambda_k. \tag{8.44}$$

The λ matrices are given in Appendix F, and the structure constants f_{ijk} are tabulated in the book by Gibson and Pollard (1976), for example. A finite SU(3) transformation on the quark triplet is then (cf (8.25))

$$\psi' = \exp(\mathrm{i}\boldsymbol{\alpha}\cdot\boldsymbol{\lambda}/2)\psi \tag{8.45}$$

which also has the 'generalised phase transformation' character, now with *eight* 'phase angles'. Higher-dimensional representations can then be built up (to accommodate the baryons, for instance) by considering wavefunctions for systems of several quarks, just as in the nuclear isospin case.

The mass splittings between members of an SU(3) multiplet which differ in strangeness quantum number can be quite substantial (e.g. $m_{\mathrm{K}} - m_{\pi} \sim 350$ MeV, $m_{\Sigma} - m_{\mathrm{N}} \sim 250$ MeV). Such differences are attributed essentially to the difference in mass between the strange and non-strange quarks $m_{\mathrm{s}} - m_{\mathrm{u}}$; the interquark forces (see Chapter 9) are believed to be *flavour-independent*. Further flavours (c, b, t?, ?, . . .) also exist, but the mass differences are now so large that it is generally not useful to think about higher-flavour groups such as SU(4), ... Further details of flavour SU(3) and its applications to hadronic spectroscopy are given in the books by Close (1979) and Lichtenberg (1978), for example.

8.2 Non-Abelian global symmetries in Lagrangian quantum field theory

As may already have begun to be apparent in Chapter 4, Lagrangian quantum field theory is a formalism which is especially well adapted for the description of symmetries. Without going into any elaborate general theory, we shall now give a few examples showing how global isospin symmetry is very easily built in, generalising in a simple way the U(1) symmetries considered in §§4.4 and 4.5. This will also prepare the way for the local (gauge) case.

Consider, for example, the Lagrangian

$$\hat{\mathscr{L}} = \bar{\hat{\psi}}_{\mathrm{u}}(\mathrm{i}\not{\partial} - m)\hat{\psi}_{\mathrm{u}} + \bar{\hat{\psi}}_{\mathrm{d}}(\mathrm{i}\not{\partial} - m)\hat{\psi}_{\mathrm{d}} \tag{8.46}$$

describing two free fermions 'u' and 'd' of equal mass m. $\hat{\mathscr{L}}$—and hence the associated Hamiltonian $\hat{\mathscr{H}}$—is certainly invariant under SU(2) unitary transformations of the form (cf (8.25)—for the difference in sign cf §4.4 in the U(1) case)

$$\begin{pmatrix} \hat{\psi}'_{\mathrm{u}} \\ \hat{\psi}'_{\mathrm{d}} \end{pmatrix} = \exp(-\mathrm{i}\boldsymbol{\alpha}\cdot\boldsymbol{\tau}/2)\begin{pmatrix} \hat{\psi}_{\mathrm{u}} \\ \hat{\psi}_{\mathrm{d}} \end{pmatrix} \tag{8.47}$$

for constant $\boldsymbol{\alpha}$; a doublet of states $|T = \frac{1}{2}, T_3 = \pm\frac{1}{2}\rangle$ is produced by $\hat{\psi}^{\dagger}_{\mathrm{u}}$ and $\hat{\psi}^{\dagger}_{\mathrm{d}}$ acting on $|0\rangle$. What are the associated conserved quantities? In the global U(1) case of §4.4, the conserved quantity $\hat{N}$ appeared in the exponent of the unitary operator $\hat{U}$ that effected the transformation $\hat{\psi} \rightarrow \hat{\psi}'$:

$$\hat{\psi}' = \hat{U}\hat{\psi}\hat{U}^{-1} \tag{8.48}$$

$$\hat{\psi}' \approx (1 - \mathrm{i}\varepsilon)\hat{\psi}, \qquad \hat{U} \approx 1 + \mathrm{i}\varepsilon\hat{N}, \qquad [\hat{N}, \hat{\psi}] = -\hat{\psi}. \tag{8.49}$$

In the present case, the single parameter ε is replaced by three $(\varepsilon_1, \varepsilon_2, \varepsilon_3)$, so that we shall obviously need *three* analogues of $\hat{N}$. Let us write

$$\hat{\psi}^{(1/2)} = \begin{pmatrix} \hat{\psi}_{\mathrm{u}} \\ \hat{\psi}_{\mathrm{d}} \end{pmatrix} \tag{8.50}$$

and

$$\hat{\psi}^{(1/2)\prime} = \hat{U}\hat{\psi}^{(1/2)}\hat{U}^{\dagger} \tag{8.51}$$

with

$$\psi^{(1/2)\prime} \approx (1 - \mathrm{i}\boldsymbol{\varepsilon}\cdot\boldsymbol{\tau}/2)\psi^{(1/2)}, \qquad \hat{U} \approx 1 + \mathrm{i}\boldsymbol{\varepsilon}\cdot\hat{\mathbf{T}}^{(1/2)} \tag{8.52}$$

for small transformations, where the three operators $\hat{\mathbf{T}}^{(1/2)}$ are Hermitian since $\hat{U}$ is unitary. Then we require

$$(1 + \mathrm{i}\boldsymbol{\varepsilon}\cdot\hat{\mathbf{T}}^{(1/2)})\hat{\psi}^{(1/2)}(1 - \mathrm{i}\boldsymbol{\varepsilon}\cdot\hat{\mathbf{T}}^{(1/2)}) = (1 - \mathrm{i}\boldsymbol{\varepsilon}\cdot\boldsymbol{\tau}/2)\hat{\psi}^{(1/2)} \tag{8.53}$$

whence (problem 8.2)

$$[\hat{\mathbf{T}}^{(1/2)}, \hat{\psi}^{(1/2)}] = -(\boldsymbol{\tau}/2)\hat{\psi}^{(1/2)}. \tag{8.54}$$

Now we need to find operators $\hat{\mathbf{T}}^{(1/2)}$ which do this—and the answer is

$$\hat{\mathbf{T}}^{(1/2)} = \int \hat{\psi}^{(1/2)\dagger}(\boldsymbol{\tau}/2)\hat{\psi}^{(1/2)}\mathrm{d}^3\boldsymbol{x} \tag{8.55}$$

as can be checked (problem 8.3) from the anticommutation relations of the $\hat{\psi}$'s. Note that if '$\boldsymbol{\tau}/2 \to 1$', (8.55) reduces to the sum of the u and d number operators again, as required for the one-parameter U(1) case. It is now possible to check that these $\hat{\mathbf{T}}^{(1/2)}$'s do indeed commute with the Hamiltonian $\hat{H}$:

$$\mathrm{d}\hat{\mathbf{T}}^{(1/2)}/\mathrm{d}t = -\mathrm{i}[\hat{\mathbf{T}}^{(1/2)}, \hat{H}] = 0 \tag{8.56}$$

so that their eigenvalues are conserved. The $\hat{\mathbf{T}}^{(1/2)}$ are, of course, precisely a *field theoretic realisation of the isospin operators* appropriate to the case $T = \frac{1}{2}$, as follows from the fact that they obey the SU(2) algebra (problem 8.4):

$$[\hat{T}_i, \hat{T}_j] = \mathrm{i}\varepsilon_{ijk}\hat{T}_k. \tag{8.57}$$

In the U(1) case, we could go further and associate the conserved operator (e.g. $\hat{Q}_{\mathrm{em}}$) with a conserved *current*:

$$\hat{Q}_{\mathrm{em}} = \int \hat{j}^{0}_{\mathrm{em}}\mathrm{d}^3\boldsymbol{x}, \qquad \hat{j}^{\mu}_{\mathrm{em}} = e\bar{\hat{\psi}}\gamma^\mu\hat{\psi} \tag{8.58}$$

where

$$\partial_\mu \hat{j}^{\mu}_{\mathrm{em}} = 0. \tag{8.59}$$

The obvious generalisation appropriate to (8.55) is

$$\hat{\mathbf{T}}^{(1/2)} = \int \hat{\boldsymbol{j}}^{0(1/2)}\mathrm{d}^3\boldsymbol{x}, \qquad \hat{\boldsymbol{j}}^{(1/2)\mu} = \bar{\hat{\psi}}^{(1/2)}\gamma^\mu(\boldsymbol{\tau}/2)\hat{\psi}^{(1/2)}. \tag{8.60}$$

Indeed one can verify from the equations of motion that

$$\partial_\mu \hat{\boldsymbol{j}}^{(1/2)\mu} = 0. \tag{8.61}$$

Thus $\hat{\boldsymbol{j}}^{(1/2)\mu}$ is a *conserved isospin current operator* appropriate to the $T = \frac{1}{2}$ (ũ, d) system.

Clearly there should be some general formalism for dealing with all this more efficiently, and it is provided by a generalisation of the steps followed, in the U(1) case, in equations (4.139)–(4.146). Suppose the Lagrangian involves a set of fields $\hat{\psi}_r$ (they could be bosons or fermions) and suppose that it is *invariant* under the small transformation

$$\delta\hat{\psi}_r = -\mathrm{i}\varepsilon T_{rs}\hat{\psi}_s \tag{8.62}$$

for some set of numerical coefficients T_{rs}. The equation (8.62) generalises (4.140). Then since $\hat{\mathscr{L}}$ is invariant under this change,

$$0 = \delta\hat{\mathscr{L}} = \frac{\partial\hat{\mathscr{L}}}{\partial\hat{\psi}_r}\delta\hat{\psi}_r + \frac{\partial\hat{\mathscr{L}}}{\partial(\partial^\mu\hat{\psi}_r)}\partial^\mu(\delta\hat{\psi}_r). \tag{8.63}$$

But

$$\frac{\partial\hat{\mathscr{L}}}{\partial\hat{\psi}_r} = \partial^\mu\left(\frac{\partial\hat{\mathscr{L}}}{\partial(\partial^\mu\hat{\psi}_r)}\right) \tag{8.64}$$

from the equations of motion. Hence

$$\partial^\mu\left(\frac{\partial\hat{\mathscr{L}}}{\partial(\partial^\mu\hat{\psi}_r)}\,\delta\hat{\psi}_r\right) = 0 \tag{8.65}$$

which is precisely a current conservation law of the form

$$\partial^\mu \hat{j}_\mu = 0. \tag{8.66}$$

Indeed, disregarding the irrelevant constant small parameter ε, the conserved current is

$$\hat{j}_\mu = -\mathrm{i}\frac{\partial\hat{\mathscr{L}}}{\partial(\partial^\mu\hat{\psi}_r)}T_{rs}\hat{\psi}_s. \tag{8.67}$$

Let us try this out on (8.46). We will condense the notation by writing

$$\hat{\mathscr{L}} = \bar{\hat{\psi}}^{(1/2)}(\mathrm{i}\not{\partial} - m)\hat{\psi}^{(1/2)} \tag{8.68}$$

and

$$\delta\hat{\psi}^{(1/2)} = (-\mathrm{i}\boldsymbol{\varepsilon}\cdot\boldsymbol{\tau}/2)\hat{\psi}^{(1/2)}. \tag{8.69}$$

So, as we knew already, there are now three ε's, and so three T_{rs}'s, namely $\frac{1}{2}(\tau_1)_{rs}$, $\frac{1}{2}(\tau_2)_{rs}$, $\frac{1}{2}(\tau_3)_{rs}$. For each one we have a current, e.g.

$$\hat{j}^{(1/2)}_{1\mu} = -\mathrm{i}\frac{\partial\hat{\mathscr{L}}}{\partial(\partial^\mu\hat{\psi}^{(1/2)})}\frac{\tau_1}{2}\hat{\psi}^{(1/2)} = \bar{\hat{\psi}}^{(1/2)}\gamma_\mu\frac{\tau_1}{2}\hat{\psi}^{(1/2)} \tag{8.70}$$

and similarly for the other τ's, so we recover (8.60). From the invariance of the Lagrangian under the transformation (8.69) there follows the conservation of an associated symmetry current. This is the quantum field theory version of Noether's theorem.

This theorem is of fundamental significance as it tells us how to relate symmetries (under transformations of the general form (8.62)) to 'current' conservation laws (of the form (8.66)), and it constructs the actual currents for us. In gauge theories, the *dynamics* is generated from a symmetry, in the sense that (as we have seen in the local U(1) of electromagnetism) the symmetry currents are the dynamical currents that drive the equations for the force field. Thus the symmetries of the Lagrangian are basic to gauge field theories.

Let us look at another example, this time involving spin-0 fields. Suppose we have three spin-0 fields all with the same mass, and take

$$\hat{\mathscr{L}} = \tfrac{1}{2}\partial_\mu\hat{\phi}_1\partial^\mu\hat{\phi}_1 + \tfrac{1}{2}\partial_\mu\hat{\phi}_2\partial^\mu\hat{\phi}_2 + \tfrac{1}{2}\partial_\mu\hat{\phi}_3\partial^\mu\hat{\phi}_3 - \tfrac{1}{2}m^2(\hat{\phi}_1^2 + \hat{\phi}_2^2 + \hat{\phi}_3^2). \tag{8.71}$$

It is obvious that $\hat{\mathscr{L}}$ is invariant under an arbitrary rotation of the three $\hat{\phi}$'s among themselves, generalising the 'rotation about the 3-axis' considered for the $\hat{\phi}_1$–$\hat{\phi}_2$ system of §4.4. An infinitesimal such rotation is

$$\hat{\boldsymbol{\phi}}' = \hat{\boldsymbol{\phi}} + \boldsymbol{\varepsilon} \times \hat{\boldsymbol{\phi}} \tag{8.72}$$

which implies

$$\delta\hat{\phi}_r = -\mathrm{i}\varepsilon_a T^{(1)}_{ars}\hat{\phi}_s \tag{8.73}$$

with

$$T^{(1)}_{ars} = -\mathrm{i}\varepsilon_{ars}. \tag{8.74}$$

There are of course three $\mathbf{T}$'s again—indeed, referring back to (8.38), we see that these ones are precisely $(T^{(1)}_a)_{rs}$, the $T = 1$ isospin matrices, evidently appropriate for the three $(2 \times 1 + 1)$-dimensional SU(2) representation, $\hat{\boldsymbol{\phi}}$. The $a = 1$ component of the conserved current in this case is, from (8.67),

$$\hat{j}^{(1)\mu}_1 = \hat{\phi}_2\partial^\mu\hat{\phi}_3 - \hat{\phi}_3\partial^\mu\hat{\phi}_2. \tag{8.75}$$

Cyclic permutations give us the other components which can be summarised as

$$\hat{\boldsymbol{j}}^{(1)\mu} = \mathrm{i}(\widetilde{\hat{\phi}}^{(1)}\mathbf{T}^{(1)}\partial^\mu\hat{\phi}^{(1)} - (\widetilde{\partial^\mu\hat{\phi}^{(1)}})\mathbf{T}^{(1)}\hat{\phi}^{(1)}) \tag{8.76}$$

where we have written

$$\hat{\phi}^{(1)} = \begin{pmatrix} \hat{\phi}_1 \\ \hat{\phi}_2 \\ \hat{\phi}_3 \end{pmatrix} \tag{8.77}$$

and $\sim$ denotes transpose. Equation (8.76) has the form expected of a bosonic spin-0 current, but with the matrices $\mathbf{T}^{(1)}$ appearing.

The general form of such SU(2) currents should now be clear. For bosons we shall have the form

$$\mathrm{i}(\hat{\phi}^{(T)\dagger}\mathbf{T}^{(T)}\partial^\mu\hat{\phi}^{(T)} - (\partial^\mu\hat{\phi}^{(T)})^\dagger\mathbf{T}^{(T)}\hat{\phi}^{(T)}) \tag{8.78}$$

where we have put the † to allow for possibly complex fields; and for fermions we shall have

$$\bar{\hat{\psi}}^{(T)}\gamma^\mu\mathbf{T}^{(T)}\hat{\psi}^{(T)} \tag{8.79}$$

where in each case $\hat{\phi}$ or $\hat{\psi}$ transforms as a $(2T + 1)$ multiplet under SU(2), i.e.

$$\hat{\psi}^{(T)\prime} = \exp(-\mathrm{i}\boldsymbol{\alpha}\cdot\mathbf{T}^{(T)})\hat{\psi}^{(T)} \tag{8.80}$$

and similarly for $\hat{\phi}^{(T)}$. The electroweak theory uses a gauge invariant version of (8.78), where $\hat{\phi}$ is the weak isospinor Higgs field; see §14.4 below.

The cases considered explicitly have all been for *free* field theories. But SU(2) invariant interactions can be easily formed. For example, the interaction $\frac{1}{2}\overline{\hat{\psi}}\boldsymbol{\tau}\hat{\psi}\cdot\hat{\boldsymbol{\phi}}$ describes SU(2) invariant interactions between a $T = \frac{1}{2}$ isospinor $\hat{\psi}$ and a $T = 1$ isovector $\hat{\boldsymbol{\phi}}$. The geometric analogy allows us to see that the 'dot product' will be invariant. Equally, the above can all be generalised to other internal symmetry groups such as SU(3), by simply replacing 'T's' by the appropriate generators 'G'.

8.3 Local non-Abelian phase invariance: non-Abelian gauge theories

Consider a global SU(2) isospinor transformation

$$\psi^{(1/2)\prime} = \exp(\mathrm{i}\boldsymbol{\alpha}\cdot\boldsymbol{\tau}/2)\psi^{(1/2)} \tag{8.81}$$

where we have now returned to ordinary wavefunctions. Invariance under this transformation amounts to the assertion that the choice of *which* two base states—(u, d), (n, p), . . .—to use is a matter of convention; any such non-Abelian phase transformation on a chosen pair produces another equally good pair. However, the choice cannot be made independently at all space–time points, only *globally*. To Yang and Mills (1954) (cf the quotation at the start of Part III) this seemed somehow an unaesthetic limitation of symmetry: 'Once one chooses what to call a proton, what a neutron, at one space–time point, one is then not free to make any choices at other space–time points'. They even suggested that this could be viewed as 'inconsistent with the localised field concept', and they therefore 'explored the possibility' of replacing this global (space–time-independent) phase transformation (8.81) by the local (space–time-dependent) one

$$\psi^{(1/2)\prime} = \exp[\mathrm{i}g\boldsymbol{\tau}\cdot\boldsymbol{\alpha}(x)/2]\psi^{(1/2)} \tag{8.82}$$

in which the phase parameters $\boldsymbol{\alpha}(x)$ are functions of $x = (t, \boldsymbol{x})$ as indicated. Notice that we have inserted a parameter g in the exponent to make the analogy with the electromagnetic U(1) case

$$\psi' = \exp[\mathrm{i}q\chi(x)]\psi \tag{8.83}$$

even stronger: g will be a coupling strength, analogous to the electromagnetic charge q. The consideration of theories based on (8.82) was the fundamental step taken by Yang and Mills (1954); see also Shaw (1955).

However, as mentioned at the start of this chapter, the original Yang–Mills attempt to get a theory of hadronic interactions by 'localising' the flavour symmetry group SU(2) turned out not to be phenomenologically viable (although a remarkable attempt was made to push the idea further by Sakurai (1960)). In the event, the successful application of a local SU(2) symmetry was to the *weak* interactions. But this is complicated by the fact that the symmetry is 'hidden', and consequently we shall delay the discussion of this application until after QCD—which *is* the theory of strong interactions, but at the quark, rather than the composite (hadronic), level. QCD is based on the local form of an SU(3) symmetry—once again it is *not* the flavour SU(3) of §8.1, but a symmetry with respect to a totally new degree of freedom, colour. This will be introduced in Chapter 9.

Although the application of local SU(2) symmetry to the weak interactions will follow that of local SU(3) to the strong, we proceed at this point with the local SU(2) case, since the group theory is simpler and more familiar. In the present section we shall give an elementary discussion of the main ideas of the non-Abelian SU(2) gauge theory which results from the demand of invariance under (8.82). We shall generally use the language of isospin when referring to the physical states and operators, bearing in mind that this will eventually mean *weak* isospin.

We shall mimic as literally as possible the discussion of electromagnetic gauge (phase) invariance in §§2.4 and 2.5. Again it is clear that no free particle wave equation can be covariant under the transformation

$$\psi^{(1/2)\prime} = \exp[\mathrm{i}g\boldsymbol{\tau}\cdot\boldsymbol{\alpha}(x)/2]\psi^{(1/2)} \tag{8.84}$$

since the gradient terms in the equation will act on the phase factor $\boldsymbol{\alpha}(x)$. However, wave equations with a suitably defined *covariant derivative* can be: physically, this means that, just as for electromagnetism, local non-Abelian phase invariance requires the introduction of a definite force field. In the electromagnetic case the covariant derivative was

$$D^\mu = \partial^\mu + \mathrm{i}qA^\mu(x). \tag{8.85}$$

The covariant derivative appropriate to the local SU(2) transformation above is

$$D^\mu = \partial^\mu + \mathrm{i}g\boldsymbol{\tau}\cdot\boldsymbol{W}^\mu(x)/2. \tag{8.86}$$

This operator acts on the two-component 'isospinor' $\psi^{(1/2)}$: the ∂^μ is understood to be multiplied by the unit 2×2 matrix. The $\boldsymbol{W}^\mu(x)$ are *three* independent gauge fields

$$\boldsymbol{W}^\mu = (W_1^\mu, W_2^\mu, W_3^\mu) \tag{8.87}$$

the generalisation of the electromagnetic field A^μ. They are called SU(2) gauge fields or, more generally, Yang–Mills fields. The term

$\boldsymbol{\tau}\cdot\boldsymbol{W}^\mu$ is then the explicit 2×2 matrix

$$\boldsymbol{\tau}\cdot\boldsymbol{W}^\mu = \begin{pmatrix} W_3^\mu & W_1^\mu - \mathrm{i}W_2^\mu \\ W_1^\mu + \mathrm{i}W_2^\mu & -W_3^\mu \end{pmatrix} \tag{8.88}$$

acting on $\psi^{(1/2)}$.

We now verify that this prescription does work. By analogy with the electromagnetic example (equation (2.32)) the key requirement is

$$(D'^\mu\psi^{(1/2)\prime}) = \exp[\mathrm{i}g\boldsymbol{\tau}\cdot\boldsymbol{\alpha}(x)/2](D^\mu\psi^{(1/2)}). \tag{8.89}$$

Consider then an *infinitesimal*† such SU(2) transformation with local parameters $\boldsymbol{\eta}(x)$. The left-hand side of this equation is

$$D'^\mu\psi^{(1/2)\prime} = (\partial^\mu + \mathrm{i}g\boldsymbol{\tau}\cdot\boldsymbol{W}'^\mu/2)[1 + \mathrm{i}g\boldsymbol{\tau}\cdot\boldsymbol{\eta}(x)/2]\psi^{(1/2)} \tag{8.90}$$

while the right-hand side is

$$[1 + \mathrm{i}g\boldsymbol{\tau}\cdot\boldsymbol{\eta}(x)/2](\partial^\mu + \mathrm{i}g\boldsymbol{\tau}\cdot\boldsymbol{W}^\mu/2)\psi^{(1/2)}. \tag{8.91}$$

We now impose equality between these two expressions to determine the required transformation law for the three $\boldsymbol{W}^\mu$. Suppose that under this infinitesimal transformation

$$\boldsymbol{W}^\mu \to \boldsymbol{W}'^\mu = \boldsymbol{W}^\mu + \delta\boldsymbol{W}^\mu. \tag{8.92}$$

Then the condition of equality is

$$[\partial^\mu + \mathrm{i}g(\boldsymbol{\tau}/2)\cdot(\boldsymbol{W}^\mu + \delta\boldsymbol{W}^\mu)]\cdot[1 + \mathrm{i}g\boldsymbol{\tau}\cdot\boldsymbol{\eta}(x)/2]\psi^{(1/2)}$$
$$= [1 + \mathrm{i}g\boldsymbol{\tau}\cdot\boldsymbol{\eta}(x)/2](\partial^\mu + \mathrm{i}g\boldsymbol{\tau}\cdot\boldsymbol{W}^\mu/2)\psi^{(1/2)}. \tag{8.93}$$

Multiplying out the terms, neglecting the term of second order involving the product of $\delta\boldsymbol{W}^\mu$ and $\boldsymbol{\eta}$ and noting that

$$\partial^\mu(\boldsymbol{\eta}\psi) = (\partial^\mu\boldsymbol{\eta})\psi + \boldsymbol{\eta}(\partial^\mu\psi) \tag{8.94}$$

we see that many terms cancel and we are left with

$$\mathrm{i}g\frac{\boldsymbol{\tau}\cdot\delta\boldsymbol{W}^\mu}{2} = -\mathrm{i}g\frac{\boldsymbol{\tau}\cdot\partial^\mu\boldsymbol{\eta}(x)}{2} + (\mathrm{i}g)^2\left[\left(\frac{\boldsymbol{\tau}\cdot\boldsymbol{\eta}(x)}{2}\right)\left(\frac{\boldsymbol{\tau}\cdot\boldsymbol{W}^\mu}{2}\right) - \left(\frac{\boldsymbol{\tau}\cdot\boldsymbol{W}^\mu}{2}\right)\left(\frac{\boldsymbol{\tau}\cdot\boldsymbol{\eta}(x)}{2}\right)\right]. \tag{8.95}$$

Using the well known identity for Pauli matrices (see problem 3.2(b))

$$\boldsymbol{\sigma}\cdot\boldsymbol{a}\boldsymbol{\sigma}\cdot\boldsymbol{b} = \boldsymbol{a}\cdot\boldsymbol{b} + \mathrm{i}\boldsymbol{\sigma}\cdot\boldsymbol{a} \times \boldsymbol{b} \tag{8.96}$$

this yields

$$\boldsymbol{\tau}\cdot\delta\boldsymbol{W}^\mu = \boldsymbol{\tau}\cdot\partial^\mu\boldsymbol{\eta}(x) - g\boldsymbol{\tau}\cdot(\boldsymbol{\eta}(x) \times \boldsymbol{W}^\mu). \tag{8.97}$$

Equating components on both sides, we deduce

†The algebra is then considerably simplified.

$$\boxed{\delta \boldsymbol{W}^\mu = -\partial^\mu \boldsymbol{\eta}(x) - g[\boldsymbol{\eta}(x) \times \boldsymbol{W}^\mu(x)].} \tag{8.98}$$

This is the way the SU(2) gauge fields $\boldsymbol{W}^\mu(x)$ transform under an infinitesimal local SU(2) gauge transformation. Compare this with the analogous transformation law for the electromagnetic gauge field $A^\mu(x)$:

$$A^\mu \rightarrow A'^\mu = A^\mu - \partial^\mu \chi \tag{8.99}$$

so that

$$\delta A^\mu = -\partial^\mu \chi. \tag{8.100}$$

This corresponds directly to the first term in $\delta \boldsymbol{W}^\mu$—three components of $\boldsymbol{W}^\mu$ and $\boldsymbol{\eta}$ corresponding to three infinitesimal χ's. The significance of the second term in $\delta \boldsymbol{W}^\mu$ is that it says that the three gauge fields $\boldsymbol{W}^\mu$ form the components of a triplet representation of SU(2)—a field with 'isospin' 1. To show this, consider how an SU(2) triplet transforms under the *global* SU(2) transformations of the last section. We use (equation (8.39))

$$\psi^{(1)\prime} = \exp(\mathrm{i}\boldsymbol{\alpha}\cdot\mathbf{T}^{(1)})\psi^{(1)} \tag{8.101}$$

where the matrices $\mathbf{T}^{(1)}$ are explicitly specified by

$$(T_i^{(1)})_{jk} = -\mathrm{i}\varepsilon_{ijk} \tag{8.38}$$

as we discussed. Under an infinitesimal SU(2) transformation $\boldsymbol{\eta}$, inserting a g for convenience, the SU(2) triplet of fields

$$\psi^{(1)} = \begin{pmatrix} \psi_1^{(1)} \\ \psi_2^{(1)} \\ \psi_3^{(1)} \end{pmatrix} \tag{8.102}$$

will transform by

$$\psi^{(1)\prime} = (1 + \mathrm{i}g\mathbf{T}^{(1)}\cdot\boldsymbol{\eta})\psi^{(1)} \tag{8.103}$$

or, in component form,

$$\psi_i^{(1)\prime} = (1 + \mathrm{i}gT_k^{(1)}\eta_k)_{ij}\psi_j^{(1)}. \tag{8.104}$$

Thus

$$\delta\psi_i^{(1)} \equiv \psi_i^{(1)\prime} - \psi_i^{(1)} = \mathrm{i}g\eta_k(-\mathrm{i}\varepsilon_{kij})\psi_j^{(1)} \tag{8.105}$$

using the explicit form for $T_k^{(1)}$ (equation (8.38)), and so, using the antisymmetry of ε_{ijk}, we obtain

$$\delta\psi_i^{(1)} = -g\varepsilon_{ikj}\eta_k\psi_j^{(1)}. \tag{8.106}$$

In vector form, this is just

$$\delta\psi^{(1)} = -g(\boldsymbol{\eta} \times \boldsymbol{\psi}^{(1)}) \tag{8.107}$$

which is precisely the form of the second term in $\delta \boldsymbol{W}^{\mu}$.

To summarise: we have shown that, for infinitesimal transformations, the relation

$$(D^{\mu\prime}\psi^{(1/2)\prime}) = [1 + \mathrm{i}g\boldsymbol{\tau}\cdot\boldsymbol{\eta}(x)/2](D^{\mu}\psi^{(1/2)}) \tag{8.108}$$

(where D^{μ} is given by (8.86)) holds true if in addition to the infinitesimal local SU(2) phase transformation on $\psi^{(1/2)}$

$$\psi^{(1/2)\prime} = [1 + \mathrm{i}g\boldsymbol{\tau}\cdot\boldsymbol{\eta}(x)/2]\psi^{(1/2)} \tag{8.109}$$

the gauge fields transform according to

$$\boldsymbol{W}^{\prime\mu} = \boldsymbol{W}^{\mu} - \partial^{\mu}\boldsymbol{\eta}(x) - g[\boldsymbol{\eta}(x) \times \boldsymbol{W}^{\mu}]. \tag{8.110}$$

In obtaining these results, the form (8.86) for the covariant derivative has been assumed, and only the infinitesimal version of (8.84) has been treated explicitly. It turns out that (8.86) is still appropriate for the finite (non-infinitesimal) transformation (8.84), but the associated transformation law for the gauge fields is then somewhat more complicated than (8.110). Since we shall not need the latter, we do not reproduce it here. With the form

$$D^{\mu} = \partial^{\mu} + \mathrm{i}g\boldsymbol{\tau}\cdot\boldsymbol{W}(x)/2 \tag{8.86}$$

for the covariant derivative, the relation (for finite $\boldsymbol{\alpha}(x)$)

$$D^{\mu\prime}\psi^{(1/2)\prime} = \exp[\mathrm{i}g\boldsymbol{\tau}\cdot\boldsymbol{\alpha}(x)/2](D^{\mu}\psi^{(1/2)}) \tag{8.111}$$

holds true, with the appropriate finite transformation law for $\boldsymbol{W}^{\mu}$.

Suppose now that we consider a Dirac equation for $\psi^{(1/2)}$:

$$(\mathrm{i}\gamma_{\mu}\partial^{\mu} - m)\psi^{(1/2)} = 0 \tag{8.112}$$

where both components of the 'isospinor' $\psi^{(1/2)}$ are four-component Dirac spinors. We assert that we can ensure *local* SU(2) covariance by replacing ∂^{μ} in this free field equation by the covariant derivative D^{μ}:

$$D^{\mu} = \partial^{\mu} + \mathrm{i}g\boldsymbol{\tau}\cdot\boldsymbol{W}^{\mu}/2. \tag{8.86}$$

Letting (for finite transformations)

$$\mathbf{U}(\boldsymbol{\alpha}(x)) \equiv \exp[\mathrm{i}g\boldsymbol{\tau}\cdot\boldsymbol{\alpha}(x)/2] \tag{8.113}$$

we have

$$\begin{aligned}\mathbf{U}(\boldsymbol{\alpha}(x))[\mathrm{i}\gamma_{\mu}D^{\mu} - m]\psi^{(1/2)} &= \mathrm{i}\gamma_{\mu}\mathbf{U}(\boldsymbol{\alpha}(x))(D^{\mu}\psi^{(1/2)}) - m\mathbf{U}(\boldsymbol{\alpha}(x))\psi^{(1/2)} \\ &= \mathrm{i}\gamma_{\mu}D^{\prime\mu}\psi^{(1/2)\prime} - m\psi^{(1/2)\prime}\end{aligned} \tag{8.114}$$

using equations (8.84) and (8.111). Thus

$$(\mathrm{i}\gamma_\mu D'^\mu - m)\psi^{(1/2)\prime} = 0 \tag{8.115}$$

if

$$(\mathrm{i}\gamma_\mu D^\mu - m)\psi^{(1/2)} = 0 \tag{8.116}$$

proving the asserted covariance. In the same way, any free particle wave equation satisfied by an 'isospinor' $\psi^{(1/2)}$—the relevant equation is determined by the Lorentz spin of the particles involved—can be made locally covariant by the use of the covariant derivative D^μ.

The essential point here, of course, is that the locally covariant form includes *interactions* between the $\psi^{(1/2)}$'s and the gauge fields $\boldsymbol{W}^\mu$, which are determined by the local phase invariance requirement (the 'gauge principle'). Indeed, we can already begin to find some of the Feynman rules appropriate to tree graphs for SU(2) gauge theories. Consider the case of an SU(2) isospinor fermion, $\psi^{(1/2)}$. The equation is

$$(\mathrm{i}\partial\!\!\!/ - m)\psi^{(1/2)} = g(\boldsymbol{\tau}/2)\cdot\boldsymbol{W}\!\!\!\!/\;\psi^{(1/2)}. \tag{8.117}$$

In lowest-order perturbation theory the one-W emission/absorption process is given by the amplitude (cf (6.22) for the electromagnetic case)

$$-\mathrm{i}\int \bar{\psi}_\mathrm{f}^{(1/2)} g(\tau_r/2) W\!\!\!\!/\;^r \psi_\mathrm{i}^{(1/2)} \mathrm{d}^4x. \tag{8.118}$$

The matrix degree of freedom in the τ's is sandwiched between the two-component isospinors $\psi^{(1/2)}$; the γ matrix in $W\!\!\!\!/$ acts on the four-component (Dirac) parts of $\psi^{(1/2)}$. The external $\boldsymbol{W}^\mu$ field is now specified by a spin-1 polarisation vector ε^μ, like a photon, and by an 'SU(2) polarisation vector' $a^r (r = 1, 2, 3)$ which tells us which of the three SU(2) $\boldsymbol{W}$ states is participating. The Feynman rule for figure 8.3 is therefore

$$-\mathrm{i}g(\tau^r/2)\gamma_\mu \tag{8.119}$$

which is to be sandwiched between spinors/isospinors u_i, $\bar{u}_\mathrm{f}$ and dotted into ε^μ and a^r. (8.119) generalises rule (ix) of §6.3 in a very economical way.

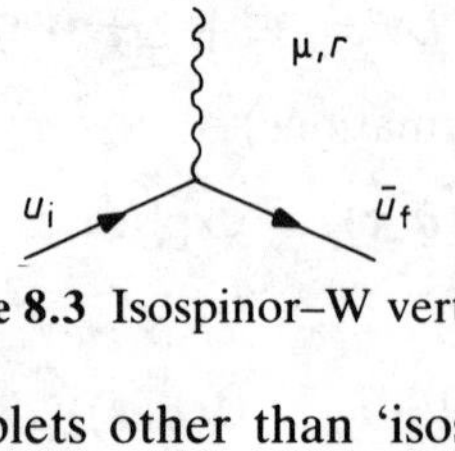

Figure 8.3 Isospinor–W vertex.

How about SU(2) multiplets other than 'isospinors'? That is, how do we make the world covariant under the general local SU(2) transformation

$$\psi^{(t)} \to \psi^{(t)\prime} = \exp[\mathrm{i}g\boldsymbol{T}^{(t)}\cdot\boldsymbol{\alpha}(x)]\psi^{(t)}. \tag{8.120}$$

In (8.120) we have introduced the notation t instead of T for the 'isospin' quantum number; this is intended to emphasise that the present 'isospin' is *not* the hadronic one, for which we retain the symbol T; t will be the symbol used for the *weak isospin* to be introduced in Chapter 14.

Comparing equations (8.120) and (8.84) for the 'isospinor' case, the answer is that in the free particle wave equation for $\psi^{(t)}$ we replace ∂^μ by

$$D^\mu = \partial^\mu + \mathrm{i}g\boldsymbol{T}^{(t)}\cdot\boldsymbol{W}^\mu. \tag{8.121}$$

This covariant derivative is of course a $(2t + 1) \times (2t + 1)$ matrix acting on the $(2t + 1)$ components of $\psi^{(t)}$. Thus the complete prescription of local SU(2) phase invariance has been obtained. The gauge fields interact with 'isomultiplets' in a *universal* way—only one g, the same for all the particles—which is prescribed by the local invariance to be simply that generated by the covariant derivatives. For the fermion vertex associated with (8.121), replace $\tau^r/2$ by $T^{(t)r}$ in (8.119).

We end this section with some comments:

(a) Suppose we are interested (as in the case of QCD) in a group other than SU(2). How does all this generalise? Whatever the group is, it will have generators G_i $(i = 1, 2, \ldots, r)$ and a multiplet transformation law

$$\psi' = \exp(\mathrm{i}g\boldsymbol{G}\cdot\boldsymbol{\alpha})\psi \tag{8.122}$$

where the 'scalar product' involves r-component vectors $\boldsymbol{G}$ and $\boldsymbol{\alpha}$ (for SU(3) there are eight generators and eight parameters α_i). The generators satisfy an algebra of the form

$$[G_i, G_j] = \mathrm{i}c_{ijk}G_k \tag{8.123}$$

where the c_{ijk} are the structure constants of the group under consideration. To make the transformation equation (8.120) local, we introduce r gauge fields $W_1^\mu, W_2^\mu, \ldots, W_r^\mu$ and define a covariant derivative

$$D^\mu = \partial^\mu + \mathrm{i}g\boldsymbol{G}^{(p)}\cdot\boldsymbol{W}^\mu \tag{8.124}$$

where p labels the representation of ψ. The $\boldsymbol{G}^{(p)}$ are some set of matrices of appropriate dimension to act on ψ. The (Yang–Mills) gauge fields $W_i^\mu (i = 1, 2, \ldots, r)$ themselves transform by

$$\delta W_i^\mu(x) = -\partial^\mu\eta_i(x) - gc_{ijk}\eta_j W_k^\mu(x) \tag{8.125}$$

for an infinitesimal transformation. Again, we see that the first term is just the expected generalisation of the electromagnetic case (equation (8.100)): the second term says that the r gauge fields $\boldsymbol{W}^\mu$ transform according to the 'regular' representation of the group, which means that the transformation coefficients are precisely the structure constants of the group.

(b) It is a remarkable fact that only one constant g (per group) is needed. This is *not* the same as in electromagnetism. There, each charged field interacts with the gauge field A^μ via a coupling whose strength is its charge (e, $-e$, $2e$, $-5e$ . . .). The crucial point is the appearance of the quadratic g^2 multiplying the *commutator* of the $\boldsymbol{\tau}$'s in the $\boldsymbol{W}^\mu$ transformation (equation (8.95)). In the electromagnetic case, there is no such commutator—the associated U(1) phase group is Abelian. As signalled by the presence of g^2, a commutator is a non-linear quantity, and the scale of quantities appearing in such commutation relations is not arbitrary. It is an instructive exercise to check that, once $\delta\boldsymbol{W}^\mu$ is given by equation (8.98)—in the SU(2) case—then the g's appearing in $\psi^{(1/2)\prime}$ (equation (8.84)) and $\psi^{(1)\prime}$ (equation (8.103)) must be the *same* as the one appearing in $\delta\boldsymbol{W}^\mu$.

(c) According to the foregoing argument, it is actually a mystery why electric charge should be quantised. Since it is the coupling constant of an Abelian group, each charged field could have an arbitrary charge from this point of view: there are no commutators to fix the scale. This is one of the motivations of attempts to 'embed' the electromagnetic gauge transformations inside a larger non-Abelian group structure. This is the case, for example, in attempts to construct so-called grand unified theories of strong, weak and electromagnetic interactions, as we shall briefly indicate in §15.3.

(d) Finally, we draw attention to the extremely important physical significance of the second term $\delta\boldsymbol{W}^\mu$ (equation (8.98)). The gauge fields themselves are not 'inert' as far as the gauge group is concerned: in the SU(2) case they have 'isospin' 1, while for a general group they belong to the regular representation of the group. This is profoundly different from the electromagnetic case, where the gauge field A^μ for the photon is of course uncharged; quite simply, $e = 0$ for a photon, and the second term in (8.98) is absent for A_μ. The fact that non-Abelian (Yang–Mills) gauge fields carry non-Abelian 'charge' degrees of freedom means that, since they are also the quanta of the force field, *they will necessarily interact with themselves*. Thus a non-Abelian gauge theory of gauge fields alone, with no 'matter' fields, has non-trivial interactions and is not a free theory.

What is the form of these interactions? The interactions between the gauge fields and the matter fields were obtained via the 'covariant derivative' prescription, which could be implemented since we knew the free particle equations of motion for the spin-0 and spin-$\frac{1}{2}$ matter fields. It would seem that we should be able to do the same thing for the spin-1 gauge fields, which should obey equations of the Maxwell form (2.19). However, it is not completely straightforward, as we now discuss.

8.4 Gauge field self-interactions: (i) heuristic

Continuing with the specific SU(2) case, we consider the 'covariantising' of the standard massless free particle equation for a vector field $\boldsymbol{W}^\mu$:

$$\Box \boldsymbol{W}^\mu - \partial^\mu \partial^\nu \boldsymbol{W}_\nu = 0. \quad (8.126)$$

Equation (8.126) is the direct generalisation of the free Maxwell equation for A^μ; each of the three SU(2) components of $\boldsymbol{W}^\mu$ obeys such a Maxwell equation. Since the $\boldsymbol{W}^\mu$ form the components of a $t = 1$ triplet as regards SU(2), the covariant derivative required for (8.126) is (cf (8.121) and (8.38))†

$$D^\mu = \partial^\mu + \mathrm{i}g\mathbf{t}^{(1)}\cdot \boldsymbol{W}^\mu. \quad (8.127)$$

This leads to the equation

$$D^2 \boldsymbol{W}^\mu - D^\mu D^\nu \boldsymbol{W}_\nu = 0. \quad (8.128)$$

But this procedure is not unique: if we had started from the perfectly equivalent free particle form

$$\Box \boldsymbol{W}^\mu - \partial^\nu \partial^\mu \boldsymbol{W}_\nu = 0 \quad (8.129)$$

we would have arrived at

$$D^2 \boldsymbol{W}^\mu - D^\nu D^\mu \boldsymbol{W}_\nu = 0 \quad (8.130)$$

and (8.130) is not the same as (8.128). It is instructive to see in detail why not. We have

$$\begin{aligned} D^\mu D^\nu \boldsymbol{W}_\nu = {} & \partial^\mu \partial^\nu \boldsymbol{W}_\nu + \mathrm{i}g[\mathbf{t}^{(1)}\cdot(\partial^\mu \boldsymbol{W}^\nu)]\boldsymbol{W}_\nu + \mathrm{i}g(\mathbf{t}^{(1)}\cdot \boldsymbol{W}^\nu)\partial^\mu \boldsymbol{W}_\nu \\ & + \mathrm{i}g(\mathbf{t}^{(1)}\cdot \boldsymbol{W}^\mu)\partial^\nu \boldsymbol{W}_\nu - g^2(\mathbf{t}^{(1)}\cdot \boldsymbol{W}^\mu)(\mathbf{t}^{(1)}\cdot \boldsymbol{W}^\nu)\boldsymbol{W}_\nu \quad (8.131) \end{aligned}$$

and

$$\begin{aligned} D^\nu D^\mu \boldsymbol{W}_\nu = {} & \partial^\mu \partial^\nu \boldsymbol{W}_\nu + \mathrm{i}g[\mathbf{t}^{(1)}\cdot(\partial^\nu \boldsymbol{W}^\mu)]\boldsymbol{W}_\nu + \mathrm{i}g(\mathbf{t}^{(1)}\cdot \boldsymbol{W}^\mu)\partial^\nu \boldsymbol{W}_\nu \\ & + \mathrm{i}g(\mathbf{t}^{(1)}\cdot \boldsymbol{W}^\nu)\partial^\mu \boldsymbol{W}_\nu - g^2(\mathbf{t}^{(1)}\cdot \boldsymbol{W}^\nu)(\mathbf{t}^{(1)}\cdot \boldsymbol{W}^\mu)\boldsymbol{W}_\nu \quad (8.132) \end{aligned}$$

whence

$$\begin{aligned} (D^\mu D^\nu - D^\nu D^\mu)\boldsymbol{W}_\nu = {} & \{\mathrm{i}g\mathbf{t}^{(1)}\cdot(\partial^\mu \boldsymbol{W}^\nu - \partial^\nu \boldsymbol{W}^\mu) \\ & - g^2(t_a^{(1)}t_b^{(1)} - t_b^{(1)}t_a^{(1)})W_a^\mu W_b^\nu\}\boldsymbol{W}_\nu \quad (8.133) \end{aligned}$$

where in the second term we have put in explicitly the SU(2) indices of the $\boldsymbol{W}$'s and $\mathbf{t}^{(1)}$'s. The different $\mathbf{t}^{(1)}$'s do not, of course, commute; in fact they obey the standard SU(2) commutation relations (8.37):

$$[t_a^{(1)}, t_b^{(1)}] = \mathrm{i}\varepsilon_{abc}t_c^{(1)}. \quad (8.134)$$

†In anticipation of the application to weak interactions, we are using lower case $\mathbf{t}^{(1)}$ matrices.

Hence (problem 8.5)

$$(D^\mu D^\nu - D^\nu D^\mu)\boldsymbol{W}_\nu = \mathrm{i}g\mathbf{t}^{(1)}\cdot(\partial^\mu \boldsymbol{W}^\nu - \partial^\nu \boldsymbol{W}^\mu - g\boldsymbol{W}^\mu \times \boldsymbol{W}^\nu)\boldsymbol{W}_\nu. \tag{8.135}$$

Noting that the 'final' $\boldsymbol{W}_\nu$ has played no role in the above, we can write the operator relation

$$(D^\mu D^\nu - D^\nu D^\mu) = \mathrm{i}g\mathbf{t}^{(1)}\cdot \boldsymbol{F}^{\mu\nu} \tag{8.136}$$

where we have introduced a very important quantity, the

$$\text{field strength tensor } \boldsymbol{F}^{\mu\nu} \equiv \partial^\mu \boldsymbol{W}^\nu - \partial^\nu \boldsymbol{W}^\mu - g\boldsymbol{W}^\mu \times \boldsymbol{W}^\nu. \tag{8.137}$$

We shall see that $\boldsymbol{F}^{\mu\nu}$ is the generalisation to non-Abelian gauge fields of the Maxwell tensor $F^{\mu\nu} = \partial^\mu A^\nu - \partial^\nu A^\mu$.

The simple prescription of replacing ∂^μ by D^μ of (8.121) has, in this case, failed to produce a unique equation of motion. We can allow for this ambiguity by introducing an arbitrary parameter δ in the field equation, which we write as

$$D^2\boldsymbol{W}^\mu - D^\nu D^\mu \boldsymbol{W}_\nu + \mathrm{i}g\delta\mathbf{t}^{(1)}\cdot\boldsymbol{F}^{\mu\nu}\boldsymbol{W}_\nu = 0. \tag{8.138}$$

It is worth pointing out here that exactly the same thing would have happened if we had considered merely the *electromagnetic* interactions of charged spin-1 particles. In that case the same steps can be followed (rather more simply, because there are no $\mathbf{t}^{(1)}$'s), and (8.136) is replaced by

$$D^\mu D^\nu - D^\nu D^\mu = \mathrm{i}qF^{\mu\nu} \tag{8.139}$$

as expected, where q is the charge. In this context, the 'δ' coupling in (8.138) is referred to as the 'ambiguous magnetic moment'. Clearly the resolution of this ambiguity is an important matter of principle as well as of practice (what is the magnetic moment of the $\mathrm{W}^\pm$?) For the photon, of course, the charge is zero and there is no such ambiguity.

We must now recall the deeper reason behind the prescription (8.121)—namely gauge invariance. This is a very powerful constraint on the *interactions* allowed in the theory. We came across an example of this in our discussion of Compton scattering of pions in §5.8.3. There we noted that, had we included only the graphs of figures 5.14 and 5.17, the amplitude at $\mathrm{O}(e^2)$ would not have been gauge invariant. And, indeed, the requirement that it should be so determined the form of the third vertex, figure (5.18). We shall now use a similar argument to fix δ in (8.138), by considering 'Compton scattering' of W's from SU(2) isospinor fermions, which we shall call quarks.

According to the rules already given, there are two diagrams, shown in figures 8.4 and 8.5. Their combined amplitude is

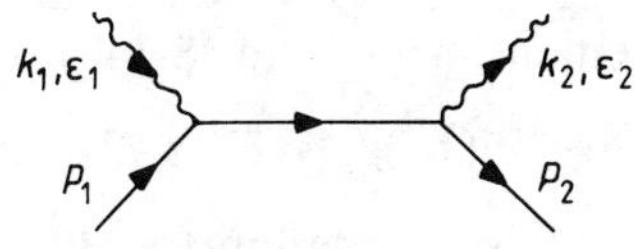

Figure 8.4 Quark–W 'Compton scattering': direct channel process.

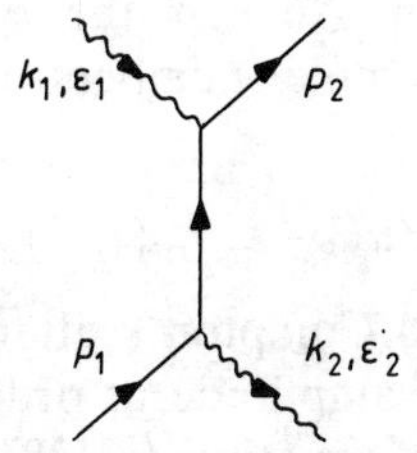

Figure 8.5 Quark–W 'Compton scattering': crossed channel process.

$$(-\mathrm{i}g)^2\bar{u}(p_2)\frac{\tau_s}{2}\not{\varepsilon}_2\frac{\mathrm{i}}{\not{p}_1+\not{k}_1-m}\frac{\tau_r}{2}\not{\varepsilon}_1 u(p_1)$$
$$+(-\mathrm{i}g)^2\bar{u}(p_2)\frac{\tau_r}{2}\not{\varepsilon}_1\frac{\mathrm{i}}{\not{p}_1-\not{k}_2-m}\frac{\tau_s}{2}\not{\varepsilon}_2 u(p_1) \quad (8.140)$$

where we have taken real polarisation vectors. What is the analogue of the '$\varepsilon_i \to k_i$' gauge invariance requirement (5.203) of QED? This latter was derived, of course, from the fact that a gauge transformation on the photon A^μ took the form

$$A^\mu \to A'^\mu = A^\mu - \partial^\mu \chi \quad (8.141)$$

so that, with the Lorentz condition, ε^μ could be replaced by $\varepsilon'^\mu = \varepsilon^\mu + \beta k^\mu$ (cf equation (5.202)) without changing the physics. But the SU(2) analogue of the gauge transformation (8.141) is, as we have seen (equation (8.98)), for infinitesimal η's,

$$W^{r\mu} \to W'^{r\mu} = W^{r\mu} - \partial^\mu \eta^r - g\varepsilon_{rst}\eta^s W^{t\mu}. \quad (8.142)$$

There is indeed a '$\partial^\mu\chi$' part—namely $\partial^\mu\eta^r$—but there is also the additional part, arising precisely from the fact that the W's carry SU(2) charge. Thus although the $\partial^\mu\eta^r$ part of (8.142) will amount to a change $\varepsilon \to \varepsilon' = \varepsilon + \beta k$ in the W polarisation vector, such a change will not be the whole content of a gauge transformation in this case. We note, however, that the extra part in (8.142) involves a factor of g. Hence, when the *full* change corresponding to (8.142) is made in a tree graph, the extra part will produce a term of higher order in g. We shall take the view that gauge invariance should hold at each order of perturbation theory separately; thus we shall demand that the tree graphs for W–Compton scattering, for example, should be invariant under $\varepsilon_i \to k_i$.

However, the replacement $\varepsilon_1 \to k_1$ in (8.140) produces the result

$$(-ig)^2 i\bar{u}(p_2)\not{\varepsilon}_2[\tau_s/2, \tau_r/2]u(p_1) \qquad (8.143)$$

where we have used the Dirac equation for the quarks of mass m, and also the mass-shell conditions $k_1^2 = k_2^2 = 0$ and $p_1^2 = p_2^2 = m^2$ (see problem 8.6). The term (8.143) is certainly *not* zero—though it would be if we were dealing with an Abelian theory like QED, for which the quark–W 'charges' are simple numbers rather than non-commuting matrices. Indeed, its value is

$$-g^2\bar{u}(p_2)\not{\varepsilon}_2\varepsilon_{rst}(\tau_t/2)u(p_1). \qquad (8.144)$$

As in the case of spin-0 Compton scattering, figures 8.4 and 8.5 are not, in fact, the complete amplitude at order g^2. To see what we have forgotten, let us return to equation (8.138) for $\boldsymbol{W}^\mu$. We write it as (cf (3.81))†

$$\Box \boldsymbol{W}^\mu - \partial^\nu\partial^\mu \boldsymbol{W}_\nu = `\hat{V}\boldsymbol{W}'^\mu \qquad (8.145)$$

where

$$\begin{aligned} `\hat{V}\boldsymbol{W}'^\mu = -ig\{&\partial^\nu[(\mathbf{t}^{(1)}\cdot \boldsymbol{W}_\nu)\boldsymbol{W}^\mu] + (\mathbf{t}^{(1)}\cdot \boldsymbol{W}^\nu)\partial_\nu \boldsymbol{W}^\mu + \delta\partial^\mu[(\mathbf{t}^{(1)}\cdot \boldsymbol{W}^\nu)\boldsymbol{W}_\nu] \\ &+ \delta\mathbf{t}^{(1)}\cdot \boldsymbol{W}^\mu\partial^\nu \boldsymbol{W}_\nu - (1+\delta)\partial^\nu[(\mathbf{t}^{(1)}\cdot \boldsymbol{W}^\mu)\boldsymbol{W}_\nu] \\ &- (1+\delta)(\mathbf{t}^{(1)}\cdot \boldsymbol{W}^\nu)\partial^\mu \boldsymbol{W}_\nu\} \end{aligned} \qquad (8.146)$$

and we have dropped terms of $O(g^2)$ which appear in the $D.D$ terms. The terms inside the { } brackets are written in such a way that they fall into three pairs, all of the form

$$`\lambda[\partial((\mathbf{t}^{(1)}\cdot \boldsymbol{W})\boldsymbol{W}) + (\mathbf{t}^{(1)}\cdot \boldsymbol{W})\partial \boldsymbol{W}]'. \qquad (8.147)$$

Consider now the lowest-order perturbation theory amplitude for 'W→ W' under the potential given above. This amplitude is

$$-i\int \boldsymbol{W}_\mu^*(\mathrm{f})\cdot`(\hat{V}\boldsymbol{W}(\mathrm{i}))'^\mu. \qquad (8.148)$$

Inserting (8.146) into (8.148) clearly gives something involving *three W's*, i.e. a triple-W vertex, shown in figure 8.6. This gives us the clue to what we were missing in figures 8.4 and 8.5: the *W exchange* process shown in figure 8.7. Our strategy is therefore clear: we obtain the 3-W vertex from (8.148), and then determine δ by the requirement that when

†The sign chosen for $\hat{V}$ here *differs* from that in the KG case (3.81). It is chosen in such a way that when the '$\mathbf{t}^{(1)}\cdot \boldsymbol{W}$' pieces in (8.146) are replaced by electromagnetic potentials, the vertex agrees in sign with that appropriate to charged vector bosons, when allowance is made for the fact that the dot product of the polarisation vectors is negative (cf equation (5.28b)).

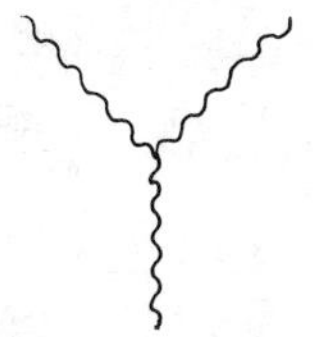

Figure 8.6 Three-W vertex.

figure 8.7 is evaluated with $\varepsilon_1 \to k_1$ it exactly cancels the corresponding contribution (8.144) from figures 8.4 and 8.5. Obtaining the 3-W vertex is less formidable than it looks. Consider the first pair in the { } bracket of (8.146). They contribute

$$-\mathrm{i}(-\mathrm{i}g)\int W_\mu^{s*}(2)\{\partial^\nu[t_{sr}^{(1)t}W_\nu^t(3)W^{r\mu}(1)] + t_{sr}^{(1)t}W^{t\nu}(3)\partial_\nu W^{r\mu}(1)\}]\mathrm{d}^4x \tag{8.149}$$

where we have written in the SU(2) indices r, s, t explicitly: the superscript index on the $\mathbf{t}^{(1)}$'s is dotted into the adjacent $W(3)$, while the lower indices are matrix indices, acting on the three components of the $W(2)$ and $W(1)$ respectively (cf (8.118)). We are labelling the W's now as '1, 2, 3', to bring out the 3-W symmetry; and also for this reason we shall consider the vertex for all momenta *in*going, i.e.

$$W^\mu \sim \varepsilon^\mu \exp(-\mathrm{i}k\cdot x). \tag{8.150}$$

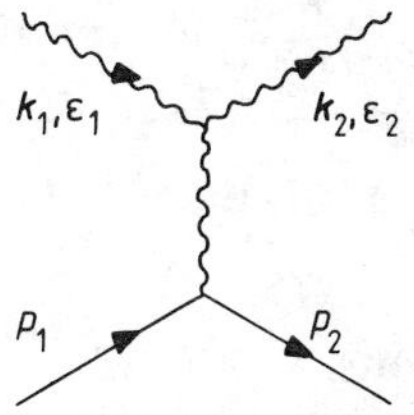

Figure 8.7 Three-W contribution to quark–W Compton scattering.

Then the first term in (8.149) can be easily evaluated by a partial integration to turn the ∂^ν onto the $W_\mu^{s*}(2)$, while in the second term ∂_ν acts on $W^{r\mu}(1)$ straightforwardly. Omitting as usual the δ^4 energy–momentum factor, we then find (problem 8.7) that (8.149) leads to

$$-g\varepsilon_{rst}a_1^r a_2^s a_3^t \varepsilon_1\cdot\varepsilon_2(k_1 - k_2)\cdot\varepsilon_3 \tag{8.151}$$

where we have used (8.38):

$$t_{sr}^{(1)t} = -\mathrm{i}\varepsilon_{tsr} = +\mathrm{i}\varepsilon_{rst}. \tag{8.152}$$

In a similar way, the other terms in (8.146) give

$$+g\delta\varepsilon_{rst}a_1^r a_2^s a_3^t(\varepsilon_1\cdot\varepsilon_3\varepsilon_2\cdot k_2 - \varepsilon_2\cdot\varepsilon_3\varepsilon_1\cdot k_1) \tag{8.153}$$

and

$$-g(1+\delta)\varepsilon_{rst}a_1^r a_2^s a_3^t(\varepsilon_2\cdot\varepsilon_3\varepsilon_1\cdot k_2 - \varepsilon_1\cdot\varepsilon_3\varepsilon_2\cdot k_1). \tag{8.154}$$

Adding all the terms up and using the 4-momentum conservation condition

$$k_1 + k_2 + k_3 = 0 \tag{8.155}$$

we obtain the vertex (suppressing the internal polarisation vectors)

$$-g\varepsilon_{rst}\{\varepsilon_1\cdot\varepsilon_2(k_1 - k_2)\cdot\varepsilon_3 + \varepsilon_2\cdot\varepsilon_3(\delta k_2 - k_3)\cdot\varepsilon_1 + \varepsilon_3\cdot\varepsilon_1(k_3 - \delta k_1)\cdot\varepsilon_2\}. \tag{8.156}$$

It is quite evident from (8.156) that the value $\delta = 1$ has a privileged role, and we strongly suspect that this will be the value selected by the symmetry of gauge invariance. To check this, we need to evaluate now the third Compton graph, figure 8.7, and set $\varepsilon_1 \to k_1$. Figure 8.7 involves an internal 'W' particle, which at this stage we are assuming is just like a photon, i.e. is massless. Its propagator can therefore be taken as

$$\frac{-\mathrm{i}g^{\mu\nu}}{q^2}\delta^{tt'} \tag{8.157}$$

where the $\delta^{tt'}$ expresses the fact that a free quantum does not change its SU(2) charge as it propagates. The amplitude for figure 8.7 is therefore

$$\{-g\varepsilon_{rst}[\varepsilon_1\cdot\varepsilon_2(k_1+k_2)_\mu + \varepsilon_{2\mu}\varepsilon_1\cdot(-\delta k_2 - k_2 + k_1) + \varepsilon_{1\mu}\varepsilon_2\cdot(k_2 - k_1 - \delta k_1)]\}$$
$$\times\left\{\frac{-\mathrm{i}g^{\mu\nu}}{q^2}\delta^{tt'}\right\}\times\left\{-\mathrm{i}g\bar{u}(p_2)\frac{\tau^{t'}}{2}\gamma_\nu u(p_1)\right\} \tag{8.158}$$

remembering that k_2 counts '$-k_2$' when considered as an ingoing momentum, and that the spin-1 propagator arises from summing on the polarisation states of particle 3 in this case. We now leave it as an exercise (problem 8.8) to verify that when $\varepsilon_1 \to k_1$ in (8.158) the resulting amplitude does exactly cancel the contribution (8.144), *provided that* $\delta = 1$. Thus with the 3-W vertex (figure 8.8)

$$\boxed{-g\varepsilon_{rst}\{\varepsilon_1\cdot\varepsilon_2(k_1 - k_2)\cdot\varepsilon_3 + \varepsilon_2\cdot\varepsilon_3(k_2 - k_3)\cdot\varepsilon_1 + \varepsilon_3\cdot\varepsilon_1(k_3 - k_1)\cdot\varepsilon_2\}} \tag{8.159}$$

the $O(g^2)$ W–Compton scattering amplitude is gauge invariant.

The verification of this non-Abelian gauge invariance to order g^2 is,

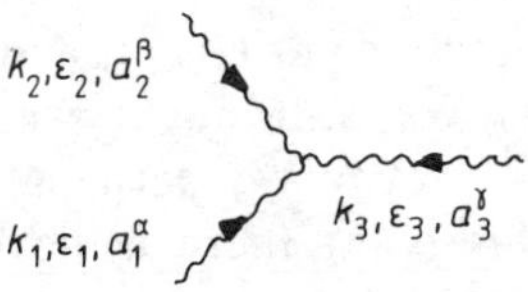

Figure 8.8 Three-W vertex with labelling appropriate to (8.159).

of course, not a proof that the entire theory will be gauge invariant if $\delta = 1$. Indeed, having obtained the 3-W vertex, we immediately have something new to check: we can see if the lowest-order W–W scattering amplitude is gauge invariant. The 3-W vertex will generate the $O(g^2)$ graphs shown in figure 8.9. The dedicated student may check that the sum of these amplitudes is *not* gauge invariant, in the sense (appropriate to tree graphs) of vanishing when any one ε is replaced by the corresponding k. Actually, this is no surprise. In obtaining the 3-W vertex we dropped an $O(g^2)$ term of order W^3 in going from (8.138) to (8.146): this will generate an $O(g^2)$ 4-W interaction, figure 8.10, when plugged into lowest-order perturbation theory, just as the analogous 'e^2A^2' term in QED generated the 4-field contact interaction of figure 5.18. We shall now simply state the required form of figure 8.10:

$$\boxed{\begin{aligned} -\mathrm{i}g^2[&\varepsilon_{rsv}\varepsilon_{tuv}(\varepsilon_1\cdot\varepsilon_3\varepsilon_2\cdot\varepsilon_4 - \varepsilon_1\cdot\varepsilon_4\varepsilon_2\cdot\varepsilon_3) \\ &+ \varepsilon_{ruv}\varepsilon_{stv}(\varepsilon_1\cdot\varepsilon_2\varepsilon_3\cdot\varepsilon_4 - \varepsilon_1\cdot\varepsilon_3\varepsilon_2\cdot\varepsilon_4) \\ &+ \varepsilon_{rtv}\varepsilon_{usv}(\varepsilon_1\cdot\varepsilon_4\varepsilon_2\cdot\varepsilon_3 - \varepsilon_1\cdot\varepsilon_2\varepsilon_3\cdot\varepsilon_4)]\end{aligned}} \tag{8.160}$$

which does, in fact, follow from (8.138) with $\delta = 1$.

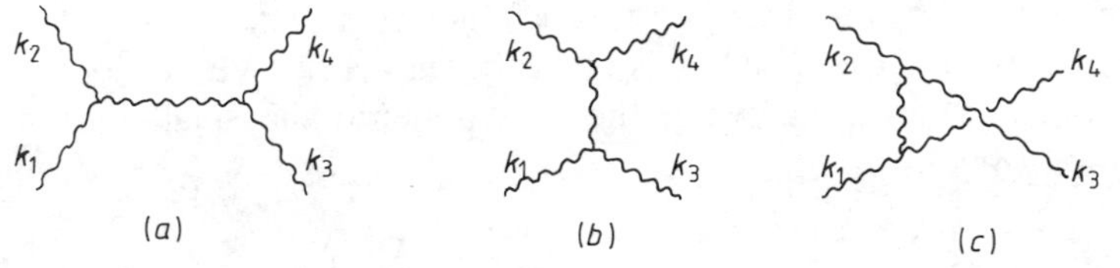

Figure 8.9 W–W scattering to order g^2: (*a*) *s* channel contribution; (*b*) *t* channel; (*c*) *u* channel.

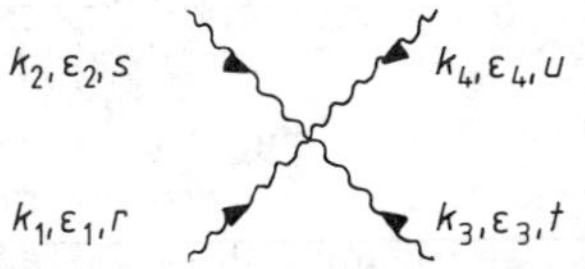

Figure 8.10 Four-W vertex with labelling appropriate to (8.160).

There are no more interactions at $O(g^2)$, and thus the theory is gauge invariant to that order†. We still might wonder about higher-order processes, such as $WW \to WWWW$; perhaps we need a 6-W vertex, even at tree level. At this point there is no denying that we need a Lagrangian: instructive though the foregoing analysis may be, to answer the general question of gauge invariance we need a gauge invariant Lagrangian which will define all the vertices for us. Let us see what it is.

8.5 Gauge field self-interactions: (ii) Lagrangian approach

The problem we face is the construction of a gauge invariant Lagrangian for the gauge fields themselves—the $\boldsymbol{W}$'s, to continue the SU(2) example. The better to advance, let us go back one step to our constant guide, QED. We gave the Lagrangian for the Maxwell (U(1)) gauge field $\hat{A}^\mu$ in §4.6. The Maxwell–Dirac Lagrangian for a particle of charge q is

$$\hat{\mathscr{L}} = \bar{\hat{\psi}}(\mathrm{i}\not\partial - m)\hat{\psi} - q\bar{\hat{\psi}}\gamma^\mu\hat{\psi}\hat{A}_\mu - \tfrac{1}{4}\hat{F}_{\mu\nu}\hat{F}^{\mu\nu}. \tag{8.161}$$

Now consider the SU(2) isospinor $\psi^{(1/2)}$. The Dirac part is

$$\hat{\mathscr{L}}_{\mathrm{D}} = \bar{\hat{\psi}}^{(1/2)}(\mathrm{i}\not\partial - m)\hat{\psi}^{(1/2)} - g\bar{\hat{\psi}}^{(1/2)}(\boldsymbol{\tau}/2)\gamma^\mu\hat{\psi}^{(1/2)}\cdot\hat{\boldsymbol{W}}_\mu. \tag{8.162}$$

The vertex (8.119) is safely accounted for in (8.162). What about the W interactions? The 'obvious' thing is to find the 'appropriate' generalisation of (8.161). Now we have already come across the SU(2) field strength tensor, which we here promote to a field operator

$$\hat{\boldsymbol{F}}^{\mu\nu} = \partial^\mu\hat{\boldsymbol{W}}^\nu - \partial^\nu\hat{\boldsymbol{W}}^\mu - g\hat{\boldsymbol{W}}^\mu \times \hat{\boldsymbol{W}}^\nu \tag{8.163}$$

which arose from considering $[D^\mu, D^\nu]$. The sceptical reader might wonder how we are sure that this is 'the' right tensor: why not just $\partial^\mu\hat{\boldsymbol{W}}^\nu - \partial^\nu\hat{\boldsymbol{W}}^\mu$ for example? This is easily answered. We are looking for an $\hat{\mathscr{L}}$ which is invariant when the W's undergo the small local transformation (8.98):

$$\hat{\boldsymbol{W}}^\mu \to \hat{\boldsymbol{W}}'^\mu = \hat{\boldsymbol{W}}^\mu - \partial^\mu\hat{\boldsymbol{\eta}} - g(\hat{\boldsymbol{\eta}} \times \hat{\boldsymbol{W}}^\mu). \tag{8.164}$$

We can easily check that the possible term

$$(\partial^\mu\hat{\boldsymbol{W}}^\nu - \partial^\nu\hat{\boldsymbol{W}}^\mu)\cdot(\partial_\mu\hat{\boldsymbol{W}}_\nu - \partial_\nu\hat{\boldsymbol{W}}_\mu) \tag{8.165}$$

is *not* invariant under (8.164)—remember that the $\hat{\boldsymbol{\eta}}$'s depend on x! The

†We should emphasise that this is only true for *massless* 'W' 's. Thus the immediate application of the foregoing results is to QCD, where the gauge group is $SU(3)_c$ and the 'W' 's are gluons. For the actual massive W's of the electroweak theory, figures 8.9 and 8.10 are not satisfactory by themselves: see §14.7.

point about $\hat{\boldsymbol{F}}^{\mu\nu}$ of (8.163) is that it transforms in a very simple way under (8.164) (see problem 8.9):

$$\hat{\boldsymbol{F}}^{\mu\nu} \rightarrow \hat{\boldsymbol{F}}'^{\mu\nu} = \hat{\boldsymbol{F}}^{\mu\nu} - g\hat{\boldsymbol{\eta}} \times \hat{\boldsymbol{F}}^{\mu\nu}. \tag{8.166}$$

(8.166) is precisely the transformation of three things that are a triplet under a *global* SU(2)—see (8.72). The problem of making an invariant out of them is therefore trivial—one takes the SU(2) dot product

$$-\tfrac{1}{4}\hat{\boldsymbol{F}}^{\mu\nu}\cdot\hat{\boldsymbol{F}}_{\mu\nu} \tag{8.167}$$

the $\frac{1}{4}$ being put in to normalise consistently with QED.

(8.167) is the answer to our problem—solved in the Yang–Mills paper. It is remarkably simple and elegant, and *defines* the theory we are talking about (together with (8.162), or whatever matter fields we have). (8.167) is invariant under the *local* transformation (8.164).

Where have our rather complicated vertices got to? For this we must unpack (8.167) a bit into

$$\begin{aligned} &-\tfrac{1}{2}(\partial_\mu \hat{\boldsymbol{W}}_\nu - \partial_\nu \hat{\boldsymbol{W}}_\mu)\cdot\partial^\mu \hat{\boldsymbol{W}}^\nu \\ &+ g(\hat{\boldsymbol{W}}_\mu \times \hat{\boldsymbol{W}}_\nu)\cdot\partial^\mu \hat{\boldsymbol{W}}^\nu \\ &- \tfrac{1}{4}g^2[(\hat{\boldsymbol{W}}^\mu\cdot\hat{\boldsymbol{W}}_\mu)^2 - (\hat{\boldsymbol{W}}^\mu\cdot\hat{\boldsymbol{W}}^\nu)(\hat{\boldsymbol{W}}_\mu\cdot\hat{\boldsymbol{W}}_\nu)]. \end{aligned} \tag{8.168}$$

The 3-W vertices are in the 'g' term, the 4-W ones in the g^2 term. The reason for the collection of terms seen in (8.159) and (8.160) can be understood as follows. Consider the 3-W vertex

$$\langle k_2, \varepsilon_2, s;\, k_3, \varepsilon_3, t|g(\hat{\boldsymbol{W}}_\mu \times \hat{\boldsymbol{W}}_\nu)\cdot\partial^\mu\hat{\boldsymbol{W}}^\nu|k_1, \varepsilon_1, r\rangle \tag{8.169}$$

for example. When each $\hat{\boldsymbol{W}}$ is expressed as a mode expansion, and the initial and final states are also written in terms of $\hat{a}$'s and $\hat{a}^\dagger$'s, the amplitude will be a vacuum expectation value of six $\hat{a}$'s and $\hat{a}^\dagger$'s. The different terms in (8.159) arise from the different ways of getting a non-zero value for this vacuum average, by manipulations similar to those in §5.2.

We have now defined all the tree graphs of an SU(2) gauge theory. The generalisation to an SU(3) one is obtained by simply replacing $\boldsymbol{\tau}/2$ by $\boldsymbol{\lambda}/2$, and ε_{rst} by f_{ijk}: thus we have the rules for the tree graphs of QCD. Vertices for representations of higher dimension that 2 for SU(2) or 3 for SU(3) would be obtained by using appropriately larger matrices, but we shall not need them, as the quarks and leptons are doublets (or singlets) under the SU(2) group of weak isospin (Chapter 14), while the quarks are triplets under SU(3) colour. The fundamental fermions thus do belong to the fundamental (smallest non-trivial) representations of these gauge groups.

We end this formal chapter with a word of warning: for non-Abelian gauge theory the Feynman rules for *loops* cannot simply be inferred from the tree graph vertices. Indeed (8.167) is not the full quantum field Lagrangian for the gauge fields, beyond tree level; additional interactions with 'ghost' particles have to be added in general (these ghosts only propagate in closed loops). We shall give a heuristic discussion of the reason for this in §15.1. For introductory treatments of the ghost Lagrangian in quantum field theory see Aitchison (1982), Mandl and Shaw (1984) or Ryder (1985).

Summary

Global non-Abelian symmetry.

Particle multiplets.

Global non-Abelian symmetries in quantum field theory: currents and conservation laws.

Local non-Abelian symmetry: non-Abelian gauge theories.

Gauge field self-interactions; field strength tensor.

Problems

8.1 Write out each of the 3×3 matrices $T_i^{(1)}$ ($i = 1, 2, 3$) whose matrix elements are given by (8.38), and verify that they satisfy the SU(2) commutation relations (8.37).

8.2 Verify, by comparing coefficients of $\boldsymbol{\varepsilon}$ on both sides of (8.53), that (8.54) follows from (8.53).

8.3 Verify that the operators $\hat{\mathbf{T}}^{(1/2)}$ given by (8.55) do satisfy the commutation relation (8.54) with $\hat{\psi}^{(1/2)}$.

8.4 Verify the commutation relations (8.57) for the operators $\hat{\mathbf{T}}^{(1/2)}$ given by (8.55).

8.5 Verify (8.135).

8.6 Verify that the replacement $\varepsilon_1 \to k_1$ in (8.140) reduces it to (8.143).

8.7 Verify that the term (8.149) leads to the vertex (8.151).

8.8 Verify that the replacement $\varepsilon_1 \to k_1$ in (8.158) reduces it to minus the quantity given in (8.144).

8.9 Verify the transformation law (8.166), given (8.163) and (8.164).

9

AN INTRODUCTION TO QUANTUM CHROMODYNAMICS

In a manner of speaking, objects are colourless

L Wittgenstein, Tractatus Logico-Philosophicus

In the previous chapter we have developed the rules for simple tree graphs in non-Abelian gauge theories. There are now many indications that the theory of the strong interactions between quarks—quantum chromodynamics, or QCD for short—is a theory of this type, in which the gauge group is an $SU(3)_c$, acting on a degree of freedom called 'colour' (indicated by the subscript c). In this chapter we introduce QCD and discuss some of the simplest experimental consequences of it. Perhaps the most remarkable thing about QCD is that, despite it being a theory of the *strong* interactions, there are situations in which it is effectively a quite *weakly* interacting theory (see §9.3). In these cases the simple lowest-order perturbation theory amplitudes (tree graphs) provide a very convincing qualitative, or even 'semi-quantitative', orientation to the data. Thus the tree graph techniques acquired for QED in earlier chapters will produce more useful physics here. Higher-order corrections are generally beyond the scope of this book, but we shall briefly indicate the sort of effects which are predicted, and the reasons—both theoretical and experimental—why it is at present difficult to subject QCD to the kind of quantitative precision tests that its distinguished ancestor continues to pass so brilliantly.

We begin by gathering together, in the next two sections, some motivations for the theory.

9.1 The colour degree of freedom

The first intimation of a new, unrevealed degree of freedom of matter came from baryon spectroscopy (Greenberg 1964). For a baryon made of three spin-$\frac{1}{2}$ quarks, the original non-relativistic quark model wavefunction took the form

$$\psi_{3q} = \psi_{\text{space}}\psi_{\text{spin}}\psi_{\text{flavour}}. \tag{9.1}$$

It was soon realised (e.g. Dalitz 1965) that the product of these space, spin and flavour wavefunctions for the ground state baryons was *symmetric* under interchange of any two quarks. For example, the Δ^{++} state mentioned in §1.2.3 is made of three u quarks (flavour symmetric) in the $J^P = \frac{3}{2}^+$ state, which has zero orbital angular momentum and is hence spatially symmetric, and a symmetric $S = \frac{3}{2}$ spin wavefunction. But we saw in §4.5 that quantum field theory requires fermions to obey the exclusion principle—i.e. the wavefunction ψ_{3q} should be *anti*symmetric with respect to quark interchange. A simple way of implementing this requirement is to suppose that the quarks carry a further degree of freedom, called colour, with respect to which the 3q wavefunction can be antisymmetrised, as follows. We introduce a colour wavefunction with 'colour' index α:

$$\psi^{\alpha} \qquad (\alpha = 1, 2, 3).$$

We are here writing the three labels as '1, 2, 3', but they are often referred to by colour names such as 'red, blue, green'; it should be understood that this is merely a picturesque way of referring to the three basic states of this degree of freedom, and has nothing to do with real colour! With the addition of this degree of freedom we can certainly form a three-quark wavefunction which is antisymmetric in colour by using the antisymmetric symbol $\varepsilon_{\alpha\beta\gamma}$, namely

$$\psi_{3q,\text{colour}} = \varepsilon_{\alpha\beta\gamma}\psi^{\alpha}\psi^{\beta}\psi^{\gamma} \tag{9.2}$$

and this must then be multiplied into (9.1) to give the full 3q wavefunction. To date, *all* known baryon states can be described this way, i.e. the symmetry of the 'traditional' space–spin–flavour wavefunction (9.1) is symmetric overall, while the required antisymmetry is restored by the additional factor (9.2). As far as meson ($\bar{q}q$) states are concerned, what was previously a $\bar{d}u$, say ($\sim \pi^+$), is now

$$\frac{1}{\sqrt{3}}(\bar{d}^1u^1 + \bar{d}^2u^2 + \bar{d}^3u^3) \tag{9.3}$$

which we write in general as $(1/\sqrt{3})\,\bar{\psi}^{\alpha}\psi^{\alpha}$. We shall shortly see the group theoretical significance of this 'neutral superposition', and of (9.2). Meanwhile, we note that (9.2) is actually the *only* way of making an antisymmetric combination of the three ψ's; it is therefore called a (colour) *singlet*. It is reassuring that there is only one way of doing this—otherwise, we would have obtained more baryon states than are physically observed. As we shall see below, (9.3) is also a colour singlet combination.

The above would seem a somewhat artificial device unless there were some physical consequences of this increase in the number of quark types—and there are. In any process which we can describe in terms of creation or annihilation of quarks, the *multiplicity* of quark types will enter into the cross section. For example, at high energies the ratio

$$R = \frac{\sigma(e^+e^- \to \text{hadrons})}{\sigma(e^+e^- \to \mu^+\mu^-)} \tag{9.4}$$

will, in the quark parton model (see §7.5), reflect the magnitudes of the individual quark couplings to the photon:

$$R = \sum_a e_a^2 \tag{9.5}$$

where a runs over all quark types. For five quarks u, d, s, c, b with respective charges $\frac{2}{3}, -\frac{1}{3}, -\frac{1}{3}, \frac{2}{3}, -\frac{1}{3}$, this yields

$$R_{\text{no colour}} = \tfrac{11}{9} \tag{9.6a}$$

and

$$R_{\text{colour}} = \tfrac{11}{3} \tag{9.6b}$$

for the two cases, as we saw in §7.5. The data, shown here again for convenience in figure 9.1, are in reasonable agreement with (9.6b) at energies well above the b threshold†. (At even higher energies, we may expect to see a rise in R above the t threshold, and also effects due to the Z^0 boson; the latter have, in fact, already been observed in the VENUS detector at KEK (Sagawa *et al* 1988).)

In Chapter 7 we also discussed the Drell–Yan process in the quark parton model; it involves the subprocess $q\bar{q} \to l\bar{l}$ which is the inverse of the one in (9.4). We mentioned that a factor of $\frac{1}{3}$ appears in this case: it arises because we must average over the nine possible initial $q\bar{q}$ combinations (factor $\frac{1}{9}$) and then sum over the number of such states that lead to the colour neutral photon, which is 3 ($\bar{q}^1q^1$, $\bar{q}^2q^2$ and $\bar{q}^3q^3$). With this factor, and using quark distribution functions consistent with deep inelastic scattering, one actually finds, however, that the cross sections are about twice as large as the predicted value, at the effective CMS energies for the parton subprocesses which are available at the CERN ISR machine. In QCD this is interpreted as being due to higher-order corrections to the free parton model—but we shall see below that the discrepancy should be less (i.e. the parton model should be better) at higher energies. Indeed, in the very similar process of W and Z production at the CERN $\bar{p}p$ collider, where the effective CMS energy is significantly higher, the Drell–Yan model with the factor $\frac{1}{3}$ gives total cross sections in good agreement with the data (§14.5.4).

9.2 The dynamics of colour

9.2.1 Colour as an SU(3) group

We now want to consider the possible dynamical role of colour—in other words, the way in which the forces between quarks depend on

†They do, however, appear to be significantly *above* the prediction (9.6b); this is actually predicted by QCD, via a higher-order correction to the free parton model—see §9.5 below.

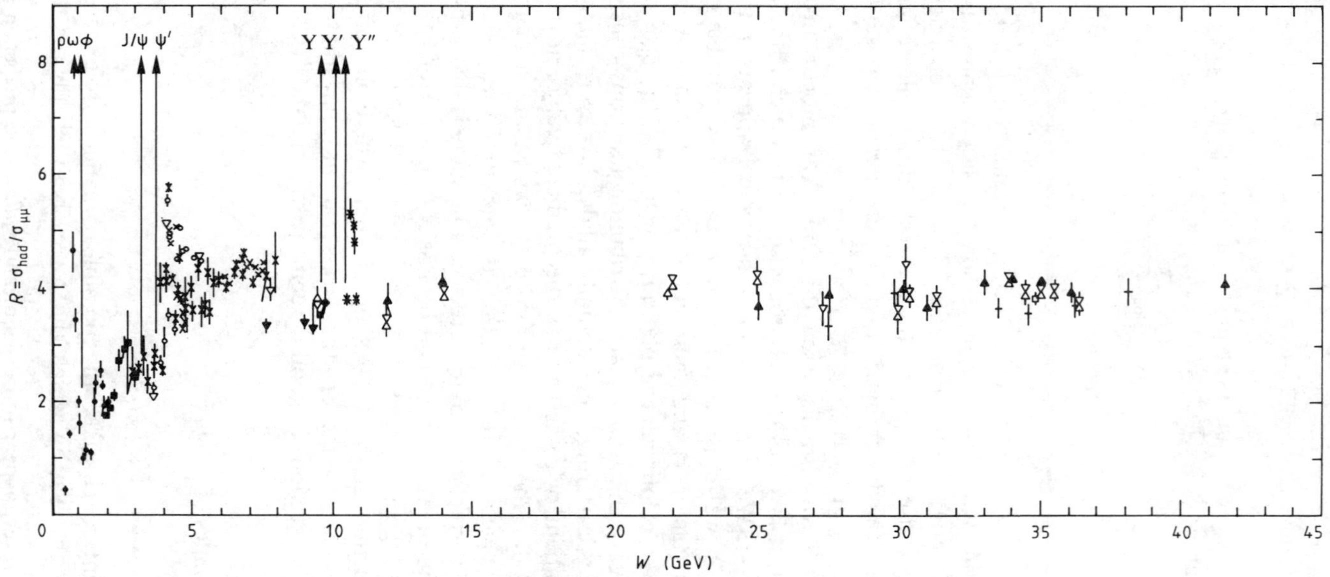

Figure 9.1 The ratio R (see equation (7.89)) (courtesy R Cashmore).

their colours. We have seen that we seem to need three different quark types for each given flavour. They must all have the same mass, or else we would observe some 'fine structure' in the hadronic levels. Furthermore, and for the same reason, 'colour' must be an exact symmetry of the Hamiltonian governing the quark dynamics. What symmetry group is involved? We shall consider how some empirical facts suggest that the answer is $SU(3)_c$.

To begin with, it is certainly clear that the interquark force must depend on colour, since we do *not* observe 'colour multiplicity' of hadronic states: for example we do not see eight other coloured π^+'s ($\bar{d}^B u^R, \bar{d}^G u^R, \ldots$) degenerate with the one 'colourless' physical π^+ whose wavefunction was given previously. The observed hadronic states are all *colour singlets*, and the force must somehow be responsible for this. More particularly, the force has to produce only those very restricted *types* of quark configuration which are observed in the hadron spectrum. Consider again the analogy drawn in §1.2 between isospin multiplets in nuclear physics and in particle physics. There is one very striking difference in the latter case: for mesons *only* $T = 0$, $\frac{1}{2}$ and 1 occur, and for baryons *only* $T = 0$, $\frac{1}{2}$, 1 and $\frac{3}{2}$, while in nuclei there is nothing in principle to stop us finding $T = \frac{5}{2}$, 3, ... states. (In fact such nuclear states are hard to identify experimentally, because they occur at high excitation energy for some of the isobars—cf figure 1.10(*c*)—where the levels are very dense.) The same restriction holds for $SU(3)_f$ also—only **1**'s and **8**'s occur for mesons, and only **1**'s, **8**'s and **10**'s for baryons. In quark terms, this of course is what is translated into the recipe: 'mesons are q̄q, baryons are qqq'. It is as if we said, in nuclear physics, that only $A = 2$ and $A = 3$ nuclei exist! Thus the quark forces must have a dramatic saturation property: apparently no q̄qq, no qqqq, qqqqq, ... states exist. Furthermore, no qq or q̄q̄ states exist either—nor, for that matter, do single q's or q̄'s.

If we assume that only colour singlet states exist, and that the interquark force depends only on colour, the fact that q̄q states are seen but qq and q̄q̄ are not gives us an important clue as to what group to associate with colour. One simple possibility might be that the three colours correspond to the components of an $SU(2)_c$ triplet '$\boldsymbol{\psi}$'. The antisymmetric, colour singlet, three-quark baryon wavefunction of (9.2) is then just the triple scalar product $\boldsymbol{\psi}_1 \cdot \boldsymbol{\psi}_2 \times \boldsymbol{\psi}_3$, which seems satisfactory. But what about the meson wavefunction? In order to discuss this, we need to know how to represent *anti*quarks in terms of these 'colour–space' wavefunctions, given that $\boldsymbol{\psi}$ represents the quarks. The answer to this is that antiquarks are described by the *complex conjugate* of the wavefunctions that describe quarks. We do not wish to digress here to give a full derivation of this fact, but we can at least note a suggestive analogy. A free spin-0 particle with 4-momentum p is described by the wavefunction $\phi \sim \exp(-\mathrm{i}p\cdot x)$; the complex conjugate of this is $\exp(+ip\cdot x)$, which describes a particle of 4-momentum $-p$ —which, by Feynman's interpretation of Chapter 3, is equivalent to an antiparticle of 4-momentum p. Thus we shall assume that if the quark colour triplet wavefunction ψ^α transforms under a colour transformation as

$$\psi^\alpha \to \psi^{\alpha\prime} = Q^\alpha_\beta \psi^\beta \tag{9.7}$$

where Q is a 3×3 unitary matrix appropriate to the $T = 1$ representation of SU(2), then the wavefunction for the 'anti' triplet is $\psi^{\alpha *}$, which transforms as

$$\psi^{\alpha *} \to \psi^{\alpha * \prime} = Q^{\alpha *}_\beta \psi^{\beta *}. \tag{9.8}$$

Given this information, we can now construct colour singlet wavefunctions for mesons, built from q̄q. Consider the quantity (cf (9.3)) $\Sigma_\alpha \psi^{\alpha *}_{(1)} \psi^\alpha_{(2)}$ where $\psi^*_{(1)}$ represents the antiquark and $\psi_{(2)}$ the quark. This may be written in matrix notation as $\psi^\dagger_{(1)}\psi_{(2)}$ where the $\psi^\dagger$ as usual denotes the transpose of the complex conjugate of the column vector ψ. Then, taking the transpose of (9.8), we find that $\psi^\dagger$ transforms by

$$\psi^\dagger \to \psi^{\dagger\prime} = \psi^\dagger Q^\dagger \tag{9.9}$$

so that the combination $\psi^\dagger_{(1)}\psi_{(2)}$ transforms as

$$\psi^\dagger_{(1)}\psi_{(2)} \to \psi^{\dagger\prime}_{(1)}\psi'_{(2)} = \psi^\dagger_{(1)} Q^\dagger Q \psi_{(2)} = \psi^\dagger_{(1)}\psi_{(2)} \tag{9.10}$$

where the last step follows since Q is unitary. Thus the product is *invariant* under (9.7) and (9.9)—that is, it is a colour singlet, as required. This is the meaning of the superposition (9.3).

All this may seem fine, but there is a problem. The three-dimensional representation of SU(2) which we are using here has a very special nature: the matrices Q which describe its transformations can be chosen to be *real*. This can be understood 'physically' if we make use of the great similarity between SU(2) and the group of rotations in three dimensions (which is the reason for the geometrical language of isospin 'rotations', and so on). We know very well how real three-dimensional vectors transform—namely by an orthogonal 3×3 matrix. It is the same in SU(2). It is always possible to choose the wavefunctions ψ to be real, and the transformation matrix Q to be real also. Since Q is, in general, unitary, this means that it must be orthogonal. But now the basic difficulty appears: there is no distinction between ψ and ψ^*! They both transform by the real matrix Q. This means that we can make SU(2) invariant (colour singlet) combinations for q̄q̄ states, and for qq states, just as well as for q̄q states—indeed they are formally identical. But such 'diquark' (or 'antidiquark') states are not found, and hence—by assumption—should *not* be colour singlets.

The next simplest possibility seems to be that the three colours correspond to the components of an $SU(3)_c$ triplet. In this case the quark wavefunction transforms as (9.7), and the antiquark wavefunction as (9.9), but now (cf (8.44))

$$Q = \exp(\mathrm{i}\boldsymbol{\alpha}\cdot\boldsymbol{\lambda}/2). \tag{9.11}$$

The proof of the invariance of $\psi^\dagger_{(1)}\psi_{(2)}$ goes through as before, and it can be shown (problem 9.1) that the antisymmetric 3q combination (9.2) is *also* an $SU(3)_c$ invariant. Thus both the proposed meson and baryon states are colour singlets. It is *not* possible to choose the $\boldsymbol{\lambda}$'s to be pure imaginary in (9.11), and thus the 3×3 Q matrices of $SU(3)_c$ cannot be real, and there is a distinction between ψ and ψ^*. Indeed, it can be shown that it is not possible to form an SU(3) colour singlet combination out of qq or $\bar{q}\bar{q}$. Thus $SU(3)_c$ seems to be a possible and economical choice for the colour group.

9.2.2 *Global $SU(3)_c$ invariance, and gluons*

As stated above, we are assuming, on empirical grounds, that the only physically observed hadronic states are colour singlets—and this now means singlets under $SU(3)_c$. What sort of interquark force could produce this dramatic result? Consider an SU(2) analogy again, the interaction of two nucleons belonging to the lowest (doublet) representation of SU(2). Labelling the states by an isospin T, the possible T values for two nucleons are $T = 1$ (triplet) and $T = 0$ (singlet). We know of an isospin-dependent force which can produce a splitting between these states—namely $V\boldsymbol{\tau}_1\cdot\boldsymbol{\tau}_2$, where the '1' and '2' refer to the two nucleons. The total isospin is $\boldsymbol{T} = \frac{1}{2}(\boldsymbol{\tau}_1 + \boldsymbol{\tau}_2)$, and we have

$$\boldsymbol{T}^2 = \tfrac{1}{4}(\boldsymbol{\tau}_1^2 + 2\boldsymbol{\tau}_1\cdot\boldsymbol{\tau}_2 + \boldsymbol{\tau}_2^2) = \tfrac{1}{4}(3 + 2\boldsymbol{\tau}_1\cdot\boldsymbol{\tau}_2 + 3) \tag{9.12}$$

whence

$$\boldsymbol{\tau}_1\cdot\boldsymbol{\tau}_2 = 2\boldsymbol{T}^2 - 3. \tag{9.13}$$

In the triplet state $\boldsymbol{T}^2 = 2$, and in the singlet state $\boldsymbol{T}^2 = 0$. Thus

$$(\boldsymbol{\tau}_1\cdot\boldsymbol{\tau}_2)_{T=1} = 1 \tag{9.14a}$$

$$(\boldsymbol{\tau}_1\cdot\boldsymbol{\tau}_2)_{T=0} = -3 \tag{9.14b}$$

and if V is positive the $T = 0$ state is pulled down. A similar thing happens in $SU(3)_c$. Suppose this interquark force depended on the quark colours via a term proportional to

$$\boldsymbol{\lambda}_1\cdot\boldsymbol{\lambda}_2. \tag{9.15}$$

Then, in just the same way, we can introduce the total colour operator

$$\boldsymbol{F} = \tfrac{1}{2}(\boldsymbol{\lambda}_1 + \boldsymbol{\lambda}_2) \tag{9.16}$$

so that

$$\boldsymbol{F}^2 = \tfrac{1}{4}(\boldsymbol{\lambda}_1^2 + 2\boldsymbol{\lambda}_1\cdot\boldsymbol{\lambda}_2 + \boldsymbol{\lambda}_2^2) \tag{9.17}$$

and

$$\boldsymbol{\lambda}_1\cdot\boldsymbol{\lambda}_2 = 2\boldsymbol{F}^2 - \boldsymbol{\lambda}^2 \tag{9.18}$$

where $\boldsymbol{\lambda}_1^2 = \boldsymbol{\lambda}_2^2 = \boldsymbol{\lambda}^2$, say. The operator $\boldsymbol{F}^2$ commutes with all components $\boldsymbol{\lambda}_1$ and $\boldsymbol{\lambda}_2$ (as $\boldsymbol{T}^2$ does with $\boldsymbol{\tau}_1$ and $\boldsymbol{\tau}_2$) and is called the quadratic Casimir operator C_2 of $SU(3)_c$ ($\boldsymbol{T}^2$ is the C_2 of SU(2)). Its eigenvalues therefore play a very important role in $SU(3)_c$, analogous to that of the total spin/angular momentum in SU(2). These eigenvalues depend, naturally, on the $SU(3)_c$ representation—indeed, they are one of the defining labels of SU(3) representations in general. The value of $\boldsymbol{F}^2$ for the singlet **1** representation is zero (no multiplicity). $\boldsymbol{F}^2$ for the three-dimensional representation **3** can be evaluated directly by hand from the $\boldsymbol{\lambda}$ matrices (Appendix F); one finds

$$\boldsymbol{F}^2(\mathbf{3}) = \tfrac{1}{4}\boldsymbol{\lambda}^2 = \tfrac{4}{3}. \tag{9.19}$$

Higher representations all give positive values for $\boldsymbol{F}^2$, and generally grow with dimensionality—for example (Lichtenberg 1978)

$$\boldsymbol{F}^2(\mathbf{8}) = 3. \tag{9.20}$$

Thus it follows (from (9.18)) that in the $SU(3)_c$ case, as in the SU(2) analogy, a force of $\boldsymbol{\lambda}_1\cdot\boldsymbol{\lambda}_2$ form will pull down the $\mathbf{1}_c$ states.

Of course, such a simple potential model does not imply that the energy difference between the $\mathbf{1}_c$ states and all 'coloured' states is infinite, as our strict 'colour singlets *only*' hypothesis would demand. Nevertheless, we can ask: what single particle exchange process (cf §5.6) between quark (or antiquark) colour triplets produces a $\boldsymbol{\lambda}_1\cdot\boldsymbol{\lambda}_2$ type of term? The answer is: the exchange of an $SU(3)_c$ **8** of particles, which (anticipating somewhat) we shall call gluons. Since colour is an exact symmetry, the quark wave equation describing the colour interactions must be $SU(3)_c$ covariant. A simple such equation is

$$(\mathrm{i}\not{\partial} - m)\psi = g_s \frac{\lambda^\alpha}{2} A^\alpha \psi \tag{9.21}$$

where g_s is a 'strong charge' and A^α ($\alpha = 1, 2, \ldots, 8$) is an octet of scalar 'gluon potentials'. This corresponds to the quark–gluon vertex

$$-\mathrm{i}g_s\lambda^\alpha/2 \tag{9.22}$$

and when we put two of these together and join them with a gluon propagator (figure 9.2), the $SU(3)_c$ structure of the amplitude will be

$$\psi_{1'}^\dagger \frac{\lambda_1^\alpha}{2}\psi_1 \delta^{\alpha\beta} \psi_{2'}^\dagger \frac{\lambda_2^\beta}{2}\psi_2 = \psi_{1'}^\dagger \frac{\boldsymbol{\lambda}_1}{2}\psi_1 \cdot \psi_{2'}^\dagger \frac{\boldsymbol{\lambda}_2}{2}\psi_2 \tag{9.23}$$

the $\delta^{\alpha\beta}$ arising from the fact that the freely propagating gluon does not change its colour. This interaction has exactly the required '$\boldsymbol{\lambda}_1 \cdot \boldsymbol{\lambda}_2$' character in the colour space.

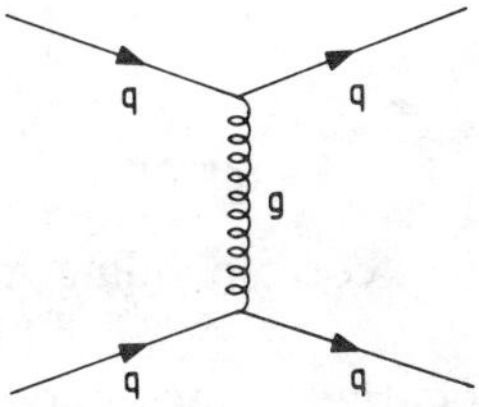

Figure 9.2 Gluon exchange between two quarks.

A natural conjecture at this stage is that these gluons (the quanta of A^α) are the uncharged partons which account for some 50% of a nucleon's momentum, according to the deep inelastic sum rule discussed in §7.3. We shall indicate how this idea is being tested in §9.4.

9.2.3 *Local SU(3)$_c$: Feynman rules for tree graphs in QCD*

It is of course tempting to suppose that the gluons introduced in (9.21) are, in fact, vector particles, like the photons of QED. (9.21) then becomes

$$(\mathrm{i}\not{\partial} - m)\psi = g_s \frac{\lambda^\alpha}{2} \not{A}^\alpha \psi \tag{9.24}$$

and the vertex (9.22) is

$$-\mathrm{i}g_s \frac{\lambda^\alpha}{2} \gamma^\mu . \tag{9.25}$$

Lying behind this temptation is the desire to make the colour dynamics as much like QED as possible, and to derive the dynamics from a gauge principle. As we have seen in the last chapter, this involves the simple but deep step of supposing that the quark world is invariant under *local* SU(3)$_c$ transformations of the form

$$\psi \rightarrow \psi' = \exp(\mathrm{i}\boldsymbol{\alpha}(x) \cdot \boldsymbol{\lambda}/2)\psi . \tag{9.26}$$

This is implemented by the replacement

$$\partial_\mu \rightarrow \partial_\mu + \mathrm{i}g_s \frac{\lambda^\alpha}{2} A^{\alpha\mu}(x) \tag{9.27}$$

which leads immediately to (9.24) and the vertex (9.25) (cf (8.119)) and figure 8.3.

Of course, it also leads to far more—in particular, it implies that the gluons are *massless vector* (spin-1) particles, and that they interact via *three-gluon* and *four-gluon vertices* which are the SU(3) analogues of the SU(2) vertices (8.148) and (8.152). The rules for the gluons are therefore

(i) propagator

$$-\mathrm{i}g^{\mu\nu}\delta^{\alpha\beta}/q^2 \tag{9.28}$$

for the gauge choice corresponding to the Lorentz condition (5.13);

(ii) three-gluon vertex (figure 9.3, momenta ingoing)

$$\begin{aligned}&-g_{\mathrm{s}}f_{\alpha\beta\gamma}[\varepsilon_1\cdot\varepsilon_2(k_1-k_2)\cdot\varepsilon_3+\varepsilon_2\cdot\varepsilon_3(k_2-k_3)\cdot\varepsilon_1\\&+\varepsilon_3\cdot\varepsilon_1(k_3-k_1)\cdot\varepsilon_2];\end{aligned} \tag{9.29}$$

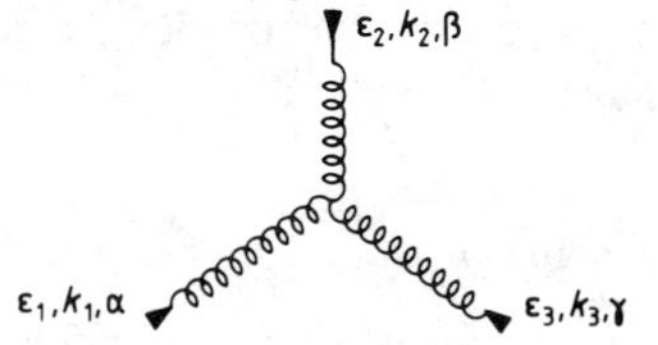

Figure 9.3 Three-gluon vertex.

(iii) four-gluon vertex (figure 9.4, momenta outgoing)

$$\begin{aligned}-\mathrm{i}g_{\mathrm{s}}^2\,[&f_{\alpha\beta\eta}f_{\gamma\delta\eta}(\varepsilon_1\cdot\varepsilon_3\varepsilon_2\cdot\varepsilon_4-\varepsilon_1\cdot\varepsilon_4\varepsilon_2\cdot\varepsilon_3)\\&+f_{\alpha\delta\eta}f_{\beta\gamma\eta}(\varepsilon_1\cdot\varepsilon_2\varepsilon_3\cdot\varepsilon_4-\varepsilon_1\cdot\varepsilon_3\varepsilon_2\cdot\varepsilon_4)\\&+f_{\alpha\gamma\eta}f_{\delta\beta\eta}(\varepsilon_1\cdot\varepsilon_4\varepsilon_2\cdot\varepsilon_3-\varepsilon_1\cdot\varepsilon_2\varepsilon_3\cdot\varepsilon_4)].\end{aligned} \tag{9.30}$$

Local $SU(3)_c$ was proposed as the underlying theory of strong interactions by Fritzsch and Gell-Mann (1972).

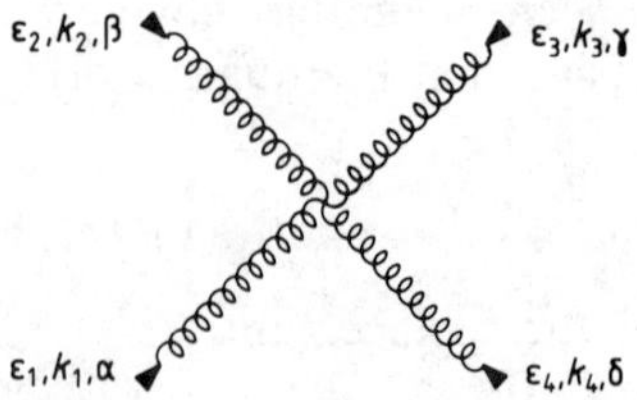

Figure 9.4 Four-gluon vertex.

Elegant and powerful as the gauge principle may be, however, any theory must ultimately stand or fall by its success, or otherwise, in explaining the experimental facts. And this brings us to a central difficulty. We have one well understood and reliable calculational procedure, namely renormalised perturbation theory. Local non-Abelian gauge theories were proved to be renormalisable by 't Hooft (1971a,b)—and it is worth noting that this is *not* true of merely globally invariant non-Abelian theories (gauge invariance is essential for renormalisability†). However, we can only use perturbation theory for relatively weak interactions, whereas QCD is supposed to be a strong interaction theory. How can our perturbative QED techniques possibly be used for QCD? Despite the considerable formal similarities between the two theories, which we have emphasised, they differ in at least one crucial respect: the fundamental quanta of QED (leptons and photons) are observed as free particles, but those of QCD (quarks and gluons) are not. It seems that in order to compare QCD with data we shall inevitably have to reckon with the complex non-perturbative strong interaction processes ('confinement') which bind quarks and gluons into hadrons—and the underlying simplicity of the QCD structure will be lost.

We must now recall from Chapter 7 the very considerable empirical success of the parton model, in which the interactions between the partons (now interpreted as quarks and gluons) were totally ignored! Somehow it does seem to be the case that in deep inelastic scattering—or more generally 'hard', high-energy, wide-angle collisions—the hadron constituents are very nearly free, and the effective interaction is *weak*. On the other hand, we are faced with an almost paradoxical situation, because we also know that the forces are indeed so *strong* that no one has yet succeeded in separating completely either a quark or a gluon from a hadron, so that they emerge as free particles. The resolution of this unprecedented mystery lies in a fundamental feature of non-Abelian gauge theories, called 'asymptotic freedom'. This property is the most compelling theoretical motivation for choosing such a theory of the strong interactions, and it establishes the circumstances in which a perturbative (tree graph) approach is useful in 'strong interaction' physics. For this reason we discuss it now, and then proceed with some simple perturbative calculations in §§9.4–9.6.

9.3 Asymptotic freedom

The essential physical idea is this: the interaction between two particles can be characterised by an effective 'charge' (coupling parameter) which is a function of the distance between them, rather than being a simple constant. In the case of non-Abelian gauge theories, the effective

†This is the real reason for the insistence on gauge invariance in §8.4.

coupling goes to zero as the separation becomes very small, and such theories are called asymptotically (i.e. as $r \to 0$) free. This is why the quarks and gluons seem to be weakly interacting at the short distances probed in hard collisions.

Let us illustrate how a distance-dependent charge can arise in a familiar context. Consider a test charge q in a polarisable dielectric medium, such as water. If we introduce another test charge $-q$ into the medium, the electric field between the two test charges will line up the water molecules (which have a permanent electric dipole moment) as shown in figure 9.5. There will be an induced dipole moment $\boldsymbol{P}$ per unit volume, and the effect of $\boldsymbol{P}$ on the resultant field is (from elementary electrostatics) the same as that produced by a volume charge equal to $-\operatorname{div}\boldsymbol{P}$. If, as is usual, $\boldsymbol{P}$ is taken to be proportional to $\boldsymbol{E}$, $\boldsymbol{P} = \chi\varepsilon_0\boldsymbol{E}$, Gauss's law will be modified from

$$\nabla\cdot\boldsymbol{E} = \rho_{\text{free}}/\varepsilon_0 \tag{9.31}$$

to

$$\nabla\cdot\boldsymbol{E} = (\rho_{\text{free}} - \operatorname{div}\boldsymbol{P})/\varepsilon_0 = \rho_{\text{free}}/\varepsilon_0 - \nabla\cdot(\chi\boldsymbol{E}) \tag{9.32}$$

where ρ_{free} refers to the test charges introduced into the dielectric. If χ is slowly varying as compared to $\boldsymbol{E}$, it may be taken as approximately constant in (9.32), which may then be written as

$$\nabla\cdot\boldsymbol{E} = \rho_{\text{free}}/\varepsilon \tag{9.33}$$

where $\varepsilon = (1 + \chi)\varepsilon_0$ is the dielectric constant of the medium, ε_0 being that of the vacuum. Thus the field is effectively reduced by the factor $(1 + \chi)^{-1} = \varepsilon_0/\varepsilon$.

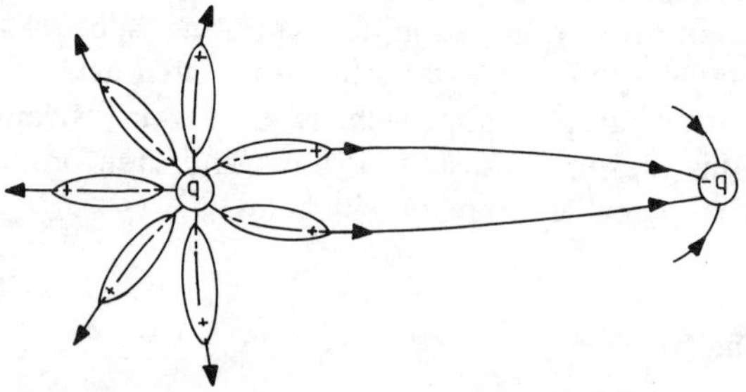

Figure 9.5 Screening of charge in a dipolar medium (from Aitchison 1985).

This is all familiar ground. Note, however, that the above treatment is essentially macroscopic, the molecules being replaced by a continuous distribution of charge density $-\operatorname{div}\mathbf{P}$. When the distance between the two test charges is as small as, roughly, the molecular diameter, the above reduction—or *screening* effect—must cease, and the field between them has the full unscreened value. In general, the electrostatic potential between two test charges q_1 and q_2 in a dielectric can be represented phenomenologically by

$$V(r) = q_1 q_2 / 4\pi\varepsilon(r) r \tag{9.34}$$

where $\varepsilon(r)$ is assumed to vary slowly from the value ε for $r \gg d$ to the value ε_0 for $r \ll d$, where d is the diameter of the polarised molecules. The situation may be described in terms of an effective charge

$$q' = q/[\varepsilon(r)]^{1/2} \tag{9.35}$$

for each of the test charges. Thus we have an effective charge which depends on the interparticle separation, as shown in figure 9.6.

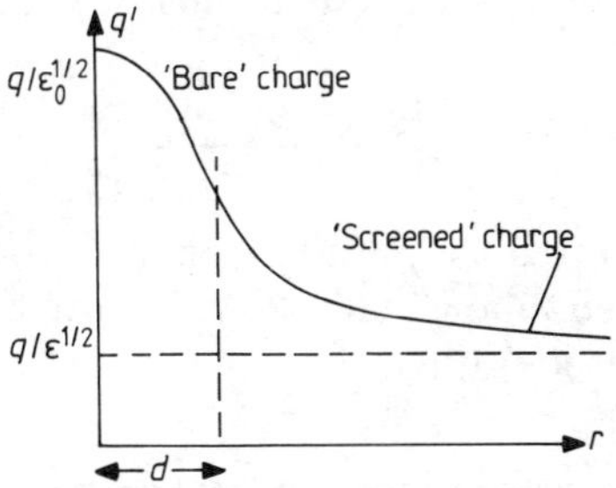

Figure 9.6 Effective (screened) charge versus separation between charges (Aitchison 1985).

Now consider the application of this idea to a quantum field theory—first of all QED—replacing the polarisable medium by the *vacuum*. We recall that the concept of vacuum polarisation was introduced in §6.10. The important idea is that, in the vicinity of a test charge *in vacuo*, charged pairs can be created. Pairs of particles of mass m can exist for a time of the order of $\Delta t \sim \hbar/mc^2$. They can spread apart a distance of order $c\Delta t$ in this time, i.e. a distance of approximately $\hbar/mc$, which is the Compton wavelength $\lambda\!\!\!^{-}_{\mathrm{C}}$. This distance gives a measure of the 'molecular diameter' we are talking about, since it is the polarised virtual pairs which now provide a *vacuum* screening effect around the original charged particle. In this analogy the bare vacuum (no virtual pairs) corresponds to the 'vacuum' used in the previous macroscopic analysis, and the physical vacuum (virtual pairs) to the polarisable dielectric. We cannot, of course, get 'outside' the physical vacuum, so that we are really always dealing with effective charges that depend on r. What, then, do we mean by the familiar symbol e? This is simply the effective charge as $r \to \infty$—or, in practice, the charge relevant for distances much larger than the particles' Compton wavelengths. It is this large distance value that is measured in Thomson scattering, for example.

In the covariant context of QED, one does not want to talk in terms of distance dependence: the appropriate analogue is a coupling strength (or dielectric constant) that depends on q^2, the invariant momentum transfer carried by the mediating photon. As usual, the large-q^2 behaviour of the one-photon amplitude will correspond to the short-distance behaviour of the interparticle potential, while the 'physical' (i.e. conventional) charge e is the limit $e(q^2)$ as $q^2 \to 0$, the large-distance region.

An indication of the q^2 dependence actually to be expected in QED can be obtained from the expressions already given in §6.10, where a vacuum polarisation contribution to $e^-\mu^-$ scattering was explicitly discussed. Equation (6.202) shows that (after infinite renormalisation) the particles can indeed be regarded as having a q^2-dependent charge given by

$$e(q^2) = e\left(1 - \frac{\alpha}{15\pi}\frac{q^2}{m^2} + \dots\right)^{1/2} \tag{9.36}$$

for small q^2, where the dots indicate higher-order corrections. Note that $e(0) = e$, as required above. It is more usual to consider the fine structure constant itself rather than the charge, so that one writes in general

$$\alpha(q^2) = \alpha[1 + \alpha F(q^2/m^2)] \tag{9.37}$$

for the effective q^2-dependent fine structure constant, with $\alpha = \alpha(0)$.

Equation (9.37) represents a *finite* change in α arising from vacuum polarisation. This change may be regarded as a renormalisation effect, using the word now in a sense quite similar to the finite renormalisations in solid state physics, mentioned in §6.10. The particular aspect of (9.37) on which we are now focusing is that it is a q^2-dependent renormalisation effect. We emphasised the measurability of this effect in §6.10.

At large $|q^2|(|q^2| \gg m^2)$, F grows logarithmically, and (see e.g. Bjorken and Drell 1964)

$$\alpha(q^2) = \alpha[1 + (\alpha/3\pi)\ln(|q^2|/m^2) + \dots]. \tag{9.38}$$

Equation (9.38) shows that the effective strength $\alpha(q^2)$ tends to increase at large $|q^2|$ (short distances). This is, after all, physically reasonable: the reduction in the effective charge caused by the dielectric constant associated with the polarisation of the vacuum disappears (the charge increases) as we pass 'inside' some typical dipole length. In the present case, that length is m^{-1} (in our standard units $\hbar = c = 1$), the electron Compton wavelength, a typical distance over which the fluctuating pairs extend.

Equation (9.38) is the lowest-order correction to α, in a form valid for $|q^2| \gg m^2$. It turns out that, in this limit, the dominant vacuum polarisation contributions can be isolated in each order of perturbation theory and summed explicitly. The result of summing these 'leading logarithms' is

$$\alpha(Q^2) = \frac{\alpha}{[1 - (\alpha/3\pi)\ln(Q^2/m^2)]} \qquad \text{for } Q^2 \gg m^2 \tag{9.39}$$

where we now introduce $Q^2 = -q^2$, a positive quantity when q is a momentum transfer. Although (9.39) might seem to be an improvement over (9.38)—since it sums an infinite number of contributions—this is not really so in practice. As Q^2 increases, (9.39) will still only differ from (9.38) by a per cent or so until $Q^2 \sim m^2 \exp(137)$, which is a region of energy ($\sim 10^{26}$ GeV) quite outside practical exploration, and one which, moreover, corresponds to a length scale far smaller than that at which hitherto neglected gravitational effects would presumably have entered (cf the discussion after (6.196)). In the energy region of physical interest, (9.39) is indistinguishable from (9.38), due essentially to the very small value of α. We shall now see that, in several respects, the situation is quite different in the non-Abelian case.

To be specific, we treat QCD in which the gauge coupling constant g_s of the local $SU(3)_c$ appears, playing a role similar to e. By analogy with QED, we introduce the 'strong' structure constant $\alpha_s = g_s^2/4\pi$ and its Q^2-dependent generalisation of $\alpha_s(Q^2)$. But there is an immediate difficulty. In QED two sources of charge can be physically separated to large distances, and the 'physical' effective charge defined in that limit ($e = e(Q^2 = 0)$). Such a separation seems not to be possible in QCD: we are *not* able to measure any such 'physical' g_s or α_s in the limit $Q^2 \to 0$. Actually, one is not *forced* to define the physical e in QED via the $Q^2 \to 0$ limit, though it is certainly physically reasonable, and convenient. Any other Q^2 would have done. In QCD, since $Q^2 = 0$ cannot be used, we introduce an arbitrary parameter into the theory, μ^2 say, with dimension [mass]^2, so that the 'reference' value of α_s is $\alpha_s(Q^2 = \mu^2)$. The question now is: how does $\alpha_s(Q^2)$ vary for $Q^2 \neq \mu^2$? In particular, what happens as $Q^2 \to \infty$?

Certainly the one-gluon exchange graph in, say, quark–quark scattering will be corrected by a vacuum polarisation diagram such as figure 9.7, involving a $q\bar{q}$ loop. This will produce just the same effect as the e^+e^- loop in QED. However, in QCD (and any non-Abelian gauge theory) there is another contribution, in the same order (α_s^2) of perturbation theory: namely, the gluon loop contribution shown in figure 9.8. The origin of this graph lies in the fact, repeatedly stressed, that in the non-Abelian case the gluons carry the gauge charge, and

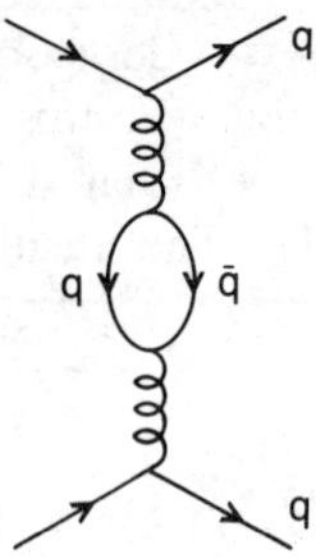

Figure 9.7 Vacuum polarisation correction to one-gluon exchange process in qq scattering.

interact with themselves. Another graph involving the gluon self-interaction, and entering at order α_s^2, is shown in figure 9.9. As yet, no simple physical argument (comparable to the 'polarisation' one for loops in QED) exists which will explain whether these kinds of gluon effects will increase or decrease $\alpha_s(Q^2)$ at large Q^2 (see, however, Hughes (1981) and Nielsen (1981)).

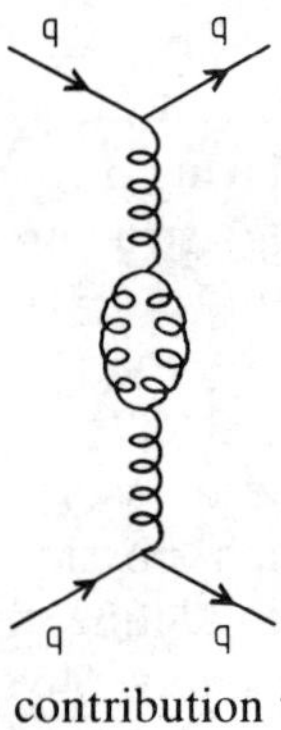

Figure 9.8 Gluon loop contribution to vacuum polarisation.

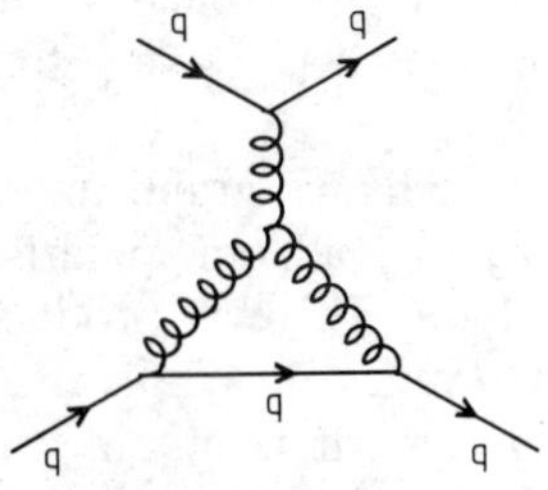

Figure 9.9 Another graph contributing at order α_s^2.

The matter can be settled by calculation, however (Politzer 1973, Gross and Wilczek 1973). It turns out that the gluon contributions do act in the opposite sense to the $q\bar{q}$ ones; they cause a *decrease* in $\alpha_s(Q^2)$ at large Q^2. The $SU(3)_c$ analogue of (9.38) is found to be

$$\alpha_s(Q^2) = \alpha_s(\mu^2)\left[1 - \frac{\alpha_s(\mu^2)}{12\pi}(33 - 2f)\ln\left(\frac{Q^2}{\mu^2}\right) + \ldots\right]. \quad (9.40)$$

The '$-2f$' contribution comes from the quark loop and is essentially the same (apart from a colour-related factor of 2) as the $f = 1$ QED expression (9.38). The crucial new point is the '33' piece, which arises from the gluons and *has the opposite sign* from the quark loop piece (its magnitude depends on the colour group: it would be 22 for $SU(2)_c$). $\alpha_s(Q^2)$ will now tend to *decrease* at large Q^2 provided $33 > 2f$, i.e. $f \leqslant 16$. The contribution of the Yang–Mills quanta is responsible for this 'antiscreening' property: no theory without Yang–Mills fields has it (Coleman and Gross 1973). Very roughly speaking, one has the picture that the effect of the gluons is to 'spread out' the QCD charge of the quarks, because the gluons can 'split apart'; scattering at large Q^2 from a spread-out charge distribution is always weaker than from a pointlike one, since for large Q^2 we get inside the area of spreading, therefore seeing less of the charge.

Once again, the dominant logarithms in higher orders can be summed, and (9.40) is then replaced by the analogue of (9.39), namely

$$\alpha_s(Q^2) = \alpha_s(\mu^2)\left[1 + \left(\frac{\alpha_s(\mu^2)}{12\pi}\right)(33 - 2f)\ln\left(\frac{Q^2}{\mu^2}\right)\right]^{-1}. \quad (9.41)$$

In this form we can see that $\alpha_s(Q^2) \to 0$ as $Q^2 \to \infty$ (for $f \leqslant 16$), *which is the property called 'asymptotic freedom'*; the approach of $\alpha_s(Q^2)$ to zero at large Q^2 is the basic reason for hoping that perturbation theory is relevant in high-Q^2 (short-distance) processes.

We note that, in contrast to the QED case, (9.41) is a significant improvement on (9.40), for two reasons. First of all, assuming that we have indeed correctly identified the dominant contributions in each order, and provided $f \leqslant 16$, (9.41) can be used to give a meaningful answer even for such large Q^2 that the perturbative series (9.40) breaks down. For $f \leqslant 16$ the denominator in (9.41) is always positive for $Q^2 > 0$; in the QED case, (9.39) cannot be used for arbitrarily high Q^2, due to the (presumably unphysical) singularity associated with the vanishing of the denominator. Secondly, comparison of QCD predictions with data (see below) indicates that typical values of α_s are in the range $0.1 \lesssim \alpha_s \lesssim 0.5$; for such an α_s the variation with Q^2—for energies greater than 1 GeV, say—is much more significant than in the QED case.

The second point is more easily seen if it is understood that, despite appearances, (9.41) actually does not involve two independent parameters μ^2 and $\alpha_s(\mu^2)$. By introducing

$$\ln \Lambda^2 = \ln \mu^2 - 12\pi/[(33 - 2f)\alpha_s(\mu^2)]$$

(9.41) can be rewritten as (problem 9.2)

$$\alpha_s(Q^2) = \frac{12\pi}{(33 - 2f)\ln(Q^2/\Lambda^2)} \tag{9.42}$$

showing that $\alpha_s(Q^2)$ depends on only one parameter in this leading log approximation. The extraction of a precise value of Λ from experiment is fraught with theoretical and experimental difficulties (Duke and Roberts 1985), but typical values lie in the range 100–200 MeV. We cannot believe (9.42) for $Q^2 \sim \Lambda^2$, for then $\alpha_s(Q^2)$ becomes very large and the perturbative approach behind the equation breaks down. Indeed, we can think of Λ as the (energy) scale at which the 'strong interactions become strong', i.e. the confining forces set in. Alternatively, in terms of distance, or range, we have a confining region of about 1 fm, suggesting a value for Λ of about 200 MeV. Figure 9.10 shows (9.42) as a function of Q for $f = 6$ and $\Lambda = 100$ MeV, and gives some indication of the expected variation of α_s with Q^2. We shall comment further on the experimental situation below.

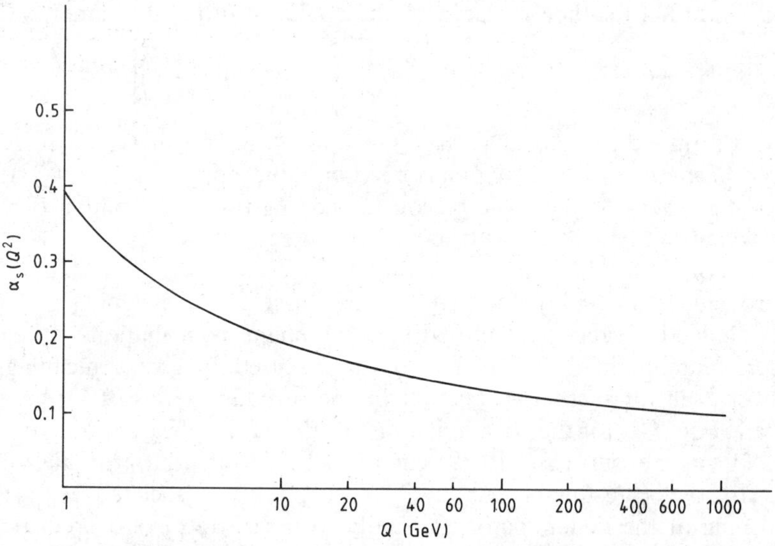

Figure 9.10 $\alpha_s(Q^2)$ versus Q for $\Lambda = 100$ MeV.

One final comment: we think of QED as being characterised by one parameter, the dimensionless fine structure constant α. By contrast, in the case of QCD the input parameter is a *mass*, namely Λ.

9.4 Hard scattering processes and QCD tree graphs

9.4.1 Two-jet events in $\bar{p}p$ collisions

In Chapter 7 we introduced the parton model and discussed how it successfully interpreted deep inelastic and large-Q^2 data in terms of the free pointlike hadronic 'partons'. This was a model rather than a theory: the theme of the rest of this chapter will be the way in which the theory of QCD both justifies the parton model and predicts observable corrections to it. In other words, the 'partons' are now to be identified precisely with the QCD quanta (quarks, antiquarks and gluons). We shall usually continue to use the language of 'partons', however, rather than—say—that of hadronic 'constituents', for the following reason. It is only at relatively low energies and/or momentum transfers that the (essentially non-relativistic) concept of a fixed number of constituents in a bound state is meaningful. At relativistic energies and short distances, pair creation and other fluctuation phenomena are so important that it no longer makes sense to think so literally of 'what the bound state is made of'—as we shall see, when we look more closely at it (larger Q^2), more and more 'bits' are revealed. In this situation we prefer 'partons' to 'constituents', since the latter term seems to us to carry with it the traditional connotation of a fixed number.

In §7.5 we briefly introduced the idea of *jets* in e^+e^- physics: well collimated sprays of hadrons, apparently created as a quark–antiquark pair, separate from each other at high speed. The dynamics at the parton level, $e^+e^- \rightarrow \bar{q}q$, was governed by QED. We also saw, in §7.4, how in hadron–hadron collisions the hadrons acted as beams of partons—quarks, antiquarks and gluons—which could produce $\bar{l}l$ pairs by the inverse process $\bar{q}q \rightarrow \bar{l}l$: this force is electromagnetic, and well described in lowest order of perturbative QED. However, it is clear that collisions between the hadronic partons should by no means be limited to QED-induced processes. On the contrary, we expect to see strong interactions between the partons, determined by QCD. In general, therefore, the data will be complicated and hard to interpret, due to all these non-perturbative strong interactions. But asymptotic freedom has taught us that at short distances we can use perturbation theory even for 'strong interactions'. Thus we might hope that the identification and analysis of short-distance parton–parton collisions will lead to direct tests of the tree graph structure of QCD.

How are short-distance collisions to be identified experimentally? The answer is: in just the same way as Rutherford distinguished the presence of a small heavy scattering centre (the nucleus) in the atom—by looking at secondary particles emerging at large angles with respect to the beam direction. For each secondary particle we can define a transverse momentum $p_T = p \sin\theta$, where p is the particle momentum and θ is the emission angle with respect to the beam axis. If hadronic matter were smooth and uniform (cf the Thomson atom), the distribution of events in p_{T} would be expected to fall off very rapidly at large p_{T} values—perhaps exponentially. This is just what is observed in the vast majority of events: the average value of p_{T} measured for charged particles at current collider energies is very low $\langle p_{\mathrm{T}}\rangle \sim 0.4$ GeV (Arnison *et al* 1983a, Banner *et al* 1983a). However, in a small fraction of collisions the emission of high-p_{T} secondaries is observed. They were first seen (Büsser *et al* 1972, 1973, Alper *et al* 1973, Banner *et al* 1983a) at the CERN ISR (CMS energies 30–62 GeV), and were interpreted in parton terms as follows. The physical process is viewed as a two-step one. In the first stage (figure 9.11) a parton from one hadron undergoes a short-distance collision with a parton from the other, leading in lowest-order perturbation theory to two wide-angle partons, emerging at high speed from the collision volume.

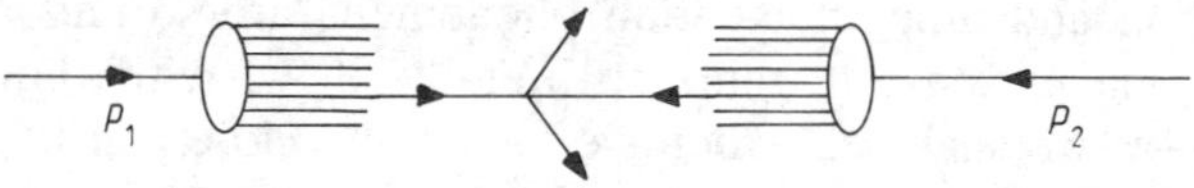

Figure 9.11 Parton–parton collision.

As the two partons separate, the effective interaction strength increases (figure 9.10) and the second stage is entered, that in which the coloured partons turn themselves—under the action of the strong colour-confining force—into colour singlet hadrons. As yet there is no detailed dynamical understanding of this second (non-perturbative) stage, which is called parton *fragmentation*. Nevertheless, we can argue that for the forces to be strong enough to produce the observed hadrons, the dominant processes in the fragmentation stage must involve small momentum transfer. Thus we have a picture in which two fairly well collimated *jets* of hadrons occur, each having a total 4-momentum approximately equal to that of the parent parton (figure 9.12). Thus jets will be the observed hadronic manifestation of the underlying QCD processes.

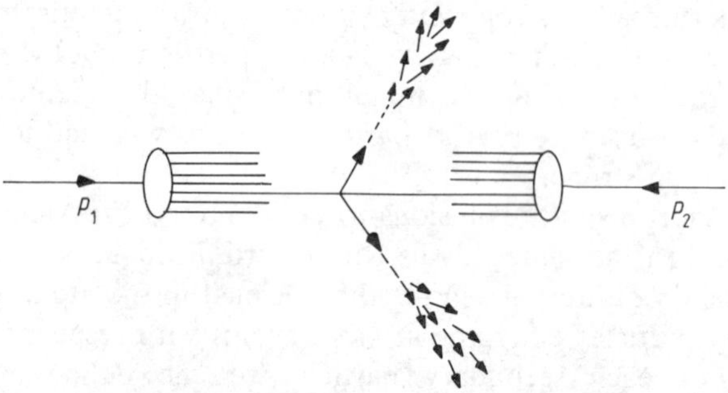

Figure 9.12 Parton fragmentation into jets.

We now face the experimental problem of picking out, from the enormous multiplicity of total events, just these hard scattering ones, in order to analyse them further. Early experiments used a trigger based on the detection of a single high-p_T particle. But it turns out that such triggering really reduces the probability of observing jets, since the probability that a single hadron in a jet will actually carry most of the jet's total transverse momentum is quite small (Jacob and Landshoff 1978; Collins and Martin 1984, Chapter 5). It is much better to surround the collision volume with an array of calorimeters which measure the total energy deposited. *Wide-angle jets* can then be identified by the occurrence of a large amount of total transverse energy deposited in a number of adjacent calorimeter cells: this is the modern 'jet trigger'. The importance of calorimetric triggers was first emphasised by Bjorken (1973), following earlier work by Berman *et al* (1971). The application of this method to the detection and analysis of wide-angle jets was first reported by the UA2 collaboration at the CERN $\bar{\text{p}}$p collider (Banner *et al* 1982). An impressive body of quite remarkably clean jet data has since been accumulated by both the UA1 and UA2 collaboration, as we now discuss.

For each event the total transverse energy ΣE_T is measured where

$$\sum E_T = \sum_i E_i \sin\theta_i. \tag{9.43}$$

E_i is the energy deposited in the ith calorimeter cell and θ_i is the polar angle of the cell centre; the sum extends over all cells. Figure 9.13 shows the observed ΣE_T distribution: it follows the 'soft' exponential form for $\Sigma E_T \lesssim 60$ GeV, but thereafter departs from it, showing clear evidence of the wide-angle collisions characteristic of hard processes.

As we shall see shortly, the majority of 'hard' events are of two-jet type, with the jets sharing the ΣE_T approximately equally. Thus a 'local' trigger set to select events with localised transverse energy $\gtrsim 30$ GeV and/or a 'global' trigger set at $\gtrsim 60$ GeV can be used. At collider energies ($\sqrt{s} \sim 500$–600 GeV) there is plenty of energy available to produce such events.

The total $\sqrt{s}$ value is important for another reason. Consider the kinematics of the two-parton collision (figure 9.11) in the $\bar{\text{p}}$p CMS. As in the Drell–Yan process of §7.4, the right-moving parton has 4-momentum

$$x_1 p_1 = x_1(P, 0, 0, P) \tag{9.44}$$

and the left-moving one

$$x_2 p_2 = x_2(P, 0, 0, -P) \tag{9.45}$$

where $P = \sqrt{s}/2$ and we are neglecting parton transverse momenta,

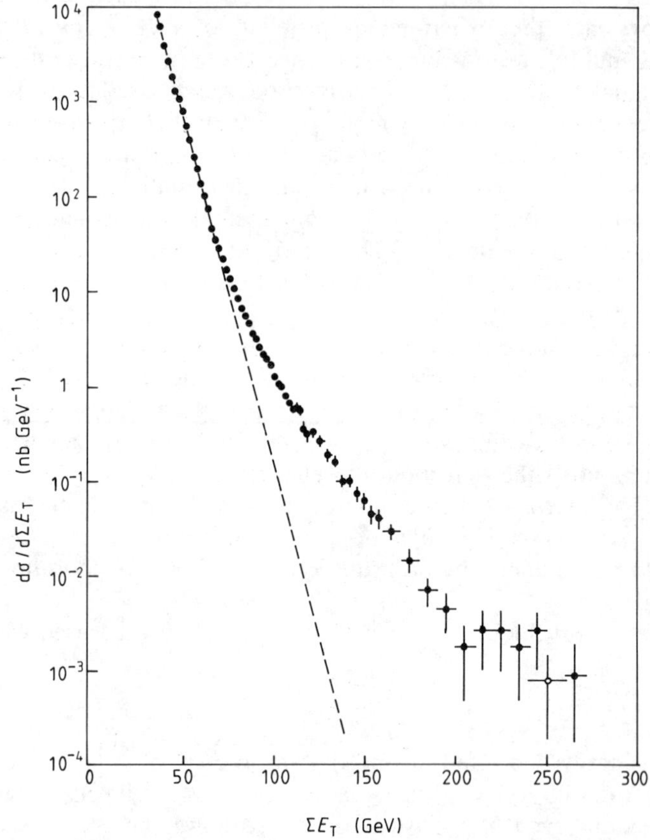

Figure 9.13 Distribution of the total transverse energy ΣE_T observed in the UA2 central calorimeter (DiLella 1985).

which are approximately limited by the observed $\langle p_T \rangle$ value (~0.4 GeV and thus negligible on the collider energy scale). Consider the simple case of 90° scattering, which requires (for massless partons) $x_1 = x_2$, equal to x say. The total outgoing transverse energy is then $2xP = x\sqrt{s}$. If this is to be greater than 50 GeV, we need $x \gtrsim 0.1$ at collider energies. The parton distribution functions are large at these small x values, due to sea quarks (§7.3) and gluons (figure 9.18 below), and thus we expect to obtain a reasonable cross section. At ISR energies ($\sqrt{s} \sim 60$ GeV), such a high ΣE_T cut would imply x's much nearer 1, where the distribution functions are very small; but the alternative of a lower ΣE_T cut will fail to pick out the jets. Indeed, the cross sections at the $\bar{p}p$ collider for inclusive jet production at 90°, as a function of jet transverse momentum, are some 10^3 times larger than those found at the ISR for the same ΣE_T—a fact which was predicted by Horgan and Jacob (1981).

What are the characteristics of jet events? When ΣE_T is large enough ($\gtrsim 150$ GeV), it is found that essentially all of the transverse energy is indeed split roughly equally between two approximately back-to-back jets. A typical such event is shown in figure 9.14. Returning to the kinematics of (9.44) and (9.45), x_1 will not in general be equal to x_2, so that—as is apparent in figure 9.14—the jets will not be collinear. However, to the extent that the transverse parton momenta can be neglected, the jets will be coplanar with the beam direction, i.e. their relative azimuthal angle will be 180°. Figure 9.15 shows a number of examples in which the distribution of the transverse energy over the calorimeter cells is analysed as a function of the jet opening angle θ and the azimuthal angle ϕ. It is strikingly evident that we are seeing precisely a kind of 'Rutherford' process, or—to vary the analogy—we might say that hadronic jets are acting as the modern counterpart of Faraday's iron filings, in rendering visible the underlying field dynamics!

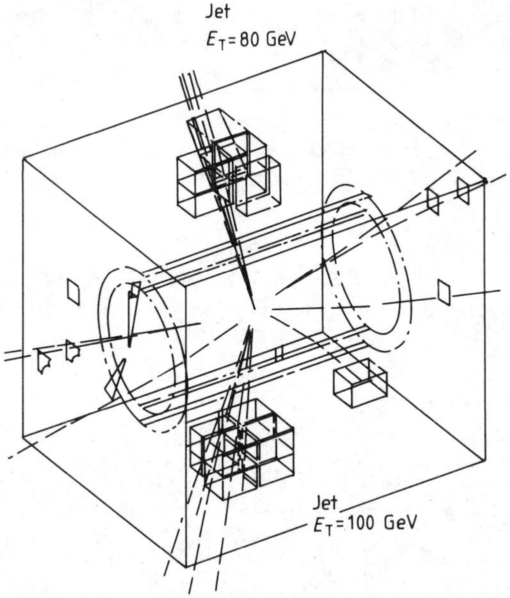

Figure 9.14 Two-jet event. Two tightly collimated groups of reconstructed charged tracks can be seen in the cylindrical central detector of UA1, associated with two large clusters of calorimeter energy depositions (Geer 1986).

We may now consider more detailed features of these two-jet events—in particular, the expectations based on QCD tree graphs (note that if we take ΣE_T as the appropriate energy scale, $\alpha_s \lesssim 0.2$). The initial hadrons provide wide-band beams of quarks, antiquarks and

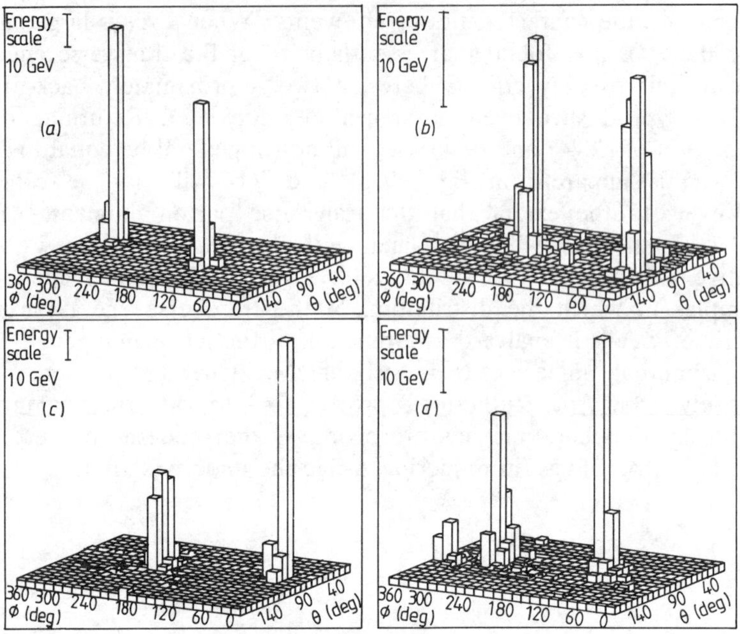

Figure 9.15 Four transverse energy distributions for events with $\Sigma E_{\mathrm{T}} >$ 100 GeV, in the θ, ϕ plane (UA2, DiLella 1985). Each bin represents a cell of the UA2 calorimeter. Note that the sum of the ϕ's equals 180° (mod 360°).

gluons; thus we shall have many parton subprocesses—$\mathrm{qq} \to \mathrm{qq}$, $\mathrm{q\bar{q}} \to \mathrm{q\bar{q}}$, $\mathrm{q\bar{q}} \to \mathrm{gg}$, $\mathrm{gg} \to \mathrm{gg}$, etc. The cross section will be given, in the parton model, by a formula of the Drell–Yan type, except that the electromagnetic annihilation cross section

$$\sigma(\mathrm{q\bar{q}} \to \mu^+\mu^-) = 4\pi\alpha^2/3q^2 \tag{9.46}$$

is replaced by the various QCD subprocess cross sections, each one being weighted by the appropriate distribution functions. At first sight this seems to be a very complicated story, with so many contributing parton processes. But a significant simplification comes from the fact that in the CMS of the parton collision, all processes involving one gluon exchange will lead to essentially the same dominant angular distribution of Rutherford type, $\sim \sin^{-4}\theta/2$, where θ is the parton CMS scattering angle (recall §1.2!). This is illustrated in table 9.1 (taken from Combridge *et al* (1977)), which lists the different relevant spin-averaged squared matrix elements $|M|^2$, where the parton differential cross section is given by (cf (5.115))

$$\frac{\mathrm{d}\sigma}{\mathrm{d}\cos\theta} = \frac{\pi\alpha_{\mathrm{s}}^2}{2\hat{s}}|M|^2. \tag{9.47}$$

Table 9.1

Subprocess	$\|M\|^2$	Dominant QCD graph
$\left.\begin{matrix} qq \to qq \\ q\bar{q} \to q\bar{q} \end{matrix}\right\}$	$\frac{4}{9}\left(\frac{\hat{s}^2 + \hat{u}^2}{\hat{t}^2}\right)$	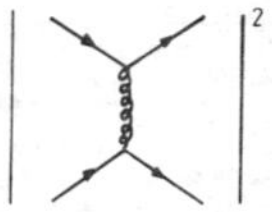
$qg \to qg$	$\frac{\hat{s}^2 + \hat{u}^2}{\hat{t}^2} + \dots$	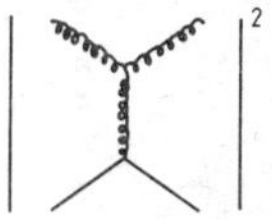
$gg \to gg$	$\frac{9}{4}\left(\frac{\hat{s}^2 + \hat{u}^2}{\hat{t}^2}\right) + \dots$	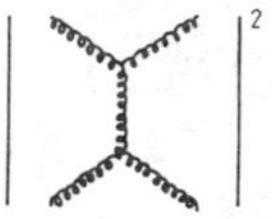

Here $\hat{s}$, $\hat{t}$ and $\hat{u}$ are the subprocess invariants, so that

$$\hat{s} = (x_1 p_1 + x_2 p_2)^2 = x_1 x_2 s \qquad \text{(cf (7.75))}. \tag{9.48}$$

Continuing to neglect the parton transverse momenta, the initial parton configuration shown in figure 9.11 can be brought to the parton CMS by a Lorentz transformation along the beam direction, the outgoing partons then emerging back-to-back at an angle θ to the beam axis, so $\hat{t} \propto (1 - \cos\theta) \propto \sin^2\theta/2$. Only the terms in $(\hat{t})^{-2} \sim \sin^{-4}\theta/2$ are given in table 9.1. We note that the $\hat{s}$, $\hat{t}$, $\hat{u}$ dependence of these terms is the same for the three types of process (and is in fact the same as that found for the 1γ exchange process $e^+e^- \to \mu^+\mu^-$—see problem 6.6, converting $d\sigma/dt$ into $d\sigma/d\cos\theta$). Figure 9.16 shows the two-jet angular distribution measured by UA1 (Arnison *et al* 1985). The broken curve is the exact angular distribution predicted by all the QCD tree graphs—it actually follows the $\sin^{-4}\theta/2$ shape quite closely.

This analysis surely constitutes compelling evidence for elementary hard scattering events proceeding via the exchange of a massless quantum. The small discrepancy between 'tree graph' theory and experiment in figure 9.16 can be accounted for by including some higher-order corrections in α_s, as shown by the full curve.

The fact that the angular distributions of all the subprocesses are so similar allows further important information to be extracted from these two-jet data. In general, the parton model cross section will have the form (cf (7.82))

$$\frac{d^3\sigma}{dx_1\, dx_2\, d\cos\theta} = \sum_{a,b} \frac{F_a(x_1)}{x_1} \cdot \frac{F_b(x_2)}{x_2} \cdot \sum_{c,d} \frac{d\sigma_{ab\to cd}}{d\cos\theta} \tag{9.49}$$

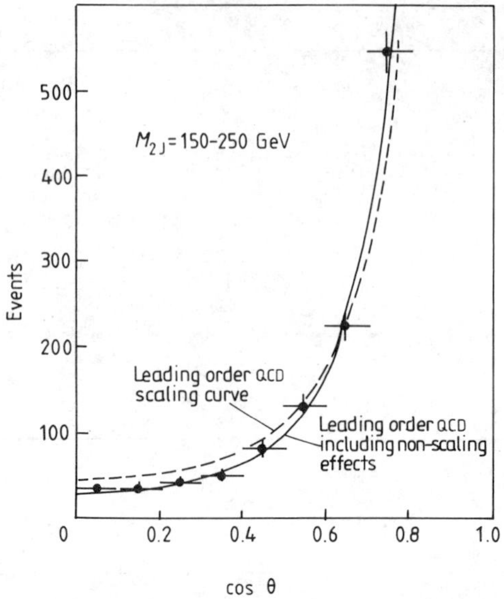

Figure 9.16 Two-jet angular distribution plotted against $\cos\theta$ (Arnison *et al* 1985).

where $F_a(x_1)/x_1$ is the distribution function for partons of type 'a' (q, $\bar{q}$ or g), and similarly for $F_b(x_2)/x_2$. Using the near identity of all $d\sigma/d\cos\theta$'s, and noting the numerical factors in table 9.1, the sums over parton types reduce to

$$\tfrac{9}{4}\{g(x_1) + \tfrac{4}{9}[q(x_1) + \bar{q}(x_1)]\}\{g(x_2) + \tfrac{4}{9}[q(x_2) + \bar{q}(x_2)]\} \quad (9.50)$$

where $g(x)$, $q(x)$ and $\bar{q}(x)$ are the gluon, quark and antiquark distribution functions. Thus effectively the weighted distribution function†

$$F(x) = g(x) + \tfrac{4}{9}[q(x) + \bar{q}(x)] \quad (9.51)$$

is measured; in fact, with the weights as in (9.50),

$$\frac{d^3\sigma}{dx_1\,dx_2\,d\cos\theta} = \frac{F(x_1)}{x_1}\cdot\frac{F(x_2)}{x_2}\cdot\frac{d\sigma_{gg\to gg}}{d\cos\theta}. \quad (9.52)$$

x_1 and x_2 are kinematically determined from the measured jet variables: from (9.48),

$$x_1x_2 = \hat{s}/s \quad (9.53)$$

where $\hat{s}$ is the invariant [mass]^2 of the two-jet system and

$$x_1 - x_2 = 2P_L/\sqrt{s} \quad (9.54)$$

†The $\frac{4}{9}$ reflects the relative strengths of the quark–gluon and gluon–gluon couplings in QCD.

with P_L the total two-jet longitudinal momentum. Figure 9.17 shows $F(x)$ obtained in the UA1 (Arnison *et al* 1984) and UA(2) (Bagnaia *et al* 1984) experiments. Also shown in this figure is the expected $F(x)$ based on fits to the deep inelastic neutrino scattering data at $Q^2 = 20\ \text{GeV}^2$ and $2000\ \text{GeV}^2$ (Abramovicz *et al* 1982a,b, 1983)—the reason for the change with Q^2 will be discussed in §9.6. The agreement is qualitatively very satisfactory. Subtracting the distributions for quarks and antiquarks as found in deep inelastic lepton scattering, UA1 deduce the gluon structure function $g(x)$ shown in figure 9.18. It is clear that gluon processes will dominate at small x—and even at larger x will be important because of the colour factors in table 9.1.

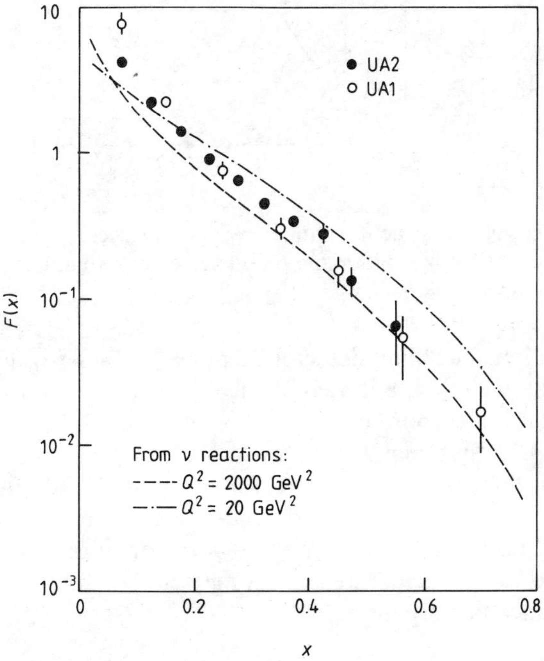

Figure 9.17 Effective structure function measured from two-jet events (Arnison *et al* 1984 and Bagnaia *et al* 1984). The broken and chain curves are obtained from deep inelastic neutrino scattering. Taken from DiLella (1985).

In concluding this important section we should note some qualifications. Parton masses have been neglected—but experiments measure hadron jets with invariant masses as low as a few GeV, comparable to heavy quark masses. Some fragmentation model is needed to relate the parton p_T to the jet transverse energy ΣE_T (e.g. QCD-inspired Monte

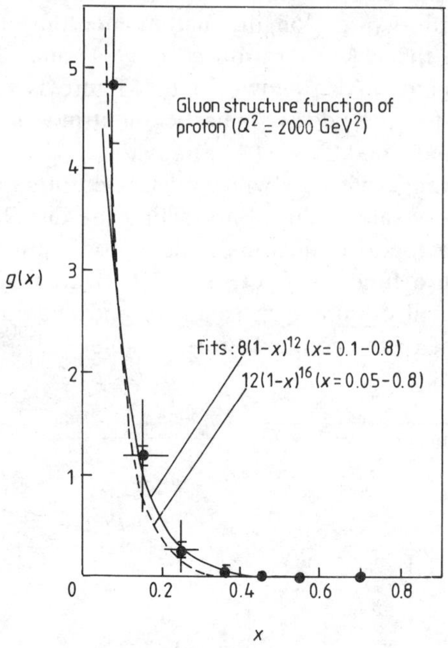

Figure 9.18 The gluon structure function $g(x)$ extracted from the effective structure function $F(x)$ by subtracting the expected contribution from the quarks and antiquarks (Geer 1986).

Carlo simulation, including detector response). Higher-order QCD effects should also be included, but a complete calculation is still missing—their effect is usually incorporated as an overall multiplicative factor $K \lesssim 2$ (Furman 1982). Systematic experimental uncertainties include energy calibration of the calorimeters, detector acceptance and integrated beam luminosity. Altogether, theory and experiment can only be compared, as regards absolute value, to within about a factor of 2. Nevertheless, it is clear that the two-jet data already strongly support the basic predictions of short-distance perturbative QCD.

9.4.2 Three-jet events

Although most of the high-ΣE_T events at the collider are two-jet events, in some 10–30% of the cases the energy is shared between three jets. An example is included as (*d*) in the collection of figure 9.15; a clearer one is shown in figure 9.19. In QCD such events are interpreted as arising from a 2 parton $\rightarrow$ 2 parton + 1 gluon process of the type gg $\rightarrow$ ggg, gq $\rightarrow$ ggq, etc. Once again, one can calculate (Kunszt and Piétarinen 1980, Gottschalk and Sivers 1980, Berends *et al* 1981) all possible contributing tree graphs, of the kind shown in figure 9.20, which should

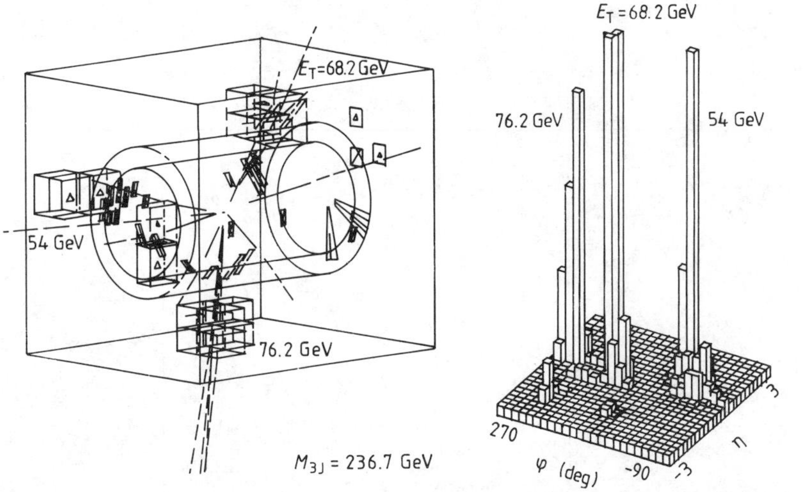

Figure 9.19 Three-jet event in the UA1 detector, and the associated transverse energy flow plot (Geer 1986).

dominate at small α_s. They are collectively known as QCD single-bremsstrahlung diagrams. Analysis of triple jets which are well separated both from each other and from the beam directions shows that the data are in good agreement with these lowest-order QCD predictions. For example, figure 9.21 shows the production angular distribution of UA2 (Appel *et al* 1986) as a function of $\cos\theta^*$, where θ^* is the angle between the leading (most energetic) jet momentum and the beam axis, in the three-jet CMS. It follows just the same $\sin^{-4}\theta^*/2$ curve as in the two-jet case (the data for which are also shown in the figure), as expected for massless quantum exchange—the particular curve is for the representative process $gg \to ggg$. Another qualitative feature is that the ratio of three-jet to two-jet events is controlled, roughly, by α_s (compare figure 9.20 with the graphs in table 9.1); precise extraction of α_s is difficult, but it appears to be in the expected range 0.15–0.25 at $Q^2 \sim 4000\ \text{GeV}^2$.

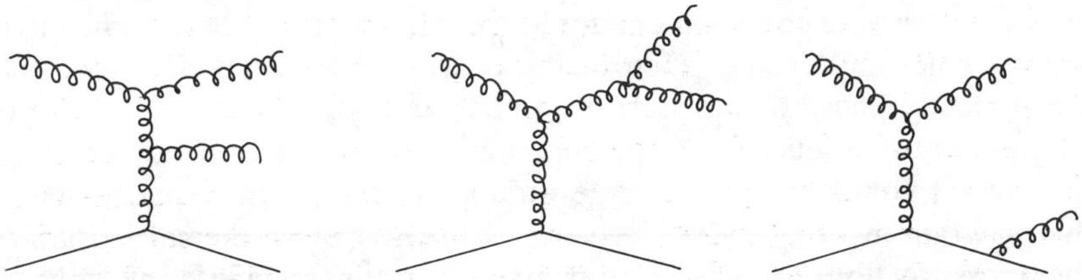

Figure 9.20 Some tree graphs associated with three-jet events.

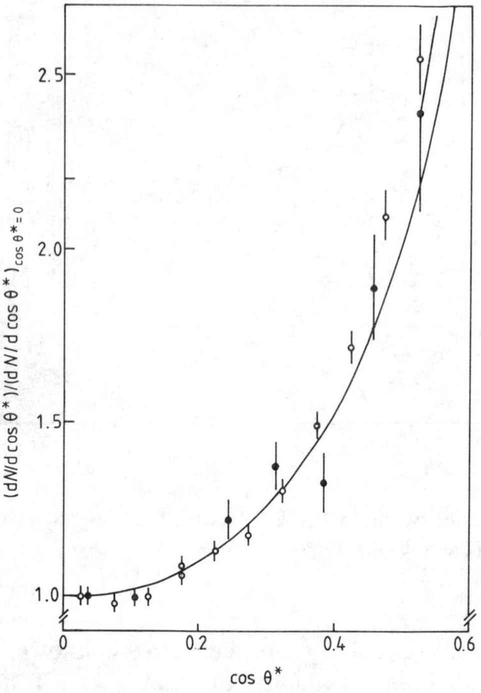

Figure 9.21 The distribution of cos $\theta^*(\bullet)$, the angle of the leading jet with respect to the beam line (normalised to unity at cos $\theta^* = 0$), for three-jet events in $\bar{p}p$ collisions (Appel *et al* 1986). The distribution for two-jet events is also shown ($\circ$). The full curve is a parton model calculation using the tree graph amplitudes for $gg \rightarrow ggg$, and cut-offs in transverse momentum and angular separation to eliminate divergences (see remarks following equation (9.70)).

Other interesting distributions concern the characteristics of the three-jet final state itself. At this point it is convenient to leave $\bar{p}p$ collisions and consider three-jet events in e^+e^- collisions instead: these originate, according to QCD, from gluon bremsstrahlung corrections to the two-jet parton model process $e^+e^- \rightarrow \gamma^* \rightarrow q\bar{q}$, as shown in figure 9.22, the existence of which was predicted by Ellis *et al* (1976b) and subsequently observed by Brandelik *et al* (1979) and Barber *et al* (1979), thus providing early encouragement for QCD. The situation here is in many ways simpler and cleaner than in the $\bar{p}p$ case: the initial state 'partons' are perfectly physical QED quanta, and their total 4-momentum is zero, so that the three jets have to be coplanar; further, there is only one type of diagram compared to the large number in the $\bar{p}p$ case, and much of that diagram involves the easier vertices of QED. Since the calculation of the cross section predicted from figure 9.22 is relevant not only to three-jet production in e^+e^- collisions, but also to QCD corrections to the *total* e^+e^- annihilation cross section (and to scaling violations in deep inelastic scattering as well), we shall now consider it in some detail.

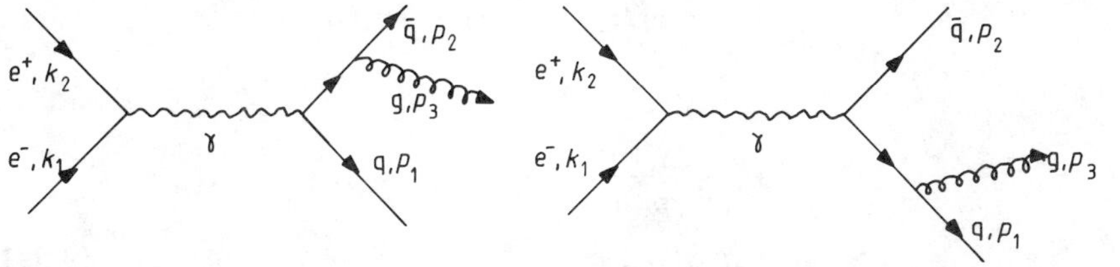

Figure 9.22 Gluon brehmsstrahlung corrections to two-jet parton model process.

The quark, antiquark and gluon 4-momenta are p_1, p_2 and p_3 respectively, as shown in figure 9.22; the e^+ and e^- 4-momenta are k_1 and k_2. The cross section is then (cf (5.90))

$$d\sigma = \frac{1}{(2\pi)^5}\delta^4(k_1 + k_2 - p_1 - p_2 - p_3)\frac{|F|^2}{2Q^2}\frac{d^3p_1}{2E_1}\frac{d^3p_2}{2E_2}\frac{d^3p_3}{2E_3} \tag{9.55}$$

where (neglecting all masses)

$$F = \frac{e_a e^2 g_s}{Q^2}\bar{v}(k_2)\gamma^\mu u(k_1)\left(\bar{u}(p_1)\gamma_\nu\frac{\lambda^\alpha}{2}\cdot\frac{(\not{p}_1 + \not{p}_3)}{2p_1\cdot p_3}\cdot\gamma_\mu v(p_2)\right.$$
$$\left. - \bar{u}(p_1)\gamma_\mu\frac{\lambda^\alpha}{2}\cdot\frac{(\not{p}_2 + \not{p}_3)}{2p_2\cdot p_3}\cdot\gamma_\nu v(p_2)\right)\varepsilon^{*\nu}(\lambda)a^\alpha \tag{9.56}$$

and $Q^2 = 4E^2$ is the square of the total e^+e^- energy, and the square of the virtual photon's 4-momentum Q, and e_a (in units of e) is the charge of a quark of type 'a'. Note the minus sign in (9.56): the antiquark coupling is $-g_s$. a^α is the colour wavefunction of the gluon ($\alpha = 1, \ldots, 8$); the colour parts of the q and q̄ wavefunctions are understood to be included in the u and v factors. Since the colour parts separate from the Dirac trace parts, we shall ignore them to begin with, and reinstate the result of the colour sum (via problem 9.3) at the end. Averaging over $e^\pm$ spins and summing over final state quark spins and gluon polarisation λ (using (5.170)), we obtain (problem 9.4)

$$\frac{1}{4}\sum_{\text{spin } s,\lambda}|F|^2 = \frac{e^4 e_a^2 g_s^2}{8Q^4}L^{\mu\nu}(k_1, k_2)H_{\mu\nu}(p_1, p_2, p_3) \tag{9.57}$$

where the lepton tensor is, as usual (equation (6.88)),

$$L^{\mu\nu}(k_1, k_2) = 2(k_1^\mu k_2^\nu + k_1^\nu k_2^\mu - k_1{\cdot}k_2 g^{\mu\nu}) \tag{9.58}$$

and the hadron tensor is

$$\begin{aligned} H_{\mu\nu}(p_1, p_2, p_3) = & \frac{1}{p_1{\cdot}p_3}[L_{\mu\nu}(p_2, p_3) - L_{\mu\nu}(p_1, p_1) + L_{\mu\nu}(p_1, p_2)] \\ & + \frac{1}{p_2{\cdot}p_3}[L_{\mu\nu}(p_1, p_3) - L_{\mu\nu}(p_2, p_2) \\ & + L_{\mu\nu}(p_1, p_2)] \\ & + \frac{p_1{\cdot}p_2}{(p_1{\cdot}p_3)(p_2{\cdot}p_3)}[2L_{\mu\nu}(p_1, p_2) + L_{\mu\nu}(p_1, p_3) \\ & + L_{\mu\nu}(p_2, p_3)]. \end{aligned} \tag{9.59}$$

Combining (9.58) and (9.59) allows complete expressions for the five-fold differential cross section to be obtained (Ellis *et al* 1976b).

As yet e^+e^- data are not extensive enough to permit such differential cross sections to be studied, and so one integrates over three angles describing the orientation (relative to the beam axis) of the production plane containing the three jets. After this integration, the (doubly differential) cross section is a function of two independent Lorentz invariant variables, which are conveniently taken to be two of the three s_{ij} defined by

$$s_{ij} = (p_i + p_j)^2. \tag{9.60}$$

Since we are considering the massless case $p_i^2 = 0$ throughout, we may also write

$$s_{ij} = 2p_i{\cdot}p_j. \tag{9.61}$$

These variables are linearly related by

$$2(p_1{\cdot}p_2 + p_2{\cdot}p_3 + p_3{\cdot}p_1) = Q^2 \tag{9.62}$$

as follows from

$$(p_1 + p_2 + p_3)^2 = Q^2 \tag{9.63}$$

and $p_i^2 = 0$. The integration yields (Ellis *et al* 1976b)

$$\frac{d^2\sigma}{ds_{13}\,ds_{23}} = \tfrac{2}{3}\alpha^2 e_a^2 \alpha_s \frac{1}{(Q^2)^3}\left(\frac{s_{13}}{s_{23}} + \frac{s_{23}}{s_{13}} + \frac{2Q^2 s_{12}}{s_{13}s_{23}}\right) \tag{9.64}$$

where $\alpha_s = g_s^2/4\pi$.

We may understand the form of this result in a simple way, as follows. It seems plausible that after integrating over the production angles, the lepton tensor will be proportional to $Q^2 g^{\mu\nu}$, all directional knowledge of the k_i having been lost. Indeed, if we use $-g^{\mu\nu}L_{\mu\nu}(p, p') = 4p{\cdot}p'$ together with (9.59), we easily find

$$-\tfrac{1}{4}g^{\mu\nu}H_{\mu\nu} = \frac{p_1\cdot p_3}{p_2\cdot p_3} + \frac{p_2\cdot p_3}{p_1\cdot p_3} + \frac{p_1\cdot p_2 Q^2}{(p_1\cdot p_3)(p_2\cdot p_3)} = \frac{s_{13}}{s_{23}} + \frac{s_{23}}{s_{13}} + \frac{2Q^2 s_{12}}{s_{13}s_{23}} \tag{9.65}$$

exactly the factor appearing in (9.64). In turn, the result may be given a simple physical interpretation. From (5.168) we note that we can replace $-g^{\mu\nu}$ by $\Sigma_{\lambda'}\varepsilon^{\mu}(\lambda')\varepsilon^{\nu *}(\lambda')$ for a virtual photon of polarisation λ', the $\lambda' = 0$ state contributing negatively. Thus effectively the result of doing the angular integration is (up to constants and Q^2 factors) to replace the lepton factor $\bar{v}(k_2)\gamma^{\mu}u(k_1)$ by $-\mathrm{i}\varepsilon^{\mu}(\lambda')$, so that F is proportional to the $\gamma^* \to \mathrm{q}\bar{\mathrm{q}}\mathrm{g}$ processes shown in figure 9.23. But these are basically the same amplitudes as the ones we already met in Compton scattering (§6.9). To compare with §6.9.2, we convert the initial state fermion (electron/quark) into a final state antifermion (positron/antiquark) by $\mathrm{p} \to -\mathrm{p}$, and then identify the variables of figure 9.23 with those of figure 6.13(a) by

$$p' \to p_1,\ k' \to p_3,\ -p \to p_2;\ s \to 2p_1\cdot p_3 = s_{13},\ t \to 2p_1\cdot p_2 = s_{12},$$
$$u \to 2p_2\cdot p_3 = s_{23}. \tag{9.66}$$

Remembering that in (6.189) the virtual γ had squared 4-momentum $-Q^2$, we see that the Compton '$\Sigma|F|^2$' of (6.189) indeed becomes proportional to the factor (9.65), as expected.

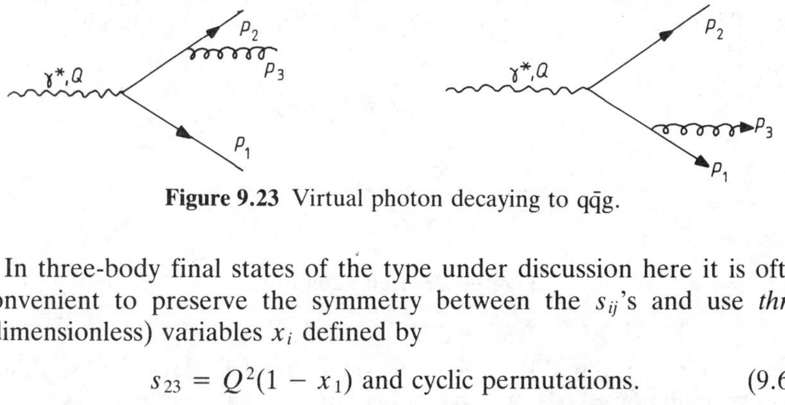

Figure 9.23 Virtual photon decaying to $\mathrm{q}\bar{\mathrm{q}}\mathrm{g}$.

In three-body final states of the type under discussion here it is often convenient to preserve the symmetry between the s_{ij}'s and use *three* (dimensionless) variables x_i defined by

$$s_{23} = Q^2(1 - x_1) \text{ and cyclic permutations.} \tag{9.67}$$

These are related by (9.62), which becomes

$$x_1 + x_2 + x_3 = 2. \tag{9.68}$$

An event with a given value of the set $\{x_i\}$ can then be plotted as a point in an equilateral triangle of height 1, as shown in figure 9.24. In order to find the limits of the allowed physical region in this x_i space, and because it will be useful subsequently, we now transform from the

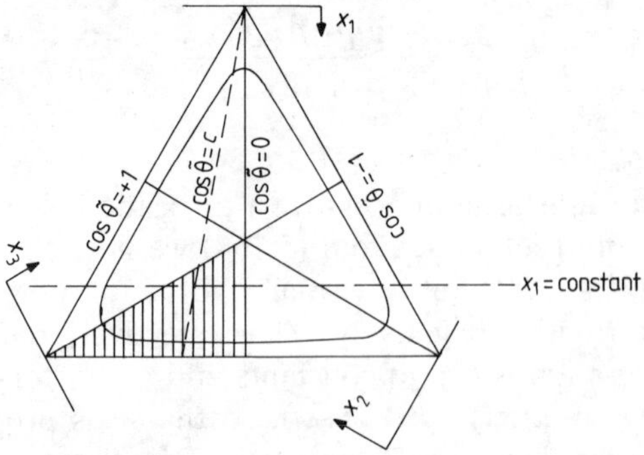

Figure 9.24 The kinematically allowed region in $\{x_i\}$ is the interior of the equilateral triangle.

overall three-body CMS to the CMS of 2 and 3 (figure 9.25). If $\tilde{\theta}$ is the angle between 1 and 3 in this system, then (problem 9.5)

$$\begin{aligned} x_2 &= (1 - x_1/2) + (x_1/2)\cos\tilde{\theta} \\ x_3 &= (1 - x_1/2) - (x_1/2)\cos\tilde{\theta}. \end{aligned} \tag{9.69}$$

The limits of the physical region are then clearly $\cos\tilde{\theta} = \pm 1$, which correspond to $x_2 = 1$ and $x_3 = 1$. By symmetry, we see that the entire perimeter of the triangle in figure 9.24 is the required boundary: physical events fall anywhere inside the triangle. (This is the massless limit of the classic Dalitz plot, first introduced by Dalitz (1953) for the analysis of $K \to 3\pi$.) Lines of constant $\tilde{\theta}$ are shown in figure 9.24.

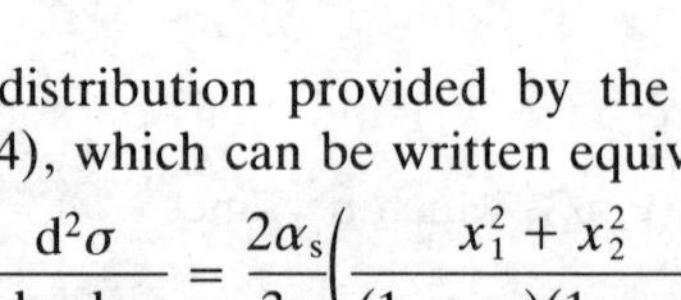

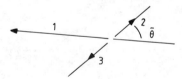

Figure 9.25 Definition of $\tilde{\theta}$.

Now consider the distribution provided by the QCD bremsstrahlung process, equation (9.64), which can be written equivalently as

$$\frac{1}{\sigma_{\text{pt}}}\frac{\mathrm{d}^2\sigma}{\mathrm{d}x_1\mathrm{d}x_2} = \frac{2\alpha_s}{3\pi}\left(\frac{x_1^2 + x_2^2}{(1-x_1)(1-x_2)}\right) \tag{9.70}$$

where σ_{pt} is the pointlike $e^+e^- \to$ hadrons total cross section of (7.88), and a factor of 4 has been introduced from the colour sum (problem 9.3). The factor in large parentheses is (9.65) written in terms of the x_i (problem 9.6). The most striking feature of (9.70) is that it is *infinite*, as x_1 or x_2, or both, tend to 1! This is a quite different infinity from the ones encountered in the loop integrals mentioned in §§5.9 and 6.10. No integral is involved here—the tree amplitude itself is becoming singular on the phase space boundary. We can trace the origin of the infinity back to the denominator factors $(p_1 \cdot p_3)^{-1} \sim (1 - x_2)^{-1}$ and $(p_2 \cdot p_3)^{-1} \sim (1 - x_1)^{-1}$ in (9.56). These become zero in two distinct configurations of the gluon momentum:

$$\text{(a) } p_3 \propto p_1 \quad \text{or} \quad p_3 \propto p_2 \quad (\text{using } p_i^2 = 0) \tag{9.71}$$

$$\text{(b) } p_3 \to 0 \tag{9.72}$$

which are easily interpreted physically. Condition (a) corresponds to a situation in which the 4-momentum of the gluon is parallel to that of either the quark or the antiquark; this is called a 'collinear divergence'† and the configuration is pictured in figure 9.26(*a*). Condition (b) corresponds to the emission of a very 'soft' gluon (figure 9.26(*b*)) and is called a 'soft divergence'. It is apparent from these figures that in either of these two cases the observed final state hadrons, after the fragmentation process, will in fact resemble a *two*-jet configuration. Such events will be found in the regions $x_1 \approx 1$ and/or $x_2 \approx 1$ of the kinematical plot shown in figure 9.24, which correspond to strips adjacent to two of the boundaries of the triangle, and to regions near the vertices of the triangle. Events inside the rounded triangular region should be mostly three-jet events.

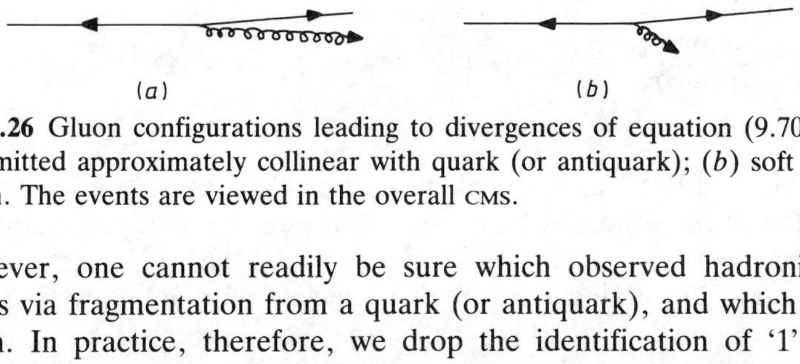

Figure 9.26 Gluon configurations leading to divergences of equation (9.70): (*a*) gluon emitted approximately collinear with quark (or antiquark); (*b*) soft gluon emission. The events are viewed in the overall CMS.

However, one cannot readily be sure which observed hadronic jet emerges via fragmentation from a quark (or antiquark), and which from a gluon. In practice, therefore, we drop the identification of '1' with quark, '2' with antiquark and '3' with gluon, and instead define x_1 as referring simply to the most energetic jet, x_2 to the next, and x_3 to the least energetic. In this way the events in the entire triangle are effectively folded onto the shaded one-sixth portion in figure 9.24, for which $x_1 > x_2 > x_3$. To isolate the three-jet events, we must keep away from the boundaries of the triangle: thus only events inside a region such as that bounded by the rounded line in figure 9.24 are considered in the remainder of this section. (Note, however, that to calculate the total cross section predicted by (9.70), we must understand how to deal

†Alternatively, a 'mass singularity', since it would be absent if either the gluon or the quark had mass.

with the boundary region: this will be discussed in the following section.) One particularly interesting distribution is that in $\cos\tilde{\theta}$, advocated (Ellis and Karliner 1979) and used (Brandelik *et al* 1980) as an early indicator for the spin of the gluon. Problem 9.7 shows that if the emitted gluon in figure 9.23 has spin 0, the expression (9.65) is changed to

$$-\tfrac{1}{4}g^{\mu\nu}H^{\mathrm{s}}_{\mu\nu} = \frac{1}{2}\left(\frac{p_1\cdot p_3}{p_2\cdot p_3} + \frac{p_3\cdot p_2}{p_1\cdot p_3} + 2\right) = \frac{x_3^2}{2(1-x_1)(1-x_2)} \tag{9.73}$$

and (9.70) becomes

$$\frac{1}{\sigma_{\mathrm{pt}}}\frac{\mathrm{d}^2\sigma^{\mathrm{S}}}{\mathrm{d}x_1\mathrm{d}x_2} = \frac{\alpha_{\mathrm{s}}'}{3\pi}\left(\frac{x_3^2}{2(1-x_1)(1-x_2)}\right) \tag{9.74}$$

where $\alpha_{\mathrm{s}}' = g_{\mathrm{s}}'^2/4\pi$ is the analogue of the QCD α_{s} for 'scalar gluons'. Clearly, for fixed x_1, the $\tilde{\theta}$ dependence of (9.70) and (9.74) will be different (see equation (9.69)). Indeed, adding together the cyclic permutations of (9.70) and (9.74) (corresponding to the folding of the Dalitz plot) and using (9.69), one finds (problem 9.8) that the angular distributions at fixed x_1, normalised to 1 at $\cos\tilde{\theta} = 0$, are

spin 1: $$\frac{\mathrm{d}\sigma}{\mathrm{d}\cos\tilde{\theta}} = \{1 + [3(2-x_1)x_1^2\cos^2\tilde{\theta}]/[4x_1^3 + (2-x_1)^3]\}/(1-\cos^2\tilde{\theta}) \tag{9.75}$$

spin 0: $$\frac{\mathrm{d}\sigma^{\mathrm{S}}}{\mathrm{d}\cos\tilde{\theta}} = \{1 + [x_1(3x_1-4)\cos^2\tilde{\theta}]/(4-3x_1^2)\}/(1-\cos^2\tilde{\theta}) \tag{9.76}$$

The forms (9.75) and (9.76) may not look very transparent. The essential physical point is that the 'scalar gluon' prediction (9.76) is *less singular* near $x_1 \sim 1$, $x_2 \sim 1$ than the vector gluon prediction (9.75): note that (9.70) is singular as both x_1 and x_2 approach 1, while (9.74) is not, since at that point $x_3 = 0$. When symmetrised over the three variables, this has the consequence that (9.70) leads to a $\tilde{\theta}$ distribution (9.75) which is singular at $\cos^2\tilde{\theta} = 1$ for $x_1 = 1$, while the corresponding 'scalar' distribution (9.76) is non-singular for $\cos^2\tilde{\theta} = 1$, $x_1 = 1$. The consequence is that (9.75) rises more steeply in $\cos^2\tilde{\theta}$ than does (9.76). Ellis and Karliner (1979) remark that for $\cos\tilde{\theta} \leqslant \frac{1}{2}$ and $x_1 \sim 0.8$–0.9, reasonable approximations are

$$\text{spin 1:}\quad \frac{\mathrm{d}\sigma}{\mathrm{d}\cos\tilde{\theta}} \simeq 1 + 2\cos^2\tilde{\theta} \tag{9.77}$$

$$\text{spin 0:}\quad \frac{\mathrm{d}\sigma}{\mathrm{d}\cos\tilde{\theta}} \simeq 1 + 0.2\cos^2\tilde{\theta}. \tag{9.78}$$

Figure 9.27 shows the early data of Brandelik *et al* (1980). Unfortunately, a very severe bias is introduced by the event selection criteria ($x_1 < 0.9$) so that the distribution is forced to zero at $\cos\tilde{\theta} \sim 0.8$, just in the region where (9.75) and (9.76) are differing substantially. The difference is thus converted, since both distributions are normalised to the same number (248) of events, to a discrepancy at small $\cos\tilde{\theta}$ values, as shown in the figure. However, the data do favour spin-1 gluons. More recent data (Behrend *et al* 1982) support this conclusion.

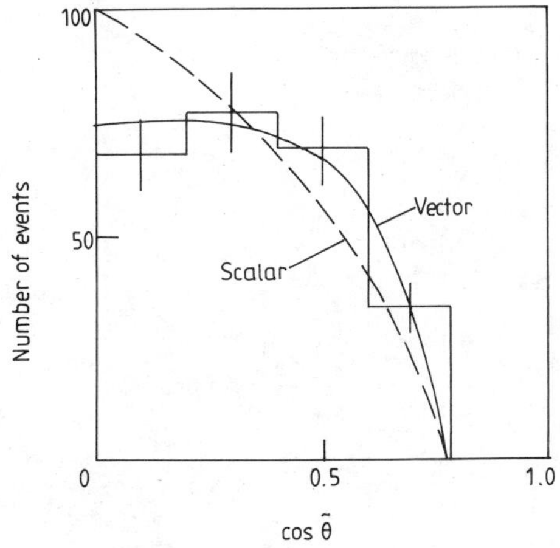

Figure 9.27 Observed (Brandelik *et al* 1980) distribution of the $e^+e^- \to$ three-jet data in the region $1 - x_1 \gtrsim 0.10$, as a function of the cosine of the Ellis–Karliner angle $\tilde{\theta}$ defined in the text. The full curve shows the QCD prediction, the broken curve the prediction for scalar gluons, both normalised to the number of observed events.

The UA2 collaboration has collected many more three-jet events than were available from e^+e^- annihilation: figure 9.28 shows the $\cos\tilde{\theta}$ distribution they have obtained for about 1000 events in the region $0.84 \leqslant x_1 \leqslant 0.96$, which is clearly more like the shape expected from (9.75). Unfortunately, of course, the parton level processes are here much more complicated than the photon-initiated one we have calculated for the e^+e^- case, and therefore harder to interpret in simple terms. However, we can make some qualitative points. We first note the interesting fact that, although the *amplitude* of (9.56) contains factors $(p_1 \cdot p_3)^{-1}$ and $(p_2 \cdot p_3)^{-1}$, the *cross section* of (9.64) does not contain $(p_1 \cdot p_3)^{-2}$ and $(p_2 \cdot p_3)^{-2}$ factors, but only $(p_1 \cdot p_3)^{-1}$ and $(p_2 \cdot p_3)^{-1}$ factors: we say that the double poles have cancelled, leaving only the single poles. This is actually a general feature of three-jet processes of bremsstrahlung type (Berends *et al* 1981) involving spin-$\frac{1}{2}$ and spin-1 particles. It is the product $[(p_1 \cdot p_3)(p_2 \cdot p_3)]^{-1}$ which leads, via (9.69), to the characteristic factor $(1 - \cos^2\tilde{\theta})^{-1}$ in (9.75). This factor will therefore occur also in the $\bar{p}p$ case, accompanied by a numerator factor of the form $1 + \alpha \cos^2\tilde{\theta}$, leading to the shape of figure 9.28. In fact it is consistent with $1 + 2\cos^2\tilde{\theta}$. Also shown in this figure is the prediction from a typical contributing QCD process (they all give similar $\cos\tilde{\theta}$ distributions). The agreement between the data and the QCD prediction is quite satisfactory.

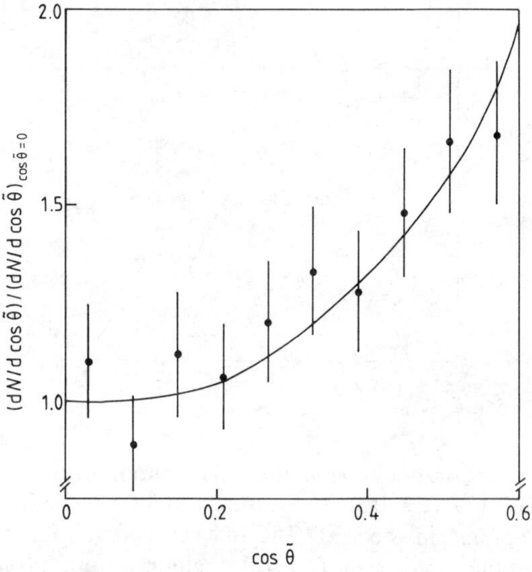

Figure 9.28 The distribution of $\cos\tilde{\theta}$ in three-jet events in $\bar{p}p$ collisions (Appel *et al* 1986), normalised to unity at $\cos\tilde{\theta} = 0$. The full curve is the $gg \to ggg$ QCD tree graph prediction, subject to the cuts mentioned in the caption to figure 9.21.

9.5 QCD prediction for the total cross section σ ($e^+e^- \to$ hadrons)

We are now in a position to take up the important question of QCD corrections to the simple parton model prediction for the total e^+e^- cross section for annihilation into hadrons (cf footnote on page 283). The basic parton model graph is shown again in figure 9.29, which has an amplitude F_γ, say. The $O(\alpha_s)$ QCD corrections to F_γ are shown in figure 9.30: we denote the amplitude for the sum of these processes by $F_{g,v}$, where 'v' stands for 'virtual', since these involve the emission and then reabsorption of gluons. The total cross section from these contributions is thus proportional to $|F_\gamma + F_{g,v}|^2$ – and this leads to a problem. The gluon loops of figure 9.30 contain, of course, the usual ('ultraviolet') divergences at large momenta. But they turn out *also* to diverge as the (virtual) gluon momenta approach zero. Such 'soft' divergences are usually called 'infrared' when they occur in loops – and they are *not* cured by renormalisation, which is relevant to the high-energy (ultraviolet) divergence of Feynman integrals. Renormalisation has nothing to offer the infrared problem.

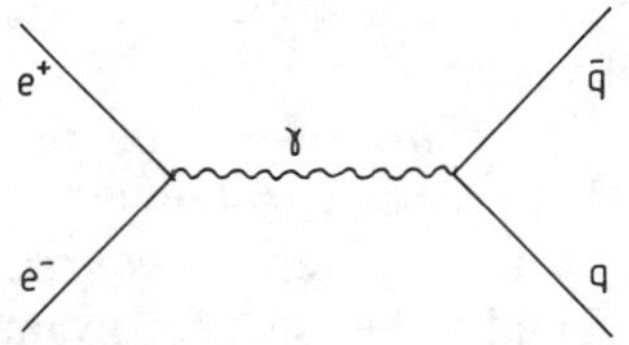

Figure 9.29 One-photon annihilation amplitude in $e^+e^- \to \bar{q}q$.

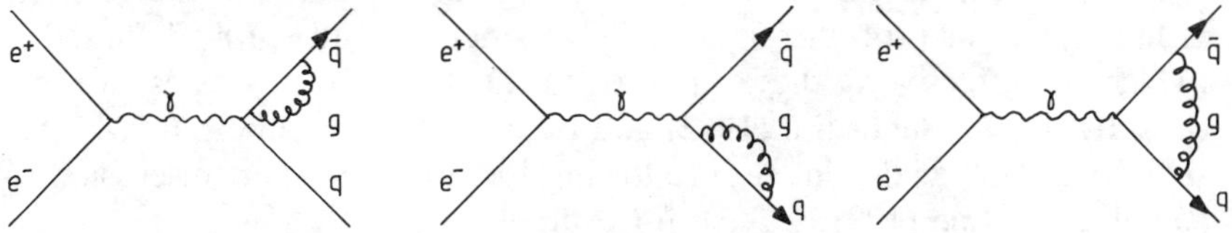

Figure 9.30 Virtual gluon corrections to figure 9.29.

In fact, the gluon loops of figure 9.30 would also, in the limit of zero quark mass, exhibit further (non-ultraviolet) divergences, arising from 'collinear' configurations of the quarks and gluons in the loops. This is, after all, not unexpected: the gluon momenta in the loops run over all possible values, including those which gave trouble in the *real* gluon emission processes of figure 9.22, discussed in the previous section. Indeed, it is the latter processes which hold the key to dealing with these troublesome divergences of figure 9.30. The cure, this time, lies in a careful analysis of what is actually meant by the total annihilation cross section to $q\bar{q}$. The point is that an outgoing quark (or antiquark) cannot be distinguished, kinematically, from one which is accompanied by a soft or collinear gluon – just as, in the appropriate kinematic regions, there are ambiguities between two-jet and three-jet events, as we saw in the previous section. Thus to $|F_\gamma + F_{g,v}|^2$ should also be added the contribution to the total cross section due to production of soft (and collinear) gluons in these dangerous kinematical regions. This will entail integrating (9.70) precisely over the area (call it ε) of figure 9.24 between the triangular boundary and the 'rounded' inner triangle, leading to a cross section (r standing for 'real')

$$\sigma_{g,r} = \sigma_{pt} \frac{2\alpha_s}{3\pi} \int\int_{\varepsilon} \frac{x_1^2 + x_2^2}{(1 - x_1)(1 - x_2)} dx_1 dx_2. \tag{9.79}$$

Clearly, as the region ε (parametrised in some way) tends to zero, (9.79) will diverge due to the vanishing denominators (soft and collinear divergences).

This refinement hardly seems to have helped: we now have *two* lots of divergences to worry about! Yet now comes the miracle. Notice that three terms appear when squaring out $|F_\gamma + F_{g,v}|^2$: one of order α^2 (from $|F_\gamma|^2$), another of order $\alpha^2\alpha_s^2$ (from $|F_{g,v}|^2$), and an *interference* term of order $\alpha^2\alpha_s$, *which is the same order as* (9.79). Thus at order $\alpha^2\alpha_s$, (9.79) must be added precisely to this interference term – and the wonderful fact is that their divergences *cancel*. The complete cross section at order $\alpha^2\alpha_s$ is found to be (see, for example, Pennington 1983)

$$\sigma = \sigma_{pt}(1 + \alpha_s/\pi). \tag{9.80}$$

This remarkable cancellation of the singularities between the real and virtual emission processes is actually a general result, proved for QED (which is effectively the same as QCD for the processes we are considering) by Kinoshita (1962) and by Lee and Nanenberg (1964). We shall meet it again in the following section. Further discussion is contained in Lee (1981, §23.5), for example.

At first sight (9.80) might appear perfectly satisfactory, since we have seen reasons for thinking that α_s is quite small (0.15–0.25, say). However, this is an illusion, which is dispelled as soon as we consider the *higher*-order processes $\sim O(\alpha^2\alpha_s^2)$. Some typical graphs contributing to this order (via, once again, interference: this time between figure 9.31 and figure 9.22) are shown in figure 9.31. It turns out (Pennington 1983)

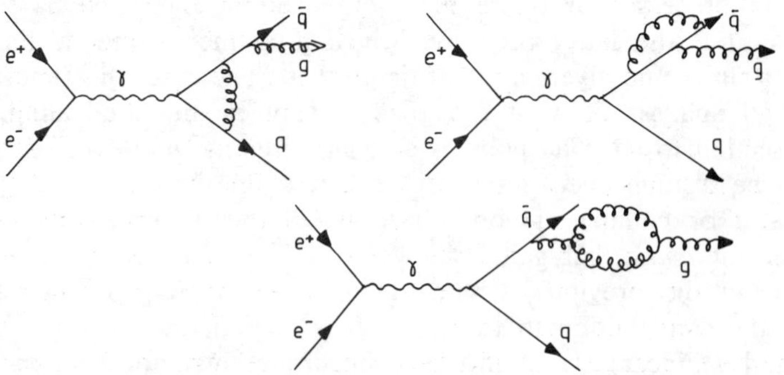

Figure 9.31 Higher-order processes contributing to $e^+e^- \to$ hadrons at the parton level.

that the result of including all such graphs amounts precisely to replacing α_s in (9.80) by the Q^2-dependent $\alpha_s(Q^2)$ given by (9.40):

$$\alpha_s(Q^2) = \alpha_s(\mu^2)\left(1 - \frac{\alpha_s(\mu^2)}{12\pi}(33 - 2f)\ln\frac{Q^2}{\mu^2} + \ldots\right). \tag{9.81}$$

Now our satisfaction is turned to gloom, since $\alpha_s(Q^2)$ can be very large (depending on Q^2), whatever $\alpha_s(\mu^2)$ may be: thus the perturbative result (9.80) with (9.81) for α_s seems after all quite useless.

Of course, there is a way out: as already stated in §9.3, a whole infinite series of such logarithms can be summed, so that (9.81) is actually the expansion in powers of $\alpha_s(\mu^2)$ of

$$\alpha_s(Q^2) = \frac{\alpha_s(\mu^2)}{1 + (\alpha_s(\mu^2)/12\pi)(33 - 2f)\ln(Q^2/\mu^2)} \tag{9.82}$$

which now *is* well behaved as $Q^2 \to \infty$—the asymptotic freedom result. Thus finally, the 'leading logarithm' prediction for $e^+e^- \to$ hadrons is

$$\sigma = \sigma_{pt}(1 + \alpha_s(Q^2)/\pi) \tag{9.83}$$

where

$$\alpha_s(Q^2) = \frac{12\pi}{(33 - 2f)\ln(Q^2/\Lambda^2)}. \tag{9.84}$$

(9.83) is a fundamental result, since it proves that the simple parton model result σ_{pt} becomes exact as $Q^2 \to \infty$, and establishes the precise form of the predicted $O(1/\ln Q^2)$ correction to the pointlike (scaling) cross section σ_{pt}.

Unfortunately, however, (9.83) is not the whole story. First, still higher-order corrections in α_s should be included, and the treatment of these is highly technical. Also, although it would be nice to use (9.82) (plus corrections) to test experimentally the Q^2-dependence of $\alpha_s(Q^2)$, we cannot believe the above 'leading logarithm improved perturbation theory' – which (9.83) represents – at low Q^2 where the Q^2 variation of $\alpha_s(Q^2)$ is substantial, and at higher $Q^2(Q^2 \gtrsim 30\ \text{GeV}^2)$ the Q^2-variation is very slow (cf figure 9.10). Furthermore, at these higher Q^2 the *weak interaction* corrections to (9.83) are important, due to the proximity of the Z^0 resonance (see §15.5.4 below). Nevertheless, the data of figure 9.1 seem to indicate at least the need for a small positive correction to σ_{pt}, and an analysis by the CELLO group (Behrend *et al* 1987) quotes $\alpha_s(34\ \text{GeV}^2) = 0.169 \pm 0.025$. One hopes that measurements at LEP will be able to confirm the predicted Q^2-dependence of α_s, either in $\sigma(e^+e^- \to \text{hadrons})$ or via the ratio $\sigma(3\ \text{jets})/\sigma(2\ \text{jets})$ which, as we have seen, is proportional to α_s.

We have learned that the first QCD correction to the free parton model causes a logarithmic violation of scaling in e^+e^- annihilation. We should also wonder whether something similar will happen to the *structure functions* introduced in §7.2, and assumed to be scale invariant functions in the free parton model. This will be the final topic in this chapter.

9.6 QCD predictions for scaling violations in deep inelastic scattering

The free parton model amplitudes we considered in Chapter 7 for deep inelastic lepton–nucleon scattering were of the form shown in figure 9.32. The obvious first QCD corrections will be due to real gluon emission by either the initial or final quark, as shown in figure 9.33, but to these we must add the one-loop virtual gluon processes of figure 9.34 in order (see below) to get rid of infrared divergences similar to those encountered in the preceding section, and also the diagrams of figure 9.35, corresponding to the presence of gluons in the nucleon. To simplify matters, we shall consider what is called a 'non-singlet structure function' F_2^{NS}, such as $F_2^{\mathrm{ep}} - F_2^{\mathrm{en}}$ in which the (flavour) singlet gluon contribution cancels out, leaving only the diagrams of figures 9.33 and 9.34.

We now want to perform, for these diagrams, calculations analogous to those of §7.2, which enabled us to find the e–N structure functions νW_2 and MW_1 from the simple parton process of figure 9.32. There are two problems here: one is to find the parton level W's corresponding to

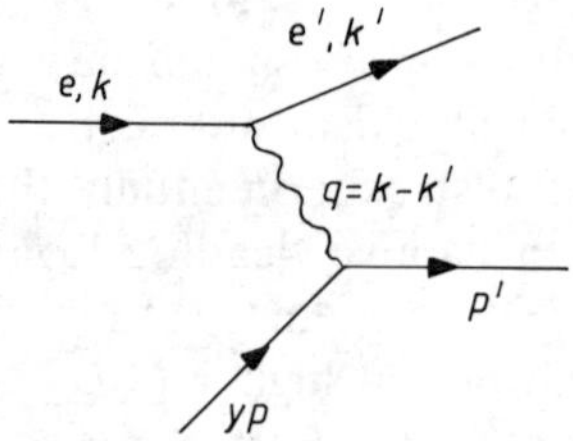

Figure 9.32 Electron–quark scattering via one–photon exchange.

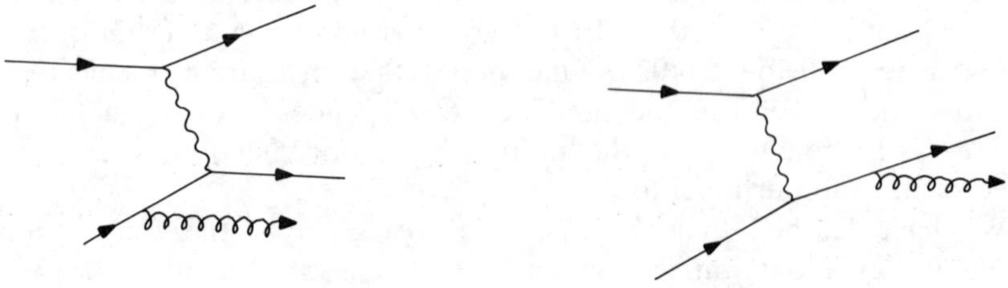

Figure 9.33 Electron–quark scattering with single-gluon emission.

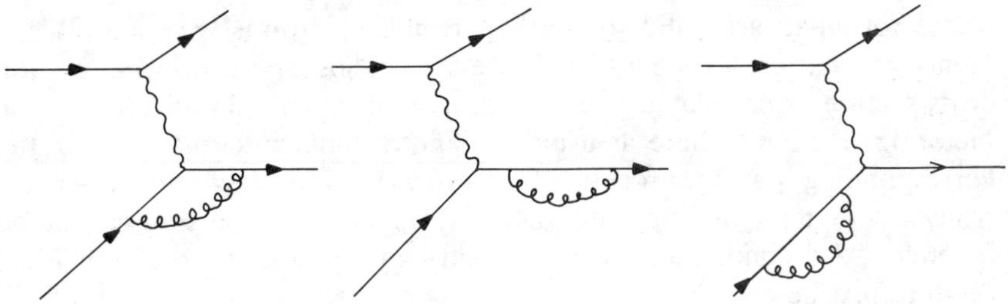

Figure 9.34 Virtual single-gluon corrections to figure 9.32.

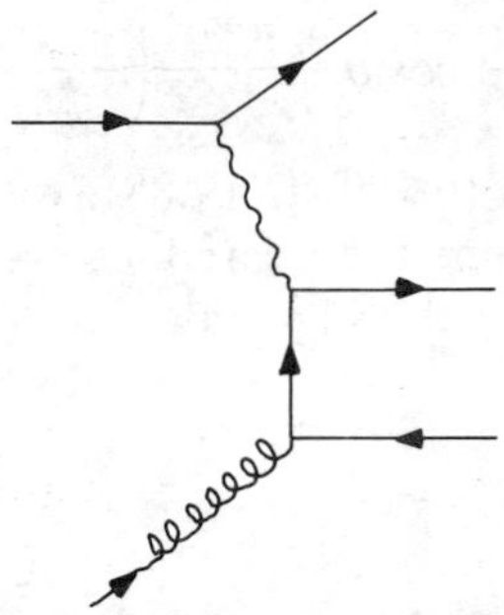

Figure 9.35 Electron–gluon scattering with $\bar{q}q$ production.

figure 9.33 (leaving aside figure 9.34 for the moment)—cf equations (7.26) and (7.27) in the case of the free parton diagram figure 9.32; the other is to relate these parton W's to observed nucleon W's via an integration over momentum fractions. In §7.2 we solved the first problem by explicitly calculating the parton level $\mathrm{d}\sigma^i/\mathrm{d}Q^2\mathrm{d}\nu$ and picking off the associated $\nu W_2^i, W_1^i$. In principle, the same can be done here, starting from the five-fold differential cross section for our $\mathrm{e}^- + \mathrm{q} \rightarrow \mathrm{e}^- + \mathrm{q} + \mathrm{g}$ process. However, a simpler—if somewhat heuristic—way is available. We note from (7.43) that in general $F_1 = MW_1$ is given by the transverse virtual photon cross section

$$W_1 = \sigma_\mathrm{T}/(4\pi\alpha^2/K) = \tfrac{1}{2}\sum_{\lambda=\pm1} \varepsilon_\mu^*(\lambda)\varepsilon_\nu(\lambda)W^{\mu\nu} \tag{9.85}$$

where $W^{\mu\nu}$ was defined in (7.2). Further, the Callan–Gross relation is still true (the photon only interacts with the charged partons, which are quarks with spin $\frac{1}{2}$), and so

$$F_2/x = 2F_1 = 2MW_1 = \sigma_\mathrm{T}/(4\pi\alpha^2/2MK). \tag{9.86}$$

These formulae are valid for both parton and proton W_1's and $W^{\mu\nu}$'s. Hence the parton level $2\hat{F}_1$ for figure 9.33 is just the transverse photon cross section as calculated from the graphs of figure 9.36, divided by the factor $4\pi^2\alpha/2\hat{M}\hat{K}$, where as usual $\hat{}$ denotes kinematic quantities in the corresponding parton process. This cross section, however, is—apart from a colour factor—just the virtual Compton cross section calculated in §6.9. Also, taking the same (Hand) convention for the individual photon flux factors,

$$2\hat{M}\hat{K} = \hat{s}. \tag{9.87}$$

Thus for the parton process of figure 9.33

$$\begin{aligned} 2\hat{F}_1 &= \hat{\sigma}_{\mathrm{T}}/(4\pi^2\alpha/2\hat{M}\hat{K}) \\ &= \frac{\hat{s}}{4\pi^2\alpha}\int_{-1}^{1} \mathrm{d}\cos\theta \cdot \frac{4}{3} \cdot \frac{\pi\alpha\alpha_{\mathrm{s}}}{\hat{s}}\left(-\frac{\hat{t}}{\hat{s}} - \frac{\hat{s}}{\hat{t}} + \frac{2\hat{u}Q^2}{\hat{s}\,\hat{t}}\right) \end{aligned} \tag{9.88}$$

where, in going from (6.189) to (9.88), we have inserted a colour factor $\frac{4}{3}$ (problem 9.3), renamed the variables $t \to u$, $u \to t$ in accordance with figure 9.36, and replaced α^2 by $\alpha\alpha_{\mathrm{s}}$.

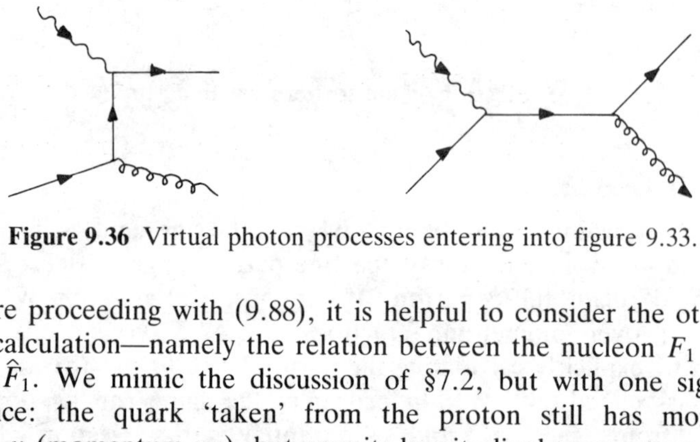

Figure 9.36 Virtual photon processes entering into figure 9.33.

Before proceeding with (9.88), it is helpful to consider the other part of the calculation—namely the relation between the nucleon F_1 and the parton $\hat{F}_1$. We mimic the discussion of §7.2, but with one significant difference: the quark 'taken' from the proton still has momentum fraction y (momentum yp), but now its longitudinal momentum must be degraded in the final state due to the gluon bremsstrahlung process we are calculating. Let us call the quark momentum after gluon emission zyp (figure 9.37). Then, assuming as in §7.2 that it stays on-shell, we have

$$q^2 + 2zyq\cdot p = 0 \tag{9.89}$$

or

$$x = yz, \qquad x = Q^2/2q\cdot p, \qquad q^2 = -Q^2 \tag{9.90}$$

and we can write (cf (7.28))

$$\frac{F_2}{x} = 2F_1 = \sum_i \int_0^1 \mathrm{d}y\, f_i(y) \int_0^1 \mathrm{d}z\, 2\hat{F}_1^i \delta(x - yz) \tag{9.91}$$

where the $f_i(y)$ are the momentum distribution functions introduced in §7.2 (we often call them $q(x)$ or $g(x)$ as the case may be) for parton type i, and the sum is over contributing partons. The reader may enjoy checking that (9.91) does reduce to (7.31) for free partons by showing that in that case $2\hat{F}_1^i = e_i^2 \delta(1 - z)$ (see Halzen and Martin (1984, §10.3) for help).

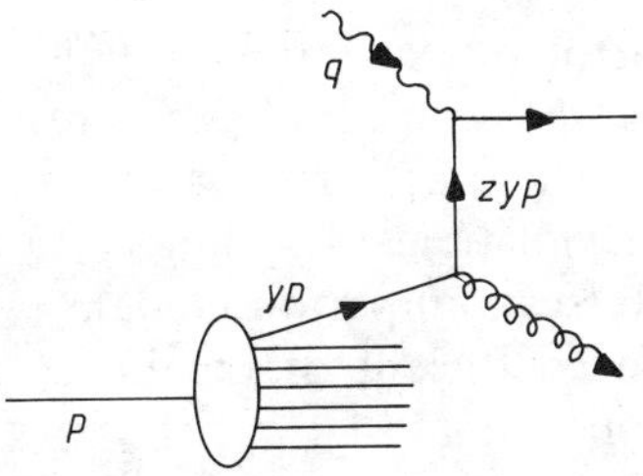

Figure 9.37 The first process of figure 9.36, viewed as a contribution to e⁻–nucleon scattering.

To proceed further with the calculation (i.e. of (9.88) inserted into (9.91)), we need to look at the kinematics of the $\gamma q \rightarrow qg$ process, in the CMS. Referring to figure 9.38, we let k, k' be the magnitudes of the CMS momenta $\boldsymbol{k}$, $\boldsymbol{k}'$. Then

$$\hat{s} = 4k'^2 = (yp + q)^2 = Q^2(1 - z)/z, \qquad z = Q^2/(\hat{s} + Q^2)$$
$$\hat{t} = (q - p')^2 = -2kk'(1 - \cos\theta) = -Q^2(1 - c)/2z, \qquad c = \cos\theta \tag{9.92}$$
$$\hat{u} = (q - q')^2 = -2kk'(1 + \cos\theta) = -Q^2(1 + c)/2z.$$

We now note that in the integral (9.88) for $\hat{F}_1$, when we integrate over

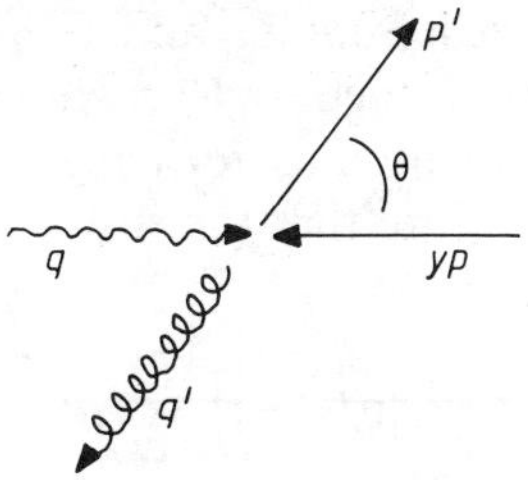

Figure 9.38 Kinematics for the parton process of figure 9.37.

$c = \cos\theta$, we shall obtain an infinite result

$$\sim \int^{1} \frac{\mathrm{d}c}{1-c} \tag{9.93}$$

associated with the vanishing of $\hat{t}$ in the forward direction. This singularity has arisen because we are assuming massless quarks: if we give the quarks a small mass μ, then the propagator factor $\hat{t}^{-1}$ becomes $(\hat{t} - \mu^2)^{-1}$ and the offending integral is

$$\sim \int^{1} \frac{\mathrm{d}c}{[(1 + 2\mu^2 z/Q^2) - c]} \tag{9.94}$$

which will produce a factor of the form $\ln(Q^2/\mu^2)$, so that μ regulates the divergence. This is the crucial physical result produced by the lowest-order QCD correction to the free parton model, since it *violates scaling*. Such logarithmic violations of scaling are a characteristic feature of corrections to the free (scaling) parton model.

We may calculate the coefficient of the $\ln Q^2$ term by retaining in (9.88) only the terms proportional to $\hat{t}^{-1}$:

$$2\hat{F}^{i} \approx \int_{-1}^{1} \frac{\mathrm{d}c}{1-c}\left(\frac{\alpha_{\mathrm{s}}}{2\pi}\cdot\frac{4}{3}\cdot\frac{1+z^2}{1-z}\right) \tag{9.95}$$

and so, for just one quark species, this QCD correction gives

$$2F_1(x) = \frac{\alpha_{\mathrm{s}}}{2\pi}\int_x^1 \frac{\mathrm{d}y}{y} f(y) \ln\left(\frac{Q^2}{\mu^2}\right) P_{\mathrm{qq}}\left(\frac{x}{y}\right) \tag{9.96}$$

where

$$P_{\mathrm{qq}}(z) = \frac{4}{3}\left(\frac{1+z^2}{1-z}\right) \tag{9.97}$$

is called a 'splitting function'. This function has an important physical interpretation: it is the probability that a quark, having radiated a gluon, is left with the fraction z of its original momentum. Similar functions arise in QED in connection with what is called the 'equivalent photon approximation' (Weizsäcker 1934, Williams 1934, Chen and Zerwas 1975). The application of these techniques to QCD corrections to the free parton model is due to Altarelli and Parisi (1977), who thereby opened the way to the above much simpler and more physical way of understanding scaling violations, which had previously been discussed only in the rather esoteric operator product formalism (Wilson 1969).

Our result so far is therefore that the 'free' quark distribution function $q(x)$, say, becomes modified to

$$q(x) + \frac{\alpha_{\mathrm{s}}}{2\pi}\ln\left(\frac{Q^2}{\mu^2}\right)\int_x^1 \frac{\mathrm{d}y}{y} q(y) P_{\mathrm{qq}}\left(\frac{x}{y}\right) \tag{9.98}$$

due to lowest-order gluon radiation. But of course, before trusting

(9.98) for physics, we have to worry about higher-order processes as well. Indeed, if we were to include just those higher-order processes which lead to $\alpha_s \to \alpha_s(Q^2)$ of equation (9.42), we would find that the additional term appears precisely to *cancel* the good asymptotic freedom result ($\alpha_s \to 0$ as $Q^2 \to \infty$), and so (9.98) could not be believed. The way out of this is to interpret (9.98) as indicating a (Q^2-dependent) change in the distribution function itself, and write

$$2F_1(x, \ln Q^2) = \int_x^1 \frac{dy}{y}[q(y) + \Delta q(y, \ln Q^2)]\delta\left(1 - \frac{x}{y}\right) \tag{9.99}$$

with

$$\Delta q(x, \ln Q^2) = \ln\left(\frac{Q^2}{\mu^2}\right)\cdot\frac{\alpha_s}{2\pi}\int_x^1 \frac{dy}{y} q(y, \ln Q^2) P_{qq}\left(\frac{x}{y}\right). \tag{9.100}$$

It can then be shown (see e.g. Altarelli 1982) that the correct 'leading order in $\alpha_s(Q^2)$' equation is

$$\frac{dq(x, \ln Q^2)}{d\ln Q^2} = \frac{\alpha_s(Q^2)}{2\pi}\int_x^1 \frac{dy}{y} q(y, \ln Q^2) P_{qq}\left(\frac{x}{y}\right) + O(\alpha_s^2(Q^2)). \tag{9.101}$$

with $\alpha_s(Q^2)$ given by (9.42). Thus, in summary, QCD does not predict the actual value of the distribution function, but it does predict how, via equation (9.101), it 'evolves' in $\ln Q^2$, given its value at one particular Q_0^2.

Equation (9.101) is known as an Altarelli–Parisi evolution equation. The right-hand side will actually have to be supplemented in general by terms (calculable from figure 9.35) in which quarks are generated from the gluon distribution; the equations must then be closed by a corresponding one describing the evolution of the gluon distribution (Altarelli 1982). Such equations can be qualitatively understood as follows. We have seen that the dependence on $\ln Q^2$ is a reflection of the fact that there is no mass scale: the distribution functions change proportionately to the fractional change in Q^2 ($d/d\ln Q^2 = Q^2 d/dQ^2$). The change in the distribution for a quark with momentum fraction x, which absorbs the virtual photon, is then given by the integral over y of the corresponding distribution for a quark with momentum fraction y, which radiated away (via a gluon) a fraction x/y of its momentum with probability $(\alpha_s/2\pi)P_{qq}(x/y)$. This probability is high for large momentum fractions: high-momentum quarks lose momentum by radiating gluons. Thus there is a predicted tendency for the distribution function to get smaller at large x and larger at small x (due to the build-up of slower partons). The gluon distribution at $Q^2 = 2000$ GeV shown in figure 9.17 was 'evolved' in this way from the measured (de Groot *et al* 1979) distribution at $Q^2 = 20$ GeV, and these qualitative features are apparent.

Actual data at different Q^2 do indeed show such a trend (figures 9.39 and 9.40). To make a more quantitative analysis, the moments of the structure functions are often considered, defined by

$$M_n(Q^2) = \int_0^1 \mathrm{d}x\, x^{n-1} q(x, \ln Q^2). \tag{9.102}$$

Taking moments of both sides of (9.101) and interchanging the order of x and y integrations, we find

$$\frac{\mathrm{d}M_n(Q^2)}{\mathrm{d}\ln Q^2} = \frac{\alpha_s(Q^2)}{2\pi}\int_0^1 \mathrm{d}y\, y^{n-1} q(y, \ln Q^2) \int_0^y \frac{\mathrm{d}x}{y}\left(\frac{x}{y}\right)^{n-1} P_{qq}\left(\frac{x}{y}\right). \tag{9.103}$$

Changing the variable to $z = x/y$ in the second integral and defining

$$\gamma_n = -4\int_0^1 \mathrm{d}z\, z^{n-1} P_{qq}(z) \tag{9.104}$$

we have

$$\frac{\mathrm{d}M_n}{\mathrm{d}\ln Q^2} = -\frac{\alpha_s(Q^2)}{8\pi}\gamma_n M_n(Q^2). \tag{9.105}$$

However, we also have from (9.42)

$$\frac{\mathrm{d}\alpha_s}{\mathrm{d}\ln Q^2} = -\frac{(33 - 2f)}{12\pi}\alpha_s^2 \tag{9.106}$$

and thus

$$\frac{\mathrm{d}\ln M_n}{\mathrm{d}\ln \alpha_s} = \frac{3}{2(33 - 2f)}\gamma_n \equiv d_n, \text{ say,} \tag{9.107}$$

which has the solution

$$\ln M_n(Q^2) = \ln M_n(Q_0^2) + d_n \ln(\alpha_s(Q^2)/\alpha_s(Q_0^2)). \tag{9.108}$$

It follows that

$$\ln M_n(Q^2) = (d_n/d_m) \ln M_m(Q^2) + \text{constant} \tag{9.109}$$

so that a plot of the logarithm of one moment against the logarithm of another should be a straight line with slope d_n/d_m calculable via (9.104) from the splitting function (9.97).

At this point we note that the function $P_{qq}(z)$ of (9.97) is singular as $z \to 1$, in such a way that the integral (9.104) for γ_n will diverge. This is the infrared divergence mentioned at the beginning of this section; it is associated with the emission of a 'soft' gluon. As discussed for the e^+e^- cross section in the preceding section, we expect that when contributions to the deep inelastic cross section from the virtual gluon diagrams of figure 9.34 are also carefully included, the divergence should be cancelled. This has been explicitly verified by Kim and Schilcher (1978) and by Altarelli *et al* (1978a,b, 1979). Alternatively, one can make the physical

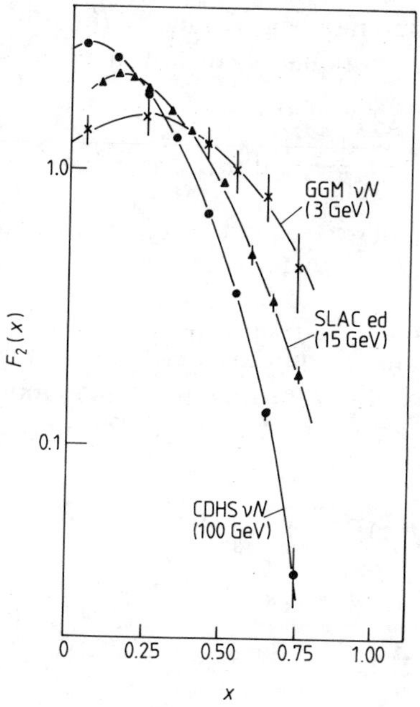

Figure 9.39 Structure function $F_2(x)$ measured in neutrino and electron scattering on nucleons at different beam energies. The functions become more peaked at small x values as the average energy, and Q^2, increases. (From Perkins D 1987 *Introduction to High Energy Physics* 3rd edn, courtesy Addison-Wesley Publishing Company.)

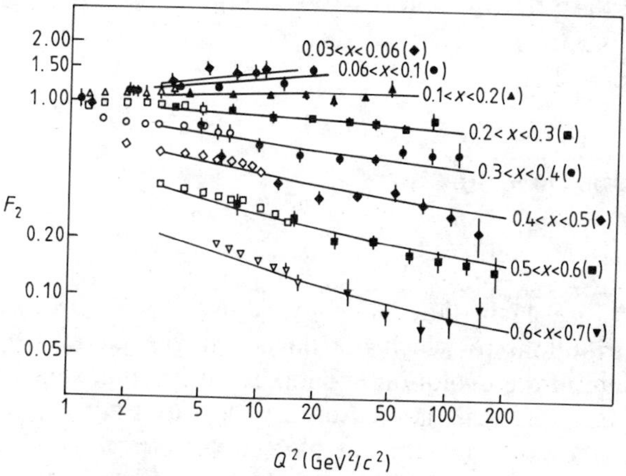

Figure 9.40 $F_2(x, Q^2)$ plotted versus Q^2 for different regions of x (de Groot *et al* 1979); compare with figure 7.2(b), which is at $x = 0.25$. The curves are from empirical fits to quark distribution functions, incorporating a Q^2 dependence associated with QCD.

argument that the net number of quarks (i.e. the number of quarks minus the number of antiquarks) of any flavour is conserved as $\ln Q^2$ varies:

$$\frac{\mathrm{d}}{\mathrm{d}\ln Q^2}\int_0^1 \mathrm{d}x\, q(x, \ln Q^2) = 0. \tag{9.110}$$

From (9.101) this implies

$$\int_0^1 \mathrm{d}x\, P_{\mathrm{qq}}^+(z) = 0 \tag{9.111}$$

where P_{qq}^+ is the complete splitting function, including the effect of the gluon loops. This fixes the contribution of these loops (which only enter, in leading log approximation, at $z \to 1$): for any function $f(z)$ regular as $z \to 1$, P_{qq}^+ is defined by

$$\int_0^1 \mathrm{d}z\, f(z) P_{\mathrm{qq}}^+(z) = \int_0^1 \mathrm{d}z [f(z) - f(1)] P_{\mathrm{qq}}(z). \tag{9.112}$$

Applying this prescription to γ_n, we find (problem 9.9)

$$d_n = \frac{4}{33 - 2f}\left(1 - \frac{2}{n(n+1)} + 4\sum_{j=2}^{n}\frac{1}{j}\right). \tag{9.113}$$

It is interesting also to calculate the d_n's for the case in which the emitted gluon has spin 0. For this we need the appropriate $P_{\mathrm{qq}}^{\mathrm{S}}(z)$. However, this can easily be extracted from the $\gamma^* \to \mathrm{q\bar{q}g'_s}$ probability calculated in (9.74), which is proportional to

$$\frac{x_3^2}{(1 - x_1)(1 - x_2)} \quad \text{instead of} \quad \frac{x_1^2 + x_2^2}{(1 - x_1)(1 - x_2)}. \tag{9.114}$$

In terms of the $\gamma^*\mathrm{q} \to \mathrm{qg}$ process, we are looking at the region $x_2 \approx 1$, while $x_1 \to z$. Thus the scalar $P_{\mathrm{qq}}^{\mathrm{S}}(z)$ function is (using $x_1 + x_2 + x_3 = 2$, $x_2 \to 1$, $x_1 \to z$)

$$P_{\mathrm{qq}}^{\mathrm{S}}(z) = \tfrac{4}{3}(1 - z) \tag{9.115}$$

and then (problem 9.10)

$$d_n^{\mathrm{S}} = \frac{4}{33 - 2f}\left(1 - \frac{2}{n(n+1)}\right). \tag{9.116}$$

We emphasise again that all of the above has been directly relevant only to distributions to which the flavour singlet gluon distributions do not contribute in the evolution equations. Such a non-singlet distribution is xF_3, for example (this is measured in ν scattering: see §11.2 below). Figure 9.41 shows a log moment plot (Bosetti *et al* 1978) from BEBC and Gargamelle data, compared with the vector gluon predictions (9.113). It is clear that the scalar prediction is ruled out: equation (9.113) predicts $d_6/d_4 = 1.29$, $d_5/d_3 = 1.456$, $d_7/d_3 = 1.76$, while

(9.116) predicts $d_6^S/d_4^S = 1.06$, $d_5^S/d_3^S = 1.12$, $d_7^S/d_3^S = 1.16$. This is certainly very encouraging—but we warn the novice against early euphoria, and refer to more advanced manuals, such as Pennington (1983).

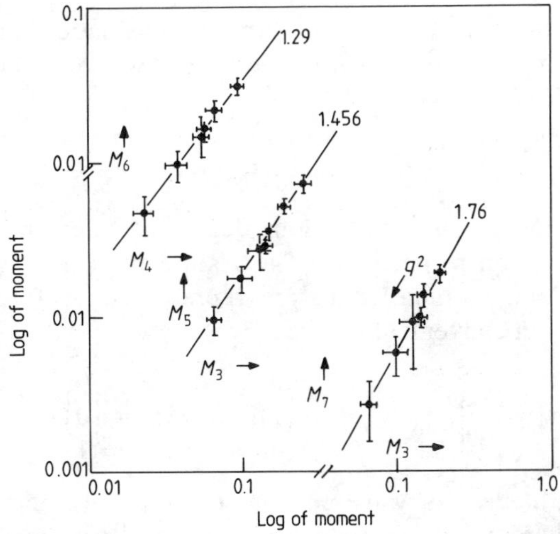

Figure 9.41 Plots of log M_m versus log M_n for various moment orders m, n, where M_n is the nth moment of xF_3 (cf (9.102)). QCD predicts the linear relations (9.109), with the d's given by (9.113). The predicted slopes are in agreement with those observed (Bosetti *et al* 1978).

Summary

Evidence for the colour degree of freedom.

The dynamics of colour, leading to QCD; gluons.

Asymptotic freedom.

Tests of tree graph QCD amplitudes in hard scattering processes in $\bar{p}p$ collisions: two-jet events, three-jet events.

Three-jet events in e^+e^- collisions.

Tests of the gluon spin.

QCD prediction for $\sigma(e^+e^- \to \text{hadrons})$ up to $O(\alpha_s(Q^2))$.

QCD predictions for scaling violations in deep inelastic scattering.

Problems

9.1 Show that the antisymmetric 3q combination of equation (9.2) is invariant under the transformation (9.7) for each colour wavefunction.

9.2 Verify equation (9.42).

9.3 This problem is concerned with the evaluation of the 'colour factor' needed for equation (9.70). The 'colour wavefunction' part of the amplitude (9.56) is

$$a^{\alpha}(c_3)\chi^{\dagger}(c_1)\frac{\lambda^{\alpha}}{2}\chi(c_2) \tag{a}$$

where c_1, c_2 and c_3 label the colour degree of freedom of the quark, antiquark and gluon respectively. The χ's are the colour wavefunctions of the quark and antiquark and are represented by three-component column vectors; a convenient choice is

$$\chi(\mathrm{r}) = \begin{pmatrix}1\\0\\0\end{pmatrix}, \qquad \chi(\mathrm{b}) = \begin{pmatrix}0\\1\\0\end{pmatrix}, \qquad \chi(\mathrm{g}) = \begin{pmatrix}0\\0\\1\end{pmatrix}$$

by analogy with the spin wavefunctions of SU(2). Indeed, the quantity $\chi^{\dagger}(\lambda^{\alpha}/2)\chi$ is quite analogous to the expression $\chi^{\dagger}\sigma_i\chi$, involving the spin matrices σ_i and spin wavefunctions χ, considered in Appendix A (see equations (A.55)–(A.57)): just as $\chi^{\dagger}\sigma_i\chi$ is a vector under SU(2) transformations, so $\chi^{\dagger}\lambda^{\alpha}\chi$ is an octet under $SU(3)_c$ transformations, and the scalar product of this with the octet wavefunction $a^{\alpha}(c_3)$ creates an $SU(3)_c$ singlet (invariant) amplitude. The cross section is obtained by forming the modulus squared of (a) and summing over colour labels c_i:

$$\sum_{c_1,c_2,c_3} a^{\alpha}(c_3)\chi^{\dagger}_r(c_1)\frac{\lambda^{\alpha}_{rs}}{2}\chi_s(c_2)\chi^{\dagger}_l(c_2)\frac{\lambda^{\beta}_{lm}}{2}\chi_m(c_1)a^{*\beta}(c_3) \tag{b}$$

where summation is understood on the matrix indices on the χ's and λ's, which have been indicated explicitly. In this form the expression is very similar to the *spin* summations considered in Chapter 6 (cf equation (6.72)). We proceed to convert (b) into a trace, and to evaluate it, as follows:

(i) Show that

$$\sum_{c_2}\chi_s(c_2)\chi^{\dagger}_l(c_2) = \delta_{sl}$$

(cf (6.80)).

(ii) Assuming the analogous result

$$\sum_{c_3}a^{\alpha}(c_3)a^{*\beta}(c_3) = \delta_{\alpha\beta}$$

show that (b) becomes

$$\tfrac{1}{4}\sum_{\alpha=1}^{8}\mathrm{Tr}(\lambda^{\alpha})^{2}.$$

(iii) Using the λ's given in Appendix F, show that

$$\sum_{\alpha=1}^{8}\mathrm{Tr}(\lambda^{\alpha})^{2} = 16$$

and hence that the colour factor for (9.70) is 4.

(iv) Why is the colour factor for (9.88) equal to $\frac{4}{3}$?

9.4 Verify equation (9.57).

9.5 Verify equation (9.69).

9.6 Verify that expression (9.65) becomes the factor in large parentheses in equation (9.70), when expressed in terms of the x_i's.

9.7 Write down the analogue of equation (9.56) for the case of a scalar (spin-0) gluon, and verify that equation (9.65) is then replaced by equation (9.73).

9.8 Verify equations (9.75) and (9.76).

9.9 Verify equation (9.113).

9.10 Verify equation (9.116).

PART IV

NOT QUITE A THEORY: PHENOMENOLOGY OF WEAK INTERACTIONS

10

INTRODUCTION TO WEAK INTERACTIONS

As we shall see in Part V, the unified electroweak gauge theory of Glashow (1961), Salam (1968) and Weinberg (1967) is in agreement with all known weak interaction phenomena. This being so, why do we not present it at this point? Remarkable, beautiful and successful though the theory certainly is, it is also quite complicated and subtle—and in some respects still rather mysterious. It seems pedagogically desirable to attempt some kind of motivation for the theory—though, of course, the fact that it agrees so well with experiment may be sufficient for some. Besides, part of the complexity resides in the nature of the weak interaction phenomena themselves. By developing weak interaction phenomenology first, following a roughly historical route, we hope to provide a more gradual entry into these complexities. In any case, the phenomenology thus acquired is still, for most purposes, a very useful approximation to the full theory, at energies well below the masses of the weak gauge bosons $W^{\pm}$ (~80 GeV) and Z^0 (~90 GeV)—see §10.1 below. In Chapter 12 we shall try to indicate why the facts of data and the constraints of theory lead one in a certain (rather peculiar) direction—the search for a gauge theory with massive gauge quanta. We shall then be prepared for the development of GSW theory in Part V.

10.1 Fermi theory and intermediate vector bosons

The first attempt at a phenomenology of weak interactions dates back to Fermi (1934a, b). For β decay he postulated an effective four-fermion point-interaction (figure 10.1). Fermi conceived of β decay as a process analogous to that of an electromagnetic transition, the electron–neutrino pair playing the role of the emitted photon. As in photon emission, the process took place, in Fermi's theory, at a single space–time point. The amplitude was assumed to involve, for the nucleons, a weak interaction analogue of the electromagnetic transition currents introduced earlier in this book; this is the *hadronic weak current* matrix element, $\langle p|\hat{j}_{\mu}^{\text{wk}}|n\rangle$. A simple Lorentz invariant amplitude is then obtained if the $e^-\nu_e$ pair also appears as a 4-vector combination $\langle e\bar{\nu}_e|\hat{j}^{\mu}_{\text{wk}}|0\rangle$, the *leptonic weak current* matrix element†, the complete matrix element being

†This possibly strange-looking current matrix element is, of course, related to one of more familiar appearance, $\langle e|\hat{j}^{\mu}_{\text{wk}}|\nu\rangle$, by the Feynman hypothesis of §3.3.

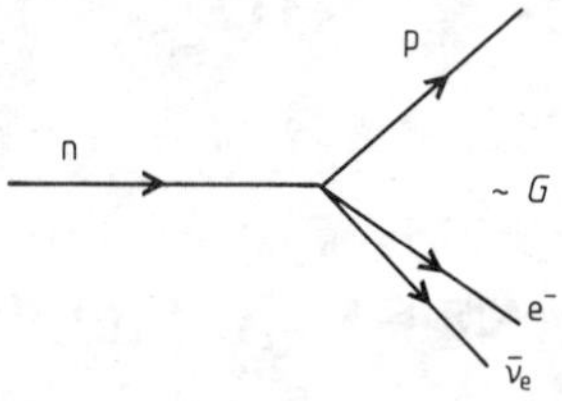

Figure 10.1 Four-fermion interaction for neutron β decay.

$$M = \langle p | \hat{j}_{\mu}^{\text{wk}} | n \rangle \langle e \bar{\nu}_e | \hat{j}_{\text{wk}}^{\mu} | 0 \rangle . \tag{10.1}$$

Because of the very low energy release, one might expect that, to a good approximation, all momentum dependence in the matrix element could be ignored, reducing it effectively to a constant, G†. The numerical value of this constant can be determined from experiments on nuclear β decay. It has dimensions of $[M]^{-2}$ and a magnitude given approximately by

$$G \simeq 1.14 \times 10^{-5} \text{ GeV}^{-2} \tag{10.2}$$

in our natural units ($\hbar = c = 1$).

The assumption of the 'current–current' form of the matrix element (10.1) can be extended to other weak processes. Our rule for dealing with antiparticles implies that the neutrino-induced reaction

$$\nu_e + n \to p + e^- \tag{10.3}$$

should exist, and have the same strength as β decay. If we assume that the *total* weak current has both hadronic (h) and leptonic (l) parts,

$$\hat{j}_{\mu}^{\text{wk}} = \hat{j}_{\mu}^{\text{h}} + \hat{j}_{\mu}^{\text{l}} \tag{10.4}$$

then a current–current form for all weak processes implies that there should exist purely leptonic weak interactions, such as μ decay,

$$\mu^- \to e^- + \bar{\nu}_e + \nu_{\mu} \tag{10.5}$$

and non-leptonic weak interactions, such as

$$\Lambda^0 \to p + \pi^- . \tag{10.6}$$

The first statement of *universality* of weak interactions was that all these processes have the same coupling strength G. In fact, as we shall see later, the notion of universality is now understood to apply at a deeper level, directly to the couplings of lepton currents, and, of Cabibbo-

†Nowadays it is conventional to call the constant appearing in μ decay 'the' Fermi constant G_F (see §10.5). Weak hadronic processes are described in terms of the underlying *quark* currents and couplings (see the following chapter).

modified quark currents, to the weak gauge bosons.

As we have seen, Fermi's vector–vector theory was motivated by analogy with the vector currents of QED. The analogy was, however, imperfect. The photon emitted in a radiative transition is the quantum of the electromagnetic force field, but it is hard to see how the corresponding $e^-\bar{\nu}_e$ pair can be the weak force quantum: for one thing, the effective mass of the pair varies from process to process. On the other hand, the remarkable success of QED encourages us to try and base theories of other interactions on as close an analogy with QED as possible. It is therefore natural to postulate the existence of a weak analogue of the photon—the intermediate vector boson (IVB)—and to suppose that weak interactions are mediated by the exchange of IVB's, as electromagnetic ones are by photon exchange. Thus the pointlike interaction of figure 10.1 is replaced by the IVB exchange, as shown in figure 10.2.

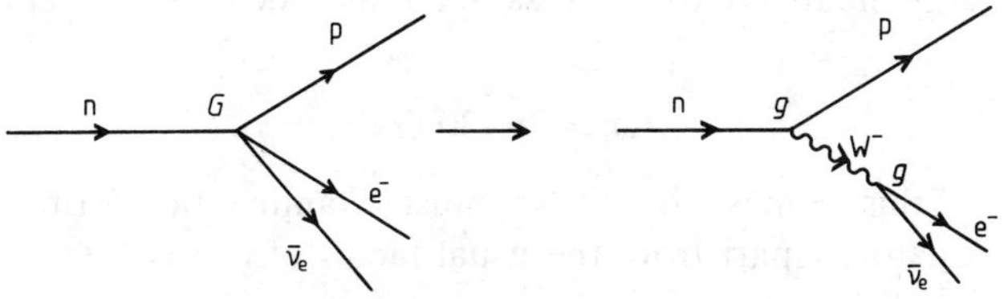

Figure 10.2 IVB exchange modification of figure 10.1.

This is the first step towards an eventual *unification* of the weak and electromagnetic forces. The path to that unification—which will be alluded to increasingly as we proceed, and will be explained in detail in Part V—has been long and hard. There are, indeed, several immediately apparent *dissimilarities* between QED and this view of weak interactions. For a start, the photon is electrically neutral, but in the weak interactions we are now considering, the hadronic and leptonic weak currents involve a change of charge by one unit†: hence (cf figure 10.2) the IVB's are charged and are denoted by $W^\pm$. They will eventually be understood, in Chapter 14, as *gauge* quanta.

A second and possibly even more striking difference is that the W's must be *massive*, unlike the photon. Since the mid-1960s, extensive experiments have been conducted to search for W's, culminating in their discovery, together with that of a neutral IVB (Z^0) associated with neutral current processes, in 1983 at the CERN $\bar{p}p$ collider (see Chapter 14). In the preceding years, the *non-observation* of W's enabled

†There are also neutral current processes: see Chapter 11.

lower limits to be placed on their mass. Obviously one way to do this was through simply trying to produce real W particles, and noting the energy at which they had still not appeared. But until the advent of the $\bar{p}p$ collider, the available CMS energy was relatively low: for example, from inclusive neutrino experiments on nuclei, via the reaction

$$\nu_\mu + A \rightarrow \mu^- + W^+ + X \tag{10.7}$$

$$W^+ \rightarrow \mu^+ + \nu_\mu \tag{10.8}$$

where X represents unobserved hadrons, a lower limit of the order

$$M_W \gtrsim 3\text{–}5\ \text{GeV} \tag{10.9}$$

was obtained, the precise value depending on the leptonic branching ratio. We now know that the W has a mass in the region of 80 GeV. In fact, a much more stringent bound than (10.9) can be obtained by considering the effect of virtual—exchanged, rather than produced—W's. Virtual W's cause deviations from pointlike behaviour (see §10.5) of high-energy neutrino total cross sections; such experiments gave a limit

$$M_W \geqslant 25\text{–}30\ \text{GeV}. \tag{10.10}$$

To see how this comes about, we must examine the form of the W boson propagator. Apart from the usual factor of i, it is

$$(-g^{\mu\nu} + q^\mu q^\nu/M^2)/(q^2 - M^2) \tag{10.11}$$

where M is the W mass. Let us attempt to make this form plausible. Recall our interpretation of the photon propagator in QED, in §5.8.1. In the presence of currents, the wave equation for the photon was assumed to have the form

$$\Box A^\mu = j^\mu_{\text{em}}. \tag{10.12}$$

The propagator (apart from the i) was just the inverse differential operator on the left-hand side of this equation, when applied to plane wave solutions:

$$-q^2 A^\mu = j^\mu_{\text{em}} \tag{10.13}$$

namely

$$-g^{\mu\nu}/q^2. \tag{10.14}$$

In fact, our treatment of QED was based on a particular gauge choice ($\partial_\mu A^\mu = 0$) for A_μ, and the form of the photon propagator in fact depends on the choice of gauge. We shall return to the problem of gauge invariance in Chapter 13. Here we wish only to motivate the form for the wave equation for a massive spin-1 particle. In a general gauge, the Maxwell equations read

$$\Box A^{\mu} - \partial^{\mu}\partial^{\nu}A_{\nu} = j^{\mu}_{\rm em}. \tag{10.15}$$

We make the natural replacement

$$\Box \to \Box + M^2 \tag{10.16}$$

to obtain

$$(\Box + M^2)W^{\mu} - \partial^{\mu}\partial^{\nu}W_{\nu} = j^{\mu}. \tag{10.17}$$

For plane wave solutions this reads

$$[(-q^2 + M^2)g^{\mu\nu} + q^{\mu}q^{\nu}]W_{\nu} = j^{\mu} \tag{10.18}$$

and the propagator is expected to correspond to the inverse operator

$$[(-q^2 + M^2)g^{\mu\nu} + q^{\mu}q^{\nu}]^{-1}. \tag{10.19}$$

How can we find this? Lorentz invariance comes to our aid. Since the only available tensors are $g_{\mu\nu}$ and $q_{\mu}q_{\nu}$, the inverse must have the form

$$Ag_{\mu\nu} + Bq_{\mu}q_{\nu}. \tag{10.20}$$

As with ordinary matrices, the inverse is defined by the requirement

$$\mathbf{M}\mathbf{M}^{-1} = \mathbf{1} \tag{10.21}$$

so we determine A and B from the requirement that

$$[(-q^2 + M^2)g^{\mu\nu} + q^{\mu}q^{\nu}](Ag_{\nu\rho} + Bq_{\nu}q_{\rho}) = \delta^{\mu}_{\rho}. \tag{10.22}$$

This leads to the desired result (problem 10.1)

$$(-g^{\mu\nu} + q^{\mu}q^{\nu}/M^2)/(q^2 - M^2) \tag{10.23}$$

which (with the i attached) we identify as the propagator for a massive spin-1 particle.

We have therefore arrived at the IVB modification of Fermi's current–current theory. The matrix element has the form (up to standard normalisation factors)

$$\mathcal{M} = j^{\rm wk}_{\mu}\frac{(-g^{\mu\nu} + q^{\mu}q^{\nu}/M^2)}{q^2 - M^2}j^{\rm wk}_{\nu}. \tag{10.24}$$

Bounds on $M_{\rm W}$ from virtual W effects were obtained by comparing data with cross sections predicted from (10.24), which contain (via the W propagator) an energy dependence absent in the 'pointlike' cross sections derived (see §10.5) from (10.1).

We saw in §5.8.2 how the range of a force is inversely proportional to the mass of the exchanged quantum. Thus the difference in mass between the photon and the IVB translates into a major qualitative difference in the associated ranges: in QED the photon is massless and the range infinite, but the limit (10.10) already implied that the range of the weak force is very small. Indeed, using the known value of M_W,

$$R_{wk} \sim 1/M_W \sim 2.5 \times 10^{-18} \text{ m}. \tag{10.25}$$

What, finally, of the *strength* of the interactions? Here we run into a difficulty if we try to compare the magnitude of Fermi's constant G_F with the fine structure constant α of QED: the latter is a dimensionless number, while the former (cf (10.2)) is not. This will turn out to be a crucial point, related to renormalisability (Chapter 12). But it would in any case seem more natural *a priori* to compare the IVB theory with QED, since both are based on the idea of an exchanged quantum. Indeed, expression (10.24) may be compared with the closely analogous one-photon exchange amplitudes derived earlier for spinless boson scattering (equation (5.72)), π^+e^- scattering (equation (6.51)) and $e^-\mu^-$ scattering (equation (6.119)). In all these latter cases, the electromagnetic current matrix elements (see, for example, equations (6.100) and (6.137)) each involve the coupling strength e appropriate to the interaction vertices in the corresponding Feynman graphs. In the same way, the W vertices and weak vector current matrix elements may be assigned a characteristic strength g (see figure 10.2)—which like e (in our units) will be *dimensionless*. We shall define g more precisely in due course, but for the moment we shall be content with the following qualitative identification for g. For typical β decay energies, and with $M_W \sim 80$ GeV, we certainly have

$$q^2 \ll M_W^2 \tag{10.26}$$

and all q dependence in the W propagator in (10.24) can be ignored. We arrive at the qualitative equivalence

$$G_F \sim g^2/M_W^2. \tag{10.27}$$

(10.27) is a fundamental relation—the exact version of it in the GSW theory is given in equation (10.93) below. It shows us *why* the Fermi constant has dimension $[\text{mass}]^{-2}$, and even, in a sense, why the weak interactions are weak! They are so (i.e. G_F is 'small') principally because M_W is so large. Indeed, out of so much apparent dissimilarity between the weak and electromagnetic interactions, perhaps some simple similarity can after all be rescued. Maybe the intrinsic strengths g and e are roughly equal:

$$g \sim e. \tag{10.28}$$

This would then lead, via (10.27), to an order-of-magnitude estimate of M_W,

$$M_W \sim e/G_F^{1/2} \sim 90 \text{ GeV} \tag{10.29}$$

which is indeed quite close to the true value. As we shall see, the GSW electroweak theory incorporates a relation of just the form (10.27)—see equation (10.93).

We shall return to these theories later. We wish now to introduce another ingredient in the theory of weak interactions—parity violation—which is *not* present in the original Fermi theory, nor in the IVB model as so far discussed.

10.2 Parity violation in weak interactions

In 1957, the experiment of Wu *et al* (1957) confirmed the prediction of Lee and Yang (1956) that parity was violated in weak interactions. What was observed was a dependence of the angular distribution of the decay electrons on the polarisation of the decaying nucleus. If we denote the polarisation vector by $\boldsymbol{S}$ and the electron momentum by $\boldsymbol{p}$, this means that, in addition to pieces depending only on dot products of momenta, the angular distribution contains a term depending on the quantity $\boldsymbol{S}\cdot\boldsymbol{p}$. To see that this represents parity violation, we must first recall the definition of vectors (polar vectors) and pseudovectors (axial vectors). A polar vector is one which transforms in the same way as the coordinate $\boldsymbol{r}$ under the parity operation P

$$\mathsf{P}:\boldsymbol{r} \rightarrow -\boldsymbol{r}. \tag{10.30}$$

Thus a vector is defined by the behaviour

$$\mathsf{P}:\boldsymbol{a} \rightarrow -\boldsymbol{a} \tag{10.31}$$

under parity. Examples are the momentum $\boldsymbol{p}$ and electromagnetic current $\boldsymbol{j}_{\text{em}}$:

$$\mathsf{P}:\boldsymbol{p} \rightarrow -\boldsymbol{p} \tag{10.32}$$

$$\mathsf{P}:\boldsymbol{j}_{\text{em}} \rightarrow -\boldsymbol{j}_{\text{em}}. \tag{10.33}$$

The vector product of two such vectors defines a pseudovector

$$\mathsf{P}:\boldsymbol{a}\times\boldsymbol{b} \rightarrow +\boldsymbol{a}\times\boldsymbol{b} \tag{10.34}$$

under parity. The most common example is angular momentum

$$\boldsymbol{L} = \boldsymbol{r}\times\boldsymbol{p}. \tag{10.35}$$

Under parity, orbital angular momentum, and by extension any angular momentum, behaves as a pseudovector

$$\mathsf{P}:\boldsymbol{L} \rightarrow +\boldsymbol{L} \tag{10.36a}$$

$$\mathsf{P}:\boldsymbol{S} \rightarrow +\boldsymbol{S}. \tag{10.36b}$$

In forming scalar products therefore, we must now distinguish between scalar, $\boldsymbol{a}\cdot\boldsymbol{b}$, and pseudoscalar, $\boldsymbol{a}\cdot\boldsymbol{b}\times\boldsymbol{c}$, quantities:

$$\mathsf{P}:\boldsymbol{a}\cdot\boldsymbol{b} \to +\,\boldsymbol{a}\cdot\boldsymbol{b} \qquad \text{scalar} \tag{10.37}$$

$$\mathsf{P}:\boldsymbol{a}\cdot\boldsymbol{b}\times\boldsymbol{c} \to -\,\boldsymbol{a}\cdot\boldsymbol{b}\times\boldsymbol{c} \qquad \text{pseudoscalar.} \tag{10.38}$$

The fact that the observed decay angular distribution in β decay contains both scalar (dot products of momenta) and pseudoscalar ($\boldsymbol{S}\cdot\boldsymbol{p}$) quantities is therefore evidence that parity is not a good quantum number for weak interactions (for further discussion, see Perkins (1987, Chapter 7)).

The Fermi theory of vector currents needs modification to account for this. After many years of painstaking experiments, and many false trails, it has been shown that the currents participating in Fermi's current–current interaction are a mixture of vector, V_μ, and axial vector, A_μ, currents. We now take this up in detail in the next two sections.

10.3 Vector and axial vector currents

In this section we wish to analyse the transformation properties under parity of (a) the *vector current* $\bar{u}\gamma_\mu u$, and (b) the *axial vector current* $\bar{u}\gamma_\mu\gamma_5 u$. A proper discussion requires a study of the Lorentz covariance of the Dirac equation (see e.g. Bjorken and Drell 1964, Aitchison 1972). Here our starting point is that $\bar{u}\gamma_\mu u$ is a Lorentz 4-vector and we want to demonstrate that the insertion of γ_5 converts this to a pseudovector. The 4×4 matrix γ_5 is defined by

$$\gamma_5 = \mathrm{i}\gamma^0\gamma^1\gamma^2\gamma^3 \tag{10.39}$$

and this can easily be shown (problem 10.2) to anticommute with the other γ matrices:

$$\{\gamma_5, \gamma_\mu\} = 0. \tag{10.40}$$

With the choice of Dirac matrices of Chapter 3, γ_5 has the explicit form†

$$\gamma_5 = \begin{pmatrix} 0 & 1 \\ 1 & 0 \end{pmatrix}. \tag{10.41}$$

Let us first consider the parity properties of the familiar electromagnetic current. As in Chapters 5 and 6 for the case of the π^+ (cf equations (5.59), (6.47), (6.100)), we introduce the electromagnetic transition current for an electron by matrix elements of the electromagnetic

†The 1's are, of course, really 2×2 unit matrices. However, we shall from now on adopt the standard convention of allowing the matrix character of the 1's to be inferred from the context.

current operator

$$\langle \mathrm{e}^-; \boldsymbol{k}', s' | \hat{j}^{\mu}_{\mathrm{em}}(0) | \mathrm{e}^-; \boldsymbol{k}, s \rangle. \tag{10.42}$$

What happens to this under parity? The electromagnetic current operator is a 4-vector transforming like x_μ under parity: $t \to +t$; $\boldsymbol{x} \to -\boldsymbol{x}$. Thus under parity

$$\mathbf{P}: \hat{j}^{\mu}_{\mathrm{em}}(0) \to \tilde{\hat{j}}^{\mu}_{\mathrm{em}}(0) \tag{10.43}$$

where

$$\hat{j}^{\mu}_{\mathrm{em}}(0) = (\hat{j}^{0}_{\mathrm{em}}(0), \hat{\boldsymbol{j}}_{\mathrm{em}}(0)) \tag{10.44}$$

and

$$\tilde{\hat{j}}^{\mu}_{\mathrm{em}}(0) = (+\hat{j}^{0}_{\mathrm{em}}(0), -\hat{\boldsymbol{j}}_{\mathrm{em}}(0)). \tag{10.45}$$

The states transform under parity by

$$|\boldsymbol{k}, s\rangle \to \eta |-\boldsymbol{k}, s\rangle \tag{10.46}$$

$$|\boldsymbol{k}', s'\rangle \to \eta |-\boldsymbol{k}', s'\rangle \tag{10.47}$$

since the spin angular momentum is a pseudovector quantity; η is the intrinsic parity and satisfies $\eta^2 = +1$. How is the parity operator represented in terms of Dirac γ matrices? To see this we write out the lowest-order expression for the current matrix element in terms of the explicit spinors:

$$\langle \mathrm{e}^-; \boldsymbol{k}', s' | \hat{j}^{\mu}_{\mathrm{em}}(0) | \mathrm{e}^-; \boldsymbol{k}, s \rangle = (-e)\mathcal{N}\mathcal{N}'\bar{u}(\boldsymbol{k}', s')\gamma^\mu u(\boldsymbol{k}, s) \tag{10.48}$$

where

$$u(\boldsymbol{k}, s) = (\omega + m)^{1/2} \begin{pmatrix} \phi^{\mathrm{s}} \\ \dfrac{\boldsymbol{\sigma}\cdot\boldsymbol{k}}{\omega + m}\phi^{\mathrm{s}} \end{pmatrix}. \tag{10.49}$$

With our present choice of γ matrices it is evident that

$$u(-\boldsymbol{k}, s) = (\omega + m)^{1/2} \begin{pmatrix} \phi^{\mathrm{s}} \\ \dfrac{-\boldsymbol{\sigma}\cdot\boldsymbol{k}}{\omega + m}\phi^{\mathrm{s}} \end{pmatrix} \tag{10.50}$$

satisfies the relation

$$u(-\boldsymbol{k}, s) = \gamma_0 u(\boldsymbol{k}, s). \tag{10.51}$$

We consider the space components of the matrix element and use equations (10.43)–(10.47) to construct the identity

$$\begin{aligned}\langle \boldsymbol{k}', s'|\hat{\boldsymbol{j}}_{\text{em}}(0)|\boldsymbol{k}, s\rangle &= \langle \boldsymbol{k}', s'|\mathbf{P}^{-1}\mathbf{P}\hat{\boldsymbol{j}}_{\text{em}}(0)\mathbf{P}^{-1}\mathbf{P}|\boldsymbol{k}, s\rangle \\ &= -\langle -\boldsymbol{k}', s'|\hat{\boldsymbol{j}}_{\text{em}}(0)|-\boldsymbol{k}, s\rangle.\end{aligned} \tag{10.52}$$

In terms of the Dirac spinors appearing in (10.48), this is obvious:

$$\begin{aligned}\bar{u}(\boldsymbol{k}', s')\boldsymbol{\gamma}u(\boldsymbol{k}, s) &= -\bar{u}(-\boldsymbol{k}', s')\boldsymbol{\gamma}u(-\boldsymbol{k}, s) \\ &= -\bar{u}(\boldsymbol{k}', s')\gamma_0\boldsymbol{\gamma}\gamma_0 u(\boldsymbol{k}, s)\end{aligned} \tag{10.53}$$

since γ_0 anticommutes with $\boldsymbol{\gamma}$. The property

$$\gamma_0\boldsymbol{\gamma}\gamma_0 = -\boldsymbol{\gamma} \tag{10.54}$$

is exactly what is needed for consistency with the requirement that, under parity,

$$\hat{\boldsymbol{j}}_{\text{em}}(0) \to -\hat{\boldsymbol{j}}_{\text{em}}(0) \tag{10.55}$$

and we see that we may represent the parity operator $\mathbf{P}$ by the matrix

$$\boxed{\mathbf{P} = \gamma_0} \tag{10.56}$$

acting on Dirac wavefunctions. This satisfies all the requirements

$$\mathbf{P}^{-1}\mathbf{P} = \mathbf{P}^{\dagger}\mathbf{P} = \mathbf{P}^2 = \mathbf{1} \tag{10.57}$$

of the parity operator. It also shows that electrons and positrons at rest have opposite intrinsic parities since

$$\gamma_0 = \begin{pmatrix} 1 & 0 \\ 0 & -1 \end{pmatrix} \tag{10.58}$$

with our choice of matrices and spinors (equations (3.60) and (3.69)).

We are now able to investigate the parity properties of other Dirac operators. As we have seen,

$$V^{\mu} = \bar{u}\gamma^{\mu}u \tag{10.59}$$

transforms as a vector under $\mathbf{P}$:

$$\tilde{V}^{\mu} = \bar{u}\gamma_0\gamma^{\mu}\gamma_0 u = (V_0, -\boldsymbol{V}). \tag{10.60}$$

There is, however, another 4-index quantity that we can construct: this involves the matrix γ_5. We define

$$A^{\mu} = \bar{u}\gamma^{\mu}\gamma_5 u. \tag{10.61}$$

Under parity

$$A^{\mu} \to \tilde{A}^{\mu} \tag{10.62}$$

where

$$\widetilde{A}^{\mu} = \bar{u}\gamma_0\gamma^{\mu}\gamma_5\gamma_0 u. \tag{10.63}$$

Thus under the parity operation the quantities

$$A^{\mu} \equiv (A_0, \boldsymbol{A}) \tag{10.64}$$

transform like an axial vector since, on using $\gamma_0\gamma_5 = -\gamma_5\gamma_0$, we can easily verify that

$$\widetilde{A}^{\mu} = (-A_0, +\boldsymbol{A}). \tag{10.65}$$

In similar fashion one can introduce

$$\text{scalar, } S = \bar{u}\mathbf{1}u \tag{10.66}$$

and

$$\text{pseudoscalar, } P = \bar{u}\gamma_5 u \tag{10.67}$$

combinations of spinors and γ matrices.

10.4 Left-handed neutrinos

A series of celebrated experiments has shown that neutrinos have the following properties:

(a) They are massless (or nearly so).
(b) There are three distinct types of neutrino each associated with its own charged lepton:

$$(\mathrm{e}^-, \nu_{\mathrm{e}}) \tag{10.68}$$

$$(\mu^-, \nu_{\mu}) \tag{10.69}$$

and

$$(\tau^-, \nu_{\tau}). \tag{10.70}$$

(c) They have spin $\frac{1}{2}$ but only the negative helicity state participates in weak interactions.

Helicity is the projection of spin along the direction of motion:

$$\lambda = \boldsymbol{S}\cdot\boldsymbol{p}/|\boldsymbol{p}|. \tag{10.71}$$

Spin-$\frac{1}{2}$ particles can occur with helicity $\lambda = \pm\frac{1}{2}$, corresponding to spin projection parallel or antiparallel to the direction of motion. By analogy with polarised light, the positive helicity state is often termed 'right-handed' and the negative helicity state, 'left-handed'. Ingenious experiments, which involved measuring correlations between helicities of the particles participating in weak interactions, established that the weak currents are vector and axial vector in nature and that only the left-handed neutrino helicity state is 'active' in weak interactions (figure 10.3) (the book by Perkins (1987) contains an account of these experiments). The fact that only one helicity state is involved shows directly that weak interactions violate parity. Helicity is a pseudoscalar quantity, and under parity

$$\mathsf{P}:\lambda \to -\lambda. \tag{10.72}$$

Hence a parity invariant theory, such as QED, necessarily requires that both helicity states participate equally in the interaction.

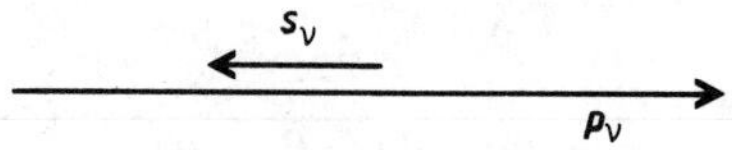

Figure 10.3 Representation of left-handed helicity state.

What we want to do now is to incorporate these facts about neutrinos into the form for the leptonic weak currents. This is done by the famous $(1 - \gamma_5)$ factor, and it was this observation that led to the postulate of the $V - A$ theory of weak interactions. There are many ways to see that this factor does the trick. Rather than set up the whole machinery of spin projection operators, the quickest and most instructive way to see how this factor arises is to go right back to the Dirac equation

$$E\omega = (\boldsymbol{\alpha}\cdot\boldsymbol{p} + \beta m)\omega \tag{10.73}$$

and *work with a different representation of the* $\boldsymbol{\alpha}$ *and* β *matrices*. This is possible since only their anticommutation relations are prescribed: clearly, however, the explicit form of the Dirac spinors depends on the choice of γ matrices. So, for this section, we must forget our explicit matrices and spinors of Chapter 3; instead, we shall work in a representation in which γ_5 is chosen to be diagonal. We choose

$$\gamma_5 = \begin{pmatrix} 1 & 0 \\ 0 & -1 \end{pmatrix}, \qquad \boldsymbol{\alpha} = \begin{pmatrix} \boldsymbol{\sigma} & 0 \\ 0 & -\boldsymbol{\sigma} \end{pmatrix}, \qquad \beta = \begin{pmatrix} 0 & 1 \\ 1 & 0 \end{pmatrix} \tag{10.74}$$

which is related to our earlier choice by a unitary transformation. We must find the form of the spinors in this new representation. Define, as before,

$$\omega = \begin{pmatrix} \phi \\ \chi \end{pmatrix} \tag{10.75}$$

so that the Dirac equation (with our new forms for $\boldsymbol{\alpha}$ and β) reduces to the two equations

$$\boldsymbol{\sigma}\cdot\hat{\boldsymbol{p}}\phi = \frac{E}{|\boldsymbol{p}|}\phi - \frac{m}{|\boldsymbol{p}|}\chi \tag{10.76a}$$

$$\boldsymbol{\sigma}\cdot\hat{\boldsymbol{p}}\chi = \frac{-E}{|\boldsymbol{p}|}\chi + \frac{m}{|\boldsymbol{p}|}\phi \tag{10.76b}$$

where we have defined

$$\hat{\boldsymbol{p}} = \boldsymbol{p}/|\boldsymbol{p}|. \tag{10.77}$$

In the limit $m = 0$, $E = |\boldsymbol{p}|$ and we see that ϕ and χ decouple and are separate helicity eigenstates:

$$\boldsymbol{\sigma}\cdot\hat{\boldsymbol{p}}\phi = +\phi, \qquad \text{positive helicity} \tag{10.78a}$$

$$\boldsymbol{\sigma}\cdot\hat{\boldsymbol{p}}\chi = -\chi, \qquad \text{negative helicity.} \tag{10.78b}$$

Thus, to project out the left-handed component of the Dirac spinor, we must select the lower components χ by a matrix of the form

$$\begin{pmatrix} 0 & 0 \\ 0 & 1 \end{pmatrix} = \tfrac{1}{2}(1 - \gamma_5) \tag{10.79}$$

in this representation. Therefore, for massless neutrinos, the combination†

$$(1 - \gamma_5)u_\nu(\boldsymbol{p}) \tag{10.80}$$

contains only a left-handed component, and although the result was derived with a specific, convenient choice of γ matrices, *this result is independent of that choice*.

What are the helicity properties of antineutrinos participating in weak interactions? The weak interactions do not conserve **P**, parity, nor do they respect invariance under **C**, charge conjugation—the operation of replacing all particles by their antiparticles. To a very good approximation, however, the weak interactions are invariant under the combined operation of **CP** (an exception is the tiny **CP** violation observed in the neutral kaon system). Thus, the active antineutrino component is the *right-handed positive helicity state*, since **C** changes ν to $\bar{\nu}$, and **P** changes λ to $-\lambda$. Remembering that a right-handed antineutrino spinor corresponds to a left-handed negative-energy neutrino solution, we see that the projection operator for a right-handed antineutrino is also $\frac{1}{2}(1 - \gamma_5)$. The antineutrino spinor will therefore appear in the combination

$$(1 - \gamma_5)v_\nu(\boldsymbol{p}). \tag{10.81}$$

†We adopt the same normalisation for massless spin-$\frac{1}{2}$ solutions as for massive ones, namely $u^\dagger u = 2E$, calling the spinors normalised in this way u rather than ω.

We can now derive the helicity properties of vector and axial vector currents. For any massless fermion we can write

$$u = \left(\frac{1+\gamma_5}{2}\right)u + \left(\frac{1-\gamma_5}{2}\right)u \tag{10.82}$$

$$\equiv u_{\mathrm{R}} + u_{\mathrm{L}}. \tag{10.83}$$

The factors

$$P_{\mathrm{R}} = \left(\frac{1+\gamma_5}{2}\right), \qquad P_{\mathrm{L}} = \left(\frac{1-\gamma_5}{2}\right) \tag{10.84}$$

are projection operators satisfying

$$P_{\mathrm{R}}^2 = P_{\mathrm{R}}, \qquad P_{\mathrm{L}}^2 = P_{\mathrm{L}},$$

$$P_{\mathrm{R}}P_{\mathrm{L}} = P_{\mathrm{L}}P_{\mathrm{R}} = 0, \qquad P_{\mathrm{L}} + P_{\mathrm{R}} = 1 \tag{10.85}$$

as can easily be verified using the result $(\gamma_5)^2 = 1$. Thus the interaction $\bar{u}\gamma_\mu u$ may be written

$$(u_{\mathrm{R}}^\dagger + u_{\mathrm{L}}^\dagger)(\gamma_0\gamma_\mu)(u_{\mathrm{R}} + u_{\mathrm{L}}) \tag{10.86}$$

and, since $(1 \pm \gamma_5)$ commutes with the product $(\gamma_0\gamma_\mu)$, we find that this reduces to

$$u_{\mathrm{R}}^\dagger(\gamma_0\gamma_\mu)u_{\mathrm{R}} + u_{\mathrm{L}}^\dagger(\gamma_0\gamma_\mu)u_{\mathrm{L}} \tag{10.87}$$

i.e. a vector interaction conserves the helicity of massless particles. Moreover, it is clear that insertion of γ_5 does not affect this argument, so the same is true for axial vector currents $\bar{u}\gamma_\mu\gamma_5 u$. Scalar and pseudoscalar interactions, on the other hand, necessarily flip the helicity of a massless particle, and hence it is correlations in helicity that distinguish the different forms of interaction.

10.5 *V–A* theory

The result of all this experimental effort is a slight, but elegant, modification of the earlier IVB model, and of Fermi's current–current theory. Instead of pure vector currents, there are now both vector, *V*, and axial, *A*, pieces. Furthermore, the leptonic weak current

$$\hat{j}_\mu^{\,\mathrm{lept}} = \hat{j}_\mu^{\,\mathrm{wk}}(\mathrm{e}) + \hat{j}_\mu^{\,\mathrm{wk}}(\mu) + \hat{j}_\mu^{\,\mathrm{wk}}(\tau) + \ldots \tag{10.88}$$

for each lepton and its (assumed massless) neutrino has the form

$$\langle \mathrm{e}^-|\hat{j}_\mu^{\,\mathrm{wk}}|\nu_{\mathrm{e}}\rangle = (g/\sqrt{2})\mathcal{N}\mathcal{N}'\bar{u}(\mathrm{e})\gamma_\mu\tfrac{1}{2}(1-\gamma_5)u(\nu_{\mathrm{e}}) \tag{10.89}$$

where we have suppressed momenta and spin labels, and labelled each spinor by a particle label $u(\mathrm{e})$, etc. Equation (10.89) defines the W coupling strength g. The $\sqrt{2}$, and the projection operator P_L rather than a simple $(1-\gamma_5)$ factor, have been adopted so as to make the notation identical to the one we shall finally arrive at in the 'standard electroweak model' described in Chapter 14.

We can now calculate a weak interaction cross section. Consider the process of figure 10.4

$$\nu_\mu e^- \to \mu^- \nu_e. \tag{10.90}$$

As indicated in the diagram, for energies such that the q^2 dependence of the W propagator can be ignored, we can replace the matrix element (cf (10.24))

$$\frac{g^2}{2}\bar{u}(\mu)\gamma_\mu\frac{(1-\gamma_5)}{2}u(\nu_\mu)\frac{(-g^{\mu\nu}+q^\mu q^\nu/M_w^2)}{q^2-M_w^2}\bar{u}(\nu_e)\gamma_\nu\frac{(1-\gamma_5)}{2}u(e) \tag{10.91}$$

by the ('Fermi') pointlike form

$$(G_F/\sqrt{2})\bar{u}(\mu)\gamma_\mu(1-\gamma_5)u(\nu_\mu)g^{\mu\nu}\bar{u}(\nu_e)\gamma_\nu(1-\gamma_5)u(e) \tag{10.92}$$

which defines the Fermi constant G_F precisely. Comparison of (10.91) and (10.92) for $q^2 \ll M_w^2$ yields the relation

$$\boxed{G_F/\sqrt{2} = g^2/8M_w^2} \tag{10.93}$$

between the parameters of the Fermi and IVB models. We shall calculate the cross section for the 'pointlike' amplitude (10.92).

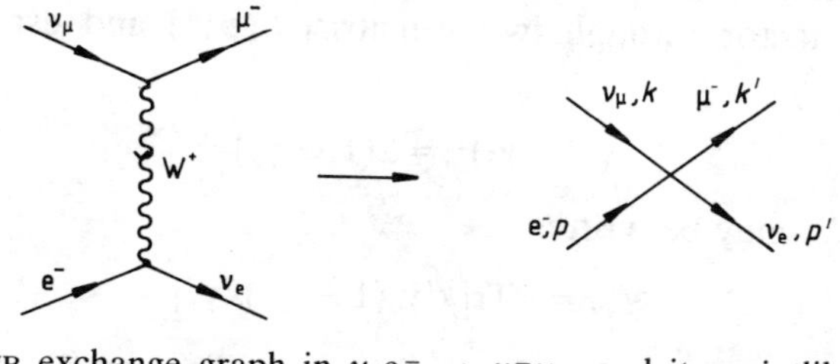

Figure 10.4 IVB exchange graph in $\nu_\mu e^- \to \mu^- \nu_e$, and its pointlike approximation.

From the general formula (5.122) for $2 \to 2$ scattering we have, neglecting all masses,

$$\frac{d\sigma}{dt} = \frac{1}{16\pi s^2}|\bar{F}|^2 \tag{10.94}$$

where $|\bar{F}|^2$ is the appropriate spin-averaged matrix element squared. In the case of neutrino–electron scattering, we must average over initial electron states for unpolarised electrons and sum over the final muon polarisation states. For the neutrinos there is no averaging over initial neutrino helicities, since only left-handed neutrinos participate in the weak interaction. Similarly, there is no sum over final neutrino helicities. However, for convenience of calculation, we can in fact sum over both helicity states of both neutrinos since the $(1 - \gamma_5)$ factors guarantee that right-handed neutrinos contribute nothing to the cross section. As for the eμ scattering example, the calculation then reduces to that of a product of traces:

$$|\bar{F}|^2 = \left(\frac{G_F^2}{2}\right)\mathrm{Tr}[\not{k}'\gamma_\mu(1 - \gamma_5)\not{k}\gamma_\nu(1 - \gamma_5)]\tfrac{1}{2}\mathrm{Tr}[\not{p}'\gamma^\mu(1 - \gamma_5)\not{p}\gamma^\nu(1 - \gamma_5)]. \quad (10.95)$$

Notice that we have neglected all lepton mass effects since these only constitute very small corrections at high energies. We define

$$|\bar{F}|^2 = \left(\frac{G_F^2}{2}\right)N_{\mu\nu}E^{\mu\nu} \quad (10.96)$$

where the $\nu_\mu \to \mu^-$ tensor $N_{\mu\nu}$ is given by

$$N_{\mu\nu} = \mathrm{Tr}[\not{k}'\gamma_\mu(1 - \gamma_5)\not{k}\gamma_\nu(1 - \gamma_5)] \quad (10.97)$$

without a $1/(2s + 1)$ factor, and the $e^- \to \nu_e$ tensor is

$$E^{\mu\nu} = \tfrac{1}{2}\mathrm{Tr}[\not{p}'\gamma^\mu(1 - \gamma_5)\not{p}\gamma^\nu(1 - \gamma_5)] \quad (10.98)$$

including a factor of $\frac{1}{2}$ for spin averaging. Since this calculation involves a couple of new features, let us look at it in some detail. By commuting the $(1 - \gamma_5)$ factor through two γ matrices ($\not{p}\gamma^\nu$) and using the result that

$$(1 - \gamma_5)^2 = 2(1 - \gamma_5) \quad (10.99)$$

the tensor $N_{\mu\nu}$ may be written

$$\begin{aligned} N_{\mu\nu} &= 2\mathrm{Tr}[\not{k}'\gamma_\mu(1 - \gamma_5)\not{k}\gamma_\nu] \\ &= 2\mathrm{Tr}(\not{k}'\gamma_\mu\not{k}\gamma_\nu) - 2\mathrm{Tr}(\gamma_5\not{k}\gamma_\nu\not{k}'\gamma_\mu). \end{aligned} \quad (10.100)$$

The first trace is the same as in our calculation of eμ scattering (cf (6.88)):

$$\mathrm{Tr}(\not{k}'\gamma_\mu\not{k}\gamma_\nu) = 4[k'_\mu k_\nu + k'_\nu k_\mu + (q^2/2)g_{\mu\nu}]. \quad (10.101)$$

The second trace must be evaluated using the result

$$\mathrm{Tr}(\gamma_5\not{a}\not{b}\not{c}\not{d}) = 4\mathrm{i}\varepsilon_{\alpha\beta\gamma\delta}a^\alpha b^\beta c^\gamma d^\delta \quad (10.102)$$

(see Appendix D). The totally antisymmetric tensor $\varepsilon_{\alpha\beta\gamma\delta}$ is just the generalisation of ε_{ijk} to four dimensions, and is defined by

$$\varepsilon_{\alpha\beta\gamma\delta} = \begin{cases} +1 & \text{for } \varepsilon_{0123} \text{ and even permutations of } 0, 1, 2, 3 \\ -1 & \text{for } \varepsilon_{1023} \text{ and for all odd permutations of } 0, 1, 2, 3 \\ 0 & \text{otherwise.} \end{cases} \tag{10.103}$$

Notice that this definition has the consequence that

$$\varepsilon_{0123} = +1 \tag{10.104a}$$

but

$$\varepsilon^{0123} = -1. \tag{10.104b}$$

We will also need to contract two ε tensors. By looking at the possible combinations, it should be easy to convince yourself of the result

$$\varepsilon_{ijk}\varepsilon_{ilm} = \begin{vmatrix} \delta_{jl} & \delta_{jm} \\ \delta_{kl} & \delta_{km} \end{vmatrix} \tag{10.105}$$

i.e.

$$\varepsilon_{ijk}\varepsilon_{ilm} = \delta_{jl}\delta_{km} - \delta_{kl}\delta_{jm}. \tag{10.106}$$

For the four-dimensional ε tensor one can show (see problem 10.3)

$$\varepsilon_{\mu\nu\alpha\beta}\varepsilon^{\mu\nu\gamma\delta} = -2!\begin{vmatrix} \delta^{\gamma}_{\alpha} & \delta^{\gamma}_{\beta} \\ \delta^{\delta}_{\alpha} & \delta^{\delta}_{\beta} \end{vmatrix} \tag{10.107}$$

where the minus sign arises from (10.104) and the 2! from the fact that the two indices are contracted.

We can now evaluate $N_{\mu\nu}$. We obtain, after some rearrangement of indices, the result for the $\nu_\mu \to \mu^-$ tensor:

$$\boxed{N_{\mu\nu} = 8[(k'_\mu k_\nu + k'_\nu k_\mu + (q^2/2)g_{\mu\nu}) - \mathrm{i}\varepsilon_{\mu\nu\alpha\beta}k^\alpha k'^\beta].} \tag{10.108}$$

For the electron tensor $E^{\mu\nu}$ we have a similar result (divided by 2):

$$E^{\mu\nu} = 4[(p'^\mu p^\nu + p'^\nu p^\mu + (q^2/2)g^{\mu\nu}) - \mathrm{i}\varepsilon^{\mu\nu\gamma\delta}p_\gamma p'_\delta]. \tag{10.109}$$

Now, in the approximation of neglecting all lepton masses,

$$q^\mu N_{\mu\nu} = q^\nu N_{\mu\nu} = 0 \tag{10.110}$$

as for the electromagnetic tensor $L_{\mu\nu}$ (6.130). Hence we may replace

$$p' = p + q \tag{10.111}$$

and drop all terms involving q in the contraction with $N_{\mu\nu}$. In the antisymmetric term, however, we have

$$\varepsilon^{\mu\nu\gamma\delta}p_\gamma(p_\delta + q_\delta) = \varepsilon^{\mu\nu\gamma\delta}p_\gamma q_\delta \tag{10.112}$$

since the term with p_δ *vanishes* because of the *antisymmetry* of $\varepsilon_{\mu\nu\gamma\delta}$. Thus we arrive at

$$\boxed{E^{\mu\nu}_{\text{eff}} = 8p^\mu p^\nu + 2q^2 g^{\mu\nu} - 4\mathrm{i}\varepsilon^{\mu\nu\gamma\delta} p_\gamma q_\delta.} \tag{10.113}$$

We must now calculate the contraction. Since we are neglecting all masses, it is easiest to perform the calculation in invariant form before specialising to the 'laboratory' frame. The usual Mandelstam variables are (neglecting all masses)

$$s = 2k\cdot p \tag{10.114a}$$

$$u = -2k'\cdot p \tag{10.114b}$$

$$t = -2k\cdot k' = q^2 \tag{10.114c}$$

satisfying

$$s + t + u = 0. \tag{10.115}$$

The result of performing the contraction

$$N_{\mu\nu}E^{\mu\nu} = N_{\mu\nu}E^{\mu\nu}_{\text{eff}} \tag{10.116}$$

may be found using the result (10.107) for the contraction of two ε tensors (see problem 10.3): the answer for $\nu\mathrm{e} \rightarrow \mu\nu$ is

$$\boxed{N_{\mu\nu}E^{\mu\nu} = 16(s^2 + u^2) + 16(s^2 - u^2)} \tag{10.117}$$

where the first term arises from the symmetric part of $N_{\mu\nu}$ similar to $L_{\mu\nu}$, and the second term from the antisymmetric part involving $\varepsilon_{\mu\nu\alpha\beta}$. We have also used

$$t = q^2 = -(s + u) \tag{10.118}$$

valid in the approximation in which we are working. Thus for

$$\nu_\mu \mathrm{e}^- \rightarrow \mu^- \nu_\mathrm{e} \tag{10.119}$$

we have

$$N_{\mu\nu}E^{\mu\nu} = +\,32s^2 \tag{10.120}$$

and with

$$\frac{\mathrm{d}\sigma}{\mathrm{d}t} = \frac{1}{16\pi s^2}\left(\frac{G_\mathrm{F}^2}{2}\right)N_{\mu\nu}E^{\mu\nu} \tag{10.121}$$

we obtain the final result

$$\boxed{\frac{\mathrm{d}\sigma}{\mathrm{d}t} = \frac{G_\mathrm{F}^2}{\pi}.} \tag{10.122}$$

All other purely leptonic processes may be calculated in an analogous fashion (see Bailin (1982) for further examples).

Since $t = -2p^2(1 - \cos\theta)$, where p is the CM momentum and θ the CM scattering angle, the total cross section is

$$\sigma = (4G_F^2/\pi)p^2. \tag{10.123}$$

As in the case of deep inelastic electron scattering, we shall, in the following chapter, be interested in neutrino 'laboratory' cross sections. A simple calculation gives $p^2 \simeq \frac{1}{2}m_e E$ (neglecting squares of lepton masses by comparison with $m_e E$), where E is the 'laboratory' energy of a neutrino incident, in this example, on a stationary electron. It follows that *the total 'laboratory' cross section for pointlike scattering rises linearly with E*†.

The process (10.119) is a difficult one to measure. As we shall see in the following chapter, much more accessible reactions in which ν's scatter from nucleons are well described in terms of pointlike interactions between the ν's and quarks. Then (10.123) is replaced by (cf (B.17))

$$\sigma = \frac{4G_F^2}{\pi}\cdot\tfrac{1}{2}ME, \qquad \sigma/E \simeq 0.3 \times 10^{-38}\ \text{cm}^2\ \text{GeV}^{-1} \tag{10.124}$$

(see problem 11.2) where M is the nucleon mass and E is the incident ν energy in the laboratory frame. Figure 10.5 shows plots of σ/E versus E for a range of ν and $\bar{\nu}$ energies (the reason for the actual magnitudes will become clear in §11.2). It was the absence of deviations from this pointlike behaviour which provided the bound on the W mass.

In the next chapter we take up the question of the hadronic weak current and deep inelastic neutrino scattering.

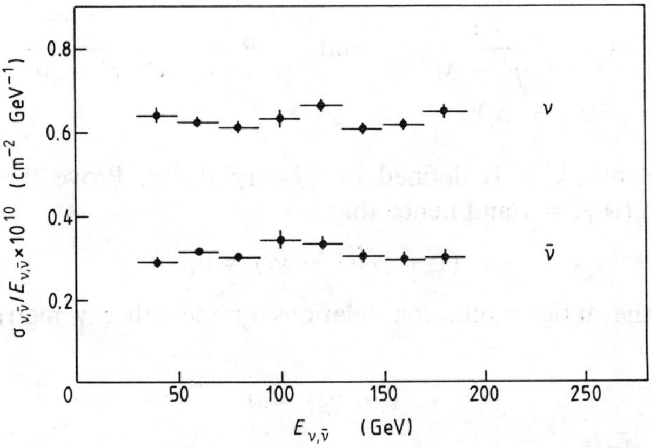

Figure 10.5 Neutrino and antineutrino total cross sections as a function of the neutrino energy (de Groot *et al* 1979).

†See also problem 10.5.

Summary

Fermi's current–current theory: hadronic and leptonic weak currents.

The IVB model and the massive spin-1 propagator.

Parity violation.

Vector and axial vector currents.

Left-handed neutrinos.

V–A theory.

Calculation of the cross section for $\nu_\mu e^- \to \mu^- \nu_e$ in the pointlike approximation.

Problems

10.1 From Lorentz invariance arguments, the inverse operator

$$[(-q^2 + M^2)g^{\mu\nu} + q^\mu q^\nu]^{-1}$$

must have the general form

$$Ag_{\mu\nu} + Bq_\mu q_\nu.$$

From the definition of an inverse, A and B must be chosen to satisfy the equation

$$[(-q^2 + M^2)g^{\mu\rho} + q^\mu q^\rho][Ag_{\rho\nu} + Bq_\rho q_\nu] = \delta^\mu_\nu.$$

Show that this requires

$$A = -\frac{1}{q^2 - M^2} \quad \text{and} \quad B = +\frac{1}{M^2}\frac{1}{q^2 - M^2}$$

(remember $g^{\mu\rho}g_{\rho\nu} = \delta^\mu_\nu$).

10.2 The matrix γ_5 is defined by $\gamma_5 \equiv i\gamma^0\gamma^1\gamma^2\gamma^3$. Prove the following properties: (i) $\gamma_5^2 = 1$ and hence that

$$(1 + \gamma_5)(1 - \gamma_5) = 0;$$

(ii) from the anticommutation relations of the other γ matrices, show that

$$\{\gamma_5, \gamma_\mu\} = 0$$

and hence that

$$(1 + \gamma_5)\gamma_0 = \gamma_0(1 - \gamma_5)$$

and

$$(1 + \gamma_5)\gamma_0\gamma_\mu = \gamma_0\gamma_\mu(1 + \gamma_5).$$

10.3 (a) Consider the two-dimensional antisymmetric tensor ε_{ij} defined by

$$\varepsilon_{12} = +1, \qquad \varepsilon_{21} = -1, \qquad \varepsilon_{11} = \varepsilon_{22} = 0.$$

By explicitly enumerating all the possibilities (if necessary), convince yourself of the result

$$\varepsilon_{ij}\varepsilon_{kl} = +1(\delta_{ik}\delta_{jl} - \delta_{il}\delta_{jk}).$$

Hence prove that

$$\varepsilon_{ij}\varepsilon_{il} = \delta_{jl} \qquad \text{and} \qquad \varepsilon_{ij}\varepsilon_{ij} = 2$$

(remember, in two dimensions, $\Sigma_i\delta_{ii} = 2$).

(b) By similar reasoning to that in part (a) of this question, it can be shown that the product of two three-dimensional antisymmetric tensors has the form

$$\varepsilon_{ijk}\varepsilon_{lmn} = \begin{vmatrix} \delta_{il} & \delta_{im} & \delta_{in} \\ \delta_{jl} & \delta_{jm} & \delta_{jn} \\ \delta_{kl} & \delta_{km} & \delta_{kn} \end{vmatrix}.$$

Prove the results

$$\varepsilon_{ijk}\varepsilon_{imn} = \begin{vmatrix} \delta_{jm} & \delta_{jn} \\ \delta_{km} & \delta_{kn} \end{vmatrix}, \qquad \varepsilon_{ijk}\varepsilon_{ijn} = 2\delta_{kn}, \qquad \varepsilon_{ijk}\varepsilon_{ijk} = 3!$$

(c) Extend these results to the case of the four-dimensional (Lorentz) tensor $\varepsilon_{\mu\nu\alpha\beta}$ (remember that a minus sign will appear as a result of $\varepsilon_{0123} = +1$ but $\varepsilon^{0123} = -1$).

10.4 The invariant amplitude for $\pi^+ \to e^+\nu$ decay may be written as

$$F = (G_F/\sqrt{2})f_\pi p^\mu \bar{u}(\nu)\gamma_\mu(1 - \gamma_5)v(e)$$

where p^μ is the 4-momentum of the pion. Evaluate the decay rate in the rest frame of the pion using the decay rate formula

$$\Gamma = (1/2M_\pi)|\bar{F}|^2\, \mathrm{dLips}(M_\pi^2;\, k_e,\, k_\nu).$$

Show that the ratio of $\pi^+ \to e^+\nu$ and $\pi^+ \to \mu^+\nu$ rates is given by

$$\frac{\Gamma(\pi^+ \to e^+\nu)}{\Gamma(\pi^+ \to \mu^+\nu)} = \left(\frac{M_e}{M_\mu}\right)^2\left(\frac{M_\pi^2 - M_e^2}{M_\pi^2 - M_\mu^2}\right)^2.$$

What factors in this formula are sensitive to the result for $|\bar{F}|^2$?

10.5 Starting from the amplitude for the process

$$\nu_\mu + e^- \to \mu^- + \nu_e$$

given by the current–current theory of weak interactions,

$$F = (G_F/\sqrt{2})\,\bar{u}(\mu)\gamma_\mu(1 - \gamma_5)u(\nu_\mu)g^{\mu\nu}\bar{u}(\nu_e)\gamma_\nu(1 - \gamma_5)u(e)$$

verify the intermediate results given in §10.5 leading to the result

$$d\sigma/dt = G_F^2/\pi$$

(neglecting all lepton masses). Hence show that the total cross section for this process rises linearly with s:

$$\sigma = G_F^2 s/\pi.$$

11

THE HADRONIC WEAK CURRENT AND NEUTRAL CURRENTS

11.1 Universality and Cabibbo theory

In our original discussion of weak interactions we were concerned with strangeness-conserving weak interactions such as neutron β decay (figure 11.1). In addition to these processes there are strangeness-changing weak interactions such as Λ β decay, illustrated in figure 11.2 (we postpone the introduction of charm until §11.4 below). Here the leptonic and hadronic weak currents still change the charge but the hadronic weak current $\Lambda \rightarrow p$ also changes strangeness.

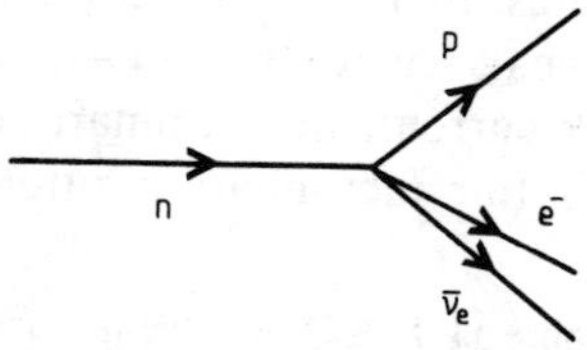

Figure 11.1 Four-fermion interaction for neutron β decay.

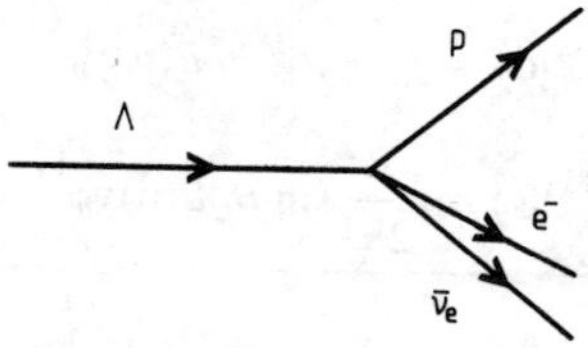

Figure 11.2 Four-fermion interaction for Λ β decay.

The notion of universality began with the observation that the coupling strength for the purely leptonic process of muon decay (figure 11.3) was the same as that for neutron β decay. In fact, detailed experimental comparison shows that this is not quite right: the hadronic strangeness-conserving weak decays are very slightly weaker than their leptonic counterparts. Moreover, the strangeness-changing hadronic

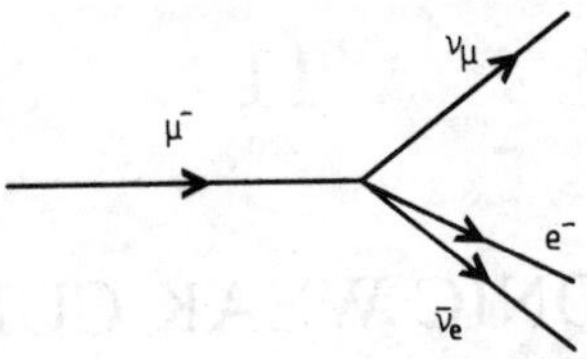

Figure 11.3 Four-fermion interaction for μ^- decay.

weak interactions are weaker still. The Cabibbo theory (Cabibbo 1963) postulated that the strength of the hadronic weak interaction is *shared* between the $\Delta S = 0$ and $\Delta S = 1$ transitions. In terms of their quark content, the amplitudes for the allowed transitions stand to one another in the ratios

$$\mathrm{e} \to \nu : g \tag{11.1a}$$

$$\mathrm{d} \to \mathrm{u} : g \cos \theta_{\mathrm{C}} \tag{11.1b}$$

$$\mathrm{s} \to \mathrm{u} : g \sin \theta_{\mathrm{C}} \tag{11.1c}$$

where θ_{C} is the Cabibbo angle. A best fit to data indicates

$$\sin \theta_{\mathrm{C}} \simeq 0.23. \tag{11.2}$$

Experiment also indicates that the space–time structure of all these charge-changing (CC) weak currents is *V–A* in character. Thus our beliefs about these weak currents are summarised in table 11.1 (in which standard spinor normalisation factors are omitted).

Table 11.1 $\Delta Q = 1$ weak currents.

$\langle \nu_{\mathrm{e}} \vert \hat{j}_\mu^{\mathrm{wk}} \vert \mathrm{e}^- \rangle$	$= \dfrac{g}{2^{1/2}} \bar{u}(\nu)\gamma_\mu \dfrac{(1-\gamma_5)}{2} u(\mathrm{e})$
$\langle u \vert \hat{j}_\mu^{\mathrm{wk}} \vert \mathrm{d} \rangle$	$= \dfrac{g}{2^{1/2}} \cos \theta_{\mathrm{C}} \bar{u}(\mathrm{u})\gamma_\mu \dfrac{(1-\gamma_5)}{2} u(\mathrm{d})$
$\langle u \vert \hat{j}_\mu^{\mathrm{wk}} \vert \mathrm{s} \rangle$	$= \dfrac{g}{2^{1/2}} \sin \theta_{\mathrm{C}} \bar{u}(\mathrm{u})\gamma_\mu \dfrac{(1-\gamma_5)}{2} u(\mathrm{s})$

In low-energy weak interaction processes, the simple structure of the quark weak current is masked by non-perturbative strong interaction effects, giving rise to form factors and consequent deviations away from the simple *V–A* form. Compare, for example, the electromagnetic current of an electron (10.48) to that of a proton (6.149) with its two form factors and anomalous magnetic moment interaction. However, just as deep inelastic electron–proton scattering gives us direct access to quark electromagnetic currents, deep inelastic neutrino scattering allows us to probe directly the quark weak currents. We therefore turn in the next section to deep inelastic neutrino scattering and the quark parton model. Before we do this, however, let us summarise our knowledge about the Cabibbo quark currents in matrix notation: the deeper significance of the matrix structure will be seen when we discuss the Glashow–Salam–Weinberg (GSW) theory in Chapter 14.

The quark weak current transition amplitudes have the form

$$j_\mu^{\text{wk,q}} \sim g_{\alpha\beta}\bar{q}_\alpha\gamma_\mu\tfrac{1}{2}(1-\gamma_5)q_\beta \tag{11.3}$$

where α, β label the 'flavour' of the quarks—u, d or s, until further notice—and $g_{\alpha\beta}$ incorporates the Cabibbo structure of the charged currents, namely

$$g_{\alpha\alpha} = 0 \tag{11.4a}$$

$$g_{\text{ud}} = (g/\sqrt{2})\cos\theta_{\text{C}} \tag{11.4b}$$

$$g_{\text{us}} = (g/\sqrt{2})\sin\theta_{\text{C}}. \tag{11.4c}$$

Consider only the flavour properties of this current and introduce a two-component quark spinor of the form

$$q = \begin{pmatrix} u \\ d\cos\theta_{\text{C}} + s\sin\theta_{\text{C}} \end{pmatrix}. \tag{11.5}$$

The Cabibbo structure may be succinctly summarised by the matrix form

$$j_{\text{wk}}^{+} = (g/\sqrt{2})(\bar{u},\ \bar{d}\cos\theta_{\text{C}} + \bar{s}\sin\theta_{\text{C}})\begin{pmatrix} 0 & 1 \\ 0 & 0 \end{pmatrix}\begin{pmatrix} u \\ d\cos\theta_{\text{C}} + s\sin\theta_{\text{C}} \end{pmatrix} \tag{11.6}$$

$$= (g/\sqrt{2})\bar{u}d\cos\theta_{\text{C}} + (g/\sqrt{2})\bar{u}s\sin\theta_{\text{C}} \tag{11.7}$$

for the charge-raising weak current amplitude j_{wk}^{+}. Notice that the matrix which *raises* charge is essentially the Pauli matrix τ_+:

$$\begin{pmatrix} 0 & 1 \\ 0 & 0 \end{pmatrix} = \frac{\tau_1 + i\tau_2}{2} = \frac{\tau_+}{2}. \tag{11.8}$$

In fact, we may consider the quarks as forming a '*weak isospin*' doublet

$$\begin{pmatrix} u \\ d_{\text{C}} \end{pmatrix} \quad \begin{matrix} Q = +\frac{2}{3} \\ Q = -\frac{1}{3} \end{matrix} \tag{11.9}$$

where d_{C} is the 'Cabibbo-rotated' d quark,

$$d_{\text{C}} = d\cos\theta_{\text{C}} + s\sin\theta_{\text{C}} \tag{11.10}$$

and j_{wk}^{+} is a 'weak-isospin-raising' transition amplitude (see Chapter 14 for detailed discussion of this weak isospin). Sometimes, the fact that only the left-handed components of all these quark currents are active in weak interactions is indicated by a subscript 'L',

$$q_{\mathrm{L}} = \begin{pmatrix} u \\ d_{\mathrm{C}} \end{pmatrix}_{\mathrm{L}} \tag{11.11}$$

and then

$$j^{+}_{\mathrm{wk}} \sim \bar{q}_{\mathrm{L}} \tau_{+} q_{\mathrm{L}}. \tag{11.12}$$

11.2 Deep inelastic neutrino scattering

It is straightforward to set up the formalism in the IVB/effective four-fermion model, for inclusive neutrino scattering (figure 11.4), assuming single-W exchange. Since no deviations from pointlike behaviour have so far been observed, we shall use the approximation, valid for $q^2 \ll M_{\mathrm{W}}^2$,

$$G_{\mathrm{F}}/\sqrt{2} = g^2/8M_{\mathrm{W}}^2 \tag{10.93}$$

to derive cross sections. The cross section will have the general form

$$\mathrm{d}\sigma^{(\nu)} \sim G_{\mathrm{F}}^2 N_{\mu\nu} W^{\mu\nu}_{(\nu)}(q, p) \tag{11.13}$$

where $N_{\mu\nu}$ is the neutrino tensor encountered in the last chapter:

$$N_{\mu\nu} = 8\{[k'_\mu k_\nu + k'_\nu k_\mu + (q^2/2) g_{\mu\nu}] - \mathrm{i}\varepsilon_{\mu\nu\alpha\beta} k^\alpha k'^\beta\}. \tag{10.108}$$

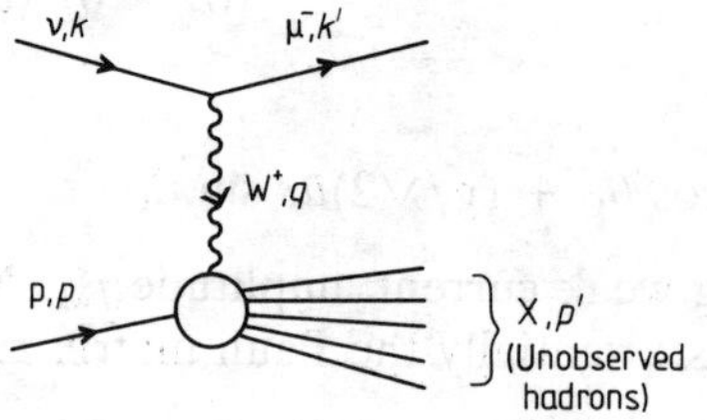

Figure 11.4 Inelastic neutrino scattering from a proton in the one-W exchange approximation.

The form of the weak hadron tensor $W^{\mu\nu}_{(\nu)}$ is deduced from Lorentz invariance. In the approximation of neglecting lepton masses, we can ignore any dependence on the 4-vector q since

$$q^\mu N_{\mu\nu} = q^\nu N_{\mu\nu} = 0. \tag{10.110}$$

Just as $N_{\mu\nu}$ contains the pseudotensor $\varepsilon_{\mu\nu\alpha\beta}$ so too will $W^{\mu\nu}_{(\nu)}$ since parity is not conserved. By analogy with the electron tensor, we define neutrino structure functions by

$$W^{\mu\nu}_{(\nu)} = (-g^{\mu\nu})W_1^{(\nu)} + \frac{1}{M^2}p^\mu p^\nu W_2^{(\nu)} - \frac{\mathrm{i}}{2M^2}\varepsilon^{\mu\nu\gamma\delta}p_\gamma q_\delta W_3^{(\nu)}. \tag{11.14}$$

In general, the structure functions depend on two variables, say Q^2 and ν, but in the Bjorken limit approximate scaling is observed, as in the electron case:

$$\left.\begin{array}{l} Q^2 \to \infty \\ \nu \to \infty \end{array}\right\} \quad x = Q^2/2M\nu, \text{ fixed} \tag{11.15a}$$

$$\nu W_2^{(\nu)}(Q^2, \nu) \to F_2^{(\nu)}(x) \tag{11.15b}$$

$$M W_1^{(\nu)}(Q^2, \nu) \to F_1^{(\nu)}(x) \tag{11.15c}$$

$$\nu W_3^{(\nu)}(Q^2, \nu) \to F_3^{(\nu)}(x) \tag{11.15d}$$

where, as with (7.19), the physics lies in the assertion that the F's are finite. This scaling can again be interpreted in terms of pointlike scattering from partons—which we shall take to have quark quantum numbers.

In the 'laboratory' frame the cross section in terms of W_1, W_2 and W_3 may be derived in the usual way from (cf equation (7.9))

$$\mathrm{d}\sigma^{(\nu)} = \left(\frac{G_\mathrm{F}}{\sqrt{2}}\right)^2 \frac{1}{4k\cdot p} 4\pi M N_{\mu\nu} W^{\mu\nu}_{(\nu)} \frac{\mathrm{d}^3k'}{2k'(2\pi)^3}. \tag{11.16}$$

In terms of 'laboratory' variables, one obtains

$$\frac{\mathrm{d}^2\sigma^{(\nu)}}{\mathrm{d}Q^2\,\mathrm{d}\nu} = \frac{G_\mathrm{F}^2}{2\pi}\frac{k'}{k}\left(W_2^{(\nu)}\cos^2(\theta/2) + W_1^{(\nu)}\,2\sin^2(\theta/2) + \frac{k+k'}{M}\sin^2(\theta/2)W_3^{(\nu)}\right). \tag{11.17}$$

For an incoming antineutrino beam, the W_3 term changes sign. It is common to use the variables

$$x = Q^2/2M\nu \tag{11.18}$$

and

$$y = \nu/k \tag{11.19}$$

in neutrino scattering and, after a routine struggle, the cross section may be written in the form (see problem 11.1)

$$\frac{\mathrm{d}^2\sigma^{(\nu)}}{\mathrm{d}x\,\mathrm{d}y} = \frac{G_\mathrm{F}^2}{2\pi}s\left(F_2^{(\nu)}\frac{1+(1-y)^2}{2} + xF_3^{(\nu)}\frac{1-(1-y)^2}{2}\right) \tag{11.20}$$

in terms of the Bjorken scaling functions, and we have assumed the relation

$$2xF_1^{(\nu)} = F_2^{(\nu)} \tag{11.21}$$

appropriate for spin-$\frac{1}{2}$ constituents.

The basic parton cross sections are eaily calculated in the same way as the νe scattering calculation of the last chapter. We obtain (see problem 11.2)

$$\nu\text{q}, \bar{\nu}\bar{\text{q}}: \quad \frac{\text{d}^2\sigma}{\text{d}x\,\text{d}y} = \frac{G_\text{F}^2}{\pi} sx\delta\left(x - \frac{Q^2}{2M\nu}\right) \tag{11.22a}$$

$$\nu\bar{\text{q}}, \bar{\nu}\text{q}: \quad \frac{\text{d}^2\sigma}{\text{d}x\,\text{d}y} = \frac{G_\text{F}^2}{\pi} sx(1-y)^2\delta\left(x - \frac{Q^2}{2M\nu}\right). \tag{11.22b}$$

The factor $(1-y)^2$ in the $\nu\bar{\text{q}}, \bar{\nu}\text{q}$ cases means that the reaction is forbidden at $y = 1$ (backwards in the CM frame). This follows from the V–A nature of the current, and angular momentum conservation, as a simple helicity argument shows (Perkins 1987).

Making the approximation of setting the Cabibbo angle to zero, we find that the contributing processes are

$$\nu\text{d} \to \text{l}^-\text{u} \tag{11.23a}$$

$$\nu\bar{\text{u}} \to \text{l}^-\bar{\text{d}} \tag{11.23b}$$

$$\bar{\nu}\text{u} \to \text{l}^+\text{d} \tag{11.23c}$$

$$\bar{\nu}\bar{\text{d}} \to \text{l}^+\bar{\text{u}}. \tag{11.23d}$$

Following the same steps as in the electron scattering case (§§7.2, 7.3) we obtain

$$F_2^{\nu\text{p}} = F_2^{\bar{\nu}\text{n}} = 2x[d(x) + \bar{u}(x)] \tag{11.24a}$$

$$F_3^{\nu\text{p}} = F_3^{\bar{\nu}\text{n}} = 2[d(x) - \bar{u}(x)] \tag{11.24b}$$

$$F_2^{\nu\text{n}} = F_2^{\bar{\nu}\text{p}} = 2x[u(x) + \bar{d}(x)] \tag{11.24c}$$

$$F_3^{\nu\text{n}} = F_3^{\bar{\nu}\text{p}} = 2[u(x) - \bar{d}(x)]. \tag{11.24d}$$

Many simple and elegant results now follow from this quark parton framework. For example, the two sum rules of (7.60a) and (7.60b) may be combined to give

$$3 = \int_0^1 \text{d}x[u(x) + d(x) - \bar{u}(x) - \bar{d}(x)] = \tfrac{1}{2}\int_0^1 \text{d}x(F_3^{\nu\text{p}} + F_3^{\nu\text{n}}) \equiv \int_0^1 \text{d}x F_3^{\nu\text{N}} \tag{11.25}$$

which is the Gross–Llewellyn Smith sum rule (1969), expressing the fact that the number of valence quarks per nucleon is three. The CDHS collaboration (de Groot *et al* 1979) quote

$$\tfrac{1}{2}\int_0^1 \text{d}x(F_3^{\nu\text{p}} + F_3^{\nu\text{n}}) = 3.2 \pm 0.5. \tag{11.26}$$

The F_2 functions for electron and neutrino scattering should be simply related in this picture. From (7.54) and (7.57) we have

$$F_2^{\mathrm{eN}} \equiv \tfrac{1}{2}(F_2^{\mathrm{ep}} + F_2^{\mathrm{en}}) = \tfrac{5}{18}x(u + \bar{u} + d + \bar{d}) + \tfrac{1}{9}x(s + \bar{s}) + \ldots \quad (11.27)$$

while (11.24a) and (11.24c) give

$$F_2^{\nu\mathrm{N}} \equiv \tfrac{1}{2}(F_2^{\nu\mathrm{p}} + F_2^{\nu\mathrm{n}}) = x(u + d + \bar{u} + \bar{d}). \quad (11.28)$$

Assuming the non-strange contributions dominate, the neutrino and electron structure functions on isoscalar targets should be approximately in the ration $\frac{18}{5}$, the reciprocal of the mean squared charge of the u and d quarks in the nucleon. Figure 11.5 shows such a comparison: the agreement is remarkable, and confirms both the ideas of the quark parton model and the fractional quark charges. The integral of $F_2^{\nu\mathrm{N}}$ ($\approx\frac{1}{2}$) has already been discussed in §7.3.

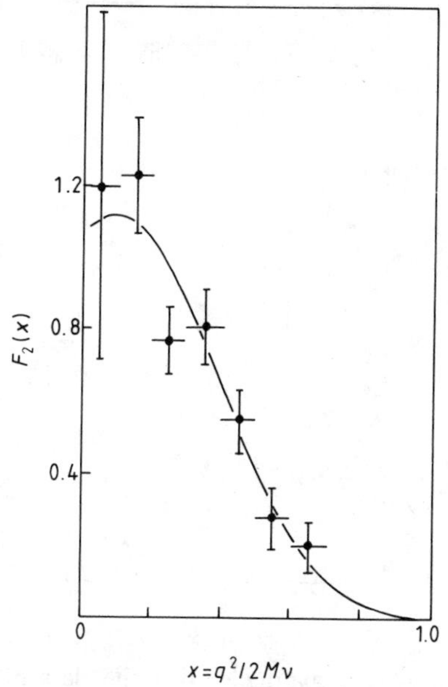

Figure 11.5 Comparison of $F_2^{\nu\mathrm{N}}$ measured in neutrino–nucleon scattering in the Gargamelle heavy-liquid bubble chamber in a PS neutrino beam at CERN, with SLAC data on F_2^{eN} from electron–nucleon scattering, in the same region of $q^2(>1)$. The data points are the neutrino results, and the curve is a fit through the electron data, multiplied by the factor $\frac{18}{5}$, which is the reciprocal of the mean squared charge of u and d quarks in the nucleon. This is a confirmation of the fractional charge assignments for the quarks. (From Perkins D 1982 *Introduction to High Energy Physics* 2nd edn, courtesy Addison-Wesley Publishing Company.).

We note from (11.22) and (11.24), after performing the integration over y (running from 0 to 1), that the total isoscalar cross sections for ν

and $\bar{\nu}$ would be in the ratio

$$\sigma^{\nu}/\sigma^{\bar{\nu}} = 3$$

if only quarks (no antiquarks) were present in the nucleon. This ratio was shown in figure 10.5 and is some way below 3, so that some antiquarks are present in the 'sea'. Quite generally, it is clear from (11.24a) and (11.24b) that neutrino reactions can separate the quark and antiquark contributions: this is because of the *V–A* interaction, which acts differently between them. Figure 11.6 shows the CDHS distributions (de Groot *et al* 1979) for $F_2^{\nu N} = x(u + d + \bar{u} + \bar{d})$ and $xF_3^{\nu N} = x(u + d - \bar{u} - \bar{d})$. From these the separate xq and $x\bar{q}$ distributions can be extracted; it is clear, for example, that the $x\bar{q}$ (sea) distribution is concentrated at small x, as we inferred earlier.

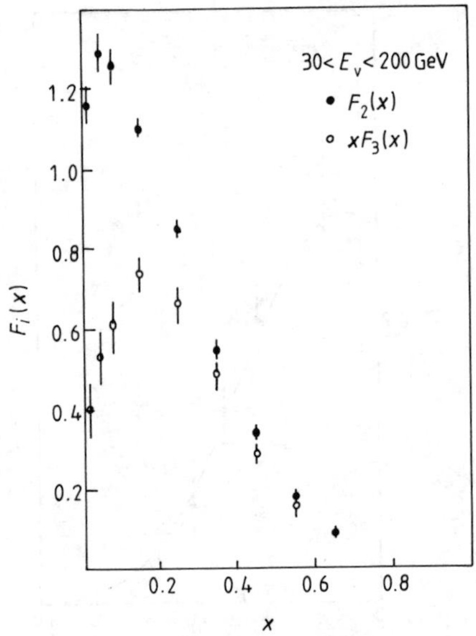

Figure 11.6 $F_2(x)$ and $xF_3(x)$ averaged over the data $30 < E_\nu < 200$ GeV (de Groot *et al* 1979).

11.3 Neutral currents

In the phenomenological framework for charged current weak interactions as so far discussed, certain types of processes are forbidden (at least in lowest order). Typical examples are the exclusive process

$$\bar{\nu}_\mu + e^- \rightarrow \bar{\nu}_\mu + e^- \tag{11.29}$$

and an analogous inclusive process

$$\nu_\mu + \mathrm{p} \rightarrow \nu_\mu + \mathrm{X} \tag{11.30}$$

which is identified experimentally as an interaction in which no μ^- appears in the final state. If such processes are to occur at the same level of strength as the charged current ones, a *neutral* weak current is needed: in the intermediate vector boson model this would correspond to the existence of a heavy, neutral weak boson, denoted by Z^0 (figure 11.7). In 1973 the discovery of such neutral current processes (see e.g. Cundy 1974) generated a whole new series of experiments. As in the case of charged currents, much detailed experimental work is necessary to determine the precise form of the neutral current interactions. If the neutral currents are assumed to be a mixture of vector and axial vector currents like their charged counterparts, they may be conveniently parametrised in terms of the left- and right-handed couplings of the leptons and quarks.

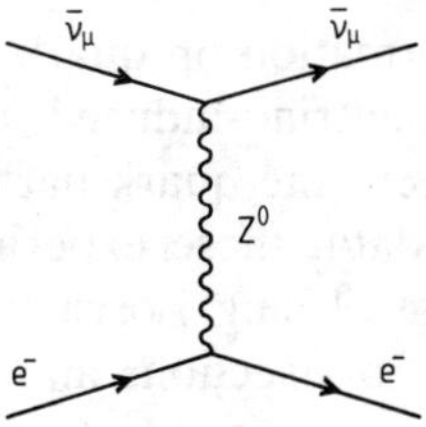

Figure 11.7 Neutral current process $\bar{\nu}_\mu \mathrm{e}^- \rightarrow \bar{\nu}_\mu \mathrm{e}^-$, mediated by neutral weak vector boson Z^0.

(1) Neutrino neutral current amplitude

$$\langle \nu | \hat{j}^{\mathrm{N}}_\mu | \nu \rangle = g_{\mathrm{N}} c^\nu \bar{u}(\nu) \gamma_\mu \frac{(1-\gamma_5)}{2} u(\nu) \qquad \nu = \nu_{\mathrm{e}}, \nu_\mu, \nu_\tau, \ldots \tag{11.31}$$

(2) Charged lepton neutral current amplitude

$$\langle l | \hat{j}^{\mathrm{N}}_\mu | l \rangle = g_{\mathrm{N}} \bar{u}(l) \gamma_\mu \left(c^l_{\mathrm{L}} \frac{(1-\gamma_5)}{2} + c^l_{\mathrm{R}} \frac{(1+\gamma_5)}{2} \right) u(l) \qquad l = \mathrm{e}^-, \mu^-, \tau^-, \ldots \tag{11.32}$$

(3) Quark neutral current amplitude

$$\langle q | \hat{j}^{\mathrm{N}}_\mu | q \rangle = g_{\mathrm{N}} \bar{u}(q) \gamma_\mu \left(c^q_{\mathrm{L}} \frac{(1-\gamma_5)}{2} + c^q_{\mathrm{R}} \frac{(1+\gamma_5)}{2} \right) u(q) \qquad q = \mathrm{u, d, s, c, b, t}, \ldots \tag{11.33}$$

In (11.31)–(11.33) we have, as before, written the amplitudes in such a way as to facilitate comparison between these phenomenological parametrisations and the GSW theory of Chapter 14. A factor g_N has been extracted, which will determine the overall strength of the neutral current processes as compared to the charged current ones; the c's determine the relative amplitudes of the various neutral current processes. We have anticipated subsequent development by extending the neutrino, lepton and quark types as indicated. For the quark neutral current only 'diagonal' neutral currents have been assumed—with no d–s quark transition, for example. This is because the decay $K_L^0 \to \mu^+\mu^-$ is highly suppressed relative to the analogous charged current decay $K^+ \to \mu^+\nu_\mu$, and this fact sets stringent limits on the magnitude of strangeness-changing neutral currents. We return to this point in the following section.

Notice that only the neutrino is assumed to couple via a purely left-handed interaction: it is up to experiment to determine the parameters c_L^l and c_R^l for the leptons, and c_L^u, c_R^u, c_L^d, c_R^d and so on for the quarks.

The major source of information on quark neutral current couplings is of course neutral current, neutrino-induced, deep inelastic scattering. In the regime of Q^2 and ν where the quark parton model provides a useful first approximation to the data, these experiments yield direct information on the quark couplings. Using isoscalar targets, as well as proton and neutron targets, total cross sections and y distributions ($d\sigma/dy$) for neutrinos and antineutrinos, one is able to make progress in separating the different couplings. Data from semi-inclusive reactions such as

$$\nu + N_{I=0} \to \nu + \pi^\pm + X \tag{11.34}$$

have also been used, as have certain exclusive channels such as

$$\nu + p \to \nu + \Delta. \tag{11.35}$$

For the current state of the art in determining these couplings, including discussions of ambiguities, effects of scaling violations and so on, the reader is referred to recent conference proceedings.

The leptonic couplings c_L^l and c_R^l, on the other hand, are best measured in purely leptonic neutral current reactions

$$\nu_\mu + e^- \to \nu_\mu + e^- \tag{11.36}$$

$$\bar{\nu}_\mu + e^- \to \bar{\nu}_\mu + e^-. \tag{11.37}$$

Additional information is obtained by combining these results with reactor experiments measuring the reaction

$$\bar{\nu}_e + e^- \to \bar{\nu}_e + e^-. \tag{11.38}$$

Again, for up-to-date analysis the reader is referred to recent conference reports.

We have indicated how the weak neutral current parameters are to be determined from experiment: the theoretical challenge is to predict and explain the results. It is a remarkable achievement that there does exist a theory of the weak interactions that not only predicted the existence of neutral currents, but also does extremely well in predicting the individual couplings. The theory is the Glashow–Salam–Weinberg (GSW) model of weak interactions which we discuss in detail in Chapter 14. Here we list some of its predictions for the neutral current couplings:

$$g_N = g/\cos\theta_W \tag{11.39}$$

and

$$c^\nu = \tfrac{1}{2} \tag{11.40a}$$

$$c_L^l = -\tfrac{1}{2} + a, \qquad c_R^l = a \tag{11.40b}$$

$$c_L^u = \tfrac{1}{2} - \tfrac{2}{3}a, \qquad c_R^u = -\tfrac{2}{3}a \tag{11.40c}$$

$$c_L^d = -\tfrac{1}{2} + \tfrac{1}{3}a, \qquad c_R^d = \tfrac{1}{3}a \tag{11.40d}$$

where

$$a = \sin^2\theta_W \tag{11.41}$$

and θ_W is the Glashow–Weinberg angle (Glashow 1961, Weinberg 1967). With the overall scale g (the parameter controlling the strength of the charged current couplings) extracted via (11.39), all these neutral current couplings are therefore expressed in terms of just one parameter, θ_W. The experimentally determined couplings are in remarkable agreement with this prediction, and give $\sin^2\theta_W \simeq 0.23$. We note that, apart from the neutrino, these neutral current couplings are apparently *not* pure V–A, in contrast to the charged currents. As implied by the absence of any distinguishing labels on the right-hand sides of the relations (11.40), in this theory all particles of a given *species* (neutral leptons: ν_e, ν_μ, ν_τ, . . .; charged leptons: e^-, μ^-, τ^-, . . .; quarks with charge $\frac{2}{3}$: u, c, t, . . .; and quarks with charge $-\frac{1}{3}$: d, s, b, . . .) are assumed to couple in the same way: for example, $c^{\nu_e} = c^{\nu_\mu} = c^{\nu_\tau} = \ldots = \frac{1}{2}$.

11.4 Charm and the Glashow–Iliopoulos–Maiani mechanism

We have seen that the flavour properties of the charged currents, with the Cabibbo mixing between d and s quarks, may be summarised by writing the hadronic weak current in matrix form. The charge-raising weak current has the form

$$j^{+}_{\text{wk}} \sim g(\bar{u}, \bar{d}_{\text{C}})\tau_{+}\begin{pmatrix} u \\ d_{\text{C}} \end{pmatrix} \tag{11.42}$$

where

$$d_{\text{C}} = d \cos\theta_{\text{C}} + s \sin\theta_{\text{C}} \tag{11.10}$$

and

$$\frac{\tau_{+}}{2} = \begin{pmatrix} 0 & 1 \\ 0 & 0 \end{pmatrix}. \tag{11.43}$$

Clearly, the charge-lowering current has the form

$$j^{-}_{\text{wk}} \sim g(\bar{u}, \bar{d}_{\text{C}})\tau_{-}\begin{pmatrix} u \\ d_{\text{C}} \end{pmatrix}, \qquad \frac{\tau_{-}}{2} = \begin{pmatrix} 0 & 0 \\ 1 & 0 \end{pmatrix}. \tag{11.44}$$

What can we say about the flavour properties of the quark neutral current? A vital clue is provided by the hypothesis that the weak currents—charged and neutral—are those appropriate to an underlying *gauge* theory: such is, after all, certainly the case for the familiar electromagnetic current. As we saw in Chapters 2 and 8, gauge theories are based on *symmetries*. The relevant symmetry here is that of 'weak isospin', which has already been mentioned. Just as for hadronic isospin, the τ_{+} and τ_{-} act as raising and lowering matrices, changing the third component of weak isospin by one unit. However, by analogy with the hadronic case, these cannot be the only relevant matrices: the full symmetry must also involve the matrix τ_3, given by the commutator

$$[\tau_{+}/2, \tau_{-}/2] = \tau_3 = \begin{pmatrix} 1 & 0 \\ 0 & -1 \end{pmatrix}. \tag{11.45}$$

As we shall see in Chapter 14, in a gauge theory based on this weak isospin there must necessarily appear a current of the form

$$j^{0}_{\text{wk}} \sim g(\bar{u}, \bar{d}_{\text{C}})\tau_3\begin{pmatrix} u \\ d_{\text{C}} \end{pmatrix} \tag{11.46}$$

which, since it involves charge-conserving transitions $u \to u$, $d_{\text{C}} \to d_{\text{C}}$, must be a weak *neutral* current.

Unfortunately, there is phenomenological evidence against this simple form, since the term $\bar{d}_{\text{C}} d_{\text{C}}$ contains *strangeness-changing* neutral currents:

$$\bar{d}_{\text{C}} d_{\text{C}} = \bar{d}d \cos^2\theta_{\text{C}} + \bar{s}s \sin^2\theta_{\text{C}} + \bar{d}s \sin\theta_{\text{C}}\cos\theta_{\text{C}} + \bar{s}d \sin\theta_{\text{C}}\cos\theta_{\text{C}}. \tag{11.47}$$

Experimentally, such strangeness-changing neutral currents seem to be highly suppressed (Review of Particle Properties 1984):

$$\text{branching fraction } (\mathrm{K_L^0} \to \mu^+\mu^-) = (9.1 \pm 1.9) \times 10^{-7}\% \qquad (11.48a)$$

to be compared with

$$\text{branching fraction } (\mathrm{K^+} \to \mu^+\nu_\mu) = (63.50 \pm 0.16)\%. \qquad (11.48b)$$

This rules out a lowest-order Z^0 $\bar{d}s$ coupling.

We emphasise that there is a problem here only if we believe that weak interactions are described by a theory based on an underlying weak isospin symmetry, which is then converted into a gauge theory. The reasons for this belief will provide the material of the following chapter, and subsequent chapters will explain the (hidden, non-Abelian) type of gauge theory involved, and its particular realisation in the GSW theory. Accepting this subsequent development, the solution to the neutral current problem is provided by the elegant hypothesis of Glashow, Iliopoulos and Maiani (1970), referred to as the GIM mechanism. We know that the weak doublet

$$\begin{pmatrix} u \\ d_C \end{pmatrix} \qquad \begin{matrix} Q = \frac{2}{3} \\ Q = -\frac{1}{2} \end{matrix} \qquad (11.9)$$

with electric charges as indicated, involves the 'Cabibbo-rotated' combination of d and s quarks

$$d_C = d\cos\theta_C + s\sin\theta_C \qquad (11.10)$$

and enters into charged current weak interactions. GIM *postulated* the existence of a second doublet, involving a then *new* $Q = \frac{2}{3}$ charge quark—the charm quark—and the orthogonal combination of d and s ($Q = -\frac{1}{3}$) quarks,

$$\begin{pmatrix} c \\ s_C \end{pmatrix} \qquad \begin{matrix} Q = +\frac{2}{3} \\ Q = -\frac{1}{3} \end{matrix} \qquad (11.49)$$

where

$$s_C = -d\sin\theta_C + s\cos\theta_C. \qquad (11.50)$$

The complete neutral current, according to GIM, is then the sum of (11.46) and an additional contribution from the new doublet:

$$j^0_{\mathrm{wk}} \sim g(\bar{u}, \bar{d}_C)\tau_3\begin{pmatrix} u \\ d_C \end{pmatrix} + g(\bar{c}, \bar{s}_C)\tau_3\begin{pmatrix} c \\ s_C \end{pmatrix}. \qquad (11.51)$$

For the $Q = -\frac{1}{3}$ quarks, this immediately yields the combination

$$\bar{d}_C d_C + \bar{s}_C s_C = \bar{d}d + \bar{s}s \qquad (11.52)$$

and the unwanted $\bar{d}s$ and $\bar{s}d$ terms have cancelled. The Z^0 thus couples directly only to $\bar{d}d$ and $\bar{s}s$—'diagonal in flavour'.

At the time GIM made their proposal, charmed states had not been observed, and neither low-energy hadron spectroscopy, nor low-energy weak interaction phenomenology, required charmed quarks. GIM attributed this non-appearance of the charm quark to the fact that it had a relatively large mass. This was not merely an easy way out: they made an estimate of its mass from an argument based on the $K_L^0 \to \mu^+\mu^-$ decay rate! The point is that although the GIM mechanism, as so far presented, certainly succeeds in cancelling $\Delta S \neq 0$ neutral current effects in lowest order, an unacceptably large *higher-order* amplitude for $K_L^0 \to \mu^+\mu^-$ (cf the experimental bound (11.48)) will result if the charm quark is too heavy as compared with the u quark.

The argument is as follows. Although the first-order (single Z^0 intermediate state) contribution to $K_L^0 \to \mu^+\mu^-$ has been explicitly cancelled, there is a non-vanishing amplitude from the two-W intermediate state graph shown in figure 11.8(*a*). Calculation of this contribution, which involves u quark exchange, leads to an amplitude roughly of order $G\alpha$, which is too large compared with the experimental rate. The GIM mechanism provides another contributing diagram from c quark exchange (figure 11.8(*b*)) which gives an additional suppression factor. Comparing couplings at the vertices, the u quark has an amplitude

$$\mathscr{A}_u \sim g^4 \sin\theta_C \cos\theta_C \tag{11.53}$$

while the c quark contributes

$$\mathscr{A}_c \sim -g^4 \sin\theta_C \cos\theta_C. \tag{11.54}$$

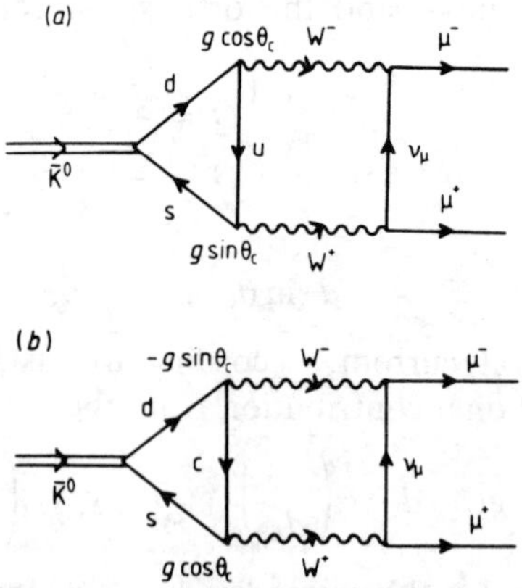

Figure 11.8 Two-W intermediate state graphs contributing to $\Delta S \neq 0$ weak neutral transition $K_L^0 \to \mu^+\mu^-$: (*a*) u exchange graph; (*b*) c exchange graph.

In the limit where u and c have equal masses, these contributions exactly cancel; for very large values of the c quark mass M_c, the c quark diagram will be very small, and the u quark diagram alone will give an unacceptable result. GIM went on to estimate the value of M_c required to give agreement with experiment. They obtained a range

$$M_c \sim 1\text{–}3\ \text{GeV} \tag{11.55}$$

which, in retrospect, is a remarkable achievement. The GIM mechanism is now incorporated into the 'standard model' of weak interactions, as will be explained in Chapter 14. For example, the neutral current coupling of the c quark is the same as that of the u quark, and those of the s and d quarks are also equal (cf the remark at the end of the preceding section).

11.5 Concluding comments

(1) It is interesting that both the GIM mechanism and the GSW theory itself were proposed before the Feynman rules for loop calculations in such theories were known. More specifically, it was not even known whether the proposed 'theory' was renormalisable, i.e. capable of yielding a sensible, consistent, perturbation expansion. Nevertheless, the important physics of the GIM mechanism—charm—has proved to be correct, as have the predictions for neutral currents. It is a tribute to the perspicacity of the originators of these theories that they were strong-minded enough to dismiss the absence of a consistent set of Feynman rules as merely a technical problem!

(2) The GIM mechanism only takes account of mixing between four flavours of quarks. To see how it may be generalised to six flavours, we write the Cabibbo combinations (11.10) and (11.50) in matrix notation as

$$\begin{pmatrix} d_C \\ s_C \end{pmatrix} = \begin{pmatrix} \cos\theta_C & \sin\theta_C \\ -\sin\theta_C & \cos\theta_C \end{pmatrix} \begin{pmatrix} d \\ s \end{pmatrix} = O_C \begin{pmatrix} d \\ s \end{pmatrix} \tag{11.56}$$

where O_C is the (orthogonal) Cabibbo rotation matrix. In this notation, it is the orthogonality of this matrix that guarantees (11.52). With six quark flavours, O_C is replaced by a 3×3 unitary matrix U_{KM} (Perkins 1987, Kobayashi and Maskawa 1973). The unitarity condition

$$U_{KM}^{\dagger} U_{KM} = 1 \tag{11.57}$$

guarantees the absence of flavour-changing neutral currents. Furthermore, it turns out that, in addition to three mixing angles (analogous to θ_C), U_{KM} contains one non-trivial phase parameter. A non-zero value for this phase corresponds to **CP** violation. Hence the theory, though not predicting the observed amount of **CP** violation, is able to incorporate it phenomenologically.

Summary

Cabibbo form of the charge-changing hadronic weak currents.

First suggestion of 'weak isospin'.

Nucleon structure functions for deep inelastic neutrino scattering.

Weak neutral currents.

Absence of strangeness-changing neutral currents; charm, and the GIM mechanism.

Indication of GSW predictions for neutral current couplings.

Problems

11.1 (a) Verify that the inclusive inelastic neutrino–proton scattering differential cross section has the form

$$\frac{d^2\sigma^{(\nu)}}{dQ^2\,d\nu} = \frac{G_F^2 k'}{2\pi\,k}\left(W_2^{(\nu)}\cos^2(\theta/2) + W_1^{(\nu)} 2\sin^2(\theta/2) + \frac{(k+k')}{M}\sin^2(\theta/2)\,W_3^{(\nu)}\right)$$

in the notation of §11.2.

(b) Using the Bjorken scaling behaviour

$$\nu W_2^{(\nu)} \to F_2^{(\nu)}, \qquad MW_1^{(\nu)} \to F_1^{(\nu)}, \qquad \nu W_3^{(\nu)} \to F_3^{(\nu)}$$

rewrite this expression in terms of the scaling functions. In terms of the variables x and y, neglect all masses and show that

$$\frac{d^2\sigma^{(\nu)}}{dx\,dy} = \frac{G_F^2}{2\pi}s[F_2^{(\nu)}(1-y) + F_1^{(\nu)}xy^2 + F_3(1-y/2)yx].$$

Remember that

$$\frac{k'\sin^2(\theta/2)}{M} = \frac{xy}{2}.$$

(c) Insert the Callan–Gross relation

$$2xF_1^{(\nu)} = F_2^{(\nu)}$$

to derive the result quoted in §11.2:

$$\frac{d^2\sigma^{(\nu)}}{dx\,dy} = \frac{G_F^2}{2\pi}sF_2^{(\nu)}\left(\frac{1+(1-y)^2}{2} + \frac{xF_3^{(\nu)}}{F_2^{(\nu)}}\frac{1-(1-y)^2}{2}\right).$$

11.2 The differential cross section for $\nu_\mu q$ scattering by charged currents has the same form (neglecting masses) as the $\nu_\mu e^- \to \mu^-\nu_e$ result of problem 10.5, namely

$$\frac{d\sigma}{dt}(\nu q) = \frac{G_F^2}{\pi}.$$

(a) Show that the cross section for scattering by antiquarks $\nu_\mu \bar{q}$ has the form

$$\frac{d\sigma}{dt}(\nu\bar{q}) = \frac{G_F^2}{\pi}(1 - y)^2.$$

(b) Hence prove the results quoted in §11.2:

$$\frac{d^2\sigma}{dx\,dy}(\nu q) = \frac{G_F^2}{\pi} sx\,\delta(x - Q^2/2M\nu)$$

and

$$\frac{d^2\sigma}{dx\,dy}(\nu\bar{q}) = \frac{G_F^2}{\pi} sx(1 - y)^2\delta(x - Q^2/2M\nu)$$

(where M is the nucleon mass).

(c) Use the parton model prediction

$$\frac{d^2\sigma}{dx\,dy} = \frac{G_F^2}{\pi} sx[q(x) + \bar{q}(x)(1 - y)^2]$$

to show that

$$F_2^{(\nu)}(x) = 2x[q(x) + \bar{q}(x)]$$

and

$$\frac{xF_3^{(\nu)}(x)}{F_2^{(\nu)}(x)} = \frac{q(x) - \bar{q}(x)}{q(x) + \bar{q}(x)}.$$

12

DIFFICULTIES WITH WEAK INTERACTION PHENOMENOLOGY

The preceding two chapters have developed the '*V–A* current–current' phenomenology of weak interactions. Many attempts were made to promote this *phenomenology* into a *theory*, or alternatively to invent a new theory of which it might be the low-energy limit. Since we now believe that the GSW theory is, in fact, the correct description of weak interactions up to currently available energies, further discussion of the difficulties with the Fermi theory, or with simple IVB theories, might seem superfluous. However, these difficulties raise several important points of principle, and an understanding of them provides valuable motivation for the GSW theory.

12.1 Violation of unitarity in the pointlike four-fermion model

In Chapter 10 we saw that the range of the weak interactions is very small; cross sections still seem to be pointlike at currently available neutrino energies, and the range corresponding to the W mass is of order 2.5×10^{-18} m. Suppose we did not know about the W: would there be anything wrong with having no intermediate vector boson at all, so that the interaction is genuinely of zero range, i.e. pointlike? Can we, in fact, promote Fermi's *phenomenological model* to the status of a *theory*? The answer is no. Let us examine why this is so, in order to gain some insight into how we may do better.

Consider the process (figure 12.1)

$$\bar{\nu}_\mu + \mu^- \rightarrow \bar{\nu}_e + e^- \qquad (12.1)$$

in the pointlike model, regarding it as a fundamental interaction, treated to lowest order in perturbation theory. A similar process has already

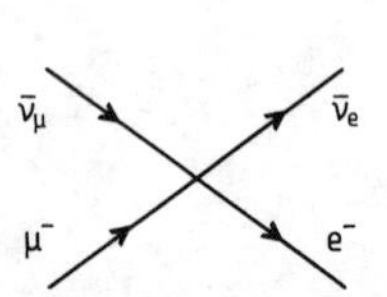

Figure 12.1 Pointlike amplitude for $\bar{\nu}_\mu + \mu^- \rightarrow \bar{\nu}_e + e^-$.

been discussed in Chapter 10. Since the troubles we shall find occur at high energies, we can simplify the expressions by neglecting the lepton masses m_e and m_μ without altering the conclusions. In this limit the invariant amplitude is (problem 12.1), up to a numerical factor,

$$F = G_F E^2(1 + \cos\theta) \tag{12.2}$$

where E is the CM energy and θ the CM scattering angle. This leads to the following behaviour of the cross section:

$$\sigma \sim G_F^2 E^2. \tag{12.3}$$

Consider now partial wave analysis of this process. For spinless particles the total cross section may be written as a sum of partial wave cross sections

$$\sigma = \frac{4\pi}{k^2}\sum_J (2J + 1)|f_J|^2 \tag{12.4}$$

where f_J is the partial wave amplitude and k is the CM momentum. It is a consequence of *unitarity*, or flux conservation (see e.g. Gasiorowicz 1974, pp 184, 383), that the partial wave amplitude may be written in terms of a phaseshift δ_J:

$$f_J = e^{i\delta_J}\sin\delta_J \tag{12.5}$$

so that

$$|f_J| \leqslant 1. \tag{12.6}$$

Thus the cross section in each partial wave is bounded by

$$\sigma_J \leqslant 4\pi(2J + 1)/k^2 \tag{12.7}$$

which falls as the CM energy rises. By contrast, we have a cross section that rises with CM energy:

$$\sigma \sim E^2. \tag{12.8}$$

Moreover, since the amplitude (equation (12.2)) only involves $(\cos\theta)^0$ and $(\cos\theta)^1$ contributions, it is clear that this rise in σ is associated with only a few partial waves, and is not due to more and more partial waves contributing to the sum in σ. Therefore, at some energy E, the unitarity bound will be violated by this lowest-order (Born approximation) expression for σ.

This is the essence of the 'unitarity disease' of the pointlike theory. To fill in all the details, however, involves a careful treatment of the appropriate partial wave analysis for the case when all particles carry spin. We shall avoid such details and instead sketch the conclusions of such an analysis. For massless spin-$\frac{1}{2}$ particles interacting via a V–A interaction we have seen that helicity is conserved. The net effect of the spin structure is to produce the $(1 + \cos\theta)$ factor in equation (12.2).

This embodies the fact that the initial state with $J_Z = -1$ (J_Z quantised along the μ^- direction in the CM system) is forbidden by angular momentum conservation to go to a $J_Z = +1$ state at $\theta = \pi$, which is the state required by the V–A interaction. Extracting this angular-momentum-conserving kinematic factor, the remaining amplitude can be regarded as that appropriate to a $J = 0$, spinless process (since there are no other factors of $\cos\theta$) so that

$$f_{\text{eff}}^{J=0} \sim G_{\text{F}}E^2. \tag{12.9}$$

The unitarity bound is therefore violated for CM energies

$$E \gtrsim G_{\text{F}}^{-1/2} \sim 300 \text{ GeV}. \tag{12.10}$$

This difficulty with the pointlike theory can be directly related to the fact that the Fermi coupling constant G_{F} is *not dimensionless*. From calculated decay rates, G_{F} is found to have the value (Review of Particle Properties 1984)

$$G_{\text{F}} \simeq (1.166\,37 \pm 0.000\,02) \times 10^{-5} \text{ GeV}^{-2} \tag{12.11}$$

and has the dimensions of $[M]^{-2}$. Given this fact, we can arrive at the form for the cross section for $\bar{\nu}_\mu \mu^- \to \bar{\nu}_e e^-$ at high energy without calculation. The cross section has dimensions of $[L]^2 = [M]^{-2}$, but must involve G_{F}^2 with dimensions $[M]^{-4}$. It must also be relativistically invariant. At energies well above lepton masses, the only invariant quantity available to restore the correct dimensions to σ is s, the square of the CM energy E, so that $\sigma \sim G_{\text{F}}^2 E^2$.

Faced with this difficulty, we make the first of many appeals we shall be making to the best theory we have, and ask: what happens in the analogous QED process? In this case we consider the process $e^+e^- \to \mu^+\mu^-$ in lowest order (figure 12.2). In Chapter 6 the total cross section for this process, neglecting lepton masses, was found to be (see problem 6.7 and equation (7.78))

$$\sigma = 4\pi\alpha^2/3E^2 \tag{12.12}$$

which obediently falls with energy as required by unitarity. In this case the coupling constant α, analogous to G_{F}, is dimensionless, so that a factor E^2 is required in the denominator to give $\sigma \sim [L]^2$.

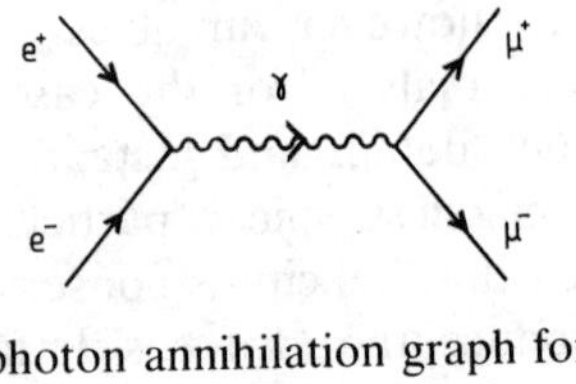

Figure 12.2 One-photon annihilation graph for $e^+e^- \to \mu^+\mu^-$.

If we accept this clue from QED, we are led to search for a theory of weak interactions that involves a dimensionless coupling constant. As can be seen from the low-q^2 correspondence

$$G_F \sim g^2/M_W^2 \tag{12.13}$$

the IVB theory, introduced in Chapter 10, is such a theory. We therefore next examine this as a candidate theory of weak interactions.

12.2 Violation of unitarity bounds in the IVB model

As the section heading indicates, matters will turn out to be fundamentally no better in the IVB model; let us see why. We begin by reconsidering the process (12.1), this time viewed as proceeding via an intermediate state consisting of one vector boson W^-, of mass M, as shown in figure 12.3. This diagram is, of course, closely analogous to the preceding one, figure 12.2; the important difference is that whereas in figure 12.2 the internal vector particle is the massless photon, in the case of figure 12.3 it is the massive W. As discussed in Chapter 10, the propagator for the massive W is

$$\frac{-g^{\mu\nu} + k^\mu k^\nu/M^2}{k^2 - M^2} \tag{12.14}$$

where k is the 4-momentum of the W, and the amplitude for figure 12.3 is

$$\mathrm{i}\frac{g^2}{2}\bar{u}(\mathrm{e})\gamma_\mu\frac{(1-\gamma_5)}{2}v(\nu_\mathrm{e})\left(\frac{-g^{\mu\nu} + k^\mu k^\nu/M^2}{k^2 - M^2}\right)\bar{v}(\nu_\mu)\gamma_\nu\frac{(1-\gamma_5)}{2}u(\mu) \tag{12.15}$$

where g is the dimensionless coupling constant of (10.89).

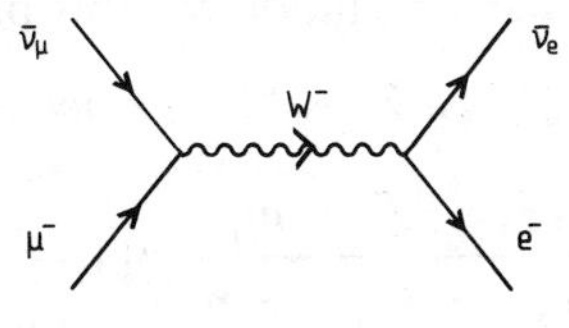

Figure 12.3 W^- intermediate state graph for $\bar{\nu}_\mu + \mu^- \rightarrow \bar{\nu}_\mathrm{e} + \mathrm{e}^-$.

We may compare (12.15) with the amplitude for figure 12.2, which is

$$\mathrm{i}e^2\bar{v}(\mathrm{e})\gamma_\mu u(\mathrm{e})\left(\frac{-g^{\mu\nu}}{k^2}\right)\bar{u}(\mu)\gamma_\nu v(\mu) \tag{12.16}$$

where k is the 4-momentum of the photon. At first sight we might conclude that the high-energy behaviour of (12.15) is going to be considerably 'worse' than that of (12.16) in view of the presence of the $k^\mu k^\nu$ factors in the numerator of (12.15). However, they turn out to be harmless, as we can see as follows. Using 4-momentum conservation,

$$\begin{aligned} k^\mu \bar{u}(\mathrm{e})\gamma_\mu(1-\gamma_5)v(\nu_\mathrm{e}) &= (p_\mathrm{e}^\mu + p_{\bar{\nu}}^\mu)\bar{u}(\mathrm{e})\gamma_\mu(1-\gamma_5)v(\nu_\mathrm{e}) \\ &= \bar{u}(\mathrm{e})\not{p}_\mathrm{e}(1-\gamma_5)v(\nu_\mathrm{e}) + \bar{u}(\mathrm{e})\not{p}_{\bar{\nu}}(1-\gamma_5)v(\nu_\mathrm{e}). \end{aligned} \tag{12.17}$$

Using the Dirac equation and $\{\gamma^\mu, \gamma_5\} = 0$, this becomes

$$m_\mathrm{e}\bar{u}(\mathrm{e})(1-\gamma_5)v(\nu_\mathrm{e}) \tag{12.18}$$

since $\not{p}_{\bar{\nu}}v(\nu_\mathrm{e}) = 0$. A similar result holds for the k^ν factor; thus the $k^\mu k^\nu$ factors have disappeared. Indeed, neglecting the lepton masses by comparison with M, the effect of the IVB is to replace the photon propagator $-g^{\mu\nu}/k^2$ by $-g^{\mu\nu}/(k^2 - M^2)$. It was this photon propagator which was really responsible, in a dynamical sense, for the fall with energy of the QED cross section (12.12), and hence we conclude that, at least for this process, the IVB modification of the four-fermion model does avoid the violation of unitarity in lowest order.

Does the IVB modification ensure that the unitarity bounds are not violated for any Born (i.e. tree graph) process? The answer is no. The unitarity-violating processes turn out to be those involving *external* W particles (rather than internal ones, as in figure 12.3). Consider, for example, the process

$$\nu_\mu + \bar{\nu}_\mu \rightarrow \mathrm{W}^+ + \mathrm{W}^- \tag{12.19}$$

proceeding via the graph shown in figure 12.4. The fact that this is experimentally a somewhat esoteric reaction is irrelevant for the subsequent argument: the proposed theory, represented by the IVB modification of the four-fermion model, will necessarily generate the amplitude shown in figure 12.4, and since this amplitude violates unitarity, the theory is unacceptable. The amplitude for this process is proportional to

$$\begin{aligned} M_{\lambda_1\lambda_2} &= g^2 \varepsilon_\mu^{-*}(k_2, \lambda_2)\varepsilon_\nu^{+*}(k_1, \lambda_1)\bar{v}(p_2)\gamma^\mu(1-\gamma_5) \\ &\quad \times \frac{(\not{p}_1 - \not{k}_1 + m_\mu)}{(p_1 - k_1)^2 - m_\mu^2}\gamma^\nu(1-\gamma_5)u(p_1) \end{aligned} \tag{12.20}$$

where the $\varepsilon^\pm$ are the polarisation vectors of the W's: $\varepsilon_\mu^{-*}(k_2, \lambda_2)$ is that associated with the outgoing W^- with 4-momentum k_2 and polarisation state λ_2, and similarly for ε_ν^{+*}.

To calculate the total cross section, we must form $|M|^2$ and sum over the three states of polarisation for each of the W's. To do this, we need the result

$$\sum_\lambda \varepsilon_\mu(k, \lambda)\varepsilon_\nu^*(k, \lambda) = -g_{\mu\nu} + k_\mu k_\nu/M^2. \tag{12.21}$$

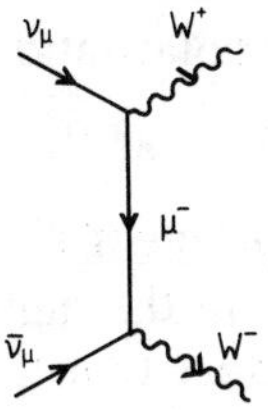

Figure 12.4 μ^- exchange graph for $\nu_\mu \bar{\nu}_\mu \to W^+W^-$.

(Note also that the right-hand side of (12.21) is the numerator of the propagator (10.11): cf (6.80) and (6.173).) (12.21) can be proved by explicitly constructing the polarisation vectors as follows. The free W wave equation is (cf (10.17))

$$(\Box + M^2)W^\mu - \partial^\mu \partial^\nu W_\nu = 0. \qquad (12.22)$$

Taking the divergence of this equation gives

$$\partial_\mu W^\mu = 0 \qquad (12.23)$$

which for the plane wave solutions

$$W^\mu = \varepsilon^\mu e^{-ik\cdot x} \qquad (12.24)$$

implies

$$k\cdot\varepsilon = 0 \qquad (12.25)$$

a covariant condition ensuring that there are just three independent ε^μ's (polarisation vectors), as we expect for a spin-1 particle. In the W rest frame, (12.25) reduces to

$$\varepsilon^0 = 0 \qquad (12.26)$$

and we may choose three independent ε's as

$$\varepsilon^\mu(k_{\text{rest}}, \lambda) = (0, \boldsymbol{\varepsilon}(\lambda)) \qquad (12.27)$$

$$\boldsymbol{\varepsilon}(\lambda = \pm 1) = \mp 2^{-1/2}(1, \pm \mathrm{i}, 0) \qquad (12.28)$$

$$\boldsymbol{\varepsilon}(\lambda = 0) = (0, 0, 1) \qquad (12.29)$$

(cf (5.26) and (5.27)) so that

$$\boldsymbol{\varepsilon}^*(\lambda)\cdot\boldsymbol{\varepsilon}(\lambda') = \delta_{\lambda\lambda'}. \qquad (12.30)$$

These states have definite spin projection $\lambda = \pm 1, 0$ along the z axis. For a general frame we can Lorentz transform $\varepsilon^\mu(k_{\text{rest}},\lambda)$ as required. For example, in a frame in which

$$k^\mu = (k^0, 0, 0, |\boldsymbol{k}|) \qquad (12.31)$$

we find

$$\varepsilon^\mu(k, \lambda = \pm 1) = \varepsilon^\mu(k_{\text{rest}}, \lambda = \pm 1) \qquad (12.32)$$

as before, but the longitudinal polarisation vector becomes

$$\varepsilon^\mu(k, \lambda = 0) = M^{-1}(|\boldsymbol{k}|, 0, 0, k^0). \tag{12.33}$$

(12.21) may now be verified (problem 12.3).

Our interest will as usual be in the high-energy behaviour of the cross section, in which regime it is clear that the $k_\mu k_\nu/M^2$ term in (12.21) will dominate the $g_{\mu\nu}$ term. It is therefore worth looking a little more closely at this term. We saw above that, in a frame in which $k^\mu = (k^0, 0, 0, |\boldsymbol{k}|)$, the transverse polarisation vectors $\varepsilon^\mu(k, \lambda = \pm 1)$ involved no momentum dependence, which was in fact carried solely in the longitudinal polarisation vector $\varepsilon^\mu(k, \lambda = 0)$. We may write this as

$$\varepsilon^\mu(k, \lambda = 0) = \frac{k^\mu}{M} + \frac{M}{(k^0 + |\boldsymbol{k}|)}\cdot(-1, \hat{\boldsymbol{k}}) \tag{12.34}$$

which at high energy tends to k^μ/M. Thus it is clear that it is the longitudinal polarisation states which are responsible for the $k^\mu k^\nu$ parts of the polarisation sum (12.21), and which will dominate real production of W's at high energy.

Concentrating therefore on the production of longitudinal W's, we are led to examine the quantity

$$\frac{g^4}{M^4(p_1 - k_1)^4}\mathrm{Tr}[(\not{k}_2(1 - \gamma_5)(\not{p}_1 - \not{k}_1)\not{k}_1\not{p}_1\not{k}_1(\not{p}_1 - \not{k}_1)\not{k}_2\not{p}_2] \tag{12.35}$$

where we have neglected m_μ, commuted the $(1 - \gamma_5)$ factors through, and used the assumed masslessness of the neutrinos, in forming $\Sigma_{\text{spins}}|M_{00}|^2$. Retaining only the leading powers of energy, we find (see problem 12.2)

$$\sum_{\text{spins}}|M_{00}|^2 \sim (g^4/M^4)(p_1\cdot k_2)(p_2\cdot k_2) = (g^4/M^4)E^4(1 - \cos^2\theta) \tag{12.36}$$

where E is the CM energy and θ the CM scattering angle. Recalling (10.93), we see that the (unsquared) amplitude must behave essentially as $G_F E^2$, precisely as in the four-fermion model (equation (12.2)). In fact, putting all the factors in, one obtains (Gastmans 1975)

$$\frac{\mathrm{d}\sigma}{\mathrm{d}\Omega} = G_F^2\frac{E^2\sin^2\theta}{8\pi^2} \tag{12.37}$$

and hence a total cross section which rises with energy as E^2, just as before. The production of longitudinal polarised W's is actually a pure $J = 1$ process, and the $J = 1$ partial wave amplitude is

$$f_1 = G_F E^2/6\pi. \tag{12.38}$$

The unitarity bound $|f_1| \leqslant 1$ is therefore violated for $E \geqslant (6\pi/G_F)^{1/2} \sim 10^3$ GeV (cf (12.10)).

Other unitarity-violating processes can easily be invented, and we have to conclude that the IVB model is, in this respect, no more fitted to be called a theory than was the four-fermion model. In the case of the latter, we argued that the root of the disease lay in the fact that G_F was not dimensionless, yet somehow this was not a good enough cure after all: perhaps (it is indeed so) 'dimensionlessness' is necessary but not sufficient (see the following section). Why is this? Returning to $M_{\lambda_1\lambda_2}$ for $\nu\bar{\nu} \to W^+W^-$ (equation (12.20)) and setting $\varepsilon_\mu = k_\mu/M$ for the *longitudinal* polarisation vectors, we see that we are involved with an effective amplitude

$$\frac{g^2}{M^2}\bar{v}(p_2)\not{k}_2(1-\gamma_5)\frac{\not{p}_1 - \not{k}_1}{(p_1-k_1)^2}\not{k}_1(1-\gamma_5)u(p_1). \tag{12.39}$$

Using the Dirac equation $\not{p}_1 u(p_1) = 0$, and $p_1^2 = 0$, this can be reduced to

$$-\frac{g^2}{M^2}\bar{v}(p_2)\not{k}_2(1-\gamma_5)u(p_1). \tag{12.40}$$

We see that the longitudinal ε's have brought in the factors M^{-2}, which are 'compensated' by the factor $\not{k}_2$, and it is this latter factor which causes the rise with energy. The longitudinal polarisation states have effectively reintroduced a dimensional coupling constant g/M.

Once more we turn in our distress to the trusty guide, QED. The analogous electromagnetic process to consider is $e^+e^- \to \gamma\gamma$. In QED there are two graphs contributing in lowest order (the 'crossed' versions of figures 6.13(*a*) and 6.13(*b*)). These are shown in figures 12.5(*a*), (*b*). The fact that there are two rather than one will turn out to be significant, as we shall see in §12.4. For the moment we just concentrate on figure 12.5(*a*) which is directly analogous to the diagram for $\nu\bar{\nu} \to W^+W^-$. The amplitude for this diagram is the same as before except that $(1-\gamma_5)/2$ is replaced by 1, $g/2^{1/2}$ by e and the ε vectors

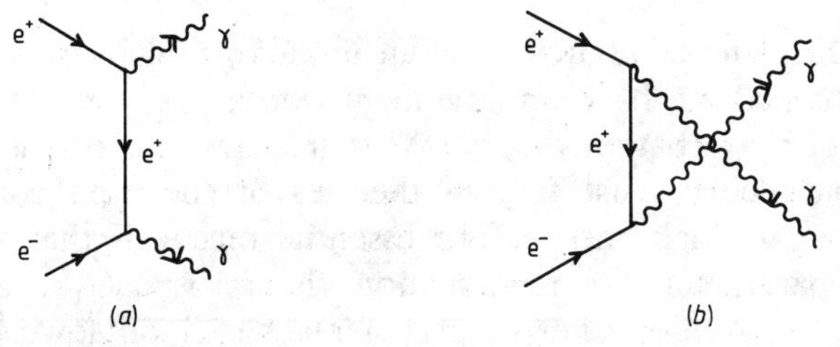

Figure 12.5 Lowest-order amplitudes for $e^+e^- \to \gamma\gamma$: (*a*) direct graph, (*b*) crossed graph.

now refer to photons:

$$M_{\lambda_1\lambda_2} = e^2\varepsilon_\mu^*(k_2, \lambda_2)\varepsilon_\nu^*(k_1, \lambda_1)\bar{v}(p_2)\gamma^\mu \frac{\not{p}_1 - \not{k}_1 + m_e}{(p_1 - k_1)^2 - m_e^2}\gamma^\nu u(p_1). \tag{12.41}$$

In the cross section we would need to sum over the photon polarisation states. For the massive spin-1 particles we used

$$\sum_\lambda \varepsilon_\mu(k, \lambda)\varepsilon_\nu^*(k, \lambda) = -g_{\mu\nu} + k_\mu k_\nu/M^2 \tag{12.21}$$

and so we would need the analogue of this result for massless photons. This is a quite non-trivial point. Clearly the answer is *not* to take the $M \to 0$ limit of the above equation, since this diverges. However, the divergent term $k_\mu k_\nu/M^2$ arises entirely from the *longitudinal* polarisation vectors, and we learnt in §5.1 that, for real photons, the longitudinal state of polarisation is absent altogether! We might well suspect, therefore, that since it was the longitudinal W's that caused the 'bad' high-energy behaviour of the IVB model, the 'good' high-energy behaviour of QED might have its origin in the absence of such states for photons. And this circumstance can, in its turn, be traced (cf §5.1) to the *gauge invariance* property of QED.

We have arrived here at an important theoretical indication that what we really need is a *gauge theory* of the weak interactions, in which the W's are gauge quanta. It must, however, be a peculiar kind of gauge theory, since, as we emphasised long ago at the start of this book, the quanta of gauge theories are massless—like photons and gluons, but not like W's. In the following chapter we shall explain how the remarkable trick of having a gauge theory with massive vector quanta is worked. But before that we discuss one further disease (related to the unitarity one) of phenomenological models of weak interactions—that of non-renormalisability.

12.3 The problem of non-renormalisability in weak interactions

The preceding line of argument about unitarity violations is open to the following objection. It is an argument conducted entirely within the famework of perturbation theory. What it shows, in fact, is simply that perturbation theory must fail, in theories of the type considered, at some sufficiently high energy. The essential reason is that the effective expansion parameter for perturbation theory is $EG_F^{1/2}$. Since $EG_F^{1/2}$ becomes large at high energy, arguments based on lowest-order perturbation theory are irrelevant. The objection is perfectly valid, and we shall take account of it by linking high-energy behaviour to the problem of renormalisability, rather than unitarity.

Renormalisation was discussed for QED in §6.10. We did not, however, give any general criterion for telling whether a given theory is renormalisable or not. There is, in fact, a simply stated criterion: if the coupling constant has negative mass dimension, the theory is non-renormalisable; if it has positive mass dimension, the theory has fewer divergences than QED and is called super-renormalisable; and if the coupling constant is dimensionless, further detailed investigation is necessary. The four-fermion model of weak interactions is an example of the first case. Let us see what goes wrong.

Consider the amplitude for the second-order process shown in figure 12.6 (the elastic scattering process $\nu_\mu e^- \to \nu_\mu e^-$ does not occur at all, at order G_F, in the four-fermion model we are considering). The amplitude is an integral over an internal momentum (cf equation (6.190)), involving two fermion propagators. Since the fermion propagator behaves like k^{-1} for large k, rather than k^{-2}, the integral will have the behaviour

$$G_F^2 \int \frac{d^4k}{k^2} \tag{12.42}$$

which is clearly divergent (see problem 6.13). At order G_F^4 we would have to include graphs such as those shown in figure 12.7; the first will diverge like

$$G_F^4 \left(\int \frac{d^4k}{k^2} \right)^3 \tag{12.43}$$

and the second like

$$G_F^4 \int \frac{d^4k_1\, d^4k_2\, d^4k_3}{k^6} \tag{12.44}$$

where k is a linear function of k_1, k_2 and k_3. If we were to attach a cut-off Λ to these integrals, as in (6.192), it is clear that we should find that the integrals would diverge as some power of Λ, unlike the situation in QED, where the divergence is always no worse than logarithmic. What is even worse, the power of the divergence gets larger, in successive orders of G_F^2. The physical reason for this is precisely the fact that G_F has negative mass dimension—every time a new factor of G_F^2 is introduced, there must be a compensating factor

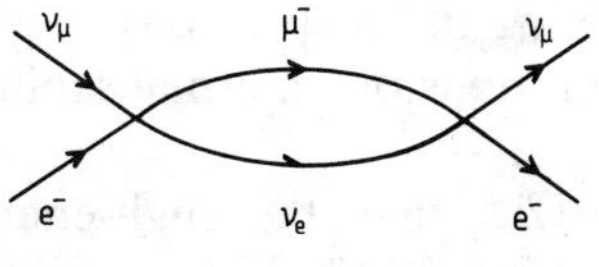

Figure 12.6 $O(G_F^2)$ contributions to $\nu_\mu e^- \to \nu_\mu e^-$.

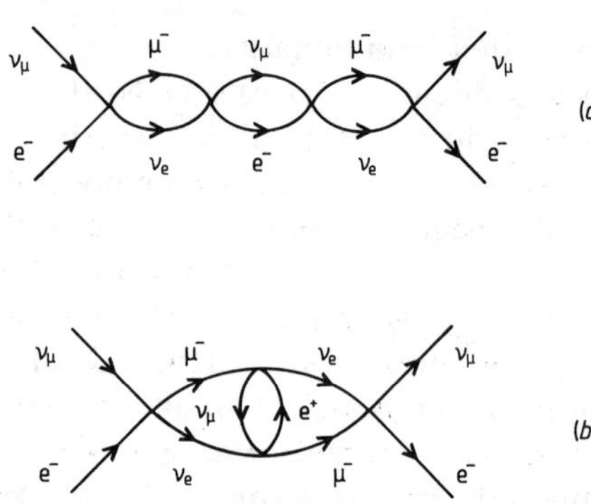

Figure 12.7 $O(G_F^4)$ contributions to $\nu_\mu e^- \rightarrow \nu_\mu e^-$.

from the momentum integral (so as to maintain the overall dimension of the amplitude) with dimension M^4. This integral is essentially $(\int d^4k/k^2)^2$, and its presence will cause the amplitude in that order of G_F^2 to diverge that much worse than at the preceding order. At each successive order, therefore, the divergence gets worse, whereas in the case of QED, with a dimensionless coupling constant, such momentum integrals are only ever logarithmically divergent. In the latter case all divergences could be tamed by the introduction of a finite number of parameters; in the four-fermion model we would need an infinite number of parameters to cope with the new divergences appearing at each order of perturbation theory.

This is therefore a good reason for rejecting the four-fermion model. Does the IVB model fare any better? In this case the coupling constant is dimensionless, just as in QED. 'Dimensionless' alone is not enough, it turns out: the IVB model is not, in fact, renormalisable. It should be reasonably clear from all the previous discussion that the key must lie in the *high-energy behaviour* of the various models (or theories): this is where the ultraviolet divergences of the integrals originate. Consider, for example, the two higher-order processes shown in figures 12.8 and 12.9—the former for the QED process $e^+e^- \rightarrow e^+e^-$ via an intermediate 2γ state, the latter for the IVB-mediated process $\nu_\mu\bar{\nu}_\mu \rightarrow \nu_\mu\bar{\nu}_\mu$. It seems plausible from the pictures that in each case the graphs must be formed by somehow 'sticking together' two of the lower-order graphs shown in figures 12.4 and 12.5. Accepting this for the moment, we are led to compare the high-energy behaviour of the processes $e^+e^- \rightarrow \gamma\gamma$ on the one hand and $\nu_\mu\bar{\nu}_\mu \rightarrow W^+W^-$ on the other. This is exactly the comparison we were making in the previous section, but now we have arrived at it from considerations of renormalisability, rather than unitarity.

Indeed, we saw in §12.2 that the high-energy behaviour of the amplitude $\nu\bar{\nu} \rightarrow W^+W^-$ (figure 12.4) grew like E^2, due to the k dependence of the longitudinal polarisation vectors, and this turns out to

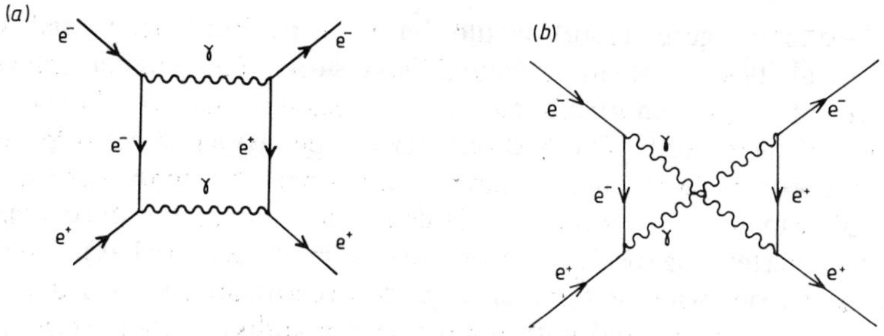

Figure 12.8 $O(e^4)$ contributions to $e^+e^- \to e^+e^-$.

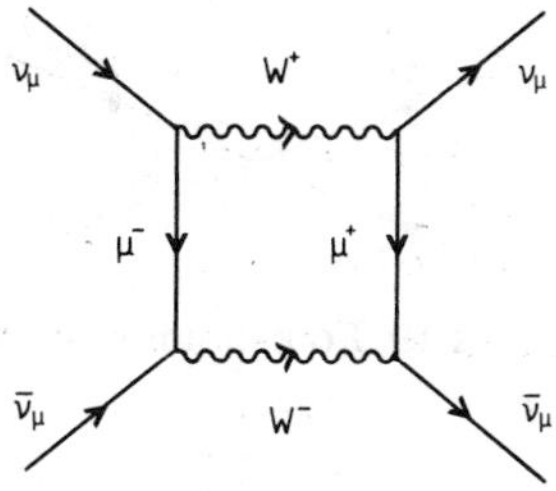

Figure 12.9 $O(g^4)$ contribution to $\nu_\mu \bar{\nu}_\mu \to \nu_\mu \bar{\nu}_\mu$.

produce, via figure 12.9, a non-renormalisable divergence. Thus the IVB model is no better in this respect that the four-fermion model. How does QED manage to survive? In this case, as we pointed out in §12.2, there are *two* graphs corresponding to figure 12.4, namely figures 12.5(a) and (b). Consider, therefore, mimicking for figures 12.5(a) and (b) the calculation we did for figure 12.4. We would obtain the leading high-energy behaviour by replacing the photon polarisation vectors by the corresponding momenta, and it can be checked (problem 12.4) that when this replacement is made (for either photon) the complete amplitude for the sum of figures 12.5(a) and (b) *vanishes*.

In physical terms, of course, this result was expected, since we knew in advance that it is always possible to choose polarisation vectors for *real* photons such that they are purely transverse (cf §5.1.1) so that no physical process can depend on a part of ε_μ proportional to k_μ. Nevertheless, the calculation is highly relevant to the question of renormalising figure 12.8. The photons in this process are not real external particles, but are instead virtual, internal ones. This has the consequence (as will be explained in more detail in Chapter 15, §15.1) that in figure 12.8 we should in general include these longitudinal ($\varepsilon_\mu \propto k_\mu$) states of the photons. The calculation of problem 12.4 then suggests that these longitudinal states are in fact harmless, provided that *both* contributions in figure 12.8 are included.

Indeed, the sum of these two contributions is not divergent. If it were, an infinite counterterm proportional to a four-point vertex

$e^+e^- \to e^+e^-$ (figure 12.10) would have to be introduced, and the original QED theory, which of course lacks such a fundamental interaction, would not be renormalisable. This is exactly what *does* happen in the case of figure 12.9. The bad high-energy behaviour of $\nu\bar{\nu} \to W^+W^-$ translates into a divergence of figure 12.9 – and this time there is no 'crossed' amplitude to cancel it. This divergence entails the introduction of a new vertex, figure 12.11, not present in the original IVB theory. Thus the theory without this vertex is non-renormalisable – and if we include it, we are landed with a four-field pointlike vertex which will generate infinitely many parameters, as in the Fermi case.

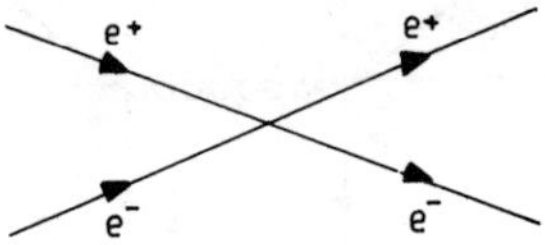

Figure 12.10 Four-point e^+e^- vertex.

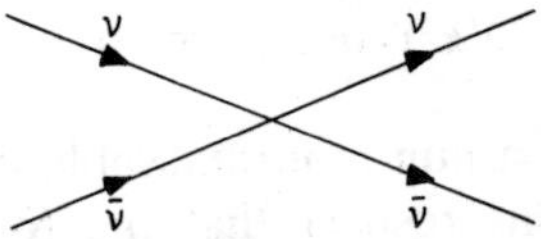

Figure 12.11 Four-point $\nu\bar{\nu}$ vertex.

Our presentation hitherto has emphasised the fact that, in QED, the bad high-energy behaviour is rendered harmless by a cancellation between contributions from figures 12.5(*a*) and (*b*) (or figures 12.8(*a*) and (*b*)). Thus one way to 'fix up' the IVB theory might be to hypothesise a new physical process, to be added to figure 12.4, in such a way that a cancellation occurred at high energies. The search for such high-energy cancellation mechanisms can indeed be pushed to a successful conclusion, given sufficient ingenuity and, arguably, a little hindsight. However, we are in possession of a more powerful principle. In QED, we have already seen, both for bosons ((5.203)) and for fermions (problem 6.10), that the vanishing of amplitudes when an ε_μ is replaced by the corresponding k_μ is due to *gauge invariance*: in other words, the potentially harmful longitudinal polarisation states are in fact harmless in a gauge-invariant theory. It is now time to pursue this clue.

Summary

Four-fermion interaction violates unitarity at high energy.

Born graphs in IVB model violate unitarity at high energy, for processes with external W's.

Longitudinal polarisation states responsible for unitarity-violating growth of amplitudes at high energy.

Gauge invariance for physical photons removes longitudinal polarisation state; no unitarity violations in QED.

Bad high-energy behaviour related to non-renormalisability.

QED graphs have acceptable high-energy behaviour due to cancellations: theory is renormalisable.

Gauge invariance as fundamental clue for constructing renormalisable vector theory.

Problems

12.1 (a) Using the representation for $\boldsymbol{\alpha}$, β and γ_5 introduced in §10.4 (equation (10.74)), massless particles are described by spinors of the form

$$u = E^{1/2}\begin{pmatrix}\phi_+\\ \phi_-\end{pmatrix} \qquad \text{(normalised to } u^\dagger u = 2E)$$

where $\boldsymbol{\sigma}\cdot\hat{\boldsymbol{p}}\phi_\pm = \pm\phi_\pm$, $\hat{\boldsymbol{p}} = \boldsymbol{p}/|\boldsymbol{p}|$. Find the explicit form of u for the case $\hat{\boldsymbol{p}} = (\sin\theta, 0, \cos\theta)$.

(b) Consider the process $\bar{\nu}_\mu + \mu^- \rightarrow \bar{\nu}_e + e^-$, discussed in §12.1, in the limit in which all masses are neglected. Following the rules of §3.4 for the antiparticles, the amplitude is proportional to

$$G_F\bar{v}(\bar{\nu}_\mu, \mathrm{R})\gamma_\mu(1-\gamma_5)u(\mu^-, \mathrm{L})\bar{u}(e^-, \mathrm{L})\gamma^\mu(1-\gamma_5)v(\bar{\nu}_e, \mathrm{R})$$

where we have explicitly indicated the appropriate helicities R or L (note that, as explained in §10.4, $(1-\gamma_5)/2$ is the projection operator for a right-handed antineutrino). In the CM frame, let the initial μ^- momentum be $(0, 0, E)$ and the final e^- momentum be $E(\sin\theta, 0, \cos\theta)$. Verify that the amplitude is proportional to $G_F E^2(1+\cos\theta)$. (Hint: evaluate the 'easy' part $\bar{v}(\bar{\nu}_\mu)\gamma_\mu(1-\gamma_5)u(\mu^-)$ first; this will show that the components $\mu = 0$, z vanish, so that only the $\mu = x, y$ components of the dot product need to be calculated.)

12.2 Verify equation (12.36).

12.3 Verify the polarisation sum (12.21) using the explicit forms (12.32) and (12.33).

PART V

THEORY OF ELECTROWEAK INTERACTIONS

In the seventh book of *The Republic*, Plato describes prisoners who are chained in a cave and can see only shadows that things outside cast on the cave wall. When released from the cave at first their eyes hurt, and for a while they think that the shadows they saw in the cave are more real than the objects they now see. But eventually their vision clears, and they can understand how beautiful the real world is. We are in such a cave, imprisoned by the limitations on the sorts of experiments we can do. In particular, we can study matter only at relatively low temperatures, where symmetries are likely to be spontaneously broken, so that nature does not appear very simple or unified. We have not been able to get out of this cave, but by looking long and hard at the shadows on the cave wall, we can at least make out the shapes of symmetries, which though broken, are exact principles governing all phenomena, expressions of the beauty of the world outside.

Weinberg (1980).

13

HIDDEN GAUGE INVARIANCE: THE U(1) CASE

Sometime in 1960 or early 1961, I learned of an idea which had originated earlier in solid state physics and had been brought into particle physics by those like Heisenberg, Nambu, and Goldstone, who had worked in both areas. It was the idea of 'broken symmetry', that the Hamiltonian and commutation relations of a quantum theory could possess an exact symmetry, and that the physical states might nevertheless not provide neat representations of the symmetry. In particular, a symmetry of the Hamiltonian might turn out to be not a symmetry of the vacuum.

As theorists sometimes do, I fell in love with this idea. But as often happens with love affairs, at first I was rather confused about its implications.

Weinberg (1980).

13.1 Introduction

In QED, the gauge theory with which we are most familiar, gauge invariance or local phase invariance is associated with a massless gauge field—the photon field A^{μ}. We took it for granted in Chapter 9 that the gauge quanta of QCD (gluons) are also massless. Is masslessness a necessary feature of all gauge fields? A simple heuristic argument (in retrospect, too simple!) would seem to suggest that it is. Recall that the introduction of the gauge field allowed us to alter the phase of the wavefunctions of all particles of a given charge independently at each space–time point, without any observable consequences. In this sense, the role of the gauge field is to 'reconcile' such phase changes, which, in principle, may refer to particles widely separated in space. But we have seen that the gauge field enters into the quantum mechanical equations as a force field (or potential), and we know that there is usually a connection between the *mass* of the quantum of a given force field and the range of the associated force. In the case of gauge fields, they must be able to perform their 'ministry of reconciliation' over arbitrarily large space–time distances, and hence we are led to interpret the associated force as having an infinite range. This argument would therefore seem to require the gauge quanta to be massless.

That gauge invariance is apparently rigorously linked to masslessness for the gauge quanta can be seen mathematically as follows. The electromagnetic potential satisfies the Maxwell equation

$$\square A^{\nu} - \partial^{\nu}(\partial_{\mu}A^{\mu}) = j^{\nu} \tag{13.1}$$

which, as discussed in §2.3, is invariant under the gauge transformation

$$A^{\mu} \to A'^{\mu} = A^{\mu} - \partial^{\mu}\chi. \tag{13.2}$$

However, if A^{μ} were to represent a *massive* field, the relevant wave equation is (cf equation (10.17))

$$(\square + M^2)A^{\nu} - \partial^{\nu}(\partial_{\mu}A^{\mu}) = j^{\nu}. \tag{13.3}$$

This equation is manifestly not invariant under (13.2), and it is precisely the mass term M^2A^{ν} that breaks the gauge invariance. Similar reasoning holds for the non-Abelian analogues of (13.1) and (13.3).

We have therefore apparently reached an impasse in any attempt to apply the gauge principle to *weak* interactions. There, we have seen that the range of the weak force is very short, and the associated vector bosons must therefore be massive. On the other hand, there was a strong suggestion at the end of Chapter 12 that, in theories involving vector quanta, gauge invariance was likely to play an essential role in ensuring renormalisability. We do seem to require some kind of gauge invariance for the case of massive vector quanta.

Remarkably enough, and notwithstanding the argument we have just given about the gauge non-invariance of mass terms, it is possible for gauge quanta to acquire mass! The gauge invariance is not lost, only hidden—it can be identified via special relations between masses and couplings—and it is for this reason that the situation has come to be described in terms of a 'hidden gauge invariance'. In order to appreciate the application of the gauge principle to generate the short-range weak forces, an understanding of how this hidden symmetry situation can come about is absolutely essential.

In order to avoid the considerable algebraic complexity of the non-Abelian case, we shall discuss in this chapter the concepts and the physics of hidden gauge invariance for the Abelian case (e.g. electromagnetism). Then, in Chapter 14, we shall generalise to the specific case of the (non-Abelian) GSW theory.

The first serious challenge to the then widely held view that electromagnetic gauge invariance required the photon to be massless was made by Schwinger (1962). Soon afterwards, Anderson (1963) pointed out that several situations in solid state physics could be interpreted in terms of an effectively massive electromagnetic field. He outlined a general framework for treating the phenomenon of the acquisition of mass by a gauge boson, and discussed its possible relevance to current attempts (Sakurai 1960) to interpret the then recently discovered vector mesons

(ρ, ω, ϕ, . . .) as the gauge quanta associated with a local extension of hadronic flavour symmetry. From his discussion, it is clear that Anderson had his doubts about the hadronic application, precisely because, as he remarked, gauge bosons can only acquire a mass if the symmetry is hidden. This has the consequence, as will be discussed in Chapter 14, that the ordinary multiplet structure to be anticipated in a theory with a non-Abelian symmetry is, in general, lost. On the other hand, flavour symmetry, if not exact, certainly leads to clearly identifiable approximately degenerate multiplets. We now believe, following Weinberg (1967) and Salam (1968), that the true application of these ideas in particle physics is to the generation of mass for the gauge quanta associated with the *weak* force. There is, however, nothing specifically relativistic about the physics of the basic mechanism involved, as Anderson showed. We shall therefore first make an apparent detour, in order to discuss some explicit examples of this mechanism at work in atomic and solid state physics. Our presentation is inspired by, and intended as an amplification of, Anderson's approach.

13.2 Screening currents and the generation of 'photon mass'

Consider the equation for a free massive vector field

$$(\Box + M^2)W^\nu - \partial^\nu(\partial_\mu W^\mu) = 0. \tag{13.4}$$

The condition

$$\partial_\mu W^\mu = 0 \tag{13.5}$$

follows from (13.4) by taking the divergence; (13.4) can therefore be written as

$$(\Box + M^2)W^\nu = 0. \tag{13.6}$$

The question we are trying to answer is this: how is it possible for the electromagnetic field A^μ, which satisfies the Maxwell equation

$$\Box A^\nu - \partial^\nu(\partial_\mu A^\mu) = j^\nu \tag{13.7}$$

to satisfy *also* an equation of the 'massive vector' type

$$(\Box + M^2)A^\nu = 0? \tag{13.8}$$

In much of the subsequent discussion, we shall be considering phenomena associated with steady currents and static magnetic fields (all time derivatives zero). We may choose the gauge condition $\operatorname{div}\boldsymbol{A} = 0$, and Maxwell's equations for $\boldsymbol{A}$ then become

$$\nabla^2\boldsymbol{A} = -\boldsymbol{j} \tag{13.9}$$

while the static version of (13.8) is (for the spatial components)

$$\nabla^2 \boldsymbol{A} = M^2 \boldsymbol{A}. \tag{13.10}$$

It is quite clear that, for (13.8) to be consistent with (13.9), the only possibility is for $\boldsymbol{j}$ to contain a part proportional to $\boldsymbol{A}$ itself:

$$\boxed{\boldsymbol{j} = -M^2 \boldsymbol{A}} \quad \text{screening current condition.} \tag{13.11}$$

The relation (13.11)—or its covariant generalisation—was emphasised by Anderson as fundamental to the whole question of gauge invariance and mass. As it stands, of course, equation (13.11) is very definitely *not* gauge invariant. In the following development we shall first be considering various situations in which such an equation can arise, and this discussion will necessarily be within the context of a special choice of gauge. Eventually, we shall show that the choice is, after all, not restrictive; we shall also discuss other choices, and explain how the gauge invariance is 'hidden'.

Why have we labelled (13.11) as the 'screening current' condition? From the Maxwell equation (in the static field case)

$$\nabla \times \boldsymbol{B} = \boldsymbol{j} \tag{13.12}$$

together with condition (13.11), and $\boldsymbol{B} = \nabla \times \boldsymbol{A}$, we deduce immediately (taking the curl of (13.12))

$$\nabla^2 \boldsymbol{B} = M^2 \boldsymbol{B} \tag{13.13}$$

where we have used $\operatorname{div} \boldsymbol{B} = 0$. Equation (13.13) could, of course, have been derived at once from the curl of (13.10), but we are now developing the argument from the Maxwell equations and the condition (13.11). The variation of magnetic field described by (13.13) is a very characteristic one, encountered in a number of contexts in solid state physics. If we consider, for simplicity, one-dimensional variation

$$\mathrm{d}^2 \boldsymbol{B}/\mathrm{d}x^2 = M^2 \boldsymbol{B} \tag{13.14}$$

in the half-plane $x \geqslant 0$, say, we have solutions of the form

$$\boldsymbol{B} = \boldsymbol{B}_0 \exp(-Mx) \tag{13.15}$$

with $\boldsymbol{B}_0$ some constant vector. The field therefore penetrates into the region $x \geqslant 0$ a distance of order M^{-1}. The range parameter M^{-1} is, in solid state physics, called the *screening length*. This expresses the fact that, in a medium such that condition (13.11) holds, the magnetic field will only penetrate the medium (from free space) a distance of the order M^{-1}: the field is 'screened out' for larger distances.

What is the physical origin of this screening? For magnetic fields the answer is provided by Lenz's law: when a magnetic field is applied to a system of charged particles, induced EMF's are set up which accelerate the particles, and the magnetic effect of the resulting currents tends to *cancel* (or screen) the applied magnetic field. On the atomic scale this is the cause of *atomic diamagnetism*, and—an example of great relevance to particle physics—on the macroscopic scale it leads to *the exclusion of magnetic flux from a superconductor*.

Before looking at some simple models of this behaviour, it will be helpful to distinguish two aspects of the problem. In typical screening situations exemplified by equation (13.13), $\boldsymbol{B}$, though 'massive', is still transverse ($\operatorname{div} \boldsymbol{B} = 0$). Similarly, equations (13.9)–(13.11) refer to a transverse $\boldsymbol{A}$. Thus the first part of the problem is to understand the origin of mass for just the transverse components of a gauge field, and this we do by appeal to the screening concept. However, we saw in §12.2 that a massive vector particle has three independent states of polarisation, while in §5.1 we learned that the massless photon has only two. Furthermore, equation (12.33) makes it clear that the 'extra' polarisation state in the $M \neq 0$ case will by no means fade away gracefully as $M \to 0$. The massive and massless cases are not smoothly related, it seems: some kind of *discontinuous change in the number of degrees of freedom is involved*. Thus the second aspect of our problem is to understand where this other degree of freedom comes from, and how it can lead to a massive longitudinal component for the vector field, while still somehow preserving gauge invariance. The eventual application to particle physics will be via a direct analogy with superconductivity; in that case, as we shall see, both aspects of the problem are related to the remarkable nature of the many-electron ground state. Specifically, a macroscopic electron pair wavefunction will be introduced, whose constant modulus (related to the electron density) will be sufficient to account for the transverse screening effect, and whose phase will be the missing 'third' degree of freedom. In the particle physics application, the role of the pair wavefunction will be played by the so-called *Higgs field*, whose function will be to produce these 'superconductivity' effects in the physical vacuum.

We now discuss some specific examples of magnetic screening.

13.2.1 One-body case: atomic diamagnetism

Consider the current associated with a single non-relativistic particle of charge q. For a free particle this would be given by the standard quantum mechanical expression

$$\boldsymbol{j} = (q/2m)[\psi^*(-\mathrm{i}\boldsymbol{\nabla}\psi) + \psi(\mathrm{i}\boldsymbol{\nabla}\psi^*)]. \tag{13.16}$$

But this is not invariant under a local phase transformation

$$\psi \to \psi' = \exp[\mathrm{i}q\chi(x)]\psi \tag{13.17}$$

so it is not gauge invariant. Indeed, if we write the wavefunction ψ as

$$\psi = |\psi| \exp(i\theta) \tag{13.18}$$

where $|\psi|$ and θ are real quantities depending on x and t, the current just becomes

$$\boldsymbol{j} = (q/m)|\psi|^2(\boldsymbol{\nabla}\theta) \tag{13.19}$$

from which it is quite clear that a local change in the phase θ will alter $\boldsymbol{j}$. Of course we know how to make $\boldsymbol{j}$ gauge invariant: we replace $\boldsymbol{\nabla}$ in equation (13.16) by the covariant derivative $\boldsymbol{D} = \boldsymbol{\nabla} - iq\boldsymbol{A}$. Then, using (13.18) we obtain

$$\boldsymbol{j} = (q/m)|\psi|^2(\boldsymbol{\nabla}\theta - q\boldsymbol{A}) \tag{13.20}$$

and we see quite explicitly that a change in $\boldsymbol{A}$ of the standard form

$$\boldsymbol{A} \to \boldsymbol{A}' = \boldsymbol{A} + \boldsymbol{\nabla}\chi \tag{13.21}$$

can be compensated by a change in θ of

$$\theta \to \theta' = \theta + q\chi. \tag{13.22}$$

Equation (13.20) is the correct expression for the probability current density (multiplied by q) for a charged particle in an electromagnetic field given by the vector potential $\boldsymbol{A}$.

We now have coupled equations for $\boldsymbol{A}$ and ψ to solve:

$$-(1/2m)(\boldsymbol{\nabla} - q\boldsymbol{A})^2\psi = E\psi \tag{13.23}$$

and

$$\nabla^2\boldsymbol{A} = -\boldsymbol{j} \tag{13.9}$$

with $\boldsymbol{j}$ given above, and we have specialised to the static field case and the gauge $\operatorname{div}\boldsymbol{A} = 0$. In general, this is a complicated problem but we may gain some insight by treating ψ as unperturbed by $\boldsymbol{A}$, in a weak field approximation. The $\boldsymbol{\nabla}\theta$ term in the current equation (13.20) will then give a contribution proportional to m_l, where m_l is the eigenvalue of the z component of orbital angular momentum L_z, since the phase of the (unperturbed) wavefunction is just $\exp(im_l\phi)$. This is the usual orbital contribution to paramagnetism. Diamagnetic screening currents may be associated with the second term in $\boldsymbol{j}$ (equation (13.20)), namely

$$\boldsymbol{j}_{sc} = (-q^2/m)|\psi|^2\boldsymbol{A}. \tag{13.24}$$

In the equation

$$\nabla^2\boldsymbol{A} = -\boldsymbol{j} \tag{13.9}$$

this current would lead to a screening length λ of the form

$$\lambda \sim (m/|\psi|^2q^2)^{1/2} \tag{13.25}$$

if, for an order-of-magnitude estimate, we could replace $|\psi|^2$ by some appropriate constant average value. Taking

$$|\psi|^2 \sim (a_0)^{-3} \tag{13.26}$$

for example, where a_0 is the Bohr radius, we find

$$\lambda \sim 300 \text{ Å} \tag{13.27}$$

if m and q are the electron mass and charge. Thus the diamagnetic screening of atomic electrons is very slight—the $\boldsymbol{A}$ field penetrates essentially unaltered.

We remind the reader that both (13.24) and (13.9) are manifestly *gauge dependent*. At this stage we are merely investigating situations in which the characteristic phenomenon of transverse screening can occur, and we are using a definite choice of gauge. Later, we shall take up the questions of different gauges, the longitudinal component of the field, and of gauge invariance.

13.2.2 Many-body case: macroscopic screening in a superconductor†

Diamagnetic screening in a single atom is a weak effect because λ is so large compared to an atomic radius. The orbitals of conduction electrons in a metal do extend over larger regions, but the wavefunctions are then, as a rule, strongly perturbed by a magnetic field, so that our approximation of using the unperturbed wavefunctions would completely fail. The diamagnetism of conduction electrons (first calculated by Landau) turns out to be quite a weak effect, rather smaller than the Pauli spin paramagnetism (though this situation can be reversed in doped semiconductors).

However, if for some reason a system were characterised by a wavefunction which was (a) 'macroscopic' in its spatial extension and (b) effectively unperturbed by the magnetic field, then, *provided λ were still macroscopic*, the diamagnetic screening effect could operate more dramatically. In fact, $\boldsymbol{B}$ could be entirely excluded from such a system except for a thin surface layer of thickness of the order of λ. This is precisely what is observed to happen in a superconductor, and it is called the Meissner effect. If a normal metal in an external applied magnetic field is cooled below its superconducting transition temperature, the magnetic flux is abruptly expelled. In physical terms, the transition must therefore be accompanied by the appearance of whatever (resistanceless) surface currents are needed to produce a magnetic flux density which, in the interior of the superconductor, exactly cancels—'perfectly screens'—the flux density of the applied magnetic field.

†We have found the book by Schrieffer (1964) particularly helpful.

We have seen how such screening can occur for a *single-particle* wavefunction. If the same ideas are to be applicable to the superconductor, the wavefunction associated with the screening current

$$j_{sc} = (-q^2/m)|\psi|^2 \mathbf{A} \tag{13.24}$$

must somehow be a macroscopic *many-body* wavefunction, since the flux expulsion is occurring as a macroscopic phenomenon, involving a macroscopic number of charge carriers. It is also necessary to assume that the macroscopic wavefunction is effectively unperturbed by the magnetic field—or, as it is sometimes phrased, that it is 'rigid'†.

We shall not discuss the second point any further here, but simply accept it as a working hypothesis. We are more concerned with the interpretation of the idea of a single macroscopic wavefunction describing many particles. We first note that if ψ were a *single*-particle wavefunction extending over a macroscopic range L, the normalisation to unity would entail $|\psi|^2 \sim L^{-3}$ and λ would be enormous. If, however, ψ were normalised to the *total* number of particles, N, then $|\psi|^2 \sim NL^{-3} \sim a_0^{-3}$ and we recover (13.26) and hence (13.27), with $\lambda \ll L$. Such a normalisation is possible if the number of particles is very large and all are in the same state, described by the wavefunction ψ; $|\psi|^2$ is then the number density.

In such a case ψ behaves as a classical field. A familiar example is the classical electromagnetic field $A^\mu(x)$ itself. This is obtained as the limit of a quantised field where there are very many photons all in the same state. It is particularly important to realise that this means that the wavefunctions for the individual particles must all have the same phase, i.e. they must be *coherent*. In the case of a superconductor, ψ will represent a type of coherent state of all the electrons.

Photons are bosons, and Bose–Einstein statistics allows arbitrarily many photons to exist in the same quantum state. But electrons are fermions, and the Exclusion Principle prevents more than one fermion from occupying any given quantum state. It would therefore seem that any kind of macroscopic wavefunction for electrons in a superconductor is impossible. The crucial observation of Cooper (1956) was that two electrons, in states above an inert Fermi sea of other electrons, would form a bound state if the force between them were attractive, whatever its magnitude‡. The way was then open to regard the electron system as

†This property may be related to the existence of an *energy gap* in the electron spectrum (see Feynman 1972, pp268–9).

‡At first sight, an attractive force between two electrons seems surprising: in fact, the coupling of the electrons to the lattice ions in a metal can provide an attractive interaction which, in certain circumstances, outweighs the Coulomb repulsion between the two electrons.

effectively a collection of Bose particles—the 'Cooper' pairs—so that macroscopic occupancy of a single quantum state could occur. In this state all the pairs have the same centre-of-mass momentum: this is the essential *coherence* property.

When a macroscopic number of particles in a many-body system are all in the same quantum state, we say there has been a 'condensation'. The Bose–Einstein condensation which occurs in a system of identical non-interacting helium atoms is an important component in the explanation of superfluidity in liquid ^{4}He. In the same way, one believes that in a superconductor a macroscopic fraction of the electrons pair together and form a condensate. Residual interactions between the pairs have the effect that, in the true ground state, some electrons are promoted out of the condensate: but there is still a macroscopic fraction $n_s(T)/n$ of electrons which carries current without dissipation and is responsible for the Meissner effect. The density of superconducting electrons, n_s, is, as indicated, a function of absolute temperature, tending to zero as T approaches the superconducting transition temperature T_c from below, and tending to n—the density of conduction electrons—for T well below T_c. The ground state below T_c is then a coherent plasma of all the superconducting pairs, and is described by a macroscopic wavefunction ψ, normalised to

$$|\psi|^2 = n_C = n_s/2 \tag{13.28}$$

where n_C is the density of Cooper pairs.

We emphasise at this point that the correct description of the superconducting ground state cannot be arrived at by using standard many-body perturbation theory based on the normal (unpaired electrons) ground state. This *non-perturbative* aspect of the true ground state is a crucial feature of both the superconductor and its particle physics analogues. The phenomenon whereby a photon (in a certain medium) propagates as if it had a mass is a non-perturbative effect, and it is this failure of perturbation theory which, mathematically, provides the loophole in the qualitative argument for masslessness with which we started this chapter.

Returning to the equation for the screening current

$$\boldsymbol{j}_{sc} = (-q^2/m)|\psi|^2\boldsymbol{A} \tag{13.24}$$

ψ now stands for the macroscopic pair wavefunction, q the charge $(-2e)$ of the Cooper pair, and m the mass $(2m_e)$ of the pair. Setting $|\psi|^2 = n_s/2$ and taking the curl of this equation, we have

$$\boldsymbol{\nabla} \times \boldsymbol{j}_{sc} = \frac{-q^2}{m}\frac{n_s}{2}\boldsymbol{B}. \tag{13.29}$$

This is known as the London equation (see e.g. London 1950). Together with Maxwell's equation in the form of Ampère's law appropriate to the static field case,

$$\nabla \times \boldsymbol{B} = \boldsymbol{j} \tag{13.12}$$

we obtain

$$\nabla^2 \boldsymbol{B} = \frac{q^2}{m}\frac{n_s}{2}\boldsymbol{B} \tag{13.30}$$

giving a screening length—a flux penetration depth—λ of the form

$$\lambda = (2m/q^2 n_s)^{1/2}. \tag{13.31}$$

With $m = 2m_e$, $q = -2e$ and $n_s \sim 4 \times 10^{28}\ \mathrm{m}^{-3}$ (roughly one conduction electron per atom), $\lambda \sim 10^{-6}$ cm; this gives the order of magnitude of the thickness of the surface layer within which the screening currents flow, and over which the flux falls to zero. As $T \to T_c$ and $n_s \to 0$, λ becomes very large and flux is not excluded.

Since the equation for a static massive vector field is

$$\nabla^2 \boldsymbol{W} = M^2 \boldsymbol{W} \tag{13.32}$$

we may also regard equation (13.30) as describing a static vector field of mass

$$M = (q^2 n_s/2m)^{1/2}. \tag{13.33}$$

Thus the superconductor does seem to provide a working physical model in which the phenomenon of a gauge field acquiring mass does occur. However, one fundamental aspect of this phenomenon has so far been evaded: namely, the question of the 'third' component of the massive vector field. Our discussion has been limited throughout, in fact, to the two *transverse* components of the field $\boldsymbol{A}$ (or $\boldsymbol{B}$). As we pointed out right at the start, a true massive vector field has three possible states of polarisation, each such degree of freedom being associated with the same mass. What has happened to the longitudinal component of the field? The answer is intimately connected to the whole question of gauge invariance. In our discussion of the superconductor, we mentioned that the macroscopic wavefunction was assumed to be 'rigid' with respect to perturbations due to the vector potential $\boldsymbol{A}$. But this was with reference to the gauge choice $\operatorname{div} \boldsymbol{A} = 0$, i.e. to a *transverse* vector potential. The screening equation (13.24) is really to be understood as applying only to the response to the transverse part of $\boldsymbol{A}$. That the macroscopic wavefunction *cannot* be rigid to longitudinal perturbations can be seen immediately by considering the case in which $\boldsymbol{A}$ is a pure gradient, which should be equivalent to no physical electromagnetic fields at all. $\boldsymbol{B}$ of course does vanish, and so must the total current, so that what we earlier called the paramagnetic contribution must, in this case, cancel the diamagnetic screening contribution (13.24) completely! In fact, the longitudinal part of $\boldsymbol{A}$ will couple strongly to longitudinal excitations of the electrons: primarily, as Bardeen (1957) first recognised, to the collective density fluctuation mode of the electron system—that is, to the plasma oscillations. When this aspect of the dynamics of the electrons is included, a fully gauge invariant description of the effectively massive $\boldsymbol{A}$ emerges (see e.g. Schrieffer 1964, §8–5).

It is not our purpose to give a detailed discussion of the theory of superconductivity†. We have been concerned merely to bring out many of the essential physical ideas, since these will be transferred to the particle physics context. In particular, we do not pursue any further here the interesting question of the longitudinal modes in a superconductor; we shall treat gauge invariance from the particle physics point of view in §13.4.

13.2.3 The 'relativistic superconductor': the Higgs model

In the preceding discussion we have considered only the static field case, and only the spatial components of $\boldsymbol{A}$. These are both clearly non-covariant restrictions. Before making any application of the foregoing ideas to particle physics, these restrictions must be removed. We try to mimic as closely as possible the procedure outlined for the superconductor. We start from the covariant equation

$$\Box A^{\nu} = j^{\nu} \tag{13.34}$$

where we have chosen to impose the Lorentz condition $\partial_{\mu} A^{\mu} = 0$. This takes the form of a massive vector boson equation if j^{ν} is proportional to A^{ν}, at least in some approximation. The covariant generalisation of the current of a charged scalar field (corresponding to a macroscopic wavefunction for some kind of 'pairs') is just

$$j^{\mu}(\phi) = \mathrm{i}q[\phi^{*}(\partial^{\mu}\phi) - (\partial^{\mu}\phi^{*})\phi] - 2q^{2}A^{\mu}|\phi|^{2} \tag{13.35}$$

which is the ordinary KG current (equation (3.20)) made gauge invariant (equation (4.196)) by the usual replacement

$$\partial^{\mu} \to D^{\mu} = \partial^{\mu} + \mathrm{i}qA^{\mu}. \tag{13.36}$$

Thus the equation for A^{ν} becomes

$$\Box A^{\nu} = \mathrm{i}q[\phi^{*}(\partial^{\nu}\phi) - (\partial^{\nu}\phi^{*})\phi] - 2q^{2}A^{\nu}|\phi|^{2}. \tag{13.37}$$

†We may remark, for example, that the simple relation (13.24) between $\boldsymbol{j}_{\mathrm{sc}}$ and $\boldsymbol{A}$ is only an approximation for real superconductors. It states that $\boldsymbol{j}_{\mathrm{sc}}$ at any given point is proportional to $\boldsymbol{A}$ at the same point, and is hence a local relation. In general $\boldsymbol{j}_{\mathrm{sc}}$ is related to the average of $\boldsymbol{A}$ around the given point.

Now in a configuration of ϕ for which the modulus of the field is effectively constant (cf (13.28)), a simple result emerges. If we take ϕ to be real, with constant value

$$\phi = \phi_0 \equiv f/2^{1/2} \tag{13.38}$$

$j^\mu(\phi)$ then becomes

$$j_0^\mu = -q^2 f^2 A^\mu \tag{13.39}$$

and the equation for A^ν is

$$\Box A^\nu = -q^2 f^2 A^\nu \tag{13.40}$$

which is the wave equation for a free vector particle of mass $M = qf$. This fundamental observation was first made, in the relativistic context, by Englert and Brout (1964), Higgs (1964) and Guralnik *et al* (1964); for a full account, see Higgs (1966).

The patient reader may well be wondering just what relevance some kind of 'relativistic superconductor' has to elementary particle physics. She may now be prepared to allow that there may be physical systems in which the electromagnetic gauge field behaves as if it had mass; but in all of our previous discussion it has been crucial that, as well as the gauge field, some *extra* matter field be present—for instance, the field of the condensed Cooper pairs. In an application of this idea to particle physics, we want to say that the gauge quanta of the weak interactions are massive when they exist as free particles flying through what we are pleased to call the vacuum! How can this be? The essential idea is to suppose that the vacuum itself has a particular property that makes this possible: namely, that it is a state in which some quantum fields—the *Higgs fields* $\hat{\phi}$—have non-vanishing expectation values

$$\langle 0|\hat{\phi}|0\rangle \neq 0. \tag{13.41}$$

Such a non-vanishing value can, in the presence of the weak force field, generate 'weak screening currents', just as the Cooper pair field ψ does for the real electromagnetic superconductor. The massless weak gauge quanta which propagate in this peculiar background—the vacuum—will then satisfy an equation of the form

$$\Box \hat{W}^\nu = \hat{j}^\nu_{\text{vac}} \tag{13.42}$$

where $\hat{j}^\nu_{\text{vac}}$ is the weak current that resides in the vacuum, due to the non-vanishing expectation value of $\hat{\phi}$. If we can arrange for $\hat{j}^\nu_{\text{vac}}$ to equal $-M^2\hat{W}^\nu$ for some M (cf (13.39)), we shall have the desired screening condition, and the vector quanta will have acquired a mass M. Such quanta will then be associated with a force of range M^{-1}: the physical vacuum will have screened out the weak force field, making it into the desired short-range phenomenon, and this despite the fact that the theory has an underlying gauge invariance. The concept of a vacuum screening current is possibly as fundamental as that of the displacement current introduced by Maxwell.

13.3 The nature of the physical vacuum: spontaneous symmetry breakdown

The vital first step along the road which leads to the possibility of vacuum screening currents was taken by Nambu (1960), who suggested that the physical vacuum of a quantum field theory is analogous to the *ground state* of an interacting many-body system—it is simply the state of minimum energy, the stable (or *equilibrium*) configuration. This concept of the vacuum was a fundamental element in our introduction to quantum field theory in Chapter 4. Now, this state of minimum energy *need not* be one in which all the quantum fields have zero average value—though that will automatically happen if we write a standard normal mode expansion such as (4.103)

$$\hat{\phi} = \int \frac{\mathrm{d}^3 k}{(2\pi)^3 2\omega_k}[\mathrm{e}^{-\mathrm{i}k\cdot x}\hat{a}(k) + \mathrm{e}^{\mathrm{i}k\cdot x}\hat{a}^\dagger(k)] \tag{13.43}$$

for all fields, and use

$$\hat{a}(k)|0\rangle = 0 \tag{13.44}$$

since then clearly

$$\langle 0|\hat{\phi}|0\rangle = 0. \tag{13.45}$$

Note that in quantum field theory, fluctuations are always possible, as we have amply experienced in earlier chapters: thus we expect that the nearest we can really get to 'nothing' is that state for which the average value of all fields is zero—hence our acceptance of (13.44) and (13.45) in Chapter 4. Such a ground state is very symmetrical (it is hard to find something more symmetrical than nothing), and we are used to quantum ground states possessing a high degree of symmetry. Nevertheless, it is possible that in the true ground state of our theory some field has a non-zero equilibrium value: this is an asymmetrical or 'lop-sided' situation, but it is precisely what we need to produce the vacuum screening current.

There are, in fact, a great many examples of this state of affairs in many-body physics. The truly stable configuration of an interacting system can be 'lop-sided' even though the basic interactions between the constituents of the system are symmetrical. An easily visualisable example is provided by a simple model of a ferromagnet, in which the Hamiltonian describing the spin–spin interaction is rotationally invariant;

nevertheless, below the ferromagnetic transition temperature the ground state is *not* rotationally invariant, but rather is one in which a particular direction in space is singled out by the alignment of the spins. *This situation, in which the ground state configuration does not display the symmetry of the Hamiltonian, is frequently described by saying that the symmetry is 'spontaneously broken'*. In a real sense, however, the symmetry is *not* broken—as it certainly would be if, by contrast, explicitly non-symmetric terms were added to the Hamiltonian. For example, adding an external magnetic field term, selecting out one direction, clearly explicitly breaks rotational invariance. In our case, however, any direction of spin alignment is equally good, and in this sense the symmetry is still there: it is only 'hidden' by the necessity of associating one of the possible directions with the ground state of our particular ferromagnet. It is for this reason that we have preferred the term 'hidden symmetry' to 'spontaneously broken symmetry' to describe this situation. At this point we can begin to understand why the true symmetry of the weak interactions is so hard to see: living inside our 'weak superconductor', trying to discover the gauge invariance associated with the weak interactions, we are in the position of Coleman's demon (Coleman 1975) living inside a ferromagnet and trying to discover rotational invariance!

In the ground state of a ferromagnet, below its transition temperature, the value of the total spin is non-zero: this is the 'spontaneous' magnetisation. In all cases known in many-body physics, such non-zero ground state equilibrium values are the result of a cooperative effect involving macroscopically many constituents. Typically, such an effect comes into play as the temperature is lowered though some critical value T_c: the system is said to undergo a *phase transition* at $T = T_c$. The onset of superconductivity is an example of just such a phase transition. The ferromagnet, however, is more easily visualised, and we now use it to illustrate how the occurrence of a non-zero equilibrium field, as T is lowered through T_c, can be modelled mathematically.

In Landau's mean field theory of phase transitions as applied to the ferromagnet (see e.g. Tilley and Tilley 1986), the relevant energy density is the Gibbs free energy density, G, which may be expanded in terms of the average spontaneous magnetisation $M(x)$ by

$$G = \alpha M^2(x) + \beta M^4(x) + \text{parts depending on } \nabla M(x). \quad (13.46)$$

The equilibrium lowest-energy configuration is found by minimising G for constant $M(x)$. In the magnetically disordered phase (above the transition temperature T_c) the constant value must equal zero: in the ordered phase we want the value to be non-zero, due to a macroscopic number of elementary spins lining up together. Such behaviour can be modelled by

$$G = \alpha M^2 + \beta M^4 \tag{13.47}$$

with $\beta > 0$ and (a) $\alpha > 0$ for $T > T_c$, (b) $\alpha < 0$ for $T < T_c$. The energy density in these two cases takes the forms shown in figure 13.1, from which it is clear that, for $T < T_c$, a minimum—i.e. position of stable equilibrium—exists with non-zero M. The field which develops a non-zero equilibrium value below the transition temperature is called the 'order parameter', corresponding to the fact that some sort of 'ordering' has occurred: in other words, some symmetry has been lost† (or hidden).

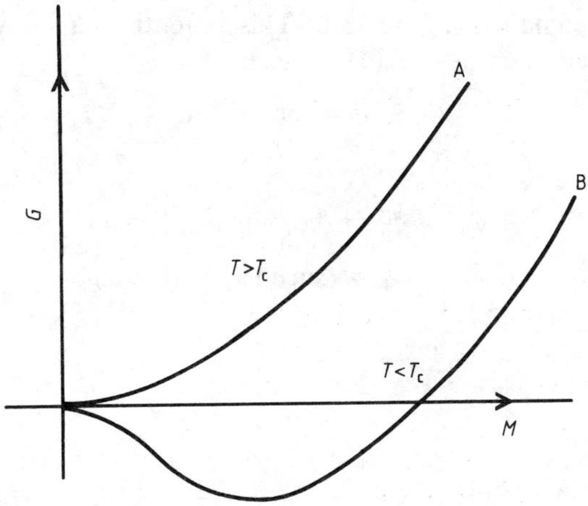

Figure 13.1 Free energy G against magnetisation M for temperature T: A, above T_c; B, below T_c.

We are more directly interested in superconductivity. There, the macroscopic Cooper pair wavefunction ψ plays the role of an order parameter: in this guise it is called the Ginsburg–Landau order parameter. Because it has to couple to the electromagnetic field, ψ must be complex (recall that electromagnetism is a phase theory!). For a 'relativistic superconductor', the analogous quantity is a complex field $\hat{\phi}$, generically called a Higgs field, which we write as (cf (4.150))

$$\hat{\phi} = \frac{1}{\sqrt{2}}(\hat{\phi}_1 - \mathrm{i}\hat{\phi}_2). \tag{13.48}$$

Just as Landau's mean field theory for the energy density of a ferromagnet was able to model the situation in which a non-zero

†It is worth pausing to reflect on the claim that *ordering* here entails a *loss* of symmetry!

magnetisation could appear 'spontaneously', so too can we devise an effective potential energy function for $\hat{\phi}$ to describe the 'hidden' symmetry phase of our 'relativistic superconductor'. We shall begin, however, by considering the case in which $\hat{\phi}$ is a classical field, $\phi = (1/\sqrt{2})(\phi_1 + i\phi_2)$.

We suppose that in the Lagrangian '$T-V$' for our (classical) field system the potential V is given by†

$$V(\phi) = -\tfrac{1}{2}\mu^2\phi^*\phi + \tfrac{1}{2}(\mu^2/f^2)(\phi^*\phi)^2 \qquad (13.49)$$

where the particular constants multiplying the two terms have been chosen for later convenience. The *signs* have been chosen so that V has the form sketched in figure 13.2. This potential has a symmetry of the kind discussed in §2.6—a U(1) symmetry:

$$\phi \to \phi' = e^{i\alpha}\phi \qquad (13.50)$$

or

$$\begin{aligned} \phi_1 \to \phi_1' &= (\cos\alpha)\phi_1 - (\sin\alpha)\phi_2 \\ \phi_2 \to \phi_2' &= (\sin\alpha)\phi_1 + (\cos\alpha)\phi_2. \end{aligned} \qquad (13.51)$$

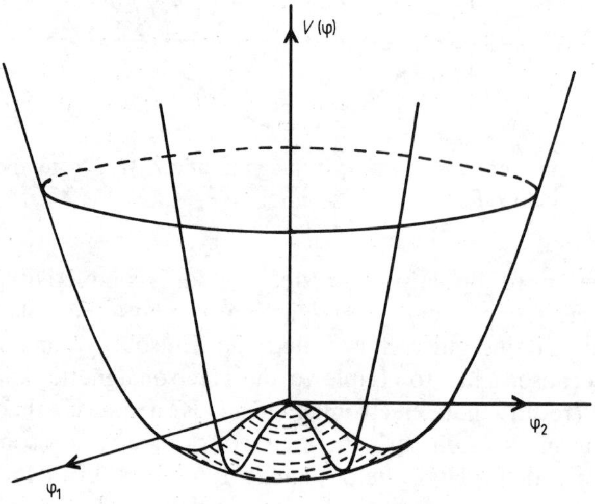

Figure 13.2 The potential (13.49), which has an equilibrium position at $\phi_1^2 + \phi_2^2 \neq 0$.

†Just such a V is assumed in Ginzburg–Landau theory (see e.g. Saint-James *et al* 1969, Tilley and Tilley 1986); for the true superconductor, replace 'vacuum' by 'ground state'. This V was first discussed in the classic paper by Goldstone (1961).

This corresponds, looking down into the 'bowl' of the potential V, to a rotation in the ϕ_1–ϕ_2 plane: the whole of V is symmetrical under this rotation. Where is the equilibrium field configuration (which will model the quantum vacuum)? The place where ϕ_1 and ϕ_2 both vanish, $\phi_1 = \phi_2 = 0$, is actually a local maximum of V, and it is clear that the fields will 'roll away' from there if given a small perturbation. The field configuration of minimum energy, which is therefore the equilibrium configuration, is an entire *locus*

$$\phi_1^2 + \phi_2^2 = f^2; \tag{13.52}$$

namely, the circle at the base of the bowl—or 'wine bottle'—of V.

For the chosen form of V, equation (13.49), there are therefore *infinitely many* possibilities for the stable configuration: any pair (ϕ_1, ϕ_2) satisfying (13.52) will do. Indeed, it is clear that there is a one-parameter family of possibilities:

$$\phi_1 = f\cos\theta, \qquad \phi_2 = f\sin\theta \tag{13.53}$$

for arbitrary θ. The entire circle (13.52) preserves the U(1) symmetry of (13.51). However, only *one* ground state—equilibrium position—is required: any point on the circle will do, for example $\phi_1 = f$, $\phi_2 = 0$, but once a particular point is chosen 'spontaneously', the U(1) symmetry is lost. We say that the U(1) symmetry is 'hidden'.

What happens in the quantum field case? Actually, it is not completely certain—but we shall assume that the naive interpretation holds. Namely, if we are faced with a potential of the form

$$\hat{V}(\hat{\phi}) = -\tfrac{1}{2}\mu^2\hat{\phi}^\dagger\hat{\phi} + \tfrac{1}{4}(\mu^2/f^2)(\hat{\phi}^\dagger\hat{\phi})^2 \tag{13.54}$$

in quantum field theory, we shall say that the state $|0\rangle$ of (13.44), for which $\langle 0|\hat{\phi}_1|0\rangle = \langle 0|\hat{\phi}_2|0\rangle = 0$, is *not* the true vacuum (it corresponds to the point $\phi_1 = \phi_2 = 0$ in figure 13.2); rather, the true vacuum $|\tilde{0}\rangle$ in this case is any state for which $\langle\tilde{0}|\hat{\phi}_1|\tilde{0}\rangle$ and $\langle\tilde{0}|\hat{\phi}_2|\tilde{0}\rangle$ lie somewhere on the circle

$$(\langle\tilde{0}|\hat{\phi}_1|\tilde{0}\rangle)^2 + (\langle\tilde{0}|\hat{\phi}_2|\tilde{0}\rangle)^2 = f^2. \tag{13.55}$$

More compactly (and cf. (13.53)), the true vacuum is any state $|\tilde{0}\rangle_\theta$ such that ${}_\theta\langle\tilde{0}|\hat{\phi}|\tilde{0}\rangle_\theta = (1/\sqrt{2})f\mathrm{e}^{\mathrm{i}\theta}$. There is clearly an infinite number of such states, parametrised by the phase angle θ. This phase is the quantity analogous to the magnetisation direction of a ferromagnet. *Any* phase is possible—the symmetry, in this sense, is still present—but the choice of a definite phase for the *actual* vacuum breaks the U(1) symmetry of (13.54) 'spontaneously' and renders it 'hidden'. A simple choice is obviously $\theta = 0$, for which

$$\langle\tilde{0}|\hat{\phi}|\tilde{0}\rangle = f/\sqrt{2}. \tag{13.56}$$

In fact, quantum fluctuations will, in general, change the value of $\langle\tilde{0}|\hat{\phi}|\tilde{0}\rangle$ away from a value lying on the minimum of the (effectively classical) potential. Nevertheless, the basic picture of a *possible* $\langle\tilde{0}|\hat{\phi}|\tilde{0}\rangle \neq 0$ seems sound. At all events, the $\hat{V}$ of equation (13.54) is taken to represent, phenomenologically, a dynamical theory in which the physical vacuum state is such that the expectation value of $\hat{\phi}$ in that state is not zero. This, as we shall see, is the crucial ingredient in the 'Higgs mechanism' for giving mass to a gauge field quantum.

It is important to realise that the physical vacuum, defined as any state $|\tilde{0}\rangle$ satisfying (13.56), represents a field configuration that cannot be arrived at by using perturbation theory based on the 'normal' vacuum $|0\rangle$, in which the expectation values of all fields are zero. The fundamental assumption that $\langle\tilde{0}|\hat{\phi}|\tilde{0}\rangle \neq 0$, which we shall ultimately make for the *weak* interactions in Chapter 14, therefore involves a hypothesis about a *non-perturbative* aspect of the physical vacuum, in respect of the weak interactions. In principle, one would eventually hope to be able to establish this property, starting from some assumed set of basic fields and interactions, in the same way as the BCS theory established the nature of the superconducting ground state as a coherent plasma of Cooper pairs. This problem has not yet been solved and may, indeed, resist solution for some time to come.

The specific application of these ideas to weak interactions will be given in Chapter 14†. The essential ingredients can best be understood by considering the simpler case of 'massive photons', i.e. let us pursue the consequences of supposing that the physical vacuum is actually a relativistic electromagnetic superconductor. We return to (13.37)—promoted to a quantum field equation—with $\hat{\phi}$ the Higgs field:

$$\Box\hat{A}^{\nu} = iq[\hat{\phi}^{\dagger}\partial^{\nu}\hat{\phi} - (\partial^{\nu}\hat{\phi}^{\dagger})\hat{\phi}] - 2q^2\hat{A}^{\nu}\hat{\phi}^{\dagger}\hat{\phi} \qquad (13.57)$$

appropriate to the gauge choice‡ $\partial_{\mu}\hat{A}^{\mu} = 0$. A naive application of perturbation theory to this equation would correspond to using the mode expansion (13.43)—but this would amount to expanding about the *unstable* point $\hat{\phi} = 0$ in figure 13.2, so that the fields will not remain 'small', and the perturbation theory will presumably diverge. It is much more plausible to develop perturbation theory in terms of small depar-

†It will be necessary there to consider the extension of the 'Higgs mechanism' to the non-Abelian case.

‡Strictly speaking, this gauge condition cannot be valid in quantum field theory, since, as remarked in §5.6, it turns out to be inconsistent with the standard commutation relations for $\hat{A}^{\mu}$ (see e.g. Aitchison 1982). We do not wish to get involved in the full quantum field theory problem of how to quantise a field with a gauge arbitrariness. The results we shall obtain with '$\partial_{\mu}\hat{A}^{\mu} = 0$' will be correct—and we shall relax this condition in §13.5 in any case.

tures from their values in the true vacuum—that is, any point on the circle (13.55). For example, we may write

$$\hat{\phi} = f/\sqrt{2} + \hat{\rho}/\sqrt{2} \tag{13.58}$$

where $\langle\tilde{0}|\hat{\rho}|\tilde{0}\rangle = 0$, consistent with (13.56).

Inserting (13.58) into (13.57), we obtain

$$\Box\hat{A}^{\nu} = -q^2f^2\hat{A}^{\nu} + \text{interactions depending on } \hat{\rho}. \tag{13.59}$$

We shall come back to the $\hat{\rho}$-dependent terms in §13.4—for the moment we must emphasise that the $-q^2f^2\hat{A}^{\nu}$ term in (13.59) has the desired screening form '$-M^2\hat{A}^{\nu}$', with $M = qf$, just as hoped for in the discussion after (13.37). Thus we see how the non-zero vacuum value of the field $\hat{\phi}$ has indeed generated a vacuum screening current, and hence a mass for the quanta of $\hat{A}^{\nu}$.

However, the argument as given so far is incomplete. The basic equation we have used,

$$\Box\hat{A}^{\nu} = \hat{j}^{\nu} \tag{13.60}$$

is *only true with the gauge choice* $\partial_{\mu}\hat{A}^{\mu} = 0$. It therefore appears that the whole analysis is only valid in this gauge, so that we have lost gauge invariance. The question we must now address is this: can we really generate a photon mass while still preserving the gauge invariance of the theory?

13.4 Gauge invariance when the symmetry is hidden

Let us remind ourselves of the essential content of the symmetry we call gauge invariance: for a charged field such as $\hat{\phi}$, a local change in its phase can be compensated by a gauge transformation on the electromagnetic potential (§§2.5, 4.6). Thus the theory has local phase invariance. The 'minimal' Lagrangian involving only $\hat{\phi}$ and $\hat{A}^{\mu}$, given in equations (4.194) and (4.195), has this property by construction. But this local gauge invariance would not be spoiled if we added on, to this minimal $\hat{\phi}$–$\hat{A}^{\mu}$ Lagrangian, the $\hat{V}$ of equation (13.54)—this is because (13.54) is also *locally* U(1) phase invariant, since it involves no derivatives of the fields. The resulting Lagrangian, together with that for the free photon field, constitutes the full field theory version of our 'relativistic superconductor':

$$\begin{aligned}\hat{\mathscr{L}} = {} & [(\partial^{\mu} + \mathrm{i}q\hat{A}^{\mu})\hat{\phi}]^{\dagger}[(\partial_{\mu} + \mathrm{i}q\hat{A}_{\mu})\hat{\phi}] - \tfrac{1}{4}\hat{F}_{\mu\nu}\hat{F}^{\mu\nu} \\ & + \tfrac{1}{2}\mu^2\hat{\phi}^{\dagger}\hat{\phi} - \tfrac{1}{2}(\mu^2/f^2)(\hat{\phi}^{\dagger}\hat{\phi})^2\end{aligned} \tag{13.61}$$

which is the Lagrangian of the *Abelian Higgs model* (Higgs 1964); the equation of motion for $\hat{A}^{\nu}$ is precisely (13.57). (13.61) is therefore locally U(1) invariant: that is, it is gauge invariant. However, the addition of $\hat{V}$ leads—we are assuming—to a non-zero vacuum expectation value for $\hat{\phi}$, and indeed to infinitely many possible vacua, the selection of *one* of which spontaneously breaks the U(1) symmetry. In equation (13.58), for example, we broke it by choosing $\hat{\phi}$ to be real. However, such a choice is not forced upon us—and we must make sure that the same physics follows if we make a different one: the gauge symmetry, though broken, has to be present in secret. We must therefore examine whether or not such choices as (13.58) destroy gauge invariance.

We shall attack the problem of gauge invariance in two stages. In this section we shall show that it is possible to choose a gauge in which $\hat{\phi}$ is real. Then, in the following one, we shall explain what happens if we do not make that choice. It will turn out that far more is involved than a merely formal check on gauge invariance: the discussion is fundamental to the question of renormalisability.

Let us therefore see what happens if we write

$$\hat{\phi} = |\hat{\phi}| \exp[\mathrm{i}\hat{\alpha}(x)] \tag{13.62}$$

where

$$|\hat{\phi}| = f/\sqrt{2} + \ldots. \tag{13.63}$$

Inserting (13.63) and (13.62) into (13.57), we find, neglecting all other interactions,

$$\Box \hat{A}^{\nu} = -q^2 f^2 (\hat{A}^{\nu} + q^{-1}\partial^{\nu}\hat{\alpha}). \tag{13.64}$$

The right-hand side of (13.64) is directly analogous to the non-relativistic current (13.20). At first sight (13.64) seems to prevent us from deducing the desired massive equation

$$(\Box + M^2)\hat{A}^{\nu} = 0. \tag{13.65}$$

But this conclusion is too hasty because (13.65) is only true in the gauge $\partial_{\mu}\hat{A}^{\mu} = 0$. Let us rather consider the complete equation for $\hat{A}^{\nu}$ before the condition $\partial_{\mu}\hat{A}^{\mu} = 0$ was imposed. We then find that the propagation of $\hat{A}^{\nu}$ through the Higgs vacuum is governed by the equation

$$\Box \hat{A}^{\nu} - \partial^{\nu}(\partial_{\mu}\hat{A}^{\mu}) = -q^2 f^2 (\hat{A}^{\nu} + q^{-1}\partial^{\nu}\hat{\alpha}). \tag{13.66}$$

But a gauge transformation on A^{ν} has the form

$$\hat{A}^{\nu} \to \hat{A}'^{\nu} = \hat{A}^{\nu} - \partial^{\nu}\hat{\chi} \tag{13.67}$$

for arbitrary $\hat{\chi}$, so we can certainly regard the whole expression $(\hat{A}^{\nu} + q^{-1}\partial^{\nu}\hat{\alpha})$ as a perfectly acceptable vector potential. Let us define

$$\hat{A}'^{\nu} \equiv \hat{A}^{\nu} + q^{-1}\partial^{\nu}\hat{\alpha}. \tag{13.68}$$

Then, since the left-hand side of the $\hat{A}^{\nu}$ equation is invariant under

(13.67), as we saw long ago, the resulting equation for $\hat{A}'^{\nu}$ is

$$\Box\hat{A}'^{\nu} - \partial^{\nu}(\partial_{\mu}\hat{A}'^{\mu}) = -q^2f^2\hat{A}'^{\nu}. \tag{13.69}$$

Thus, in a configuration in which for stability

$$\hat{\phi} = (f/2^{1/2})\exp[\mathrm{i}\hat{\alpha}(x)] + \ldots \tag{13.70}$$

the field $\hat{A}'^{\nu}$ satisfies (neglecting other interactions)

$$(\Box + M^2)\hat{A}'^{\nu} - \partial^{\nu}(\partial_{\mu}\hat{A}'^{\mu}) = 0 \tag{13.71}$$

with $M = qf$, which is nothing but the standard wave equation for a free vector particle of mass M. It is here obeyed, we emphasise, by a vector field $\hat{A}'^{\nu}$ which, as we have just seen, has a perfect right to be regarded as an electromagnetic vector potential!

From equation (13.71) we can derive, as in equations (13.4)–(13.6), the familiar spin-1 condition

$$\partial_{\mu}\hat{A}'^{\mu} = 0 \tag{13.72}$$

so that (13.71) reduces to

$$(\Box + M^2)\hat{A}'^{\nu} = 0 \tag{13.73}$$

after all. Moreover, this analysis—in particular equation (13.68)—shows quite explicitly how a degree of freedom, namely the phase of the Higgs field $\hat{\phi}$, is being surrendered to make the gauge field massive.

This is the answer to the question about the origin of the third degree of freedom. In the case of the real superconductor, the phase of the pair wavefunction ψ is associated with the local centre-of-mass momentum of the pairs (all of which, we recall, are coherently in the same state of CM momentum). Oscillations in this phase correspond physically to plasma oscillations, which represent precisely the mode to which the longitudinal part of the vector potential is coupled.

In order to interpret the relationship

$$\hat{A}'^{\nu} = \hat{A}^{\nu} + q^{-1}\partial^{\nu}\hat{\alpha} \tag{13.68}$$

as a gauge transformation on $\hat{A}^{\nu}$, we must, to be consistent with (13.72), regard $\hat{A}^{\mu}$ as being in a gauge specified (cf footnote on page 410) by

$$\partial_{\mu}\hat{A}^{\mu} = -q^{-1}\Box\hat{\alpha}. \tag{13.74}$$

In going from the situation described by $\hat{A}^{\mu}$ and $\hat{\alpha}$ to one specified by $\hat{A}'^{\mu}$ alone via (13.68), we have chosen a gauge function

$$\hat{\chi} = -\hat{\alpha}/q \tag{13.75}$$

so that, recalling the form of the associated local phase change

$$\hat{\phi} \to \hat{\phi}' = \exp[\mathrm{i}q\hat{\chi}(x)]\hat{\phi} = \exp[-\mathrm{i}\hat{\alpha}(x)]\hat{\phi} \tag{13.76}$$

we see, comparing (13.76) with (13.62), that in this gauge the phase of $\hat{\phi}$ is reduced to zero. We have therefore established that our earlier special choice of a real value for $\hat{\phi}$ was not, in fact, a real restriction, since if we start with a non-zero phase for $\hat{\phi}$ we can always make a particular choice of gauge for the vector potential so as to reduce the phase of $\hat{\phi}$ to zero†.

We may now ask: what happens if we start with a certain phase $\hat{\alpha}$ for $\hat{\phi}$, but do *not* make use of the gauge freedom in $\hat{A}^{\nu}$ to reduce $\hat{\alpha}$ to zero? We shall see that the equation of motion, and hence the propagator, for the vector particle *depends on the choice of gauge*; furthermore, Feynman graphs involving quanta corresponding to the degree of freedom associated with the phase field $\hat{\alpha}$ will have to be included for a consistent theory, even though this must be an unphysical degree of freedom, as follows from the fact that a gauge can be chosen in which this field vanishes. That the propagator is gauge dependent should, on reflection, come as a relief. After all, if the massive vector boson generated in this way were *simply* described by the wave equation (13.71), all the troubles with massive vector particles detailed in Chapter 12 would be completely unresolved. As we shall see, a different choice of gauge from that which renders $\hat{\phi}$ real has precisely the effect of ameliorating the bad high-energy behaviour associated with (13.71). This is ultimately the reason for the wonderful fact already stated in §12.4: *massive vector theories, in which the vector particles acquire mass through the vacuum screening current mechanism, are renormalisable* ('t Hooft 1971b).

We shall consider alternative gauges somewhat further in §13.5: for the moment we consider the gauge in which $\hat{\phi}$ is real and make a brief comment about the interactions present in the theory. Hitherto we have concentrated on the problem of generating a mass for a *free* spin-1 particle, while starting from a gauge invariant theory. Thus, by setting

†A subtle but important qualification must be made here in the case of a real superconductor, where $\hat{\phi}$ is replaced by the pair wavefunction ψ. The argument given implicitly assumes that the phase of ψ (i.e. $\arg \psi$) is unique. Actually, this need not be so. It is customary to assume that ψ is single-valued, but this still allows $\arg \psi$ to be arbitrary to within a multiple of 2π. Whether this refinement matters depends on whether the 'superconductor'—real or pedagogical—is simply connected or not (i.e. on whether it has any holes in it). A hole may occur either as a consequence of the physical geometry of the superconductor or—in a type II superconductor—as a region in which a filament of magnetic flux exists. The fact that the phase arbitrariness in ψ is limited to a multiple of 2π can be shown to lead to quantisation of the flux passing through such holes. For further discussion of the possible relevance of this point to particle physics, see §15.3. Incidentally, in superconductivity the choice of gauge which makes the pair wavefunction real is called the 'London gauge'.

$|\hat{\phi}| = f\sqrt{2}$, we arrived at (13.73). In this case, of the two degrees of freedom present in $\hat{\phi}$, one—the phase, corresponding to 'motion round the circle' in figure 13.2—has been transferred to $\hat{A}^\nu$ to render it massive, while the other—'radial' oscillations in figure 13.2—has been lost since $|\hat{\phi}|$ has been taken as constant. However, the physics of the radial oscillations can soon be uncovered by returning to (13.58) and inserting it into (13.57); we obtain

$$(\Box + M^2)\hat{A}^\nu - \partial^\nu(\partial_\mu \hat{A}^\mu) = -2q^2 f\hat{\rho}(x)\hat{A}^\nu - q^2\hat{\rho}^2(x)\hat{A}^\nu. \quad (13.77)$$

This is the equation of motion for a massive vector field $\hat{A}^\nu$ *coupled in a specific way to the neutral field* $\hat{\rho}(x)$. The quanta of $\hat{\rho}$ (called Higgs bosons) are the particle aspect of the 'radial' degree of freedom of $\hat{\phi}$. We expect that it will make sense to develop the perturbation theory of (13.77), since this amounts to looking at small oscillations of $\hat{\phi}$ about its *stable* value.

Since this 'electromagnetic superconducting vacuum' has been only a pedagogical excercise, by way of preparation for the physically interesting application to weak interactions, we need not pursue the precise forms of the couplings and the associated Feynman graphs any further. But the following general point can be made. The breaking of manifest gauge invariance by the phase choice

$$\hat{\phi} = [f + \hat{\rho}(x)]/2^{1/2} \quad (13.58)$$

has therefore, from the gauge invariant starting point (cf (13.66))

$$\Box\hat{A}^\nu - \partial^\nu(\partial_\mu \hat{A}^\mu) = \hat{j}^\nu \quad (13.78)$$

produced an *apparently* gauge non-invariant result (13.77). As we have just argued, however, different phase and gauge choices *do* lead to the same physics. Thus it is permissible to draw a general conclusion from (13.77), namely, a fundamental consequence of the original gauge invariance is the appearance of specific relations between the parameters of the theory (in the gauge used above, $M = qf$, and the couplings of $\hat{A}^\nu$ to $\hat{\rho}$ are also determined by q and f). Verifying such relations, then, is the way we must experimentally identify the hidden weak symmetry of the world in which we live.

It has been shown explicitly (see e.g. Llewellyn Smith 1974) in several more realistic examples of field theories that such relations among the parameters of the theory are precisely those required to bring about cancellations among graphs with bad high-energy behaviour, and to render the massive vector boson theory renormalisable. It is generally believed that the *only* renormalisable massive vector theories are those in which there is a hidden gauge invariance. Hence, this appears to be the only way to make a *theory* of the weak interaction phenomenology discussed earlier.

13.5 't Hooft's gauges

We must now at last grasp the nettle and consider what happens if, in the parametrisation

$$\hat{\phi} = |\hat{\phi}| e^{i\hat{\alpha}(x)}$$

we do not choose the gauge (recall the second footnote on page 410)

$$\partial_\mu \hat{A}^\mu = -q^{-1}\Box\hat{\alpha}. \tag{13.74}$$

This was the gauge that enabled us to transform away the phase degree of freedom and reduce the equation of motion for the electromagnetic field to that of a massive vector boson. Instead of using the modulus and phase as the two independent degrees of freedom for the complex Higgs field $\hat{\phi}$, we choose to parametrise $\hat{\phi}$, quite generally, by the decomposition (compare (13.58))

$$\hat{\phi}(x) = 2^{-1/2}[f + \hat{\chi}_1(x) + i\hat{\chi}_2(x)] \tag{13.80}$$

where the vacuum values of $\hat{\chi}_1$ and $\hat{\chi}_2$ are zero. Substituting this form for $\hat{\phi}$ into the master equation for $\hat{A}^\nu$

$$\Box\hat{A}^\nu - \partial^\nu(\partial_\mu\hat{A}^\mu) = iq[\hat{\phi}^\dagger(\partial^\nu\hat{\phi}) - (\partial^\nu\hat{\phi}^\dagger)\hat{\phi}] - 2q^2\hat{A}^\nu\hat{\phi}^\dagger\hat{\phi} \tag{13.81}$$

leads to an equation of motion of the form

$$(\Box + M^2)\hat{A}^\nu - \partial^\nu(\partial_\mu\hat{A}^\mu) = -M\partial^\nu\hat{\chi}_2 + q(\hat{\chi}_2\partial^\nu\hat{\chi}_1 - \hat{\chi}_1\partial^\nu\hat{\chi}_2) - q^2\hat{A}^\nu(\hat{\chi}_1^2 + 2f\hat{\chi}_1 + \hat{\chi}_2^2) \tag{13.82}$$

with $M = qf$. At first sight this just looks like the equation of motion of an ordinary massive vector field $\hat{A}^\nu$ coupled to a rather complicated current. However, this certainly cannot be right, as we can see by a count of the degrees of freedom. In the previous gauge we had four degrees of freedom, counted either as two for the original massless $\hat{A}^\nu$ plus one each for $\hat{\alpha}$ and $\hat{\rho}$, or as three for the massive $\hat{A}'^\nu$ and one for $\hat{\rho}$. If we take this new equation at face value, there seem to be three degrees of freedom for the massive field $\hat{A}^\nu$, and one for each of $\hat{\chi}_1$ and $\hat{\chi}_2$, making *five* in all. Actually, we know perfectly well that we can make use of the freedom of gauge choice to set $\hat{\chi}_2$ to zero, say, reducing $\hat{\phi}$ to a real quantity and eliminating a spurious degree of freedom: we have then returned to the form (13.58). In terms of (13.82), the consequence of the unwanted degree of freedom is quite subtle, but it is basic to all gauge theories. The difficulty appears when we try to calculate the propagator for $\hat{A}^\nu$ from equation (13.82). The operator on the left-hand side can be simply inverted, as done in §10.1, to yield, apparently, the standard massive vector boson propagator

$$(-g^{\mu\nu} + k^\mu k^\nu/M^2)/(k^2 - M^2). \tag{13.83}$$

However, the current on the right-hand side is rather peculiar: instead of having only terms corresponding to $\hat{A}^\nu$ coupling to two or three particles, there is also a term involving only one field. This is the term $-M\partial^\nu\hat{\chi}_2$, which tells us that $\hat{A}^\nu$ actually couples directly to the scalar field χ_2 via the gradient coupling $(-M\partial^\nu)$. In momentum space this corresponds to a coupling strength $-ik^\nu M$ and a vertex as in figure 13.3.

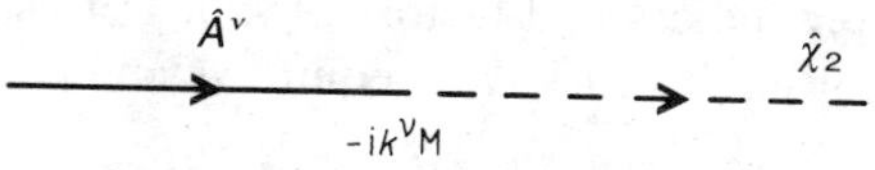

Figure 13.3 $\hat{A}^\nu$–$\hat{\chi}_2$ coupling.

Clearly, for a scalar particle, the momentum 4-vector is the only quantity that can couple to the vector index of the vector boson. The existence of this coupling shows that the propagators of $\hat{A}^\nu$ and $\hat{\chi}_2$ are necessarily mixed: the complete vector propagator must be calculated by summing the infinite series shown diagrammatically in figure 13.4. This complication is, of course, completely eliminated by the gauge choice $\hat{\chi}_2 = 0$. However, we are interested in pursuing the case $\hat{\chi}_2 \neq 0$.

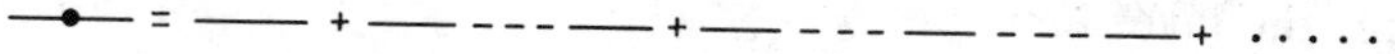

Figure 13.4 Series for the full $\hat{A}^\nu$ propagator.

In figure 13.4. the only unknown factor is the propagator for $\hat{\chi}_2$, for which we have to make a specific model for the scalar field. Actual specific models for $\hat{\phi}$, in which $\hat{\phi}$ develops a non-zero vacuum as above, show that $\hat{\chi}_2$ is a massless field with propagator $1/k^2$. Now that all the elements of the diagrams are known, we can formally sum the series by generalising the well known result

$$(1 - x)^{-1} = 1 + x + x^2 + x^3 + \dots . \tag{13.84}$$

Diagrammatically, we rewrite the propagator of figure 13.4 as in figure 13.5 and perform the sum. Inserting the expressions for the propagators and vector–scalar coupling, and keeping track of the indices, we finally

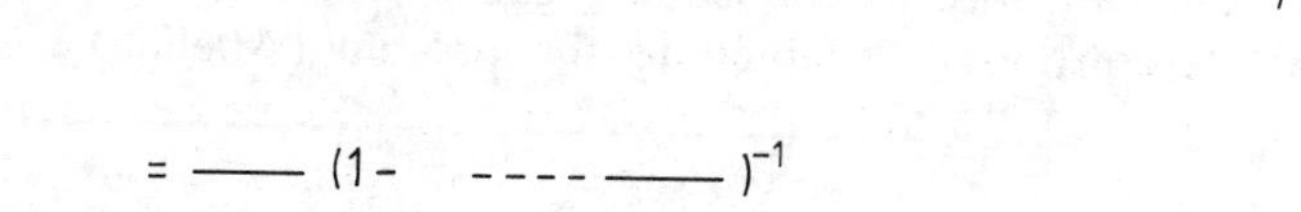

Figure 13.5 Formal summation of the series in figure 13.4.

arrive at the result (see problem 13.1)

$$\frac{-g^{\mu\lambda} + k^\mu k^\lambda/M^2}{k^2 - M^2}(\delta_\lambda^\nu - k^\nu k_\lambda/k^2)^{-1} \qquad (13.85)$$

for the full propagator. We therefore need to calculate the inverse operator

(after rearranging indices and factors of k^2). Let us try to do this by the method we followed in §10.1. We require A and B such that

$$(-k^2 g^{\mu\lambda} + k^\mu k^\lambda)(A g_{\lambda\nu} + B k_\lambda k_\nu) = \delta_\nu^\mu. \qquad (13.87)$$

Multiplying out, we find

$$-k^2 A \delta_\nu^\mu - \cancel{Bk^2 k^\mu k_\nu} + A k^\mu k_\nu + \cancel{Bk^2 k^\mu k_\nu} = \delta_\nu^\mu \qquad (13.88)$$

which is impossible for any A! This operator has *no* inverse. Thus all is not as it seems in equation (13.82).

At this point, the alert reader may be thoroughly rattled, since she recalls the equation of motion for electrodynamics, namely

$$\Box \hat{A}^\nu - \partial^\nu(\partial_\mu \hat{A}^\mu) = \hat{j}^\nu \qquad (13.78)$$

from which it appears that the *photon* propagator in momentum space requires the inverse

$$(-k^2 g^{\mu\nu} + k^\mu k^\nu)^{-1}. \qquad (13.89)$$

As we have just seen, this does not exist! These diseases are indeed identical, and the remedy the same: in all gauge theories, in order to define the propagators of the gauge quanta, a choice of gauge must be made. In QED we have so far always worked with the (covariant) Lorentz condition $\partial_\mu \hat{A}^\mu = 0$. The inversion of the operator equation

$$\Box \hat{A}^\nu = \hat{j}^\nu \qquad (13.60)$$

is then trivial. More general gauge conditions are also possible, leading to different forms for the photon propagator, as we shall see.

Let us return to the 'hidden gauge invariance' case and the equation

$$(\Box + M^2)\hat{A}^\nu - \partial^\nu(\partial_\mu \hat{A}^\mu) = -M\partial^\nu \hat{\chi}_2 + q(\hat{\chi}_2 \partial^\nu \hat{\chi}_1 - \hat{\chi}_1 \partial^\nu \hat{\chi}_2)$$
$$- q^2 \hat{A}^\nu(\hat{\chi}_1^2 + 2f\hat{\chi}_1 + \hat{\chi}_2^2). \qquad (13.82)$$

The difficulty over the $\hat{A}^\nu$ propagator must be overcome by a choice of gauge. A clever way to choose the gauge was suggested by 't Hooft (1971b). His proposal, adapted to the present (Abelian) case, was to set†

†The second footnote as on page 410 applies here. However, at the classical (tree graph) level at which we shall in fact work, our results are correct.

$$\boxed{\partial_\mu \hat{A}^\mu = M\xi\hat{\chi}_2} \tag{13.90}$$

in equation (13.82), where ξ is an arbitrary parameter. This is still manifestly covariant, and furthermore, it effectively reduces the degrees of freedom by one. With this choice we obtain

$$(\Box + M^2)\hat{A}^\nu - \partial^\nu(\partial_\mu \hat{A}^\mu)(1 - 1/\xi)$$
$$= q(\hat{\chi}_2\partial^\nu\hat{\chi}_1 - \hat{\chi}_1\partial^\nu\hat{\chi}_2) - q^2\hat{A}^\nu(\hat{\chi}_1^2 + 2f\hat{\chi}_1 + \hat{\chi}_2^2). \tag{13.91}$$

The operator appearing on the left-hand side now *does* have an inverse (see problem 13.2) and yields the general form for the gauge boson propagator

$$\left[-g^{\mu\nu} + \frac{(1-\xi)k^\mu k^\nu}{k^2 - \xi M^2}\right](k^2 - M^2)^{-1}. \tag{13.92}$$

This propagator is very remarkable†. The standard massive vector boson propagator

$$(-g^{\mu\nu} + k^\mu k^\nu/M^2)(k^2 - M^2)^{-1} \tag{13.83}$$

is seen to correspond to the limit $\xi \to \infty$, and in this gauge all the high-energy diseases described in Chapter 12 appear to threaten renormalisability (in fact, it can be shown that there is a consistent set of Feynman rules for this gauge, and all the required cancellations do occur). For any finite ξ, however, the high-energy behaviour of the gauge boson propagator is actually $\sim 1/k^2$, which is as good as the *renormalisable* theory of QED (in Lorentz gauge). Note, however, that there seems to be another pole in this propagator (equation (13.92)) at $k^2 = \xi M^2$: this is surely unphysical since it depends on the arbitrary parameter ξ. A full treatment ('t Hooft 1971b) shows that this pole is always cancelled by an exactly similar pole in the propagator for the $\hat{\chi}_2$ field itself. These finite-ξ gauges are called *R gauges* (since they are 'manifestly renormalisable') and typically involve unphysical Higgs fields such as $\hat{\chi}_2$. The infinite-ξ gauge is known as the *U gauge* (U for unitary) since only physical particles appear in this gauge. For tree diagram calculations, of course, it is easiest to use the U gauge Feynman rules: the technical difficulties with this gauge choice only enter in loop calculations, for which the R gauge choice is easier.

Notice that in our master formula for the gauge boson propagator

$$\left[-g^{\mu\nu} + \frac{(1-\xi)k^\mu k^\nu}{k^2 - \xi M^2}\right](k^2 - M^2)^{-1} \tag{13.92}$$

†A vector boson propagator of similar form was first introduced by Lee and Yang (1962), but their discussion was not within the framework of a spontaneously broken theory so that Higgs particles were not present, and the physical limit was obtained *only* as $\xi \to 0$.

the limit $M \to 0$ may be taken (compare the remarks about such a limit on page 384). This yields the massless vector boson (photon) propagator in an arbitrary gauge

$$\boxed{[-g^{\mu\nu} + (1 - \xi)k^{\mu}k^{\nu}/k^{2}]k^{-2}.} \tag{13.93}$$

The gauge which we have used in all our QED calculations corresponds to the case of $\xi = 1$ in this formula; it is called the Feynman gauge.

We now proceed with the generalisation of these ideas to the (non-Abelian) GSW electroweak theory.

Summary

Apparent impossibility of massive vector particle theory exhibiting gauge invariance.

The screening current condition.

Screening currents in an atom.

Screening currents in a superconductor
- macroscopic wavefunction;
- coherent occupancy of a single quantum state;
- condensation;
- non-perturbative aspect.

Higgs model (relativistic superconductor)
- Higgs field providing vacuum screening currents.

The physical vacuum
- the vacuum as ground state;
- spontaneous symmetry breakdown in ground state;
- ferromagnet;
- superconductor; U(1) symmetry spontaneously broken;
- $\langle\tilde{0}|\hat{\phi}|\tilde{0}\rangle \neq 0$;
- vacuum screening current from $\langle\tilde{0}|\hat{\phi}|\tilde{0}\rangle \neq 0$.

Gauge invariance
- Higgs model;
- choice of phase of $\hat{\phi}$ breaks gauge invariance;
- restored by change in $\hat{A}^{\nu}$;
- possible to choose $\hat{\phi}$ real;
- dependence of gauge boson propagator on choice of gauge;
- such a theory is renormalisable.

't Hooft gauges
gauges other than '$\hat{\phi}$ = real' necessitate extra contributions from 'unphysical Higgs field' $\hat{\chi}_2$;
difficulty with gauge boson propagator in general case;
't Hooft gauge and general form of propagator;
photon propagator.

Problems

13.1 Verify equation (13.82).

13.2 By considering plane waves for A^μ in the usual way, find the momentum space inverse of the operator

$$(\Box + M^2)g^{\nu\mu} - (1 - 1/\xi)\partial^\nu\partial^\mu$$

appearing (after a slight change in indices) in (13.91); and hence verify equation (13.92).

14

THE GLASHOW–SALAM–WEINBERG GAUGE THEORY OF ELECTROWEAK INTERACTIONS

14.1 Introduction

In Chapters 10 and 11 we gave an introduction to the low-energy phenomenology of weak interactions, and explained in Chapter 12 why it could not be regarded as a theory. The application of the gauge principle—invariance under certain local phase transformations—to weak interactions produces, at last, a true theory. We shall consider the specific theory associated principally with the names of Glashow, Salam and Weinberg, but now often referred to as 'the standard model' (i.e. of the electroweak interactions). This is a non-Abelian gauge theory in which the local phase invariance is 'hidden', in the sense of Chapter 13. It is therefore a somewhat more complicated theory than one with an unbroken (or manifest) non-Abelian symmetry, which is why we discussed QCD first. In particular, the Feynman rules for the GSW theory will need to include prescriptions for graphs involving the scalar ('Higgs') field associated with the generation of mass for the gauge quanta.

To provide motivation, we begin by reviewing the way in which we seem to be led by the desire to turn the low-energy phenomenology into a theory. In the first place, the known V–A structure of the quark and lepton currents requires us to conclude that the weak field quanta must be vector particles. As we have seen in Chapter 10, these must be massive, because the weak force is known to be short-ranged, and they are also charged. The only known renormalisable theories involving charged massive vector bosons are those in which the bosons are the quanta associated with a gauge symmetry, which must be of the hidden variety in order for the bosons to acquire mass. It is remarkable that if massive charged W's exist, a hidden gauge invariance has to be invoked in order to make sense of even their *electromagnetic* interactions.

A further, very suggestive, experimental hint of the likely relevance of some gauge principle is the empirical fact that a 'universal' strength can be associated with the weak interactions of both leptons and quarks. In §11.1 we saw that after allowance for the Cabibbo rotation, the weak charged current of the quarks had the same overall coupling strength g as the weak charged current of the leptons (see table 11.1). Such a 'universality' is precisely what we should expect from a gauge theory of weak interactions, in which the 'weak charge' g plays a role analogous to e in U(1) electromagnetic gauge transformations in QED.

The remaining question is then: what is the relevant symmetry group of local phase transformations, i.e. the relevant *weak gauge group*? Several possibilities were suggested, but it is now very well established that the one originally proposed by Glashow (1961), subsequently treated as a hidden gauge symmetry by Weinberg (1967) and by Salam (1968), and later extended by other authors, produces a theory which is in agreement with all currently known data. We shall not give a critical review of the experimental evidence, but instead proceed directly to an outline of the GSW theory, introducing elements of the data as required.

14.2 Weak isospin and hypercharge: the SU(2) × U(1) group of the electroweak interactions

An important clue to the symmetry group involved in the weak interactions is provided by considering the transitions induced by these interactions. This is somewhat analogous to discovering the multiplet structure of atomic levels and hence the representations of the rotation group, a prominent symmetry of the Schrödinger equation, by studying electromagnetic transitions. However, there is one very important difference, which is worth emphasising right at the start, between the 'weak multiplets' we shall consider and those of conventional 'manifest' symmetry groups. This aspect of hidden symmetry was not apparent in the Abelian example discussed in Chapter 13. We saw in Chapter 8 how a *non-Abelian* global symmetry reveals itself in the existence of multiplets of states which are degenerate in mass. However, in that discussion we implicitly assumed that the symmetry was not 'hidden': when it is, the degenerate multiplet structure will, in general, disappear entirely.

What can go wrong with the argument for multiplets that we gave in Chapter 8? To understand this, we must use the field theory formalism of §8.2, in which the generators of the symmetry, say $\hat{T}_i$, are Hermitian field operators, and the states are created by operators acting on the vacuum. Thus consider two states $|A\rangle$, $|B\rangle$:

$$|A\rangle = \hat{\phi}^{\dagger}_A|0\rangle, \qquad |B\rangle = \hat{\phi}^{\dagger}_B|0\rangle \tag{14.1}$$

where $\hat{\phi}^{\dagger}_A$ and $\hat{\phi}^{\dagger}_B$ are related to each other by (cf (8.54))

$$[\hat{T}, \hat{\phi}^{\dagger}_A] = \hat{\phi}^{\dagger}_B \tag{14.2}$$

for some generator $\hat{T}$ of a symmetry group, such that

$$[\hat{T}, \hat{H}] = 0. \tag{14.3}$$

(14.2) is equivalent to

$$\hat{U}\hat{\phi}^{\dagger}_{A}\hat{U}^{-1} \approx \hat{\phi}^{\dagger}_{A} + \mathrm{i}\varepsilon\hat{\phi}^{\dagger}_{B} \tag{14.4}$$

for an infinitesimal transformation $\hat{U} \approx 1 + \mathrm{i}\varepsilon\hat{T}$. Thus $\hat{\phi}^{\dagger}_{A}$ is 'rotated' into $\hat{\phi}^{\dagger}_{B}$ by $\hat{U}$, and the operators will create states related by the symmetry transformation. We want to see what are the assumptions necessary to prove that

$$E_A = E_B \quad \text{where} \quad \hat{H}|A\rangle = E_A|A\rangle, \quad \hat{H}|B\rangle = E_B|B\rangle. \tag{14.5}$$

We have

$$E_B|B\rangle = \hat{H}|B\rangle = \hat{H}\hat{\phi}^{\dagger}_{B}|0\rangle = \hat{H}(\hat{T}\hat{\phi}^{\dagger}_{A} - \hat{\phi}^{\dagger}_{A}\hat{T})|0\rangle. \tag{14.6}$$

Now if

$$\hat{T}|0\rangle = 0 \tag{14.7}$$

we can rewrite the right-hand side of (14.6) as

$$\begin{aligned}\hat{H}\hat{T}\hat{\phi}^{\dagger}_{A}|0\rangle &= \hat{T}\hat{H}\hat{\phi}^{\dagger}_{A}|0\rangle \quad \text{using (14.3)} = \hat{T}\hat{H}|A\rangle = E_A\hat{T}|A\rangle \\ &= E_A\hat{T}\hat{\phi}^{\dagger}_{A}|0\rangle = E_A(\hat{\phi}^{\dagger}_{B} + \hat{\phi}^{\dagger}_{A}\hat{T})|0\rangle \quad \text{using (14.2)} \\ &= E_A|B\rangle \quad \text{if (14.7) holds;}\end{aligned} \tag{14.8}$$

whence, comparing (14.8) with (14.6), we see that

$$E_A = E_B \quad \text{if (14.7) holds.} \tag{14.9}$$

Remembering that $\hat{U} = \exp(\mathrm{i}\alpha\hat{T})$, we see that (14.7) is equivalent to

$$|0\rangle' = \hat{U}|0\rangle = |0\rangle. \tag{14.10}$$

Thus *a multiplet structure will emerge provided that the vacuum is left invariant under the symmetry transformation*. The 'hidden' symmetry situation arises in the contrary case—that is, when the application of the generator $\hat{T}$ to the vacuum gives, not zero as in (14.7), but another possible vacuum! We can get a visualisation of how this can arise by looking back at the wine-bottle potential of figure 13.2, which is so essential to the idea of hidden symmetry. As we emphasised in §13.3, in the quantum field theory version of that classical potential there is an infinity of *equally possible vacuum* states $|\tilde{0}\rangle$, which are such that $(\langle\tilde{0}|\hat{\phi}_1|\tilde{0}\rangle)^2 + (\langle\tilde{0}|\hat{\phi}_2|\tilde{0}\rangle)^2 = f^2$. Clearly a symmetry operator which rotates the complex field $\hat{\phi} = (1/\sqrt{2})(\hat{\phi}_1 - \mathrm{i}\hat{\phi}_2)$ by a simple phase effectively moves the system from one of these vacua to another. Instead of 'killing $|0\rangle$' as in (14.7), such an operator merely shifts $|\tilde{0}\rangle$ to the adjacent $|\tilde{0}\rangle'$. In this case we cannot get beyond (14.6) and are unable to deduce (14.9).

It is easy to relate the above discussion to our previous treatment of hidden symmetry, which emphasised the necessity of some $\hat{\phi}$ such that $\langle\tilde{0}|\hat{\phi}|\tilde{0}\rangle \neq 0$. If we take the vacuum expectation value of equation

(14.2), we find

$$\langle 0|\hat{T}\hat{\phi}_A^\dagger|0\rangle - \langle 0|\hat{\phi}_A^\dagger\hat{T}|0\rangle = \langle 0|\hat{\phi}_B^\dagger\,|0\rangle \tag{14.11}$$

so that, if (14.7) holds, then $\langle 0|\hat{\phi}_B^\dagger|0\rangle = 0$ as expected. But if (14.7) does not hold, we are in a 'Higgs' vacuum $|\tilde{0}\rangle$, the left-hand side of (14.11) does not vanish, and we can have $\langle\tilde{0}|\hat{\phi}_B^\dagger|\tilde{0}\rangle \neq 0$. Of course, this is really just a slightly more formal way of stating the physics modelled by figure 13.2.

Thus a 'spontaneously broken' symmetry is well and truly hidden, and this is surely why, in retrospect, the weak symmetry group took so long to discover.

Nevertheless, as we shall see, vestiges of the weak symmetry group—specifically, the relations it requires between otherwise unrelated masses and couplings (see §13.4)—are accessible to experiment. Moreover, despite the fact that members of a multiplet of a global symmetry which is hidden will, in general, no longer have even approximately the same mass, the concept of a multiplet is still useful. This is because when the symmetry is made a *local* one, we shall find (in §14.3) that the associated gauge quanta still mediate interactions between members of a given global symmetry multiplet, just as in the manifest local non-Abelian symmetry example of QCD. Now, the leptonic transitions associated with the weak charged currents are, as we saw in Chapters 10 and 11, $\nu_e \leftrightarrow e$, $\nu_\mu \leftrightarrow \mu$, etc. This suggests that these pairs should be regarded as *doublets* under some group. Further, the quark current introduced in §11.1 had a suggestively simple form when written in terms of the left-handed doublet q_L:

$$q_L = \begin{pmatrix} u \\ d_C \end{pmatrix}_L \tag{14.12}$$

and

$$j_{wk}^+ \sim \bar{q}_L \tau_+ q_L. \tag{14.13}$$

The simplest possibility is therefore to suppose that, in both cases, a 'weak SU(2) group' is involved, called 'weak isospin'. We emphasise once more that this weak isospin is distinct from the hadronic isospin of Chapter 8, which is part of $SU(3)_f$. We use the symbols t, t_3 for the quantum numbers of weak isospin, and make the specific lepton assignments

$$\boxed{t = \tfrac{1}{2}, \qquad t_3 = \begin{cases} +\tfrac{1}{2} \\ -\tfrac{1}{2} \end{cases} \qquad \begin{pmatrix} \nu_e \\ e^- \end{pmatrix}_L, \begin{pmatrix} \nu_\mu \\ \mu^- \end{pmatrix}_L, \ldots} \tag{14.14}$$

where the dots indicate further '*generations*', such as $(\nu_\tau, \tau^-)_L$ and possibly others, yet to be discovered. For quarks, we assign, in a corresponding sequence of generations,

$$t = \tfrac{1}{2}, \qquad t_3 = \begin{cases} +\tfrac{1}{2} \\ -\tfrac{1}{2} \end{cases} \qquad \begin{pmatrix} u \\ d_{\mathrm{C}} \end{pmatrix}_{\mathrm{L}}, \begin{pmatrix} c \\ s_{\mathrm{C}} \end{pmatrix}_{\mathrm{L}}, \ldots \tag{14.15}$$

where the other generations involve the b and, presumably, t quarks, and so on. As in §11.1, the subscript 'L' indicates that only the left-handed parts of the wavefunctions enter into these weak transitions, as explained in §§10.4 and 11.1. For this reason, the weak isospin group is usually referred to as $\mathrm{SU(2)_L}$, to show that the weak isospin assignments and corresponding transformation properties apply only to these left-handed parts: for example, under an $\mathrm{SU(2)_L}$ transformation

$$\begin{pmatrix} \nu_{\mathrm{e}} \\ \mathrm{e}^- \end{pmatrix}_{\mathrm{L}} \to \begin{pmatrix} \nu_{\mathrm{e}} \\ \mathrm{e}^- \end{pmatrix}'_{\mathrm{L}} = \exp(\mathrm{i}\boldsymbol{\alpha} \cdot \boldsymbol{\tau}/2) \begin{pmatrix} \nu_{\mathrm{e}} \\ \mathrm{e}^- \end{pmatrix}_{\mathrm{L}}. \tag{14.16}$$

Notice that, as anticipated for hidden symmetry, these doublets all involve pairs of particles which are by no means mass degenerate.

Making this $\mathrm{SU(2)_L}$ into a local phase invariance (following the logic of Chapter 8) will entail the introduction of three gauge fields, transforming as a $t = 1$ multiplet (a triplet) under the group. Because (as with ordinary $\mathrm{SU(2)_f}$ hadronic isospin) the members of a weak isodoublet differ by one unit of charge, the two gauge fields associated with transitions between doublet members will have charge ± 1. The quanta of these fields will, of course, be the familiar $\mathrm{W}^{\pm}$ bosons mediating the charged current transitions, and associated with the weak-isospin-raising and lowering operators $t_{\pm}$. What about the third gauge boson of the triplet? This will be electrically neutral, and a very economical and appealing idea would be to associate this neutral vector particle with the photon—thereby *unifying* the weak and electromagnetic interactions. A model of this kind was originally suggested many years ago by Schwinger (1957). Of course, the W's must somehow acquire mass, while the photon remains massless. Schwinger arranged this by introducing appropriate couplings of the vector bosons to additional scalar and pseudoscalar fields. These couplings were arbitrary and no prediction of the W masses could be made. We now believe, as emphasised in §13.1, that the W mass must arise from the 'Higgs mechanism', generalised to the case of spontaneous breakdown of a non-Abelian gauge symmetry, and as we shall shortly see, this *does* constrain the W mass.

Apart from the question of the W mass in Schwinger's model, we now know (see Chapter 11) that there exist *neutral current* weak interactions, in addition to those of the charged currents. We must also include these in our emerging gauge theory, and an obvious suggestion is to have these currents mediated by the neutral member W^0 of the $SU(2)_L$ gauge field triplet. Such a scheme was indeed proposed by Bludman (1958), again pre-Higgs, so that W masses were put in 'by hand'. In this model, however, the neutral currents will have the same pure left-handed V–A structure as the charged currents: it is now known that the neutral currents are *not* pure V–A. Furthermore, the attractive feature of including the photon, and thus unifying weak and electromagnetic interactions, has been lost.

A key contribution was made by Glashow (1961); similar ideas were also advanced by Salam and Ward (1964). Glashow suggested enlarging the Schwinger–Bludman SU(2) schemes by the inclusion of an additional U(1) gauge group, resulting in an '$SU(2)_L \times U(1)$' group structure. The new Abelian U(1) group is associated with a weak analogue of hypercharge—'weak hypercharge'—just as $SU(2)_L$ was associated with 'weak isospin'. Indeed, Glashow proposed that the Gell-Mann–Nishijima relation for charges should also hold for these weak analogues, giving

$$eQ = e(t_3 + y/2) \tag{14.17}$$

for the electric charge Q (in units of e) of the t_3 member of a weak isomultiplet, assigned a weak hypercharge y. Clearly, therefore, the lepton doublets, (ν_e, e^-), etc, then have $y = -1$, while the quark doublets, (u, d_C), etc, have $y = +\frac{1}{3}$. Now, when *this* group is gauged, everything falls marvellously into place: the charged vector bosons appear as before, but there are now *two* neutral vector bosons, which between them will be reponsible for the weak neutral current processes, and for electromagnetism. We must of course demonstrate that the hidden symmetry realisation of the $SU(2)_L \times U(1)$ gauge group can in fact accommodate a massive neutral boson and, at the same time, a massless photon. To describe the way in which the intertwining of weak and electromagnetic interactions comes about will be the work of the following sections of this chapter. We end this section by giving further weak quantum number assignments.

The most important omission so far concerns the *right-handed* parts of the fermion wavefunctions. Although neutrinos may only exist in one helicity state, electrons and muons certainly have both. However, there do not seem to be any weak transitions that couple their right-handed components

$$\left(\frac{1+\gamma_5}{2}\right)\psi_e, \left(\frac{1+\gamma_5}{2}\right)\psi_\mu, \ldots \tag{14.18}$$

to other states. It is therefore natural to assume that all right-handed components are *singlets* under the weak isospin group. With the help of the weak charge formula (equation (14.17)), this then produces the assignments shown in table 14.1. Other assignments will follow later.

Table 14.1 Weak isospin and hypercharge assignments for fermions.

	t	t_3	y	Q
ν_e, ν_μ	1/2	1/2	−1	0
e_L, μ_L	1/2	−1/2	−1	−1
e_R, μ_R	0	0	−2	−1
u_L, c_L	1/2	1/2	1/3	2/3
$(d_C)_L, (s_C)_L$	1/2	−1/2	1/3	−1/3
μ_R, c_R	0	0	4/3	2/3
$(d_C)_R, (s_C)_R$	0	0	−2/3	−1/3

14.3 SU(2)$_L$ × U(1) as a local symmetry: charged current vertices

From now on we shall primarily be concerned with the application of the theory to leptons, as in the original Weinberg–Salam model. By this time, the reader should need no further instruction on how to make the global SU(2)$_L$ invariance

$$\begin{pmatrix} \nu_e \\ e^- \end{pmatrix}_L \to \begin{pmatrix} \nu_e \\ e^- \end{pmatrix}'_L = \exp(i\boldsymbol{\alpha}\cdot\boldsymbol{\tau}/2)\begin{pmatrix} \nu_e \\ e^- \end{pmatrix}_L \tag{14.19}$$

into a local one; nor for the even simpler U(1)$_y$ phase invariance. We merely introduce the covariant derivative appropriate to the particular representation, specified by t and y, namely (using quantum fields now)

$$\partial^\mu \to \hat{D}^\mu = \partial^\mu + ig\boldsymbol{t}^{(t)}\cdot\hat{\boldsymbol{W}}^\mu + i(g'/2)y\hat{B}^\mu. \tag{14.20}$$

The fundamental constants g and g' are the weak 'charges' of the SU(2)$_L$ and U(1)$_y$ parts of the gauge group respectively: the factor of $\frac{1}{2}$ in the g' term is purely conventional. The three $\hat{\boldsymbol{W}}^\mu$ are the SU(2)$_L$ gauge fields, while $\hat{B}^\mu$ is the new U(1)$_y$ gauge field. Thus for the left-handed electron-type doublet of (14.14), which we denote l_e, we have

$$\left.\begin{array}{l}\text{left-handed}\\ \text{leptons}\\ t=\frac{1}{2}\\ y=-1\end{array}\right\}\ \partial^\mu \to \hat{D}^\mu = \partial^\mu + ig\frac{\boldsymbol{\tau}\cdot\hat{\boldsymbol{W}}^\mu}{2} - \frac{ig'}{2}\hat{B}^\mu \tag{14.21}$$

and similarly for the muon doublet l_μ. For the right-handed part of the electron wavefunction e_R, which is left untouched by SU(2)$_L$ transformations since $t = 0$, we have

$$\left.\begin{array}{l}\text{right-handed}\\ \text{leptons}\\ t = 0\\ y = -2\end{array}\right\} \partial^\mu \to \hat{D}^\mu = \partial^\mu - \mathrm{i}g'\hat{B}^\mu \tag{14.22}$$

and similarly for μ_R.

Just as in QED and QCD, the isodoublet–W interaction follows immediately we introduce the covariant derivative into the Dirac equation: it is

$$-\mathrm{i}g\,\bar{\hat{l}}_e(\tau^a/2)\gamma_\mu \hat{l}_e \hat{W}^{a\mu} \tag{14.23}$$

or

$$-\mathrm{i}g[\bar{\hat{l}}_e(\boldsymbol{\tau}/2)\gamma_\mu \hat{l}_e]\cdot\hat{\boldsymbol{W}}^\mu \tag{14.24}$$

and similarly for l_μ, etc. The degree of freedom associated with the index a, running over the values 1, 2, 3 for the SU(2)$_L$ indices of the $\boldsymbol{W}^\mu$ triplet, may be written in a number of equivalent notations, such as

$$W^{a\mu},\ \boldsymbol{W}^\mu,\ \omega^a W^\mu \tag{14.25}$$

where, in the last form, we have introduced an 'SU(2)$_L$ polarisation vector' ω^a. In this notation, the SU(2)$_L$ structure of the interaction (14.24) is summarised by the product

$$(\boldsymbol{\tau}/2)\cdot\boldsymbol{\omega} \tag{14.26}$$

which we may write as

$$\frac{1}{2^{1/2}}\left(\tau_+\frac{\omega^1 - \mathrm{i}\omega^2}{2^{1/2}} + \tau_-\frac{\omega^1 + \mathrm{i}\omega^2}{2^{1/2}}\right) + \frac{\tau_3}{2}\omega^3 \tag{14.27}$$

where $\tau_\pm = (\tau^1 + \mathrm{i}\tau^2)/2$ are the raising and lowering operators for isodoublets. The matrix τ_+ picks out the transitions like $e^- \to \nu_e$, in which we either absorb a W^+ or emit a W^-. The combination $\omega^- = (\omega^1 - \mathrm{i}\omega^2)/2^{1/2}$ creates a W^- or destroys a W^+. It follows that the $e^- \to \nu_e + W^-$ vertex is

$$-\frac{\mathrm{i}g}{2^{1/2}}\bar{u}(\nu)\gamma_\mu\frac{1-\gamma_5}{2}u(e) \tag{14.28}$$

which, of course, is precisely the vertex which, in the regime $q^2 \ll M_W^2$, produces the phenomenological form given in table 11.1. This theory therefore reproduces the low-energy charged current phenomenology, with the specific identification

$$G_F/2^{1/2} = g^2/8M_W^2 \tag{10.93}$$

previously made.

The neutral current part and the electromagnetic current are not so straightforward. They must clearly correspond in some way to the $W^{3\mu}$ and B^{μ} parts of the interaction. One gauge field must acquire mass (to reproduce the observed short-range neutral weak force) while the other must be massless (for the photon). In order to understand the precise physical interpretation of the neutral fields, we must now discuss the detailed application of the Higgs mechanism to this specific non-Abelian theory.

14.4 Hidden $SU(2)_L \times U(1)$ gauge invariance

In §§13.3 and 13.4 we explained how the existence of a charged (complex) scalar field with a non-zero expectation value in the ground state of the world—the physical 'vacuum'—could give the U(1) gauge boson a mass, thereby 'hiding' the gauge invariance of the theory. An Abelian U(1) gauge group was considered there, merely for the sake of illustrating this 'Higgs mechanism' in its simplest context, but we do not want a real photon to have a non-zero mass. We must now generalise this analysis to the non-Abelian case, as required by the GSW theory; this case was discussed by Englert and Brout (1964) and (in greater detail) by Kibble (1967).

We shall not attempt a general discussion: we will limit ourselves to the particular way in which the GSW symmetry is hidden. Let us begin by briefly recalling how the U(1) case went. We started with a massless gauge field $\hat{A}^{\mu}$ with two degrees of freedom; we introduced the complex (Higgs) scalar field $\hat{\phi}$, with two more degrees of freedom; one of these latter degrees of freedom was 'eaten' (cf equation (13.68)) by the massless gauge field $\hat{A}^{\mu}$ to make a massive vector field $\hat{A}'^{\mu}$, while the other remained on as a massive scalar field $\hat{\rho}$ (although we did not explicitly demonstrate this last point in Chapter 13). In our present case we have to make *three* gauge fields massive—the $\hat{W}^{\pm}$ and one associated with the neutral weak currents. We must therefore introduce some scalar field with at least three degrees of freedom. However, we must also ensure that some effective U(1) gauge symmetry remains which is *unbroken*, in order that 'its' gauge field remains massless and can represent the photon.

It turns out that the simplest way to do all this is to generalise the charged scalar field of Chapter 13 to a (complex) scalar *doublet*, transforming as a $t = \frac{1}{2}$ multiplet under $SU(2)_L$. This was the specific proposal of Weinberg and Salam. Such a doublet field has two components—the $t_3 = +\frac{1}{2}$ and the $t_3 = -\frac{1}{2}$ states—where each component, as with any spinor, is complex. Thus our Higgs field may be written in the form

$$t = \tfrac{1}{2}, \qquad \hat{\phi} = \begin{pmatrix} 2^{-1/2}(\hat{\phi}_1 + i\hat{\phi}_2) \\ 2^{-1/2}(\hat{\phi}_3 + i\hat{\phi}_4) \end{pmatrix}, \qquad t_3 = \begin{cases} +\tfrac{1}{2} \\ -\tfrac{1}{2} \end{cases} \tag{14.29}$$

in which *four* real scalar fields appear. We have still to decide on the weak hypercharge assignment for $\hat{\phi}$, but we defer this for the moment.

The basic mechanism by which a non-zero vacuum expectation value of $\hat{\phi}$ gives a mass to three of the (originally massless) gauge bosons is substantially the same as the U(1) case. In essence, we want *the current induced in the vacuum* to contain a part *proportional to the gauge field*, so that an equation for a massless vector field, of the form

$$\Box \hat{A}^{\mu} - \partial^{\mu}\partial_{\nu}\hat{A}^{\nu} = \hat{j}^{\mu} \tag{14.30}$$

together with the vacuum screening current condition

$$\hat{j}^{\mu} = -M^2\hat{A}^{\mu} + \text{interactions} \tag{14.31}$$

yields the equation for a massive vector field

$$(\Box + M^2)\hat{A}^{\mu} - \partial^{\mu}\partial_{\nu}\hat{A}^{\nu} = 0 \tag{14.32}$$

which describes the free propagation of the gauge field in this peculiar vacuum. More precisely, we recall from the U(1) 'superconducting vacuum' example that it was the choice of the *phase* of the scalar field's vacuum expectation value that broke (or 'hid') the manifest gauge invariance of the theory. Actually, of course, gauge invariance is maintained in the sense that a change in phase of $\hat{\phi}$ can always be compensated by a gauge transformation on the gauge field $\hat{A}^{\mu}$, with no change in the physics. Indeed, in equations (13.74)–(13.76) we showed that it was possible to choose $\hat{\phi}$ to be real, in which case the corresponding gauge field $\hat{A}'^{\mu}$ satisfied the standard massive spin-1 equation (14.32). The phase of $\hat{\phi}$ was precisely the degree of freedom absorbed by the original massless $\hat{A}^{\mu}$ to generate a massive field (cf equation (13.68)).

In the present case 'the phase' of the complex isospinor (14.29) is a more complicated object than the single phase field which appeared in the U(1) case. An appropriate generalisation of an expression of the type

$$\hat{\phi} = |\hat{\phi}| \exp[i\hat{\alpha}(x)] \tag{14.33}$$

is

$$\hat{\phi} = \exp[i(\boldsymbol{\tau}/2)\cdot\hat{\boldsymbol{\alpha}}(x)]\begin{pmatrix} 0 \\ \hat{P}(x) \end{pmatrix} \tag{14.34}$$

in which *three* real phase fields $\hat{\boldsymbol{\alpha}} = (\hat{\alpha}^1, \hat{\alpha}^2, \hat{\alpha}^3)$ appear, together with the real field $\hat{P}(x)$. Notice that if $\hat{\boldsymbol{\alpha}}$ and $\hat{P}$ were independent of x, this equation would simply correspond to the statement that an arbitrary $\hat{\phi}$ can be regarded as arising from a certain isospin rotation, parametrised by $\hat{\boldsymbol{\alpha}}$, performed on a basic isospinor lined up along the '−3 direction'. The dependence of $\hat{\boldsymbol{\alpha}}$ on x corresponds to taking *independent* isospin rotations at each space–time point, precisely as sanctioned by our fundamental principle of *local* phase invariance. It is these three fields $\hat{\boldsymbol{\alpha}}(x)$ that take the place, in $SU(2)_L$, of the single field $\hat{\alpha}(x)$ in the U(1) case, and they will give a mass to three of the four gauge fields required to make $SU(2)_L \times U(1)$ a local symmetry.

We must now introduce the currents associated with the complex isospinor $\hat{\phi}$, whose vacuum values will produce the desired screening effects. In the U(1) example of Chapter 13, the relevant current was

$$\hat{j}^{\mu}(\hat{\phi}) = iq[\hat{\phi}^{\dagger}(\partial^{\mu}\hat{\phi}) - (\partial^{\mu}\hat{\phi})^{\dagger}\hat{\phi}] \tag{14.35}$$

which we made gauge invariant by replacing ∂^{μ} by the covariant derivative (cf (13.35) and (13.36))

$$\partial^{\mu} \rightarrow \hat{D}^{\mu} \equiv \partial^{\mu} + iq\hat{A}^{\mu}. \tag{14.36}$$

The symmetry group of the GSW theory is $SU(2)_L \times U(1)_y$. Thus corresponding to the $U(1)_y$, there will be a weak hypercharge analogue of (14.35):

$$\hat{j}^{\mu}_{y}(\hat{\phi}) = i(g'/2)y[\hat{\phi}^{\dagger}(\partial^{\mu}\hat{\phi}) - (\partial^{\mu}\hat{\phi})^{\dagger}\hat{\phi}] \tag{14.37}$$

where we have left unspecified, at this stage, the weak hypercharge of the Higgs field. The SU(2) generalisation of (14.37) was considered in §8.2: from equation (8.78) the $SU(2)_L$ current appropriate to the weak isospinor $\hat{\phi}$ is

$$\hat{j}^{a\mu}(\hat{\phi}) = ig\left(\hat{\phi}^{\dagger}\frac{\tau^a}{2}(\partial^{\mu}\hat{\phi}) - (\partial^{\mu}\hat{\phi})^{\dagger}\frac{\tau^a}{2}\hat{\phi}\right). \tag{14.38}$$

As indicated by the notation, these are only the $\hat{\phi}$ contributions to the total weak isospin current: there will also be contributions for all particles that carry isospin. For example, the leptons will contribute terms of the form

$$\bar{\hat{l}}\,\gamma^{\mu}(\tau^a/2)\,\hat{l} \tag{14.39}$$

and similarly for the quarks. In addition, the W's themselves will contribute a self-interaction term (cf §§8.4, 8.5) which we may write as $\hat{j}^{a\mu}(W)$. However, we need not discuss these other contributions at present since we are interested only in the vacuum screening currents, and only the Higgs field $\hat{\phi}$ will be assumed to have a non-zero vacuum value. These currents are, however, not invariant under local $SU(2)_L \times U(1)$ phase transformations, and to remedy this we must replace ∂^{μ} by the covariant derivative appropriate to the $t = \frac{1}{2}\,\hat{\phi}$ field with hypercharge y:

$$\hat{D}^\mu = \partial^\mu + ig(\boldsymbol{\tau}/2)\cdot \hat{\boldsymbol{W}}^\mu + i(g'/2)y\hat{B}^\mu. \tag{14.40}$$

This produces the currents

$$\hat{j}^{a\mu}(\hat{\phi}) = ig\left(\hat{\phi}^\dagger\frac{\tau^a}{2}(\partial^\mu\hat{\phi}) - (\partial^\mu\hat{\phi})^\dagger\frac{\tau^a}{2}\hat{\phi}\right) - \frac{g^2}{2}\hat{\phi}^\dagger\hat{\phi}\hat{W}^{a\mu} - gg'y\hat{\phi}^\dagger\frac{\tau^a}{2}\hat{\phi}\hat{B}^\mu \tag{14.41}$$

and

$$\hat{j}^\mu_y(\hat{\phi}) = \frac{ig'}{2}y[\hat{\phi}^\dagger(\partial^\mu\hat{\phi}) - (\partial^\mu\hat{\phi})^\dagger\hat{\phi}] - gg'y\hat{\phi}^\dagger\frac{\boldsymbol{\tau}}{2}\hat{\phi}\cdot\hat{\boldsymbol{W}}^\mu - \frac{g'^2}{2}y^2\hat{\phi}^\dagger\hat{\phi}\hat{B}^\mu \tag{14.42}$$

where we have used the relation

$$\tau^a\tau^b + \tau^b\tau^a = 2\delta^{ab} \tag{14.43}$$

specific to the 2 × 2 matrices of SU(2).

To find the vacuum screening currents, we must now insert the vacuum value of $\hat{\phi}$ into these equations. In the U(1) case we chose

$$\hat{\phi} = (f/2^{1/2})\exp[i\hat{\alpha}(x)] + \ldots \tag{14.44}$$

where (cf (13.63)) the dots represented a field with vanishing vacuum expectation value, which did not contribute to the equation describing free propagation of the gauge field $\hat{A}^\mu$ in the Higgs vacuum. In the SU(2)$_L$ × U(1)$_y$ case, Weinberg (1967) suggested the form

$$\hat{\phi} = \exp[i(\boldsymbol{\tau}/2)\cdot\hat{\boldsymbol{\alpha}}(x)]\begin{pmatrix}0\\ f/2^{1/2}\end{pmatrix} + \ldots \tag{14.45}$$

with f a real constant and $\hat{\boldsymbol{\alpha}}(x) \neq 0$ (in general). Now in the case of U(1) phase invariance, we showed explicitly that, by a suitable choice of gauge, we could without loss of generality set the phase $\hat{\alpha}(x)$ to zero. The same is true in the non-Abelian case, so we may simplify our equations by the choice

$$\hat{\phi} = \begin{pmatrix}0\\ f/2^{1/2}\end{pmatrix} + \ldots \tag{14.46}$$

that is

$$\langle\tilde{0}|\hat{\phi}|\tilde{0}\rangle = \begin{pmatrix}0\\ f/2^{1/2}\end{pmatrix} \tag{14.47}$$

without altering the physics. Remember, however, that in a general 't Hooft gauge (see §13.5) there are always some unphysical Higgs fields appearing in the equations and the Feynman rules. Here our discussion is entirely in the U gauge, where such unphysical fields are absent and the physics consequently most transparent. Inserting this form for $\hat{\phi}$ (equation (14.46)) into equations (14.41) and (14.42) leads to the vacuum screening currents which must give mass to the $W^{\pm}$ and one of the neutral fields.

Since the neutral sector is a little more involved than the charged sector, let us for the moment ignore the $\hat{B}^{\mu}$ field and the hypercharge current and look at the effect of (14.46) on the $SU(2)_L$ current

$$\hat{j}^{a\mu}(\hat{\phi}) = \mathrm{i}g\left(\hat{\phi}^{\dagger}\frac{\tau^a}{2}(\partial^{\mu}\hat{\phi}) - (\partial^{\mu}\hat{\phi})^{\dagger}\frac{\tau^a}{2}\hat{\phi}\right) - \frac{g^2}{2}\hat{\phi}^{\dagger}\hat{\phi}\hat{W}^{a\mu}. \tag{14.48}$$

We obtain immediately the result, for the current in the Higgs vacuum,

$$\hat{j}^{a\mu}_{\mathrm{vac}} = -g^2 f^2 \hat{W}^{a\mu}/4 + \ldots \tag{14.49}$$

which, with the identification

$$\boxed{M_{\mathrm{W}} = gf/2} \tag{14.50}$$

looks just like a mass term as before. In fact, as repeatedly emphasised in Chapter 8, the field equation for $\hat{W}^{a\mu}$ is more complicated than the Abelian case since there must always be W self-interaction terms $\hat{j}^{a\mu}(\hat{W})$. In the spontaneously broken $SU(2)_L$ case we are considering, the presence of the *additional* (vacuum screening) current due to the Higgs field will lead to an equation of the form

$$(\square + M_{\mathrm{W}}^2)\hat{W}^{a\mu} - \partial^{\mu}\partial_{\nu}\hat{W}^{a\nu} = \hat{j}^{a\mu}(\hat{W}) \tag{14.51}$$

where we have reintroduced the self-interaction current. In the U gauge in which we are working it is indeed correct to identify M_{W} via equation (14.50) as the mass of the $W^{\pm}$ bosons of the $SU(2)_L \times U(1)$ GSW theory.

However, this is not the whole story since we must deal with the neutral current and electromagnetic current components of the gauge fields. Here we must consider the y value of the $\hat{\phi}$ field. Unlike a real superconductor, we do not allow the particle physics vacuum to give an electrically charged field a non-zero value. Thus we require that the component of $\hat{\phi}$ with non-zero vacuum value has zero charge. With the choice

$$\langle\tilde{0}|\hat{\phi}|\tilde{0}\rangle = \begin{pmatrix} 0 \\ f/2^{1/2} \end{pmatrix}. \tag{14.47}$$

it is the $t_3 = -\frac{1}{2}$ component, and the relation

$$eQ = e(t_3 + y/2) \tag{14.52}$$

therefore fixes $y(\phi) = 1$. Let us look again at the contribution of $\hat{\phi}$ to the complete $SU(2)_L \times U(1)_y$ weak isospin current (equation (14.41) with $y = 1$)

$$\hat{j}^{a\mu}(\hat{\phi}) = ig\left(\hat{\phi}^\dagger \frac{\tau^a}{2}(\partial^\mu \hat{\phi}) - (\partial^\mu \hat{\phi})^\dagger \frac{\tau^a}{2}\hat{\phi}\right) - \frac{g^2}{2}\hat{\phi}^\dagger \hat{\phi} \hat{W}^{a\mu} - gg' \hat{\phi}^\dagger \frac{\tau^a}{2}\hat{\phi}\hat{B}^\mu. \tag{14.53}$$

We can now understand the full significance of our choice of $\langle \tilde{0}|\hat{\phi}|\tilde{0}\rangle$ above. It implies, firstly, that the $\hat{B}^\mu$ term in $\hat{j}^{a\mu}$ does not contribute to the vacuum screening current for $a = 1, 2$. This follows at once from the fact that τ^1 and τ^2 have no diagonal elements so that

$$(0, f)(\tau^1 \text{ or } \tau^2)\begin{pmatrix} 0 \\ f \end{pmatrix} = 0. \tag{14.54}$$

Thus our previous analysis, supposedly only for $SU(2)_L$, in fact is true for the $a = 1, 2$ components of the $SU(2)_L \times U(1)$ theory, and $M_W = gf/2$ is to be interpreted as the physical $W^\pm$ mass. For $a = 3$, however, we find

$$\hat{j}^{3\mu}(\hat{\phi}) = -(g^2 f^2/4)\hat{W}^{3\mu} + (gg' f^2/4)\hat{B}^\mu + \ldots. \tag{14.55}$$

This implies that instead of a massive equation of the form

$$\Box \hat{W}^{3\mu} - \partial^\mu \partial_\nu \hat{W}^{3\nu} = -M^2 \hat{W}^{3\mu} + \hat{j}^{3\mu}(\hat{W}) \tag{14.56}$$

we shall now have

$$\Box \hat{W}^{3\mu} - \partial^\mu \partial_\nu \hat{W}^{3\nu} = -M^2 \hat{W}^{3\mu} + MM' \hat{B}^\mu + \hat{j}^{3\mu}(\hat{W}) \tag{14.57}$$

where we have defined $M' = g'f/2$ and $M = gf/2$.

The corresponding equation for the U(1) gauge field $\hat{B}^\mu$ is found by considering the hypercharge current (equation (14.42) with $y = 1$)

$$\hat{j}^\mu_y(\hat{\phi}) = i(g'/2)[\hat{\phi}^\dagger(\partial^\mu \hat{\phi}) - (\partial^\mu \hat{\phi})^\dagger \hat{\phi}] - gg' \hat{\phi}^\dagger (\boldsymbol{\tau}/2)\hat{\phi} \cdot \hat{\mathbf{W}}^\mu - (g'^2/2)\hat{\phi}^\dagger \hat{\phi} \hat{B}^\mu. \tag{14.58}$$

The current for the $\hat{B}^\mu$ equation is therefore

$$\hat{j}^\mu_y(\hat{\phi}) = +(gg' f^2/4)\hat{W}^{3\mu} - (g'^2 f^2/4)\hat{B}^\mu + \ldots \tag{14.59}$$

and $\hat{B}^\mu$ satisfies an equation of the form

$$\Box \hat{B}^\mu - \partial^\mu \partial_\nu \hat{B}^\nu = -M'^2 \hat{B}^\mu + MM' \hat{W}^{3\mu} \tag{14.60}$$

where, *in vacuo*, there are no other terms on the right-hand side, since $\hat{B}^\mu$ is an Abelian gauge field and has no self-coupling.

We see that the new feature of the specific $SU(2)_L \times U(1)$ group structure and our choice of $\langle \tilde{0}|\hat{\phi}|\tilde{0}\rangle$ is that the gauge fields $\hat{B}^\mu$ and $\hat{W}^{3\mu}$ satisfy *coupled* equations. More precisely, there is a term which is linear in $\hat{B}^\mu$ on the right-hand side of equation (14.57) for $\hat{W}^{3\mu}$, and a corresponding term in $\hat{W}^{3\mu}$ on the right-hand side of the equation for $\hat{B}^{\mu}$, (14.60). Normally, such linear terms (as in the U(1) vacuum screening current case) could be brought to the left of the equations, and reinterpreted as mass terms; but here it is precisely these mass terms that are coupled. Physically, this means that the fields $\hat{W}^{3\mu}$ and $\hat{B}^{\mu}$ do not have a definite mass, and are not the physical fields. To obtain these latter, we must find the linear combinations of $\hat{B}^{\mu}$ and $\hat{W}^{3\mu}$ that diagonalise the mass terms. This is very easy to do. From equations (14.55) and (14.59) we see that the mass terms cancel completely in the particular linear combination

$$g'\hat{W}^{3\mu} + g\hat{B}^{\mu}; \tag{14.61}$$

in fact, (14.61) satisfies an equation of the form

$$(\Box\delta^{\mu}_{\nu} - \partial^{\mu}\partial_{\nu})(g'\hat{W}^{3\nu} + g\hat{B}^{\nu}) = g'\hat{j}^{3\mu}(\hat{W}) \tag{14.62}$$

appropriate to a massless particle—it will be the long-lost photon—coupled (via the self-interaction terms) to the charged W's.

It is customary to normalise the combination (14.61) slightly differently. We introduce the *Glashow–Weinberg angle* θ_{W}, defined in terms of the fundamental constants g and g' by

$$\tan\theta_{\mathrm{W}} = g'/g \tag{14.63a}$$

$$\cos\theta_{\mathrm{W}} = g/(g^2 + g'^2)^{1/2} \tag{14.63b}$$

$$\sin\theta_{\mathrm{W}} = g'/(g^2 + g'^2)^{1/2}. \tag{14.63c}$$

The electromagnetic field is identified as

$$\hat{A}^{\mu} = \sin\theta_{\mathrm{W}}\hat{W}^{3\mu} + \cos\theta_{\mathrm{W}}\hat{B}^{\mu} \tag{14.64}$$

and satisfies (in the absence of all other fields) the equation

$$(\Box\delta^{\mu}_{\nu} - \partial^{\mu}\partial_{\nu})\hat{A}^{\nu} = \sin\theta_{\mathrm{W}}\hat{j}^{3\mu}(\hat{W}). \tag{14.65}$$

The 'orthogonal' combination

$$\hat{Z}^{\mu} = \cos\theta_{\mathrm{W}}\hat{W}^{3\mu} - \sin\theta_{\mathrm{W}}\hat{B}^{\mu} \tag{14.66}$$

correspondingly obeys a massive wave equation of the form

$$\Box\hat{Z}^{\mu} - \partial^{\mu}\partial_{\nu}\hat{Z}^{\nu} = -(f^2/4)(g'^2 + g^2)\hat{Z}^{\mu} + \cos\theta_{\mathrm{W}}\hat{j}^{3\mu}(\hat{W}) \tag{14.67}$$

showing that the theory also contains a weak neutral current mediated by a neutral vector boson Z^0 of mass

$$\boxed{M_{\mathrm{Z}} = \tfrac{1}{2}f(g'^2 + g^2)^{1/2} = M_{\mathrm{W}}/\cos\theta_{\mathrm{W}}.} \tag{14.68}$$

With these identifications, we can now write the charge-non-changing part of the general $SU(2)_L \times U(1)$ covariant derivative (for a weak isomultiplet of dimension $2t + 1$ and hypercharge y) as

$$\hat{D}^\mu = \partial^\mu + \mathrm{i}g \sin\theta_W \hat{A}^\mu(t_3^{(t)} + y/2) + \mathrm{i}(g/\cos\theta_W)\hat{Z}^\mu[\cos^2\theta_W t_3^{(t)} - (y/2)\sin^2\theta_W]. \quad (14.69)$$

Since we have the relation

$$Q = (t_3 + y/2) \quad (14.17)$$

we therefore identify from the $\hat{A}^\mu$ term the specific connection between the coupling strengths of the weak and electromagnetic interactions predicted by the GSW theory:

$$\boxed{g \sin\theta_W = e.} \quad (14.70)$$

We end this section with two comments.

(a) The reader may well be wondering how we can be sure about identifying the physical photon with that particular combination (14.61) of the fields $\hat{W}^{3\mu}$ and $\hat{B}^\mu$ which ended up massless. The answer lies in the choice

$$\langle\tilde{0}|\hat{\phi}|\tilde{0}\rangle = \begin{pmatrix} 0 \\ f/2^{1/2} \end{pmatrix}. \quad (14.47)$$

Any choice for the Higgs field vacuum value $\langle\tilde{0}|\hat{\phi}|\tilde{0}\rangle$ which is such that it explicitly breaks the relevant symmetry operation will inevitably lead to a corresponding massive gauge boson. However, if our original symmetry group is larger than just a U(1) phase invariance, it may happen that a part of this symmetry group is not broken by our choice for $\langle\tilde{0}|\hat{\phi}|\tilde{0}\rangle$. In other words, $\langle\tilde{0}|\hat{\phi}|\tilde{0}\rangle$ may be left invariant by some of the group transformations, and the massless gauge bosons associated with the local version of these particular transformations will not acquire mass. In the case of equation (14.47), since $y(\phi) = 1$, we have the relation

$$\left(\frac{y}{2} + \frac{\tau_3}{2}\right)\langle\tilde{0}|\hat{\phi}|\tilde{0}\rangle = 0 \quad (14.71)$$

which implies that $\langle\tilde{0}|\hat{\phi}|\tilde{0}\rangle$ is invariant under the transformation

$$\langle\tilde{0}|\hat{\phi}|\tilde{0}\rangle \to \langle\tilde{0}|\hat{\phi}|\tilde{0}\rangle' = \exp[\mathrm{i}\alpha(y/2 + t_3^{(1/2)})]\langle\tilde{0}|\hat{\phi}|\tilde{0}\rangle. \quad (14.72)$$

But from the weak Gell-Mann–Nishijima formula (14.17) we can see that $(y/2 + t_3^{(1/2)})$ is just equal to the *charge* operator for $t = \frac{1}{2}$ particles. Thus the particular transformation, selected from $SU(2)_L \times U(1)_y$ transformations, which leaves $\langle\tilde{0}|\hat{\phi}|\tilde{0}\rangle$ invariant is nothing other than the electromagnetic U(1) phase invariance. Thus the corresponding gauge boson that is left massless is indeed the photon.

(b) The second comment is a final reminder to the reader that all our analysis has been specific to the unitary gauge case. In a general gauge, the massive vector boson propagator in this hidden symmetry situation will have the general form

$$\left(-g^{\mu\nu} + \frac{(1-\xi)k^\mu k^\nu}{k^2 - \xi M^2}\right)(k^2 - M^2)^{-1} \tag{13.92}$$

in which the presence of the unphysical pole at $k^2 = \xi M^2$ signifies that the Feynman rules will involve an unphysical Higgs particle. Our equations (14.51) and (14.67) for the W and Z bosons are for the special U gauge choice $\xi \rightarrow \infty$.

14.5 Simple phenomenological implications

At this point we interrupt the theoretical development in order to consider some phenomenological implications of the GSW theory as so far presented.

14.5.1 W and Z masses

The theory has introduced two new parameters, the vacuum expectation value of the Higgs field f, and one of g and g', which are related to the assumed-known constant e by equation (14.70). From the equation for the W mass

$$M_{\mathrm{W}} = gf/2 \tag{14.50}$$

and the connection to the low-energy phenomenology via

$$G_{\mathrm{F}}/2^{1/2} = g^2/8M_{\mathrm{W}}^2 \tag{10.93}$$

we can determine f in terms of the known Fermi constant G_{F}:

$$G_{\mathrm{F}}/2^{1/2} = 1/2f^2 \tag{14.73}$$

giving

$$f \simeq 246\ \mathrm{GeV}. \tag{14.74}$$

As yet, no theory is available to predict this number. The other parameter introduced so far may be taken to be θ_{W}. This is determined from neutral current processes involving Z^0 exchange, the couplings for which are given via the covariant derivative (14.69)—see §14.5.3 below. Fits to neutrino data (Review of Particle Properties 1984) give

$$\sin^2\theta_{\mathrm{W}} = 0.226 \pm 0.004. \tag{14.75}$$

Equation (14.50) can then be written as

$$M_{\mathrm{W}} = \left(\frac{\pi\alpha}{\sqrt{2}G_{\mathrm{F}}}\right)^{1/2} \frac{1}{\sin\theta_{\mathrm{W}}} \simeq \frac{37.28}{\sin\theta_{\mathrm{W}}}\,\mathrm{GeV} \simeq 78.4 \pm 0.7\,\mathrm{GeV}. \tag{14.76}$$

Equation (14.68) gives

$$M_{\mathrm{Z}} = M_{\mathrm{W}}/\cos\theta_{\mathrm{W}} \simeq 89.1 \pm 0.7\,\mathrm{GeV}. \tag{14.77}$$

These precise predictions of the GSW theory indicate the power of the hidden symmetry in determining many apparently unrelated parameters (in this instance, masses), in terms of only a few basic parameters.

However, there is a complication—numerically a minor one at present, but ultimately of great interest. Equations (14.76) and (14.77) represent only the *tree level* predictions for the masses. They are, in fact, the mass parameters that appear in the field equations (14.51) and (14.67)—or, alternatively, in the basic Lagrangian—and hence in the corresponding propagators. But we have learned in Chapter 6 that loop corrections can modify these propagators. Indeed, since the GSW theory *is* renormalisable, these modifications can be precisely calculated, and they have been (Marciano 1979, Antonelli and Maiani 1981, Dawson *et al* 1981, Sirlin and Marciano 1981, Llewellyn Smith and Wheater 1982). Obviously we cannot discuss the details of such calculations here. Nevertheless, the discussion of Chapters 6 and 9 gives us some access to the results. We learned that in renormalisable theories, such as QED and QCD, one-loop corrections typically produce (after the subtraction procedures involved in infinite renormalisation) finite changes in masses and coupling constants, which depend logarithmically on the energy variable relative to some fixed energy point. The same is true in GSW theory. Thus, in extracting the value of $\sin^2\theta_{\mathrm{W}}$ from experiment, one has to be precise about the method of renormalisation that one is adopting, and the energy point which is used in a *definition* of the relevant quantities—as, for example, one conventionally chooses $q^2 = 0$ for the definition of α. To the accuracy at which only leading logarithmic corrections (in the sense of §9.3, for example) are retained, it can be shown (Marciano 1979, Antonelli and Maiani 1981, Dawson *et al* 1981) that formulae (14.76) and (14.77) are correct if interpreted in terms of 'running' parameters α and $\sin\theta_{\mathrm{W}}$.

One convenient *definition* of θ_{W} is actually (Sirlin and Marciano 1981, Marciano and Sirlin 1984)

$$\sin^2\theta_{\mathrm{W}} \equiv 1 - (M_{\mathrm{W}}^2/M_{\mathrm{Z}}^2) \tag{14.78}$$

i.e. the reference energy is taken to be M_{W}. Then (14.76) becomes

$$M_{\mathrm{W}} = \left(\frac{\pi\alpha}{\sqrt{2}G_{\mathrm{F}}}\right)^{1/2} \cdot \frac{1}{\sin\theta_{\mathrm{W}}} \cdot [1 + \delta_\alpha + \delta_M] \tag{14.79}$$

where δ_α represents the finite vacuum polarisation correction in α, when going from $q^2 = 0$ (where $\alpha \simeq (137)^{-1}$ is conventionally defined) to $q^2 = M_W^2$, and δ_M represents all effects of virtual W's and Z's, and Higgs bosons (see §14.7). Including three light lepton and quark families gives

$$\alpha(M_W) \simeq (128)^{-1} \tag{14.80}$$

so that

$$\delta_\alpha \simeq +0.03 \tag{14.81}$$

which is, in fact, the largest effect. The full one-loop result† is that (14.76) becomes

$$M_W \simeq 38.6 \text{ GeV}/\sin\theta_W. \tag{14.82}$$

The value given in (14.75) is actually appropriate to $\sin^2\theta_W$ defined at M_W, and so we finally obtain

$$M_W \simeq 81.2 \pm 0.7 \text{ GeV} \tag{14.83a}$$

$$M_Z \simeq 92.3 \pm 0.7 \text{ GeV}. \tag{14.83b}$$

14.5.2 Charged currents, and W width

Charged current (or, more generally, flavour-changing) vertices were already discussed briefly in §14.3. In the low-energy limit, the lowest-order amplitudes for four-fermion processes involving W exchange will take the form already used in §10.5:

$$\frac{e^2}{2\sin^2\theta_W} j_\mu^C j^{C\mu} = \frac{4G_F}{2^{1/2}} j_\mu^C j^{C\mu} \tag{14.84}$$

where, for the lepton doublets, the charge current amplitude is

$$j_\mu^C = \bar{u}(\nu_e)\gamma_\mu \frac{1-\gamma_5}{2} u(e^-) \tag{14.85}$$

for example, and similarly for (ν_μ, μ^-), (ν_τ, τ^-), ... For the quarks, we have already introduced the Cabibbo–GIM mixing mechanism, whereby d and s are replaced by

$$\begin{pmatrix} d_C \\ s_C \end{pmatrix} = \begin{pmatrix} \cos\theta_C & \sin\theta_C \\ -\sin\theta_C & \cos\theta_C \end{pmatrix} \begin{pmatrix} d \\ s \end{pmatrix}. \tag{14.86}$$

The b quark with charge $-\frac{1}{3}$, and its still unconfirmed generation partner t with charge $+\frac{2}{3}$, can be handled (Kobayashi and Maskawa 1973) by enlarging the mixing scheme (14.86) to a 3×3 arrangement involving new combinations d', s' and b', which can then accommodate

†Assuming $m_t = 36$ GeV and a Higgs mass μ (see §14.7) $= M_Z$, for no particular reason.

(without predicting magnitudes) a parametrisation of **CP**-violating effects (see §11.5).

This charged current coupling determines the width for $W \to e^- \bar{\nu}_e$ as (problem 14.1)

$$\Gamma(W \to e^- \bar{\nu}_e) = \frac{1}{12}\frac{g^2}{4\pi}M_W = \frac{G_F}{2^{1/2}}\frac{M_W^3}{6\pi} \simeq \frac{23\,\text{MeV}}{\sin^3\theta_W} \simeq 214\,\text{MeV} \tag{14.87}$$

using (14.75) for $\sin\theta_W$. The widths to $\mu^-\bar{\nu}_\mu$, $\tau^-\bar{\nu}_\tau$ are the same, and those for decay to $\bar{u}d$, $\bar{c}s$ and $\bar{t}b$ are each three times as much† (from a colour factor), neglecting all Cabibbo-type mixing. The total W width for this number of fermions will therefore be about 2.5 GeV, while the branching ratio for $W \to e\nu$ is

$$B(e\nu) = \Gamma(W \to e\nu)/\Gamma(\text{total}) \simeq 9\%. \tag{14.88}$$

14.5.3 *Neutral currents, and Z^0 width*

In the GSW theory, weak neutral current processes are mediated by Z°-exchange. The associated weak current transition amplitudes, describing the coupling of leptons and quarks to the Z^0, can be immediately deduced from the $\hat{Z}_\mu$ part of the neutral covariant derivative (14.69):

$$\partial_\mu + ie\hat{A}_\mu Q + \frac{ie}{\sin\theta_W\cos\theta_W}\hat{Z}_\mu(t_3^{(t)} - \sin^2\theta_W Q). \tag{14.89}$$

If we assume that all neutrinos are purely left-handed, with $t = \frac{1}{2}$ and $t_3^{(1/2)} = +\frac{1}{2}$, then we obtain

$$\langle\nu|\hat{j}_\mu^N|\nu\rangle = \frac{e}{\sin\theta_W\cos\theta_W}\frac{1}{2}\,\bar{u}(\nu)\gamma_\mu\frac{1-\gamma_5}{2}u(\nu). \tag{14.90}$$

For the other (massive) fermions, we shall have a left-handed coupling proportional to

$$(t_3^{(1/2)} - \sin^2\theta_W Q)(1-\gamma_5)/2 \tag{14.91}$$

and a right-handed coupling proportional to

$$-\sin^2\theta_W Q(1+\gamma_5)/2 \tag{14.92}$$

since the right-handed components are assumed to be weak isosinglets. From (14.69) and (14.70) we may write generally for the massive fermions

$$\langle f|\hat{j}_\mu^N|f\rangle = \frac{e}{\sin\theta_W\cos\theta_W}\bar{u}(f)\gamma_\mu\left(c_L^f\frac{1-\gamma_5}{2} + c_R^f\frac{1+\gamma_5}{2}\right)u(f) \tag{14.93}$$

†But the (expected) channel $\bar{t}b$ will be affected, if open, by phase space.

with

$$c_{\rm L}^f = t_3^{(1/2)} - \sin^2\theta_{\rm W} Q \tag{14.94a}$$

$$c_{\rm R}^f = -\sin^2\theta_{\rm W} Q \tag{14.94b}$$

or equivalently

$$\langle f|\hat{j}_\mu^{\rm N}|f\rangle = \frac{e}{\sin\theta_{\rm W}\cos\theta_{\rm W}}\bar{u}(f)\gamma_\mu\left(t_3^{(1/2)}\frac{1-\gamma_5}{2} - \sin^2\theta_{\rm W} Q\right)u(f). \tag{14.95}$$

These relations are just the ones given earlier, in §11.3, with the identification $g_{\rm N} = g/\cos\theta_{\rm W} = e/\sin\theta_{\rm W}\cos\theta_{\rm W}$; for example, the negatively charged leptons with $t_3^{(1/2)} = -\frac{1}{2}$ have relative strengths

$$c_{\rm L}^{{\rm e},\mu,\ldots} = -\tfrac{1}{2} + a, \qquad c_{\rm R}^{{\rm e},\mu,\ldots} = a \tag{14.96}$$

while the quarks with $Q = \frac{2}{3}$ and $t_3^{(1/2)} = +\frac{1}{2}$ have strengths

$$c_{\rm L}^{{\rm u,c},\ldots} = \tfrac{1}{2} - \tfrac{2}{3}a, \qquad c_{\rm R}^{{\rm u,c},\ldots} = -\tfrac{2}{3}a \tag{14.97}$$

and those with $Q = -\frac{1}{3}$ and $t_3^{(1/2)} = -\frac{1}{2}$ have strengths

$$c_{\rm L}^{{\rm d,s},\ldots} = -\tfrac{1}{2} + \tfrac{1}{3}a, \qquad c_{\rm R}^{{\rm d,s},\ldots} = +\tfrac{1}{3}a \tag{14.98}$$

with $a = \sin^2\theta_{\rm W}$. The neutrinos have strength $c_{\rm L}^\nu = \frac{1}{2}$, $c_{\rm R}^\nu = 0$, assuming they are massless. The alternative notation

$$\langle f|\hat{j}_\mu^{\rm N}|f\rangle = \frac{e}{\sin\theta_{\rm W}\cos\theta_{\rm W}}\bar{u}(f\gamma_\mu\left(\frac{c_V - c_A\gamma_5}{2}\right)u(f) \tag{14.99}$$

is sometimes used: comparing (14.99) and (14.95), we find

$$c_V = t_3 - 2\sin^2\theta_{\rm W} Q, \tag{14.100}$$

$$c_A = t_3. \tag{14.101}$$

For q^2 such that the momentum dependence of the Z° propagator can be neglected, matrix elements for processes involving four fermions will reduce, as usual, to an effective pointlike current–current form

$$\frac{e^2}{\sin^2\theta_{\rm W}\cos^2\theta_{\rm W} M_{\rm Z}^2} j_\mu^{\rm N} j^{{\rm N}\mu} \tag{14.102}$$

where

$$j_\mu^{\rm N} = \bar{u}(f)\gamma_\mu\left(c_{\rm L}\frac{1-\gamma_5}{2} + c_{\rm R}\frac{1+\gamma_5}{2}\right)u(f) \tag{14.103}$$

and $c_{\rm L} = \frac{1}{2}$, $c_{\rm R} = 0$ for massless neutrinos, while $c_{\rm L}$, $c_{\rm R}$ are given by (14.94a) and (14.94b) for the massive fermions. Since, in this model, $M_{\rm Z}\cos\theta_{\rm W} = M_{\rm W}$, while also $e/\sin\theta_{\rm W} = g$, the factor multiplying the j's in (14.102) is just

$$g^2/M_{\rm W}^2 = 4G_{\rm F}2^{1/2} \tag{14.104}$$

where we have used (10.93).

From (14.102), (14.104) and (14.84), we see that the effective pointlike current–current amplitude for four-fermion processes involving both $W^{\pm}$ and Z^0 exchange can be written as

$$(4G_F/2^{1/2})(j_{\mu}^{C}j^{C\mu} + 2\rho j_{\mu}^{N}j^{N\mu}) \tag{14.105}$$

where

$$\rho = M_W^2/(M_Z^2 \cos^2\theta_W). \tag{14.106}$$

The ratio of the neutral to charged current contribution is determined by the relationship between the $W^{\pm}$ and Z^0 masses, which is a consequence of the assumed form of $\langle\tilde{0}|\hat{\phi}|\tilde{0}\rangle$. With the choice (14.47), ρ is of course 1, from (14.68). An alternative to the choice (14.47) is developed in problem 14.2 and is seen to lead to $\rho \neq 1$.

The width for $Z^0 \to \nu\bar{\nu}$ can be found from (14.87) by replacing $g/2^{1/2}$ by $g/2\cos\theta_W$, and M_W by M_Z, giving

$$\Gamma(Z^0 \to \nu\bar{\nu}) = \frac{1}{24}\frac{g^2}{4\pi}\frac{M_Z}{\cos^2\theta_W} = \frac{G_F}{2^{1/2}}\frac{M_Z^3}{12\pi} \simeq \frac{91\ \text{MeV}}{(\sin 2\theta_W)^3} \simeq 155\ \text{MeV} \tag{14.107}$$

using (14.75). Massive lepton pairs couple with both c_L and c_R terms, leading to

$$\Gamma(Z^0 \to f\bar{f}) = \left(\frac{|c_L^f|^2 + |c_R^f|^2}{6}\right)\frac{g^2}{4\pi}\frac{M_Z}{\cos^2\theta_W} \tag{14.108}$$

where the e^+e^- mode, for example, will have

$$c_L^e = -\tfrac{1}{2} + \sin^2\theta_W, \qquad c_R^e = \sin^2\theta_W \tag{14.109}$$

as before. The values $c_L^{\nu} = \frac{1}{2}$, $c_R^{\nu} = 0$, in (14.108) reproduce (14.107). With (14.75) for $\sin^2\theta_W$, we find

$$\Gamma(Z^0 \to e^+e^-) \simeq 81\ \text{MeV}. \tag{14.110}$$

Quark pairs will couple as in (14.97), (14.98), with a colour factor 3 and Cabibbo-type factors, as for the W. The branching ratios are (problem 14.3)

$$\Gamma(Z^0 \to e^+e^-) : \Gamma(Z^0 \to \nu\bar{\nu}) : \Gamma(Z^0 \to u\bar{u}) : \Gamma(Z^0 \to d\bar{d}) \simeq 1:2:3.5:4.5, \tag{14.111}$$

and the total width is estimated as

$$\Gamma_Z \simeq 30.4\,\Gamma(Z^0 \to e^+e^-) \simeq 2.46\ \text{GeV} \tag{14.112}$$

for three generations, assuming $\Gamma(Z^0 \to t\bar{t}) \simeq 0.25\Gamma(Z^0 \to u\bar{u})$. QCD corrections increase these estimates by a factor $(1 + \alpha_s/\pi) \simeq 1.06$ (cf. (9.80)). The branching ratio for the e^+e^- mode is

$\Gamma(Z^0 \to e^+e^-)/\Gamma_Z \simeq 0.3$.

Cross sections for lepton–lepton scattering proceeding via Z^0 exchange can be calculated (for $q^2 \ll M_Z^2$) from (14.102) by the methods of §§10.5 and 11.2: examples are

$$\nu_\mu e^- \to \nu_\mu e^- \tag{14.113}$$

and

$$\bar{\nu}_\mu e^- \to \bar{\nu}_\mu e^- \tag{14.114}$$

as shown in figure 14.1. Now for an equivalent charged current process such as $\nu_\mu e^- \to \mu^- \nu_e$, in the same pointlike limit, we would replace g^2/M_W^2 by (cf (14.28) and (10.91)) $g^2/2M_W^2$. The neutrino coupling, however, introduces a factor of $\frac{1}{2}$ into the neutral current processes (14.113) and (14.114), so that if only the V–A part c_L in (14.103) contributed, the cross sections for (14.113) and (14.114) could be obtained from that for $\nu_\mu e^- \to \mu^- \nu_e$ as calculated in §10.5 by multiplying by $|c_L|^2$. However, since the neutral current for the electron is not pure V–A, as was the charged current, we expect to see terms involving both $|c_L|^2$ and $|c_R|^2$, and possibly an interference term. The cross section for (14.113) is found to be ('t Hooft 1971c)

$$d\sigma/dy = (2G_F^2 E m_e/\pi)[|c_L^e|^2 + |c_R^e|^2(1-y)^2 - \tfrac{1}{2}(c_R^{e*}c_L^e + c_L^{e*}c_R^e)ym_e/E] \tag{14.115}$$

where E is the energy of the incident neutrino in the 'laboratory' system, and $y = (E - E')/E$, as before, where E' is the energy of the outgoing neutrino in the 'laboratory' system†. (14.115) may be compared with the $\nu_\mu e^- \to \mu^- \nu_e$ (charged current) cross section of (10.122) by noting that $t = -2m_e Ey$: the $|c_L^e|^2$ term agrees with the pure V–A result (10.122), while the $|c_R^e|^2$ term involves the same $(1-y)^2$ factor discussed for $\nu\bar{q}$ scattering in §11.2. The interference term is negligible for $E \gg m_e$. The cross section for the antineutrino process (14.114) is found from (14.115) by interchanging c_L^e and c_R^e.

A third lepton–lepton process is experimentally available,

$$\bar{\nu}_e e^- \to \bar{\nu}_e e^-. \tag{14.116}$$

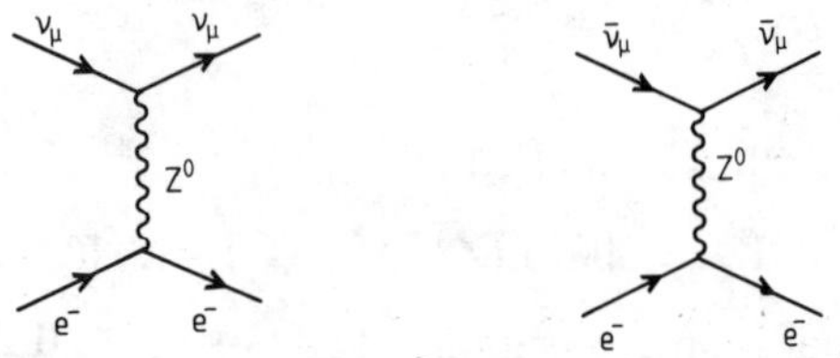

Figure 14.1 Neutrino–electron graphs involving Z^0 exchange.

†In the kinematics, lepton masses have been neglected wherever possible.

In this case there is a single W intermediate state graph to consider as well as the Z^0 one, as shown in figure 14.2. The cross section for (14.116) turns out to be given by an expression of the form (14.115), but with the replacements

$$c_L^e \rightarrow \tfrac{1}{2} + a, \qquad c_R^e \rightarrow a. \tag{14.117}$$

We emphasise once more that all these cross sections are determined in terms of the Fermi constant G_F and only one further parameter, $\sin^2\theta_W$. Current fits to the data will be found in the latest Review of Particle Properties.

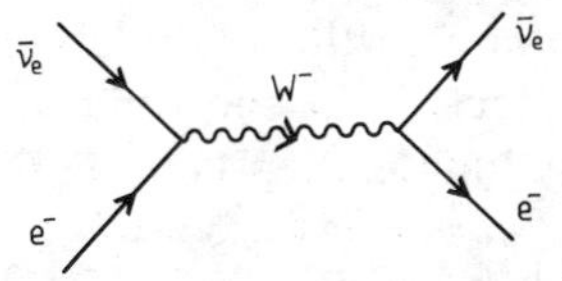

Figure 14.2 One-W annihilation graph in $\bar{\nu}_e e^- \rightarrow \bar{\nu}_e e^-$

14.5.4 *Production cross sections for W and Z*

The existence of $W^{\pm}$ and Z^0 particles, with masses and branching ratios given above, is clearly a crucial test of the GSW theory. In order to produce such heavy particles, one needs at least 100 GeV available CMS energy. This suggests colliding beam machines. The cleanest way to form Z^0 is certainly via the process

$$e^+e^- \rightarrow Z^0 \rightarrow \text{all} \tag{14.118}$$

shown in figure 14.3. However, no e^+e^- colliding beam machines of sufficient energy will be available until LEP and SLC come into operation (but see §14.5.6 for evidence that the effect of the Z^0 is seen at PETRA in $e^+e^- \rightarrow \mu^+\mu^-$).

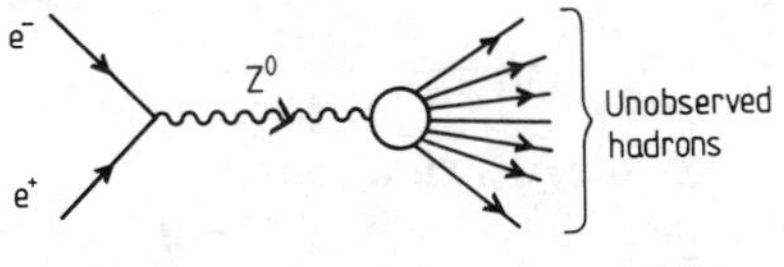

Figure 14.3 One-Z^0 annihilation graph in $e^+e^- \rightarrow Z^0 \rightarrow$ all.

First, however, we discuss W and Z^0 production in the CERN $\bar{p}p$ collider, where they were, in fact, first discovered (UA1 collaboration, Arnison *et al* 1983a, b; UA2 collaboration, Banner *et al* 1983b, Bagnaia *et al* 1983). Indeed, the possibility of such discoveries was the principal motivation for transforming the CERN SPS into a $\bar{p}p$ collider using the stochastic cooling technique (Rubbia *et al* 1977, Staff of the CERN $\bar{p}p$ project 1981). Estimates of W and Z^0 production in $\bar{p}p$ collisions may be obtained (see, for example, Quigg 1977) from the parton model, in a way analogous to that used for the Drell–Yan process in §7.4, with γ replaced by W or Z^0, as shown in figure 14.4 (cf figure 7.12), and for two-jet cross sections in §9.4.1. As in (9.48), we denote by $\hat{s}$ the subprocess invariant

$$\hat{s} = (x_1 p_1 + x_2 p_2)^2 = x_1 x_2 s \tag{14.119}$$

for massless partons. With $\hat{s}^{1/2} = M_W \sim 80$ GeV and $s^{1/2} = 540$ GeV $\bar{p}p$ collider energy, we see that the x's are typically ~ 0.15, so that the valence q's in the proton and $\bar{q}$'s in the antiproton will dominate (at $\sqrt{s} = 1.6$ TeV, appropriate to the Fermilab Tevatron, $x \simeq 0.05$ and the sea quarks will be expected to contribute). The parton model cross section $p\bar{p} \to W^\pm$ + anything is then (setting $\theta_C \approx 0$)

$$\sigma(p\bar{p} \to W^\pm + X) = \tfrac{1}{3}\int_0^1 dx_1 \int_0^1 dx_2 \hat{\sigma}(x_1, x_2)\begin{Bmatrix} u(x_1)\bar{d}(x_2) + \bar{d}(x_1)u(x_2) \\ \bar{u}(x_1)d(x_2) + d(x_1)\bar{u}(x_2) \end{Bmatrix} \tag{14.120}$$

where the $\frac{1}{3}$ is the same colour factor as in the Drell–Yan process, and the subprocess cross section $\hat{\sigma}$ is (problem 14.4)

$$\begin{aligned}\hat{\sigma} &= 4\pi^2\alpha\cdot(1/4\sin^2\theta_W)\delta(\hat{s} - M_W^2) \\ &= \pi 2^{1/2} G_F M_W^2 \delta(x_1 x_2 s - M_W^2).\end{aligned} \tag{14.121}$$

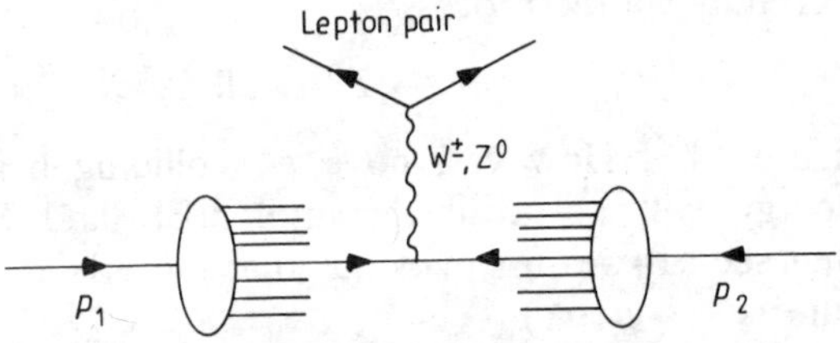

Figure 14.4 Parton model amplitude for $W^\pm$ or Z^0 production in $\bar{p}p$ collisions.

QCD corrections to (14.120) must also be considered. Leading logarithms will make the distributions in (14.120) Q^2-dependent, and they should be evaluated at $q^2 = M_W^2$. There will also be further (non-leading logarithm) corrections, which are conventionally accounted for by a multiplicative factor K (Altarelli *et al* 1979, Parisi 1980), which is of order 1.5 in this case. (14.120) then gives, for the total cross sections at $\sqrt{s} = 540$ GeV,

$$\sigma(W) = \sigma(W^+) + \sigma(W^-) \simeq 3.6 \times 10^{-33}\ \text{cm}^2. \tag{14.122}$$

A similar calculation can be done for the Z^0, and one finds

$$\sigma(Z^0) \simeq 10^{-33}\ \mathrm{cm}^2. \tag{14.123}$$

Multiplying (14.122) and (14.123) by the branching ratios (see (14.88) and (14.112)) gives

$$\sigma(p\bar{p} \to W + X \to e\nu X) \simeq 0.36\ \mathrm{nb} \tag{14.124}$$

$$\sigma(p\bar{p} \to Z^0 + X \to e^+e^-X) \simeq 0.03\ \mathrm{nb} \tag{14.125}$$

at $\sqrt{s} = 540$ GeV (1 nb $= 10^{-33}$ cm^2). More precise estimates are given in Altarelli *et al* (1984, 1985).

The total cross section for $p\bar{p}$ is about 70 mb at these energies: hence (14.124) represents $\sim 10^{-8}$ of the total cross section, and (14.125) is 10 times smaller. The rates could of course be increased by using the $q\bar{q}$ modes of W and Z^0, which have bigger branching ratios. But the detection of these is very difficult, being very hard to distinguish from conventional two-jet events produced via the mechanism discussed in §9.4, which has a cross section some 10^3 higher than (14.124). W and Z^0 would appear as slight shoulders on the edge of a very steeply falling invariant mass distribution, similar to that shown in figure 7.13, and the calorimetric jet energy resolution capable of resolving such an effect will be hard to achieve. Thus despite the unfavourable branching ratios, the leptonic modes provide the better signatures, as discussed further in §14.6 below.

In the future, the process (14.118) is likely to be the preferred one for Z^0 production. Since the width of the Z^0 is expected to be considerably wider than the experimental energy resolution of an e^+e^- machine at these energies ($\sim$50 MeV), we can use a Breit–Wigner form for the peak cross section

$$\sigma_{Z^0} = \frac{4\pi}{s_0} 3B(Z^0 \to e^+e^-) \tag{14.126}$$

where $s_0 = M_Z^2$ and B is the branching ratio $\Gamma(Z^0 \to e^+e^-)/\Gamma(Z^0 \to \mathrm{all})$. It follows that, at the Z^0 peak,

$$\frac{\sigma_{Z^0}}{\sigma_{\mathrm{QED}}} = \frac{9}{\alpha^2} B(Z^0 \to e^+e^-) \tag{14.127}$$

which should be a factor of about 5000. With luminosities $\sim 10^{32}$ cm^{-2} s^{-1}, this implies that something like five background-free Z^0 events per second at the peak should be seen. (Radiative corrections may reduce the estimate by a factor of about 2; see, for example, Ellis (1979) and Llewellyn Smith (1979).) The consequent high statistics should enable detailed determinations of the weak neutral current couplings to fermions.

The reaction $e^+e^- \to W^+W^-$ is of great interest, because three diagrams contribute importantly (figure 14.5) and there are definite relations, predicted by the underlying gauge symmetry, between the coupling constants involved in these diagrams. These relations turn out to imply (Alles *et al* 1977, Sushkov *et al* 1975) considerable cancellation among the diagrams, so that the cross sections would be much larger if, for example, the $W^\pm$ did not have the particular magnetic moment predicted by the theory (see §14.7 below), or if the ZWW vertex were different. This cross section, then, directly tests the gauge structure.

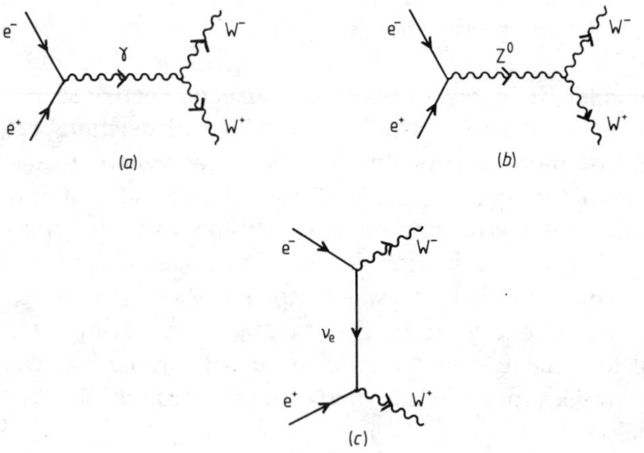

Figure 14.5 Diagrams contributing to $e^+e^- \to W^+W^-$ in lowest order: (*a*) one-γ annihilation; (*b*) one-Z^0 annihilation; (*c*) ν_e exchange.

14.5.5 Charge asymmetry in $W^\pm$ decay

At energies such that the simple valence quark picture of (14.120) is valid, the W^+ is created in the annihilation of a left-handed u quark from the proton and a right handed $\bar{d}$ quark from the $\bar{p}$, according to the fundamental *V–A* couplings introduced in §14.3. In the $W^+ \to e^+\nu_e$ decay, a right-handed e^+ and left-handed ν_e are emitted. Referring to figure 14.6, we see that angular momentum conservation allows e^+ production parallel to the direction of the antiproton, but forbids it parallel to the direction of the proton. Similarly, in $W^- \to e^-\bar{\nu}_e$, the e^- is emitted preferentially parallel to the proton (these considerations are exactly similar to those mentioned in §11.2 with reference to νq and $\bar{\nu} q$ scattering). The actual distribution can be inferred from (11.26b)—which is also a process of the form $f_1\bar{f}_2 \to f_3\bar{f}_4$—and is $\sim(1+\cos\theta)^2$, where θ is the angle between the e^- and the p (for $W^- \to e^-\bar{\nu}_e$) or the e^+ and the $\bar{p}$ (for $W^+ \to e^+\nu_e$).

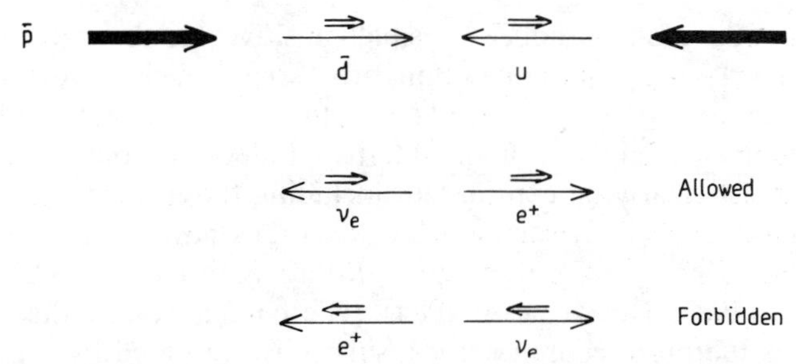

Figure 14.6 Preferred direction of leptons in W^+ decay.

14.5.6 Electroweak interference effects

Although the energy of the PETRA e^+e^- storage ring at DESY is insufficient to create real Z^0's, the effect of the 'tail' of the Z^0 can be seen, via interference effects between the standard QED (one-γ) process of figure 14.7(a) and the GSW (one-Z^0) process of figure 14.7(b). The main effect accessible at present is the forward–backward asymmetry defined as

$$A = (N_F - N_B)/(N_F + N_B) \tag{14.128}$$

where N_F (N_B) are the numbers of events with a positive lepton in the forward (backward) hemisphere, with respect to the incident e^+. From the Born graphs of figure 14.7, A is found to be (Budny 1975)

$$A = (-\tfrac{3}{2}c_A^e c_A^\mu \operatorname{Re}\chi + 3c_V^e c_V^\mu c_A^e c_A^\mu |\chi|^2)/R_{\mu\mu} \tag{14.129}$$

where

$$R_{\mu\mu} = 1 - 2c_V^e c_V^\mu \operatorname{Re}\chi + (c_V^{e2} + c_A^{e2})(c_V^{\mu 2} + c_A^{\mu 2})|\chi|^2 \tag{14.130}$$

and

$$\chi = s/[4\sin^2\theta_W \cos^2\theta_W (M_Z^2 - s - iM_Z\Gamma_Z)] \tag{14.131}$$

s being $(p_{e^+} + p_{e^-})^2$ as usual. For $\sin^2\theta_W \simeq 0.25$ we see from (14.100) that the c_V's are ≈ 0, so that the effect is controlled essentially by the first term in (14.129). At $\sqrt{s} = 29$ GeV it gives $A \approx -0.063$.

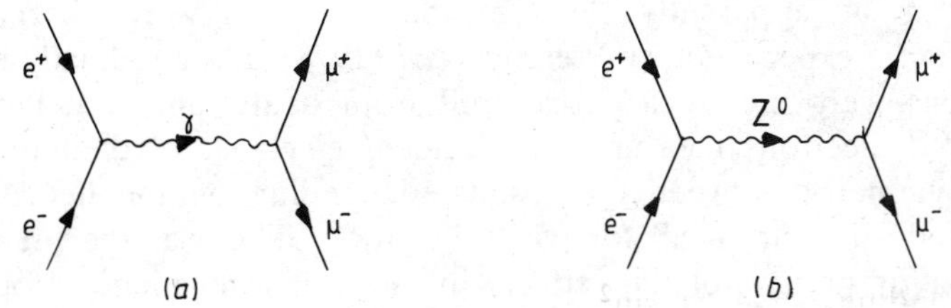

Figure 14.7 (a) One-γ and (b) one-Z^0 graphs contributing to $e^+e^- \to \mu^+\mu^-$.

However, QED alone produces a small positive A, through interference between 1γ and 2γ annihilation processes (which have different charge conjugation parity), as well as between initial and final state bremsstrahlung corrections to figure 14.7(a). Indeed, *all* one-loop radiative effects must clearly be considered, including those affecting $\sin^2\theta_W$, (unless *defined* via (14.78)), and such calculations have been performed (Wetzel 1983, Lynn and Stuart 1985, Böhm and Hollik 1982, 1984, Berends *et al* 1983, Decker *et al* 1984). The comparison of theory and experiment is therefore complicated; a survey is presented by Cashmore *et al* (1986). As an illustration, in figure 14.8 we show A versus s (note that (14.129) predicts $A \propto s$, approximately, for $s \ll M_Z^2$), taken from Adeva *et al* (1985), the broken curve being the GSW prediction with $\sin^2\theta_W = 0.22$ (defined by M_W/M_Z) and $M_Z = 93$ GeV.

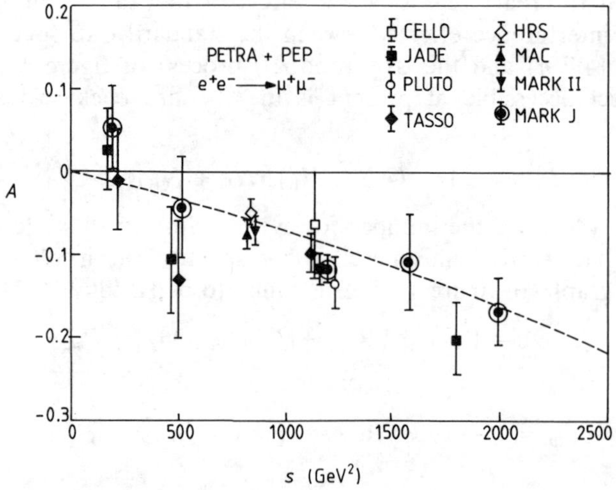

Figure 14.8 A versus s (from Adeva *et al* 1985).

The asymmetry A is not, in fact, direct evidence for parity violation in $e^+e^- \to \mu^+\mu^-$, since we see from (14.129) that it is even under $c_A \to -c_A$, whereas a true parity-violating effect would involve terms odd (linear) in c_A. However, electroweak-induced parity violation effects in an apparently electromagnetic process were observed in a remarkable experiment by Prescott *et al* (1978). Longitudinally polarised electrons were inelastically scattered from deuterium, and the flux of scattered electrons measured for incident electrons of definite helicity. An asymmetry between the results, depending on the helicities, was observed—a clear signal for parity violation. This was the first demonstration of parity-violating effects in an 'electromagnetic' process; the corresponding value of $\sin^2\theta_W$ is in agreement with that determined from ν data.

14.6 Discovery of the $W^\pm$ and Z^0 at the $p\bar{p}$ collider, and their properties

As already indicated in §14.5.4, the best signatures for W and Z production in $p\bar{p}$ collisions are provided by the leptonic modes

$$p\bar{p} \to W^\pm X \to e^\pm \nu X \tag{14.132}$$

$$p\bar{p} \to Z^0 X \to e^+e^- X. \tag{14.133}$$

(14.132) has the larger cross section, by a factor of 10 (cf (14.124) and (14.125)), and was observed first (UA1, Arnison *et al* 1983a; UA2, Banner *et al* 1983a). However, the kinematics of (14.133) is simpler and so the Z^0 discovery (UA1, Arnison *et al* (1983); UA2, Bagnaia *et al* 1983) will be discussed first.

The signature for (14.133) is of course an isolated, and approximately back-to-back, e^+e^- pair with invariant mass peaked around 92 GeV (cf (14.83)). Very clean events can be isolated by imposing a modest transverse energy cut—the e^+e^- pairs we require are coming from the decay of a massive relatively slowly moving Z^0. Figure 14.9 shows one of the original UA1 events illustrating this. Figure 14.10 shows the transverse energy distribution of a candidate Z^0 event from the first UA2 sample. Figure 14.11 shows (Geer 1986) the invariant mass distribution for a later sample of 14 UA1 events in which both electrons have well measured energies, together with the Breit–Wigner resonance curve appropriate to $M_Z = 93\ \mathrm{GeV}/c^2$, with experimental mass resolution folded in. The full result for the Z^0 mass was

$$M_Z = 93.0 \pm 1.4\ (\text{stat.}) \pm 3.2\ (\text{syst.})\ \mathrm{GeV}. \tag{14.134}$$

The corresponding UA2 result (di Lella 1986), based on 13 well measured pairs, was

$$M_Z = 92.5 \pm 1.3\ (\text{stat.}) \pm 1.5\ (\text{syst.})\ \mathrm{GeV}. \tag{14.135}$$

In both cases the systematic error reflects the uncertainty in the absolute calibration of the calorimeter energy scale. Clearly the agreement with (14.83b) is excellent.

Cross sections for (14.133) are listed in table 14.2 (di Lella 1986). Within the rather larger errors, the two experiments are consistent with each other and with the expectation (14.123).

The total Z^0 width Γ_Z is an interesting quantity. If we assume that, for any fermion family additional to the three known ones, only the neutrinos are significantly less massive than $M_Z/2$, we have

$$\Gamma_Z \simeq (2.6 + 0.16\Delta N_\nu)\ \mathrm{GeV} \tag{14.136}$$

from §14.5.3, where ΔN_ν is the number of additional light neutrinos (i.e. beyond ν_e, ν_μ and ν_τ). Thus (14.136) can be used as an important measure of such neutrinos (i.e. generations?) if Γ_Z can be determined

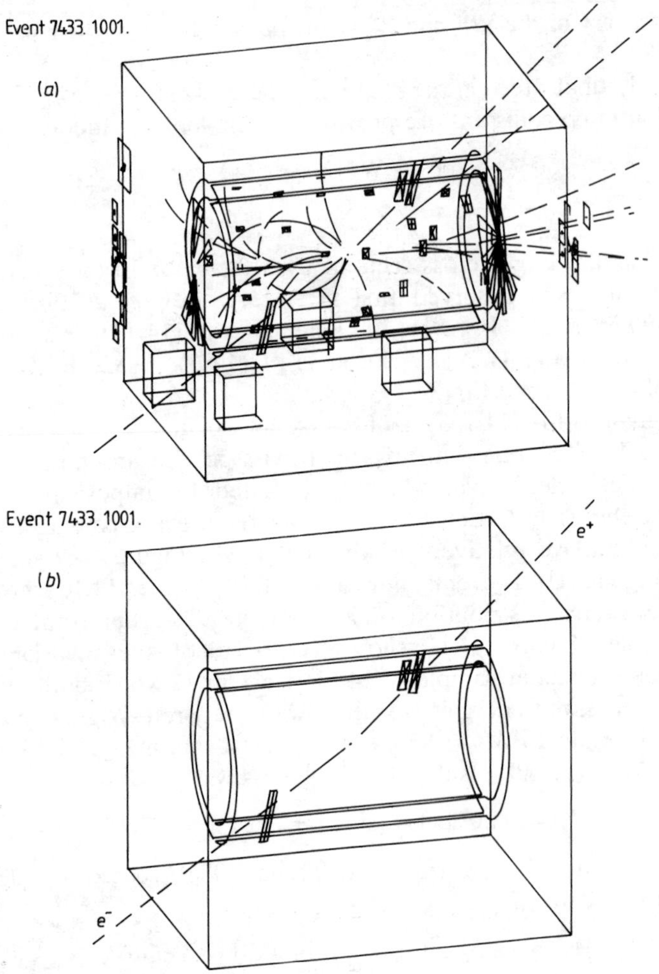

Figure 14.9 Event display from UA1 (Arnison *et al* 1983b). In (a) all reconstructed vertex-associated tracks and all calorimeter hits are displayed. In (b) thresholds are raised to $p_T > 2$ GeV/c for charged tracks and to $E_T > 2$ GeV for calorimeter hits; only the electron pair survives these mild cuts.

accurately enough. At the time of writing, the mass resolution in the $\bar{p}p$ experiments is of the same order as the total expected width (~3 GeV/c^2), so that (14.136) is difficult to use directly. The advent of LEP and the SLC, with energy resolution in the region of 50 MeV/c^2, will provide crucial new data bearing on (14.136).

We turn now to the $W^{\pm}$. In this case an invariant mass plot is impossible, since we are looking for the $e\nu$ ($\mu\nu$) mode, and cannot measure the ν's. However, it is clear that—as in the case of $Z^0 \rightarrow e^+e^-$ decay—slow moving massive W's will emit isolated electrons with high

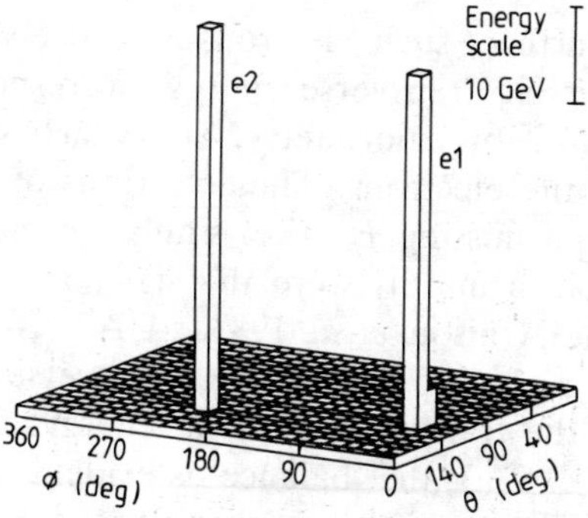

Figure 14.10 The cell transverse energy distribution for a $Z^0 \rightarrow e^+e^-$ event (UA2, Banner *et al* 1983b) in the θ, ϕ plane, where θ and ϕ are the polar and azimuthal angles relative to the beam axis.

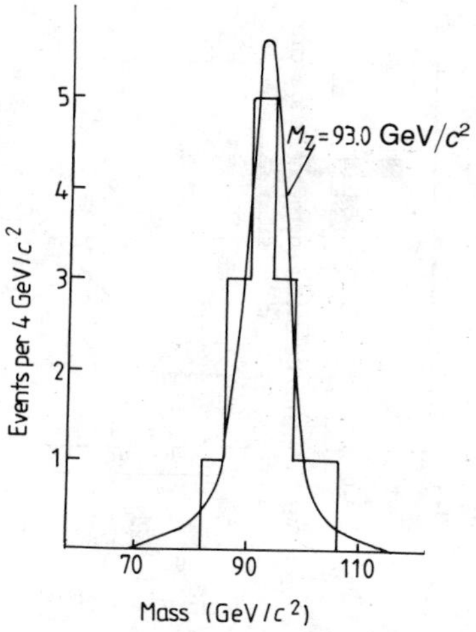

Figure 14.11 Invariant mass distribution for 14 well measured $Z^0 \rightarrow e^+e^-$ decays (UA1, Geer 1986).

Table 14.2 $Z \rightarrow e^+e^-$ cross sections (di Lella 1986).

	σ_Z^e(pb)	
Source	$\sqrt{s}$ = 546 GeV	$\sqrt{s}$ = 630 GeV
UA1	$41 \pm 20 \pm 6$	$85 \pm 23 \pm 13$
UA2	$110 \pm 39 \pm 9$	$52 \pm 19 \pm 4$
Theory	42^{+13}_{-6}	51^{+16}_{-10}

The second error is the systematic uncertainty.

transverse energy. Further, such electrons should be produced in association with large *missing* transverse energy (corresponding to the ν's), which can be measured by calorimetry, and which should balance the transverse energy of the electrons. Thus electrons of high E_T accompanied by balancing high missing E_T (i.e. similar in magnitude to that of the e^- but opposite in azimuth) were the signatures used for the early event samples (UA1, Arnison *et al* 1983a; UA2, Banner *et al* 1983b). Figure 14.12 shows a plot of the missing transverse energy parallel to the electron, versus the transverse electron energy, from the first UA1 paper (Arnison *et al* 1983a); the 'balance' is evident. Figure 14.13 shows one of the lone high-transverse-energy electrons from the first UA2 paper (Banner *et al* 1983b), illustrating the cleanness of the isolation of the e^- in the calorimetric system.

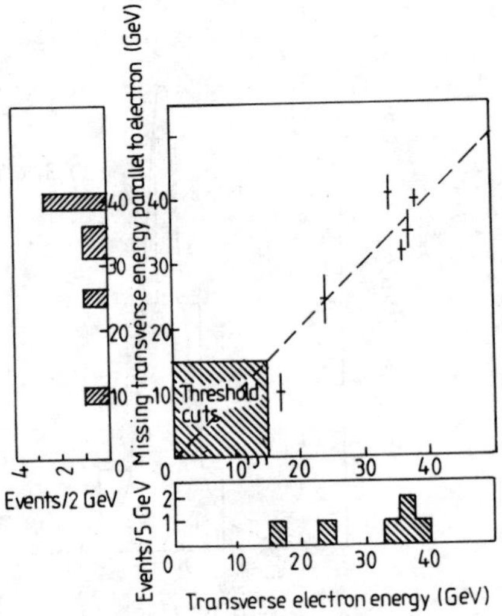

Figure 14.12 Missing transverse energy component parallel to the electron, versus transverse electron energy, for original six $W \rightarrow e\nu$ events without jets (UA1, Arnison *et al* 1983a).

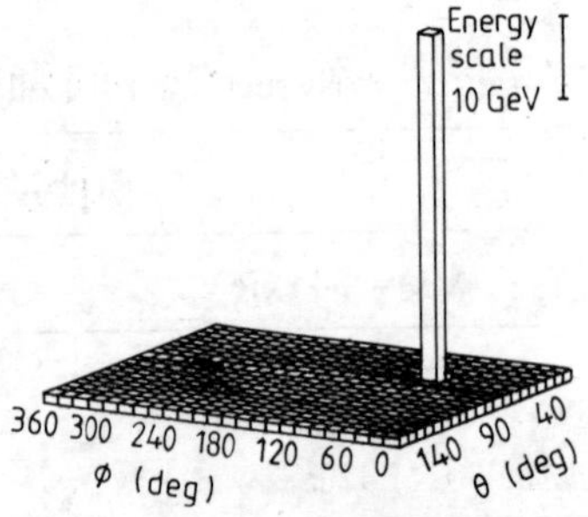

Figure 14.13 The cell distribution as a function of polar angle θ and azimuth ϕ for an early $W \rightarrow e\nu$ event candidate (UA2, Banner *et al* 1983b).

The determination of the mass of the W is not quite so straightforward as that of the Z, since we cannot construct directly an invariant mass plot for the eν pair: only the missing transverse momentum (or energy) can be attributed to the ν, since some unidentified longitudinal momentum will always be lost down the beam pipe. We note from figure 14.12, however, that the transverse energy spectra for both leptons have end-points around 40 GeV (as confirmed in the more extensive later data), which strongly suggests the decay of an object of mass $\sim$80 GeV/c^2. In fact, the distribution of events in p_{eT}, the transverse momentum of the e$^-$, should show a pronounced peak at $p_{\mathrm{eT}} \approx \frac{1}{2}M_{\mathrm{W}}$, as is easily seen from the following argument. Consider the decay of a W at rest (figure 14.14). We have $|\boldsymbol{p}_{\mathrm{e}}| = \frac{1}{2}M_{\mathrm{W}}$ and $|\boldsymbol{p}_{\mathrm{eT}}| = \frac{1}{2}M_{\mathrm{W}}\sin\theta$. Thus the transverse momentum distribution is given by

$$\frac{\mathrm{d}\sigma}{\mathrm{d}p_{\mathrm{eT}}} = \frac{\mathrm{d}\sigma}{\mathrm{d}\cos\theta}\frac{\mathrm{d}\cos\theta}{\mathrm{d}p_{\mathrm{eT}}} = \frac{\mathrm{d}\sigma}{\mathrm{d}\cos\theta}\left(\frac{2p_{\mathrm{eT}}}{M_{\mathrm{W}}}\right)\left(\tfrac{1}{4}M_{\mathrm{W}}^2 - p_{\mathrm{eT}}^2\right)^{-1/2} \quad (14.137)$$

and the last factor in (14.137) is peaked at $p_{\mathrm{eT}} = \frac{1}{2}M_{\mathrm{W}}$. This peak will be smeared by the width, and transverse motion, of the W. Early

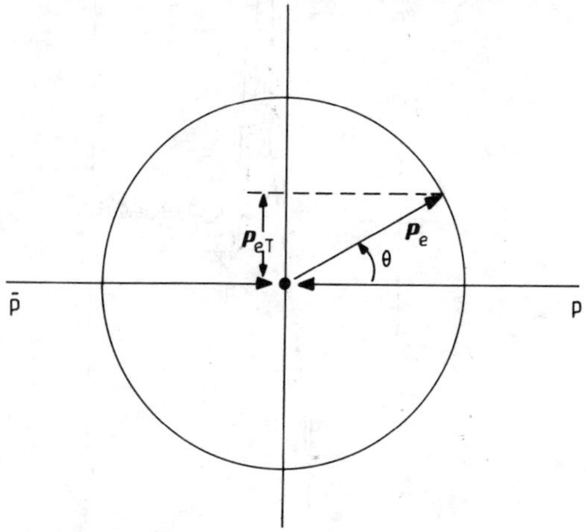

Figure 14.14 Kinematics of W $\rightarrow$ eν decay.

determinations of M_{W} used (14.137), but sensitivity to the transverse momentum of the W can be much reduced (Barger *et al* 1983) by considering instead the distribution in 'transverse mass', defined by

$$M_{\mathrm{T}}^2 = (E_{\mathrm{eT}} + E_{\nu\mathrm{T}})^2 - (\boldsymbol{p}_{\mathrm{eT}} + \boldsymbol{p}_{\nu\mathrm{T}})^2 \simeq 2p_{\mathrm{eT}}p_{\nu\mathrm{T}}(1 - \cos\phi) \quad (14.138)$$

where ϕ is the azimuthal separation between p_{eT} and $p_{\nu\mathrm{T}}$. A Monte Carlo simulation is used to generate M_{T} distributions for different values of M_{W}, and the most probable value is found by a maximum likelihood fit. Figure 14.15 shows the M_{T} distribution for a sample of 86 UA1 events for which both p_{eT} and $p_{\nu\mathrm{T}}$ exceed 30 GeV/c, with the fit corresponding to $M_W = 83.5\ \text{GeV}/c^2$. The quoted results were

$$\text{UA1 (Geer 1986)} \quad M_W = 83.5 \pm {}^{1.1}_{1.0}\ (\text{stat.}) \pm 2.8\ (\text{syst.})\ \text{GeV} \tag{14.139}$$

$$\text{UA2 (di Lella 1986)} \quad M_W = 81.2 \pm 1.1\ (\text{stat.}) \pm 1.3\ (\text{syst.})\ \text{GeV} \tag{14.140}$$

the systematic errors again reflecting uncertainty in the absolute energy scale of the calorimeters. The two experiments also quote (Geer 1986, di Lella 1986)

$$\left.\begin{array}{ll}\text{UA1} & \Gamma_W < 6.5\ \text{GeV} \\ \text{UA2} & \Gamma_W < 7.0\ \text{GeV}\end{array}\right\} 90\%\ \text{c.l.} \tag{14.141}$$

Once again, the agreement between the experiments, and of both with (14.83a) and (14.112), is excellent.

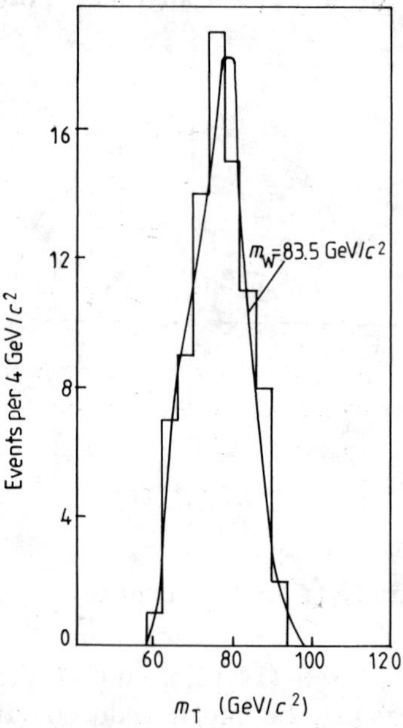

Figure 14.15 Electron–neutrino transverse mass distribution for a subsample of 86 events in which both the electron and the neutrino have a transverse energy greater than 30 GeV (UA1, Arnison *et al* 1986).

Cross sections for $W \to e\nu$ are given in table 14.3 (di Lella 1986); they are consistent with (but systematically higher than) the theoretical predictions (Altarelli *et al* 1984, 1985), but the latter have substantial uncertainties associated with ignorance of the precise form of the

Table 14.3 $W \to e\nu$ cross sections (di Lella 1986).

Source	σ_W^e(nb) $\sqrt{s}$ = 546 GeV	σ_W^e(nb) $\sqrt{s}$ = 630 GeV	$r = \frac{\sigma_W^e(630 \text{ GeV})}{\sigma_W^e(546 \text{ GeV})}$
UA1	$0.55 \pm 0.08 \pm 0.09$	$0.63 \pm 0.05 \pm 0.09$	1.15 ± 0.19
UA2	$0.50 \pm 0.09 \pm 0.05$	$0.53 \pm 0.06 \pm 0.05$	1.06 ± 0.23
Theory	$0.36^{+0.11}_{-0.05}$	$0.45^{+0.14}_{-0.08}$	1.26 ± 0.02

The second error is the systematic uncertainty.

structure functions, and with higher-order QCD effects.

Using (14.78), a value of $\sin^2\theta_W$ can be obtained which is free of the systematic error in the energy scale. The results are (di Lella 1986)

$$\text{UA1} \quad \sin^2\theta_W = 0.194 \pm 0.031 \text{ (stat.)} \tag{14.142}$$

$$\text{UA2} \quad \sin^2\theta_W = 0.229 \pm 0.030 \text{ (stat.)} \tag{14.143}$$

with weighted average

$$\sin^2\theta_W = 0.212 \pm 0.022 \text{ (stat.)} \tag{14.144}$$

in excellent agreement with the ν scattering value (14.75). Equations (14.139) and (14.140) can also be used to extract $\sin^2\theta_W$ from the combined W and Z data of each experiment. This reduces the statistical error in (14.144), but reintroduces the systematic error: the results are in excellent agreement with (14.144).

By using the definition (14.78), we are implicitly assuming that the ρ parameter of (14.106), evaluated at M_W^2, is equal to unity. This can be checked by combining (14.106) and (14.82) to form

$$\rho \equiv M_W^2/(M_Z^2\cos^2\theta_W) = M_W^2/[M_Z^2(1 - (38.6/M_W)^2)]. \tag{14.145}$$

Quoted results are (di Lella 1986)

$$\text{UA1} \quad \rho = 1.028 \pm 0.037 \text{ (stat.)} \pm 0.019 \text{ (syst.)} \tag{14.146}$$

$$\text{UA2} \quad \rho = 0.996 \pm 0.033 \text{ (stat.)} \pm 0.009 \text{ (syst.)} \tag{14.147}$$

Although one-loop radiative corrections have been included in the above, as explained in §14.5.1, the present statistical and systematic efforts are such that the need for such corrections cannot, yet, be demonstrated (however, the more accurate UA2 data incline to a preference for the corrected values). It will be of great interest to check the details of the radiative corrections when data from LEP and the SLC become available.

Finally, figure 14.16 shows (Arnison *et al* 1986) the angular distribution of the charged lepton in $W \to e\nu$ decay (see §14.5.5); θ_e^* is the

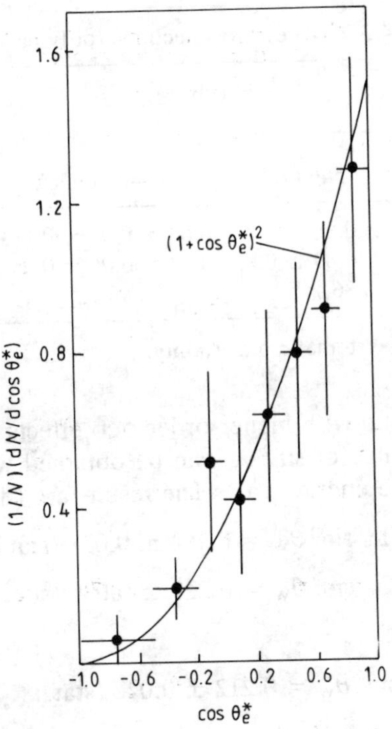

Figure 14.16 The W decay angular distribution of the emission angle θ_e^* of the positron (electron) with respect to the antiproton (proton) beam direction, in the rest frame of the W.; 75 events; background subtracted and acceptance corrected.

$e^+(e^-)$ angle in the W rest frame, measured with respect to a direction parallel (antiparallel) to the $\bar{p}(p)$ beam. The expected form $(1 + \cos\theta_e^*)^2$ is followed very closely.

In summary, these remarkable and beautiful results constitute an extraordinarily convincing confirmation of the principal expectations of the GSW theory, as outlined in §14.5.

We now consider some further aspects of the theory.

14.7 Feynman rules for tree graphs in GSW theory and the Higgs sector

We have already derived, in §§14.2–14.5, many of the vertices and propagators needed to construct tree graphs of this theory. The charged and neutral currents have been given; the vector boson propagators have the general form (13.92), but we shall always work in the gauge $\xi = 1$. As in the case of QCD, there are also the trilinear and quadrilinear gauge boson vertices to consider. These refer only to the non-Abelian $SU(2)_L$ part of the electroweak group, which has just the same group structure as the non-Abelian model considered in Chapter 8. In particular, the 3-W vertex has the form

$$-g\varepsilon_{abc}\omega_1^a\omega_2^b\omega_3^c[(\varepsilon_1\cdot\varepsilon_2)(k_1-k_2)\cdot\varepsilon_3+(\varepsilon_2\cdot\varepsilon_3)(k_2-k_3)\cdot\varepsilon_1 + (\varepsilon_3\cdot\varepsilon_1)(k_3-k_1)\cdot\varepsilon_2] \quad (14.148)$$

where $SU(2)_L$ 'polarisation vectors' ω_i have been introduced for the W's.

Consider the specific case in which the boson labelled with subscript '3' is also the neutral one with $SU(2)_L$ index 3 as well:

$$\omega_3^c=\delta^{c3}. \quad (14.149)$$

Now from (14.64) and (14.66) we have

$$W^{3\mu}=\sin\theta_W A^\mu+\cos\theta_W Z^\mu \quad (14.150)$$

so that this 3-W coupling implies the physical γWW vertex

$$\mathrm{i}e(\omega_1^-\omega_2^+-\omega_1^+\omega_2^-)[(\varepsilon_1\cdot\varepsilon_2)(k_1-k_2)\cdot\varepsilon_\gamma+(\varepsilon_2\cdot\varepsilon_\gamma)(k_2-k_\gamma)\cdot\varepsilon_1 + (\varepsilon_\gamma\cdot\varepsilon_1)(k_\gamma-k_1)\cdot\varepsilon_2] \quad (14.151)$$

where we have introduced the combinations

$$\omega^\pm=(\omega^1\pm \mathrm{i}\omega^2)/2^{1/2}. \quad (14.152)$$

The term $\omega_1^-\omega_2^+$ corresponds to the absorption of a W^+ (particle 1) and absorption of a W^- (particle 2)—or emission of W^-, W^+. This is the *unique* electromagnetic coupling to the W's, according to the GSW theory; in particular, the magnetic moment of the W's is predicted by (14.151) (cf §8.4).

The 4-W vertex is the same as (8.160). This is not the end of the story, however. A nasty surprise is in store for the dedicated reader who, at this point, might venture to calculate the amplitude for WW scattering—analogous to the 'W–W' amplitude discussed in §8.4, but this time involving *massive* vector particles. She will find that the sum of the tree graphs shown in figure 14.17 *still* carries the disease gauge invariance set out to cure—namely, their sum violates unitarity at high energy, due to non-cancellation of longitudinal polarisation states. This seems to raise, once again, the question of renormalisability. However, all is in fact taken care of, since these are no ordinary, unsophisticated massive vector particles. Their mass has been contrived in such a way as to *respect* gauge invariance—in the hidden sense of Chapter 13. This was done by insisting on the presence of a Higgs scalar field $\hat{\phi}$, and it turns out that there is a coupling of the vector particles to the Higgs field which is of order g, and consequently an additional graph of order g^2 contributes to WW scattering (figure 14.18). This graph restores acceptable high-energy behaviour.

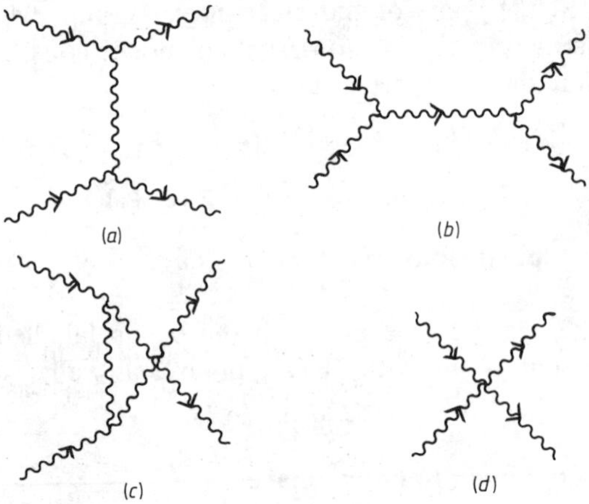

Figure 14.17 Graphs contributing to WW scattering at order g^2: (*a*) t channel; (*b*) s channel; (*c*) u channel; (*d*) four-W vertex.

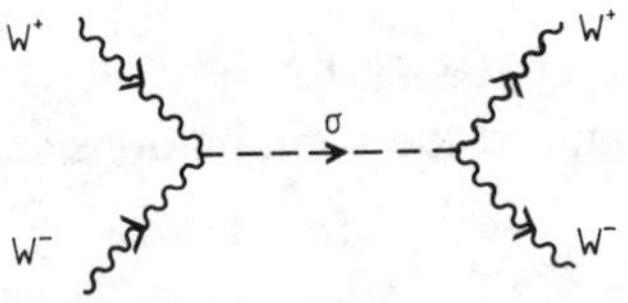

Figure 14.18 Exchange of Higgs scalar σ in WW scattering.

This leads us to a more detailed discussion of the Higgs sector of the theory. The way in which a Higgs coupling, such as the one shown in figure 14.18, arises is precisely analogous to our discussion in §13.4 of the coupling of the U(1) gauge field $\hat{A}^\mu$ and the *deviation* $\hat{\rho}$ of the Higgs field from its vacuum expectation value. In the case of $SU(2)_L \times U(1)$, the three *phase* degrees of freedom of the complex Higgs doublet have been 'eaten' to give the $W^\pm$ and Z^0 mass, and the remaining physical degree of freedom of the Higgs field corresponds to fluctuations σ about the vacuum expectation value f.

Since this is a new piece of physics not discussed in Chapter 13, we begin with the simpler U(1) case. Let us examine the properties of the Higgs field in the *absence* of any gauge field. By an appeal to the example of spontaneous magnetisation and the mathematical model for its free energy density, we suggested in Chapter 13 that there must be an analogous 'potential' function for the complex Higgs field $\hat{\phi}$

$$\hat{V}(\hat{\phi}) = -\tfrac{1}{2}\mu^2\hat{\phi}^\dagger\hat{\phi} + \tfrac{1}{2}(\mu^2/f^2)(\hat{\phi}^\dagger\hat{\phi})^2 \qquad (13.54)$$

which was used to make plausible the possibility of $\langle\tilde{0}|\hat{\phi}|\tilde{0}\rangle \neq 0$. We now want to take it rather more seriously, and use it to discuss the dynamics of the $\hat{\phi}$ field. To do this, we shall *assume* that the complete Lagrangian for $\hat{\phi}$ is

$$\hat{\mathscr{L}}_{\phi} = \hat{T}-\hat{V} = \tfrac{1}{2}\partial_{\mu}\hat{\phi}^{\dagger}\partial^{\mu}\hat{\phi} + \tfrac{1}{2}\mu^{2}\hat{\phi}^{\dagger}\hat{\phi} - \tfrac{1}{2}(\mu^{2}/f^{2})(\hat{\phi}^{\dagger}\hat{\phi})^{2}. \quad (14.153)$$

The Euler–Lagrange equations then give

$$(\Box - \mu^{2})\hat{\phi} = -(2\mu^{2}/f^{2})(\hat{\phi}^{\dagger}\hat{\phi})\hat{\phi} \quad (14.154)$$

which is almost the equation for a scalar field of mass μ with a self-interaction—but not quite! The *sign* of the μ^2 term is wrong for that interpretation: a genuine mass term would appear in $\hat{V}$ as $+\frac{1}{2}\mu^{2}\hat{\phi}^{\dagger}\hat{\phi}$, not $-\frac{1}{2}\mu^{2}\hat{\phi}^{\dagger}\hat{\phi}$, and correspondingly in $\hat{\mathscr{L}}_{\phi}$. With such a genuine mass term the potential has only the single minimum at $\hat{\phi} = 0$, so that $\langle 0|\hat{\phi}|0\rangle$ would be zero: the change in sign of the 'mass' term $\hat{\phi}^{\dagger}\hat{\phi}$ is entirely analogous to changing from one side of T_{c} to the other in our magnetisation example (see figure 13.1). It is thus the characteristic signal for the hidden symmetry situation.

In our case we wish to insist on $\langle\tilde{0}|\hat{\phi}|\tilde{0}\rangle \neq 0$, so the $\hat{\phi}^{\dagger}\hat{\phi}$ term will not correspond to a mass term. How do we interpret this state of affairs? As we have remarked, in the spontaneously broken phase the remaining degree of freedom of the Higgs field lies in the fluctuations of $\hat{\phi}$ away from $\langle\tilde{0}|\hat{\phi}|\tilde{0}\rangle$. In the gauge in which $\hat{\phi}$ is real, with

$$\langle\tilde{0}|\hat{\phi}|\tilde{0}\rangle = f/2^{1/2} \quad (13.56)$$

for real f, we therefore set

$$\hat{\phi} = [f + \hat{\rho}(x)]/2^{1/2} \quad (13.58)$$

and recalculate $\hat{V}$ in terms of f and $\hat{\rho}$. We find

$$\hat{V} = +\tfrac{1}{2}\mu^{2}\hat{\rho}^{2} + \tfrac{1}{8}(\mu^{2}/f^{2})\hat{\rho}^{4} + \tfrac{1}{2}(\mu^{2}/f)\hat{\rho}^{3} - \tfrac{1}{8}\mu^{2}f^{2}. \quad (14.155)$$

Notice that an amazing thing has happened: the quadratic $\hat{\rho}^2$ term now has the *correct* sign for a mass term! The field $\hat{\rho}$ now obeys the equation of motion

$$(\Box + \mu^{2})\hat{\rho} = -\frac{3}{2}\frac{\mu^{2}}{f}\hat{\rho}^{2} - \frac{\mu^{2}}{2f^{2}}\hat{\rho}^{3} \quad (14.156)$$

appropriate to a genuine scalar particle of mass μ, plus certain self-interactions. Thus the significance of the parameter μ in the Higgs potential $\hat{V}(\hat{\phi})$ is that it is indeed a *mass*; but that of a field describing fluctuations of $\hat{\phi}$ away from its vacuum value.

To summarise: the physical Higgs field, representing fluctuations of $\hat{\phi}$ from $\langle\tilde{0}|\hat{\phi}|\tilde{0}\rangle$, will have the free propagator

$$\mathrm{i}/(p^{2} - \mu^{2}) \quad (14.157)$$

appropriate to a scalar particle of mass μ. In the GSW case, this will be a new parameter of the theory: the mass of the physical Higgs boson. We now return to the electroweak problem, and examine the coupling of this Higgs particle to the gauge fields. In terms of $\hat{\phi}$, the scalar contributions to the weak isospin and hypercharge currents were

$$\hat{j}^{a\mu}(\hat{\phi}) = \mathrm{i}g\left(\hat{\phi}^{\dagger}\frac{\tau^a}{2}(\partial^\mu\hat{\phi}) - (\partial^\mu\hat{\phi})^{\dagger}\frac{\tau^a}{2}\hat{\phi}\right) - \frac{g^2}{2}\hat{\phi}^{\dagger}\hat{\phi}\hat{W}^{a\mu} - gg'\hat{\phi}^{\dagger}\frac{\tau^a}{2}\hat{\phi}\hat{B}^\mu. \tag{14.41}$$

and

$$\hat{j}^{\mu}_{y}(\hat{\phi}) = \frac{\mathrm{i}g'}{2}(\hat{\phi}^{\dagger}(\partial^\mu\hat{\phi}) - (\partial^\mu\hat{\phi})^{\dagger}\hat{\phi}) - gg'\hat{\phi}^{\dagger}\frac{\boldsymbol{\tau}}{2}\hat{\phi}\cdot\hat{\boldsymbol{W}}^\mu - \frac{g'^2}{2}\hat{\phi}^{\dagger}\hat{\phi}\hat{B}^\mu. \tag{14.42}$$

Setting, in U gauge, $\hat{\boldsymbol{\alpha}}(x) = 0$ (cf 14.34)), we now insert

$$\hat{\phi} = \begin{pmatrix} 0 \\ [f + \hat{\sigma}(x)]/2^{1/2} \end{pmatrix} \tag{14.158}$$

into these equations and obtain

$$\hat{j}^{a\mu}(\hat{\phi}) = \left(-\frac{g^2\hat{W}^{a\mu}}{4} + \frac{gg'}{4}\hat{B}^\mu\delta^{a3}\right)(f^2 + 2f\hat{\sigma} + \hat{\sigma}^2) \tag{14.159}$$

$$\hat{j}^{\mu}_{y}(\hat{\phi}) = \left(-\frac{g'^2}{4}\hat{B}^\mu + \frac{gg'}{4}\hat{W}^{3\mu}\right)(f^2 + 2f\hat{\sigma} + \hat{\sigma}^2). \tag{14.160}$$

We repeat that the *full* currents will include many other terms, in general, but we are here specifically interested in the couplings of the gauge fields to the scalar field $\hat{\sigma}$. From these equations, it is not difficult to show (problem 14.4) that the physical fields $W^{\pm}$, Z^0 and γ satisfy the equations (neglecting all other contributions, including self-couplings)

$$(\Box + M_{\mathrm{W}}^2)\hat{W}^{\pm\mu} - \partial^\mu\partial_\nu\hat{W}^{\pm\nu} = -\frac{e}{\sin\theta_{\mathrm{W}}}M_{\mathrm{W}}\hat{W}^{\pm\mu}\hat{\sigma} - \frac{M_{\mathrm{W}}^2}{f^2}\hat{W}^{\pm\mu}\hat{\sigma}^2 \tag{14.161}$$

$$(\Box + M_{\mathrm{Z}}^2)\hat{Z}^\mu - \partial^\mu\partial_\nu\hat{Z}^\nu = -\frac{2e}{\sin 2\theta_{\mathrm{W}}}M_{\mathrm{Z}}\hat{Z}^\mu\hat{\sigma} - \frac{M_{\mathrm{Z}}^2}{f^2}\hat{Z}^\mu\hat{\sigma}^2 \tag{14.162}$$

$$\Box\hat{A}^\mu - \partial^\mu\partial_\nu\hat{A}^\nu = 0 \tag{14.163}$$

where the 0 in the photon equation signifies that the photon does not couple directly to the neutral Higgs scalar $\hat{\sigma}$.

Now at last we have the equations we need. By considering the matrix element for W to scatter in the 'external σ field', in the usual way† we deduce the forms of the WWσ vertex

†But recall the footnote on p.274.

deduce the forms of the WWσ vertex

$$\boxed{\text{WW}\sigma \text{ vertex} \quad \frac{\mathrm{i}e}{\sin\theta_{\mathrm{W}}} M_{\mathrm{W}} \varepsilon_{\mathrm{i}} \cdot \varepsilon_{\mathrm{f}}^{*}} \tag{14.164}$$

and similarly derive the ZZσ vertex

$$\boxed{\text{ZZ}\sigma \text{ vertex} \quad \frac{\mathrm{i}2e}{\sin 2\theta_{\mathrm{W}}} M_{\mathrm{Z}} \varepsilon_{\mathrm{i}} \cdot \varepsilon_{\mathrm{f}}^{*}.} \tag{14.165}$$

There are also four-point vertices coming from the $\hat{\sigma}^2$ terms on the right-hand sides of equations (14.161) and (14.162).

With these couplings of the Higgs field to the gauge bosons—determined in terms of the same parameters as before—and with the Higgs self-couplings as in (14.156), it might appear that we have finally obtained all the rules for tree graphs in the GSW theory. Unfortunately, this is still not quite the case, and the reason is sufficiently important to merit a new section.

14.8 The problem of the fermion masses

What problem? Simply this: *fermion masses spoil invariance under all gauge transformations, such as SU(2)$_L$, which act on only the left-handed component of the fermion wavefunctions*. This seems to be a disaster for the whole of the beautiful framework we have been developing!

Let us first see why fermion masses cause this problem. The standard logic of the gauge field approach to dynamics is to start from the free field equation

$$(\mathrm{i}\partial\!\!\!/ - m)\hat{\psi} = 0 \tag{14.166}$$

and make this consistent with local phase invariance by replacing ∂^μ by the appropriate covariant derivative. Now the covariant derivatives postulated for fermions in the GSW theory are *different* for the left-handed

$$\hat{\psi}_{\mathrm{L}} \equiv \tfrac{1}{2}(1 - \gamma_5)\hat{\psi} \tag{14.167}$$

and right-handed

$$\hat{\psi}_{\mathrm{R}} = \tfrac{1}{2}(1 + \gamma_5)\hat{\psi} \tag{14.168}$$

parts of the field: the former transforms as part of an SU(2)$_{\mathrm{L}}$ doublet, while the latter is a singlet. However, the mass term in our free Dirac equation can be written

$$m\hat{\psi} = m\hat{\psi}_{\mathrm{L}} + m\hat{\psi}_{\mathrm{R}} \tag{14.169}$$

and since $\hat{\psi}_{\mathrm{L}}$ and $\hat{\psi}_{\mathrm{R}}$ transform *differently* under $\mathrm{SU(2)_L}$, there is no hope of making equation (14.166) locally $\mathrm{SU(2)_L}$ covariant: it is analogous to having pieces with opposite parity in a wave equation, which of course implies that parity cannot be a good symmetry of the equation. There seems no escape from the conclusion that such a 'left-handed' symmetry can only be a true symmetry for massless fermions, such as neutrinos (if indeed they are massless).

Needless to say, there is a way out, albeit rather ugly. The escape lies in the fact that the symmetry is hidden, and there exists a scalar field with $\langle\tilde{0}|\hat{\phi}|\tilde{0}\rangle \neq 0$. The idea is to introduce a coupling between the originally massless fermions and the Higgs field, such that when the massless fermions propagate in a vacuum with background $\langle\tilde{0}|\hat{\phi}|\tilde{0}\rangle \neq 0$, they develop a mass. This is reminiscent of the mechanism which gave the gauge fields mass, but with one vital difference: namely, there is nothing, in the case of these fermion fields, to fix the fermion–Higgs coupling strength, except that it must produce the observed masses. Each new fermion mass is therefore a new parameter in the GSW model as it now stands; and the attempt to find a mechanism which *predicts* these masses remains an important, unsolved problem.

For simplicity, we shall suppose that the neutrinos are actually massless, and limit ourselves to just giving a mass to the $t_3 = -\frac{1}{2}$ members of the lepton isodoublets. The form of the desired couplings may be seen by separating the Dirac field $\hat{\psi}$ into $\hat{\psi}_{\mathrm{L}}$ and $\hat{\psi}_{\mathrm{R}}$ in the massive Dirac equation. We apply the projection operator

$$P_{\mathrm{R}} = \tfrac{1}{2}(1 + \gamma_5) \tag{14.170}$$

from the left and remember that

$$\{\gamma_\mu, \gamma_5\} = 0 \tag{14.171}$$

to derive

$$\mathrm{i}\not{\partial}\,\hat{\psi}_{\mathrm{L}} = m\hat{\psi}_{\mathrm{R}} \tag{14.172a}$$

and similarly

$$\mathrm{i}\not{\partial}\,\hat{\psi}_{\mathrm{R}} = m\hat{\psi}_{\mathrm{L}}. \tag{14.172b}$$

Thus the free particle massless equations

$$\mathrm{i}\not{\partial}\,\hat{\psi}_{\mathrm{L}} = 0 \tag{14.173a}$$

and

$$\mathrm{i}\not{\partial}\,\hat{\psi}_{\mathrm{R}} = 0 \tag{14.173b}$$

must be modified by introducing an interaction term involving the isodoublet $\hat{\phi}$ on the right-hand side, in such a way as to yield the above massive equations, (14.172a) and (14.172b), after we set

$$\hat{\phi} \to \langle\tilde{0}|\hat{\phi}|\tilde{0}\rangle = \begin{pmatrix} 0 \\ f/2^{1/2} \end{pmatrix}. \tag{14.174}$$

The recipe which works is to introduce the following 'Yukawa' interactions:

$$\mathrm{i}\not\partial\,\hat{\psi}_{\mathrm{R}} = g_{\mathrm{f}}\hat{\phi}^{\dagger}\hat{l} \tag{14.175a}$$

$$\mathrm{i}\not\partial\,\hat{l} = g_{\mathrm{f}}\hat{\phi}\hat{\psi}_{\mathrm{R}} \tag{14.175b}$$

where $\hat{l}$ is the left-handed fermion doublet. It is clear that these equations are *globally* $\mathrm{SU(2)_L}$ covariant—both sides transform in the same way under $\mathrm{SU(2)_L}$ transformations—and furthermore they can both be made *locally* covariant by the introduction of the appropriate covariant derivative. Substituting $\langle\tilde{0}|\hat{\phi}|\tilde{0}\rangle$ of (14.174) for $\hat{\phi}$, these equations reduce to the correct massive equations (14.172a) and (14.172b) if

$$g_{\mathrm{f}} = \frac{m2^{1/2}}{f} = \frac{e}{2^{1/2}\sin\theta_{\mathrm{W}}}\left(\frac{m}{M_{\mathrm{W}}}\right). \tag{14.176}$$

We can now find the trilinear Yukawa-like Higgs–fermion vertices from equations (14.175) in the usual way. In addition, there will also be the usual fermion couplings to the gauge fields which we have ignored in this discussion.

We must conclude this description of the GSW theory by emphasising that this prescription for fermion masses is one of the least satisfactory aspects of the theory. It is an entirely *ad hoc* procedure, requiring a different 'g_{f}' for each different massive fermion ($g_{\mathrm{e}}, g_{\mu}, \ldots, g_{\mathrm{u}}, g_{\mathrm{d}}, \ldots$). Furthermore, these g_{f}'s are unusually small as compared with other coupling constants we have met—which of course is just another way of saying the fermions are unusually light in mass, as compared with the weak scale of M_{W}, or of f. There is as yet no satisfactory theory of the origin of these disparate scales. Nevertheless, without such Higgs–fermion couplings the theory with massive leptons would not have been locally $\mathrm{SU(2)_L \times U(1)}$ gauge invariant. This would have meant (cf §12.2) that we could find some tree graph, or sum of tree graphs, that would violate unitarity at high energies—and, as in §12.3, this would ultimately imply that the theory is not renormalisable. Indeed one can show explicitly that the graph of figure 14.19 is needed to cancel such violations in the process $\mathrm{e^+e^- \to W^+W^-}$, and that, in general, such cancellations require just the above couplings, proportional to the lepton masses.

Phenomenological expectations regarding the Higgs particle H are much less precise than those concerning the W and $\mathrm{Z^0}$. The mass of H is largely unknown. The parameter μ in the Higgs potential $\hat{V}$ gives the *tree level* mass of H via (14.155), but radiative corrections will change this to m_{H}, say. If μ vanishes (or is small), radiative corrections can be a

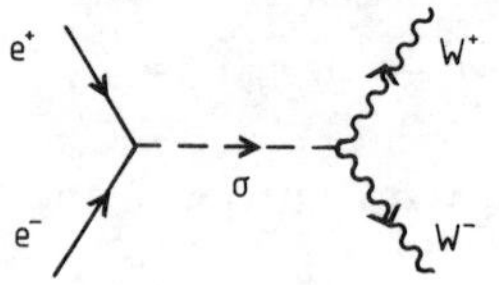

Figure 14.19 Graph with an intermediate Higgs scalar σ, involving the fermion–Higgs coupling, and contributing to $e^+e^- \rightarrow W^+W^-$.

significant source of mass, and place (Linde 1976, Weinberg 1976) a lower bound on m_H:

$$m_H \gtrsim 5 - 10 \text{ GeV} \tag{14.177}$$

using (14.75) for $\sin^2\theta_W$. Larger values of μ, for fixed values of the vacuum value f determined by (14.74), correspond (cf (14.156)) to larger values of the Higgs self-interactions, which could become strong if $m_H \sim O(1 \text{ TeV})$ (Lee *et al* 1977, Veltman 1977). In the latter case it would seem reasonable to expect a complicated Higgs spectroscopy, along the lines of the strongly interacting meson sector of QCD.

In the case of a single Higgs particle, as discussed above and in §14.7, the couplings are all determined in terms of the phenomenologically known mass parameters. It is clear that the strongest couplings are to the largest mass particles (e.g. Z^0, W and t), so these would provide the most favourable production rates. The radiative decay of toponium, $t\bar{t} \rightarrow H + \gamma$ (Wilczek 1977) and—if the mass allows—various modes of Z^0 decay involving H, should be possible ways to detect such an H. Further discussion is given by Ellis *et al* (1976a) and Lee *et al* (1977); see also the Proceedings of the 1978 LEP Summer Study (LEP 1979) and Ellis (1981).

Summary

No multiplet structure for a hidden non-Abelian symmetry.

Weak isospin and hypercharge.

Covariant derivatives for local $SU(2)_L \times U(1)_y$.

Charged current vertices.

The choice of Higgs field, and of $\langle\tilde{0}|\hat{\phi}|\tilde{0}\rangle$.

Associated vacuum screening currents.

Generation of $W^{\pm}$ masses.

W^3–B mixing, and physical Z^0, γ.

Phenomenological implications
- $W^\pm$, Z^0 masses;
- charged currents;
- W width;
- neutral currents;
- Z^0 width;
- production cross sections for $W^\pm$ and Z^0;
- charge asymmetry in $W^\pm$ decay;
- electroweak interference effects.

Discovery of $W^\pm$ and Z^0, and comparison of properties with predictions.

Couplings of $W^\pm$ and Z^0 to Higgs scalar σ.

Fermion mass problem.

The Higgs particle.

Problems

14.1 Verify equation (14.87).

14.2 This is an example to illustrate the effect of choosing different forms of $\langle\tilde{0}|\hat{\phi}|\tilde{0}\rangle$.

(a) Suppose we took $\hat{\phi}$ to be a triplet ($t = 1$) representation of $SU(2)_L$. In the simple gauge in which $\hat{\phi}$ is real, we might then try

$$\langle\tilde{0}|\hat{\phi}|\tilde{0}\rangle = \begin{pmatrix} 0 \\ 0 \\ f \end{pmatrix}$$

say, with f real. This non-vanishing component of $\langle\tilde{0}|\hat{\phi}|\tilde{0}\rangle$ has $t_3 = -1$, and to make the charge zero we must assign $y = 2$. Repeat the calculations of §14.4 for this case, and show that the photon and Z^0 fields are still given by (14.64) and (14.66), but that the Z^0 mass is now

$$M_Z = 2^{1/2} M_W / \cos\theta_W.$$

What would the parameter ρ of (14.106) be for this model?

(b) Alternatively, we might set $y(\phi) = 0$ and take

$$\langle\tilde{0}|\hat{\phi}|\tilde{0}\rangle = \begin{pmatrix} 0 \\ f \\ 0 \end{pmatrix}$$

this component having $t_3 = 0$. Show that only W^1 and W^2 acquire mass in this case.

14.3 Verify equation (14.111).

14.4 Verify equations (14.161)–(14.163).

PART VI

BEYOND THE TREES

15

FOUR LAST THINGS

This book has been almost entirely concerned with obtaining and applying the Feynman rules for graphs which represent the amplitudes, calculated in perturbation theory, for various processes to occur according to the QED, QCD and GSW theories. For the most part, our calculations have been limited to the lowest-order—or *tree*—graphs. In this final chapter we shall give a very brief introduction to four topics—ghosts in non-Abelian gauge theories, grand unified theories, confinement, and lattice gauge theories—each of which, in different ways, extends the foregoing framework.

15.1 Ghost loops and unitarity in non-Abelian gauge theories

We have indicated at various points in earlier chapters how amplitudes for higher-order processes—ones involving *loops*—are to be calculated. For our present purposes, the essential steps are: (i) draw the Feynman graph; (ii) label each line with a 4-momentum, ensuring 4-momentum conservation at each vertex (this will entail the introduction of a variable 4-momentum—not that of any of the external particles—associated with each independent loop in the graph); (iii) associate wavefunctions with external lines, propagators with internal lines, and factors with vertices, as determined by the particles and their interactions; (iv) write these elements down in the natural order (initial state $\rightarrow$ intermediate state $\rightarrow$ final state) and, for closed loops, integrate over the loop momenta, and sum over all internal and Lorentz degrees of freedom; (v) for each closed fermion loop, multiply by -1. According to all that has been said so far, therefore, there would seem to be no particular difficulty in writing down amplitudes for arbitrarily complicated Feynman graphs in non-Abelian gauge theories, once the rules for all the tree graphs have been discovered. This, however, is by no means the case: the problem of finding the complete Feynman rules for such theories resisted solution for many years. Although our primary aim has been to derive the rules for tree graphs, and use them in practical cases, we think it important to attempt some explanation of the problem associated with loops in non-Abelian gauge theories, and of its resolution, if only to counteract any misleading impression of simplicity we may have inadvertently given in earlier chapters.

The problem is one which is endemic in all gauge theories (manifest as well as hidden); it concerns amplitudes involving gauge quanta as internal lines, and it has to do with apparent *violations of unitarity*. We must emphasise that this unitarity violation is quite distinct from the one discussed in Chapter 12 in connection with the high-energy behaviour of Born graphs in massive vector theories. The problem we shall be dealing with arises even for massless (manifestly symmetric) gauge theories, and in what follows we shall deal explicitly with this case. Briefly, the problem is one of *unwanted degrees of freedom*, introduced by treating a massless vector field covariantly. The sum of Feynman graphs contributing to the total amplitude for a physical process, up to some given order in perturbation theory, must satisfy unitarity to that order. Unitarity determines the imaginary part of the amplitude via a relation of the form

$$\mathrm{Im}\,\langle \mathrm{f}|F|\mathrm{i}\rangle = \int \sum_n \langle \mathrm{f}|F|n\rangle\langle n|F^\dagger|\mathrm{i}\rangle\, \mathrm{d}\rho_n \tag{15.1}$$

where $\langle \mathrm{f}|F|\mathrm{i}\rangle$ is the amplitude for the process $\mathrm{i}\rightarrow \mathrm{f}$, and the sum is over a complete set of physical intermediate states $|n\rangle$, which can enter at the given energy; $\mathrm{d}\rho_n$ represents the phase space element for the general intermediate state $|n\rangle$. Consider the possibility of gauge quanta appearing in these intermediate states. Since unitarity deals only with physical states, such quanta can have only the two degrees of freedom (polarisations) allowed for a physical massless gauge field (cf §5.1.2). Now part of the power of the 'Feynman rules' approach to perturbation theory is that it is manifestly covariant. But there is no completely covariant way of selecting out just the two physical components of a massless polarisation vector ε_μ, from the four originally introduced for covariance. In fact, when gauge quanta appear as virtual particles in intermediate states in Feynman graphs, they will not be restricted to having only two polarisation states (as we shall see explicitly in a moment). Hence there is a real chance that when the imaginary part of such graphs is calculated, a contribution from the unphysical polarisation states will be found, which has no counterpart at all in the physical unitarity relation, so that unitarity will not be satisfied. This would be a disease even more fundamental than non-renormalisability.

Instead of tackling the complexity of the general case, we illustrate the trouble with some simple examples. Consider a two-particle amplitude whose imaginary part has a combination from a state containing one particle and one photon (figure 15.1):

$$\mathrm{Im}\,\langle \mathrm{f}|F|\mathrm{i}\rangle = \int \sum \langle \mathrm{f}|F|\gamma \mathrm{A}\rangle\langle \gamma \mathrm{A}|F^\dagger|\mathrm{i}\rangle\, \mathrm{d}\rho_2 \tag{15.2}$$

where we have called the intermediate particle 'A', and $\mathrm{d}\rho_2$ is the

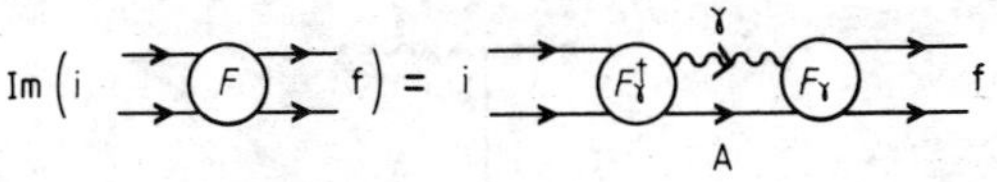

Figure 15.1 Contribution to the imaginary part of a $2 \to 2$ amplitude, involving one photon in the intermediate state.

two-body phase space for the γA state. For simplicity, the A's are taken to be spinless. The one-photon amplitudes involved in the right-hand side of (15.2) must have the general form $\varepsilon_\mu F^\mu_{(\gamma)}$, where ε_μ is the photon polarisation vector. The remaining sum in (15.2) is over the photon polarisations, so that, suppressing unnecessary labels, (15.2) takes the form

$$\operatorname{Im} F = \int F^\mu_{(\gamma)} \left(\sum_{\lambda=1,2} \varepsilon_\mu(k, \lambda) \varepsilon_\nu(k, \lambda) \right) F^{\nu\dagger}_{(\gamma)} \, \mathrm{d}\rho_2. \tag{15.3}$$

Since only physical states $|n\rangle$ can contribute in the unitarity summation in (15.2), the sum on λ in (15.3) is over the two transverse polarisation states only, with of course $k^2 = 0$. For later convenience, we are now using real polarisation vectors: in the frame $k = (|\boldsymbol{k}|, 0, 0, |\boldsymbol{k}|)$, we take $\varepsilon(k, \lambda = 1) = (0, 1, 0, 0)$ and $\varepsilon(k, \lambda = 2) = (0, 0, 1, 0)$.

We now wish to find out whether or not a result of the form (15.3) will hold when the F's represent some suitable Feynman graphs. We first note that we want the unitarity relation (15.3) to be satisfied order by order in perturbation theory: that is to say, when the F's on both sides are expanded in powers of the coupling strengths (as in the usual Feynman graph expansion), the coefficients of corresponding powers on each side should be equal. Since each emission or absorption of a photon produces one power of e, the right-hand side of (15.3) involves at least the power e^2. Thus the lowest-order process in which (15.3) may be 'tested' is the one shown in figure 15.2. The question is: does the imaginary part of the amplitude for figure 15.2 have the form (15.3) required by unitarity?

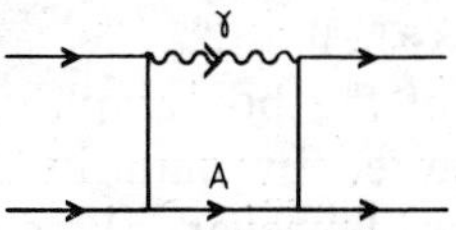

Figure 15.2 Lowest-order Feynman graph for which the unitarity relation (15.3) can be checked.

Since each $F_{(\gamma)}$ on the right-hand side of (15.3) must now be precisely of order e, we expect each of them to be the lowest-order (tree) graph shown in figure 15.3. Let us explain how this comes about. For simplicity, we choose to work in the so-called Feynman gauge ($\xi = 1$ in

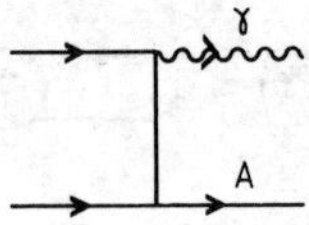

Figure 15.3 Tree graph entering into the imaginary part of the amplitude for figure 15.2.

(13.93)), so that the photon propagator takes the familiar and simple form (ignoring the i)

$$-\ g^{\mu\nu}/k^2. \tag{15.4}$$

According to the rules for loops already given, the amplitude $B^{(1)}$ for figure 15.2 will be proportional to the product of the propagators for the scalar particles and the photon propagator (15.4), together with appropriate vertex factors, the whole being integrated over the one independent loop momentum, which we may take to be k. The extraction of the imaginary part of a Feynman amplitude is a technical matter, for which, indeed, rules exist (see e.g. Eden *et al* 1966, §2.9). In the present case the result is that, to compute the imaginary part of figure 15.2, one replaces the photon propagator (15.4) by

$$\pi(-\ g_{\mu\nu})\delta(k^2)\theta(k_0); \tag{15.5}$$

that is, the propagator is replaced by a condition stating that, in evaluating the imaginary part, the photon's mass is constrained to be zero (instead of being a free variable of integration), and its energy is to be positive. In addition, the propagator for the intermediate A particle in figure 15.2 is also replaced by the corresponding mass-shell condition. These two conditions have the effect of converting the integral over k into a standard two-body phase space integral for the 'photon + A' intermediate state, so that finally

$$\operatorname{Im} B^{(1)} = \int T^{\mu}_{(\gamma)}(-\ g_{\mu\nu})T^{\nu}_{(\gamma)}\ \mathrm{d}\rho_2 \tag{15.6}$$

where $T^{\mu}_{(\gamma)}$ is indeed the tree graph shown in figure 15.3, with all external legs being on-mass-shell.

So, the imaginary part of the box graph of figure 15.2 does seem to have the form (15.3) required by unitarity, with $|n\rangle$ the intermediate state of $1\gamma + 1\mathrm{A}$. In fact, however, there is an essential difference between (15.6) and (15.3). Comparing these two equations, we see that the place of the factor $-\ g_{\mu\nu}$ in (15.6) is taken in (15.3) by the photon polarisation sum

$$P_{\mu\nu} = \sum_{\lambda=1,2} \varepsilon_\mu(k,\lambda)\varepsilon_\nu(k,\lambda). \tag{15.7}$$

Thus we have to investigate whether this difference matters.

To proceed further, it is helpful to have an explicit expression for the sum over physical polarisation states $P_{\mu\nu}$ which enters into (15.3). We already encountered this quantity in §12.2 in discussing the calculation of cross section for $e^+e^- \to \gamma\gamma$. There we saw that we could not simply take the $M \to 0$ limit of the corresponding polarisation sum for a massive vector particle, since that diverged. Nevertheless, we might still think of calculating the necessary sum in $P_{\mu\nu}$ by brute force, using two ε's specified by the conditions

$$\varepsilon^{\mu}(k, \lambda)\varepsilon_{\mu}(k, \lambda') = -\delta_{\lambda\lambda'}, \qquad \varepsilon\cdot k = 0. \tag{15.8}$$

The trouble is that conditions (15.8) *do not fix the ε's uniquely if $k^2 = 0$*. Indeed, it is precisely the fact that any given ε_μ satisfying (15.8) can be replaced by $\varepsilon_\mu + \lambda k_\mu$ that both reduces the degrees of freedom to two (as we saw in §5.1.2), and evinces the essential arbitrariness in the ε_μ specified only by (15.8). In order to calculate (15.7), we need to put another condition on ε_μ, in order to fix it uniquely. A standard choice (see e.g. Taylor 1976, pp 14–15) is to supplement (15.8) with the further condition

$$t\cdot\varepsilon = 0 \tag{15.9}$$

where t is some 4-vector. This certainly fixes ε_μ, and enables us to calculate (15.7), but of course now two further difficulties have appeared; namely, the physical results depend on t_μ and, have we not lost Lorentz covariance, because the theory now involves a special 4-vector t_μ?

Setting these questions aside for the moment, we can calculate (15.7) using the conditions (15.8) and (15.9), finding (problem 15.1)

$$\begin{aligned} P_{\mu\nu} &\equiv \sum_{\lambda=1,2} \varepsilon_\mu(k, \lambda)\varepsilon_\nu(k, \lambda) \\ &= -g_{\mu\nu} - [t^2 k_\mu k_\nu - k\cdot t(k_\mu t_\nu + k_\nu t_\mu)]/(k\cdot t)^2. \end{aligned} \tag{15.10}$$

But only the *first* term on the right-hand side of (15.10) is to be seen in (15.6). A crucial quantity is clearly

$$\begin{aligned} U_{\mu\nu}(k, t) &\equiv -g_{\mu\nu} - P_{\mu\nu} \\ &= [t^2 k_\mu k_\nu - k\cdot t(k_\mu t_\nu + k_\nu t_\mu)]/(k\cdot t)^2. \end{aligned} \tag{15.11}$$

We note that whereas

$$k^\mu P_{\mu\nu} = k^\nu P_{\mu\nu} = 0 \tag{15.12}$$

(from the condition $k\cdot\varepsilon = 0$), the same is *not* true of $k^\mu U_{\mu\nu}$—in fact,

$$k^\mu U_{\mu\nu} = -k_\nu \tag{15.13}$$

where we have used $k^2 = 0$. It follows that $U_{\mu\nu}$ may be regarded as including polarisation states for which $\varepsilon \cdot k \neq 0$. In physical terms, therefore, a photon appearing internally in a Feynman graph has to be regarded as existing in more than just the two polarisation states available to an external photon (cf §5.6). $U_{\mu\nu}$ characterises the contribution of these unphysical polarisation states.

The discrepancy between (15.6) and (15.3) is then

$$\operatorname{Im} B^{(1)}_{\text{unphys}} = \int T^{\mu}_{(\gamma)} [U_{\mu\nu}(k, t)] T^{\nu}_{(\gamma)} \, \mathrm{d}\rho_2. \tag{15.14}$$

It follows that this unwanted contribution will, in fact, vanish if

$$k_{\mu} T^{\mu}_{(\gamma)} = 0 \tag{15.15}$$

where k is the 4-momentum of the photon. In general, there will be more than one box graph (of the type shown in figure 15.2) present, and thus more than one corresponding tree graph; condition (15.15) must hold for the *sum* of all tree graphs.

We can easily imagine a generalisation to arbitrary one-photon amplitudes of the type appearing in the full unitarity relation of 15.1. For such amplitudes, the condition

$$\boxed{k_{\mu} F^{\mu}_{(\gamma)} = 0} \tag{15.16}$$

is called a *Ward identity*. If (15.16) is true for the amplitude $F^{\mu}_{(\gamma)}$, unitarity will be obeyed by Feynman graphs in the $\mathrm{i} \to \mathrm{f}$ amplitude, as far as the $\gamma\mathrm{A}$ intermediate state is concerned. Furthermore, the questions about t dependence and Lorentz non-covariance are then satisfactorily disposed of. Condition (15.16) is just the one we introduced in §5.8.3, pointing out that it followed from gauge invariance. Thus all is well when the intermediate state $|n\rangle$ in (15.1) contains just one photon.

A crucial new feature appears when we consider an intermediate state with two photons. Consider, to be specific, the box graph shown in figure 15.4 for $\mathrm{e^+e^-} \to \mathrm{e^+e^-}$, with a two-photon intermediate state. This amplitude is of order e^4, but it is not the only graph with a 2γ intermediate state at this order—there is also figure 15.5. To be consistent, we must include both; let us call their sum $B^{(2)}$. Proceeding as before, we calculate the contribution to $\operatorname{Im} B^{(2)}$ by putting both photons on-mass-shell. Labelling the photon momenta and polarisations as k_1, λ_1 and k_2, λ_2, we shall find a term in $\operatorname{Im} B^{(2)}$ of the form

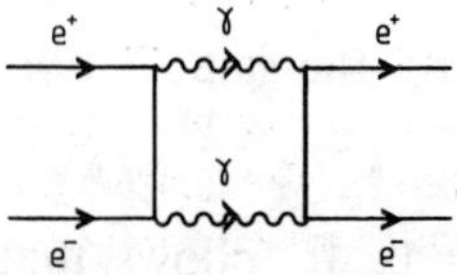

Figure 15.4 Two-photon intermediate state graph in $\mathrm{e^+e^-} \to \mathrm{e^+e^-}$.

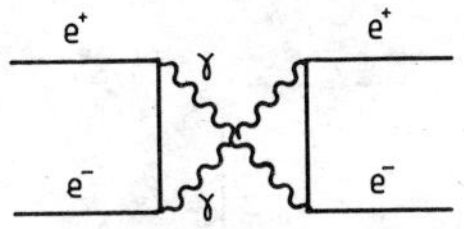

Figure 15.5 Additional two-photon graph in $e^+e^- \to e^+e^-$.

$$\int T^{\mu\nu}_{(\gamma\gamma)}\, T^{\mu'\nu'\dagger}_{(\gamma\gamma)} [U_{\mu\mu'}(k_1, t_1)][U_{\nu\nu'}(k_2, t_2)]\, \mathrm{d}\rho \tag{15.17}$$

where each $T^{\mu\nu}_{(\gamma\gamma)}$ is now the sum of the two tree graphs of figure 15.6, and $\mathrm{d}\rho$ is the 2γ phase space factor. There will of course also be the 'right' contribution to $\operatorname{Im} B^{(2)}$ involving just the transverse polarisations for both photons ('$P \cdot P$'), and also mixed contributions ('$P \cdot U$').

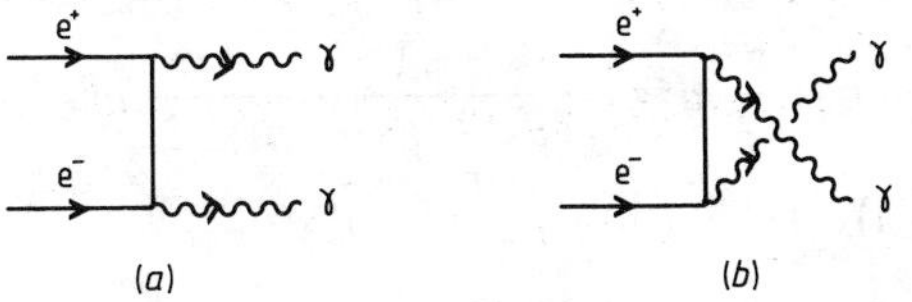

Figure 15.6 Tree graphs for $e^+e^- \to \gamma\gamma$: (*a*) direct; (*b*) crossed.

For unitarity to be correctly satisfied with only physical intermediate states, (15.17) must vanish (as must the mixed contributions). In the unwanted piece (15.17), neither photon is in a polarisation state satisfying $\varepsilon \cdot k = 0$; for it to vanish, we require

$$k_{1\mu} T^{\mu\nu}_{(\gamma\gamma)} = 0 \tag{15.18}$$

(and similarly for k_2). At first sight (15.18) looks no different from (15.15): the sum of the tree graphs in figure 15.6 must vanish when the polarisation vector of one photon is replaced by its momentum. Note, however, that (15.18) must be true for each photon, *even when the other has* $\varepsilon \cdot k \neq 0$. This is the new feature of the 2γ intermediate state, and it will turn out to be of crucial significance in the non-Abelian case.

Actually, it can be verified explicitly that (15.18) is true for each photon, whatever the state of the other. This result holds for arbitrary Feynman graphs with 2γ intermediate states, which will, therefore, not violate unitarity, and the same is true for arbitrary numbers of photons. With non-Abelian gauge quanta (gluons), the situation turns out to be different. Consider a theory with the non-Abelian gauge symmetry of SU(2). In checking unitarity (in lowest order) for box graphs involving a two-gluon intermediate state, we shall again be led to the condition (15.18) if, as before, we choose the gauge $\xi = 1$ for the gluon propagator. $T^{\mu\nu}_{(gg)}$ will be the sum of all relevant tree graphs, and these are shown in figure 15.7: there are now three, because of the existence of

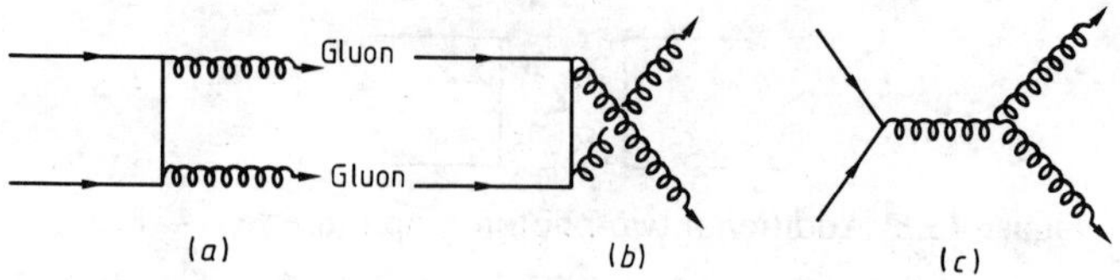

Figure 15.7 Tree graphs in fermion–antifermion scattering.

the three-gluon vertex, absent in QED. Correspondingly, the 'box' (one-loop) graphs must now include the one shown in figure 15.8. For definiteness, we take the solid lines to represent 'isospinor' fermions. These three tree graphs together yield

$$\begin{aligned}\varepsilon_{1\mu}\varepsilon_{2\nu}T^{\mu\nu}_{(gg)} &= g^2\bar{v}(p_2)\,\frac{\tau^\beta}{2}\,\not\varepsilon_2 a_2^\beta\,\frac{1}{(\not p_1 - \not k_1 - m)}\,\frac{\tau^\alpha}{2}\,a_1^\alpha\not\varepsilon_1 u(p_1)\\ &+ g^2\bar{v}(p_2)\,\frac{\tau^\alpha}{2}\,a_1^\alpha\not\varepsilon_1\,\frac{1}{(\not p_1 - \not k_2 - m)}\,\frac{\tau^\beta}{2}\,a_2^\beta\not\varepsilon_2 u(p_1)\\ &+ (-\,\mathrm{i})g^2\varepsilon^{\gamma\alpha\beta}[(p_1 + p_2 + k_1)^\nu g^{\mu\rho}\\ &+ (-\,k_2 - p_1 - p_2)^\mu g^{\rho\nu} + (-\,k_1 + k_2)^\rho g^{\mu\nu}]\\ &\times\,\varepsilon_{1\mu}a_1^\alpha a_2^\beta\varepsilon_{2\nu}\,\frac{-1}{(p_1 + p_2)^2}\,\bar{v}(p_2)\,\frac{\tau^\gamma}{2}\,\gamma_\rho u(p_1)\end{aligned} \tag{15.19}$$

where we have written the gluon polarisation vectors as a product of a Lorentz 4-vector ε_μ and an 'SU(2) polarisation vector' a^α. Now replace ε_1, say, by k_1. Using the Dirac equation and the mass-shell conditions $p_1^2 = p_2^2 = m^2$, $k_1^2 = k_2^2 = 0$, the first two terms reduce to

$$g^2\bar{v}(p_2)\not\varepsilon_2[\tau^\alpha/2,\ \tau^\beta/2]u(p_1)a_1^\alpha a_2^\beta \tag{15.20}$$

$$= \mathrm{i}g^2\varepsilon^{\alpha\beta\gamma}\bar{v}(p_2)\not\varepsilon_2\,\frac{\tau^\gamma}{2}\,u(p_1)a_1^\alpha a_2^\beta \tag{15.21}$$

where the second line follows from the SU(2) algebra of the τ's. The third term in (15.19) gives

$$-\,\mathrm{i}g^2\varepsilon^{\alpha\beta\gamma}\bar{v}(p_2)\not\varepsilon_2\,\frac{\tau^\gamma}{2}\,u(p_1)a_1^\alpha a_2^\beta \tag{15.22}$$

$$+\,\mathrm{i}g^2\,\frac{\varepsilon^{\alpha\beta\gamma}}{2k_1\cdot k_2}\,\bar{v}(p_2)\not k_1\,\frac{\tau^\gamma}{2}\,u(p_1)k_2\cdot\varepsilon_2 a_1^\alpha a_2^\beta. \tag{15.23}$$

We see that the part (15.22) certainly does cancel (15.21), but there remains the piece (15.23), *which only vanishes if* $k_2\cdot\varepsilon_2 = 0$. But this is not sufficient to guarantee the absence of all unphysical contributions to the imaginary part of the two-gluon box graphs, as the preceding discussion shows. We conclude that *loop diagrams involving two (or more) gluons, if constructed according to the simple rules for tree diagrams, will violate unitarity.*

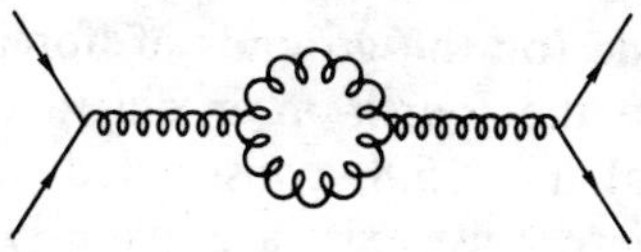

Figure 15.8 Additional fourth-order graph in $f\bar{f} \to f\bar{f}$, present in the non-Abelian case.

The correct rule for such loops must be such as to satisfy unitarity. Since there seems no other way in which the offending piece in (15.23) can be removed, we must infer that the rule for loops will have to involve some extra term, or terms, over and above the simple tree-type constructions, which will cancel the contributions of unphysical polarisation states. To get an intuitive idea of what such extra terms might be, we return to expression (15.11) for the sum over unphysical polarisation states $U_{\mu\nu}$, and make a specific choice for t. We take $t_\mu = \bar{k}_\mu$, where the 4-vector $\bar{k}$ is defined by $\bar{k} = (-|\boldsymbol{k}|, \boldsymbol{k})$, and $\boldsymbol{k} = (0, 0, |\boldsymbol{k}|)$. This choice obviously satsifies $t \cdot \varepsilon(k, \lambda = 1 \text{ or } 2) = 0$. Then

$$U_{\mu\nu}(k, \bar{k}) = (k_\mu \bar{k}_\nu + k_\nu \bar{k}_\mu)/(2|\boldsymbol{k}|^2). \tag{15.24}$$

Unitarity requires

$$\int T^{\mu\nu}_{(\mathrm{gg})} T^{\mu'\nu'\dagger}_{(\mathrm{gg})} \frac{(k_{1\mu}\bar{k}_{1\mu'} + k_{1\mu'}\bar{k}_{1\mu})}{2|\boldsymbol{k}_1|^2} \frac{(k_{2\nu}\bar{k}_{2\nu'} + k_{2\nu'}\bar{k}_{2\nu})}{2|\boldsymbol{k}_2|^2} \,\mathrm{d}\rho \tag{15.25}$$

to vanish, but it does not. Let us work in the CM frame of the two gluons, with $k_1 = (|\boldsymbol{k}|, 0, 0, |\boldsymbol{k}|)$, $k_2 = (|\boldsymbol{k}|, 0, 0, -|\boldsymbol{k}|)$, $\bar{k}_1 = (-|\boldsymbol{k}|, 0, 0, |\boldsymbol{k}|)$, $\bar{k}_2 = (-|\boldsymbol{k}|, 0, 0, -|\boldsymbol{k}|)$, and consider for definiteness the contractions with the $T^{\mu\nu}_{(\mathrm{gg})}$ term. These are $T^{\mu\nu}_{(\mathrm{gg})} k_{1\mu} k_{2\nu}$, $T^{\mu\nu}_{(\mathrm{gg})} k_{1\mu} \bar{k}_{2\nu}$, etc. Such quantities can be calculated from expression (15.19) by setting $\varepsilon_1 = k_1$, $\varepsilon_2 = k_2$ for the first, $\varepsilon_1 = k_1$, $\varepsilon_2 = \bar{k}_2$ for the second, and so on. We have already obtained the result of putting $\varepsilon_1 = k_1$. From (15.23) it is clear that a term in which ε_2 is replaced by k_2 as well as ε_1 by k_1, will vanish, since $k_2^2 = 0$. A typical non-vanishing term is of the form $T^{\mu\nu}_{(\mathrm{gg})} k_{1\mu} \bar{k}_{2\nu}/2|\boldsymbol{k}|^2$. From (15.23) this reduces to

$$-\mathrm{i}g^2 \frac{\varepsilon^{\alpha\beta\gamma}}{2k_1 \cdot k_2} \bar{v}(p_2) \not{k}_1 \frac{\tau^\gamma}{2} u(p_1) a_1^\alpha a_2^\beta \tag{15.26}$$

using $k_2 \cdot \bar{k}_2 / 2|\boldsymbol{k}|^2 = -1$. We may write (15.26) as

$$j^\gamma_\mu \frac{-g^{\mu\nu}\delta^{\gamma\eta}}{(k_1 + k_2)^2} \mathrm{i}g\varepsilon^{\alpha\beta\eta} a_1^\alpha a_2^\beta k_{1\nu} \tag{15.27}$$

where

$$j^\gamma_\mu = g\,\bar{v}(p_2)\gamma_\mu(\tau^\gamma/2)u(p_1) \tag{15.28}$$

in the SU(2) current associated with the fermion–antifermion pair.

The unwanted terms of the form (15.27) can be eliminated if we adopt the following rule (on the grounds of 'forcing the theory to make sense'). In addition to the fourth-order diagrams of figures 15.4, 15.5 (generalised to gluons) and 15.8, constructed according to the simple 'tree' prescriptions, there must exist a previously unknown fourth-order contribution, *only present in loops*, such that it has an imaginary part which is non-zero in the same physical region as the two-gluon intermediate state, and moreover is of just the right magnitude to cancel all the contributions to (15.25) from terms like (15.27). Now (15.27) has the appearance of a one-gluon intermediate state amplitude. The $f\bar{f} \rightarrow g$ vertex is represented by the current (15.28), the gluon propagator appears in Feynman gauge $\xi = 1$, and the rest of the expression would have the interpretation of a coupling between the intermediate gluon and two scalar particles with SU(2) polarisations a_1^{α}, a_2^{β}. Thus (15.27) can be interpreted as the amplitude for the tree graph shown in figure 15.9, where the dotted lines represent the scalar particles. It seems plausible, therefore, that the fourth-order graph we are looking for has the form shown in figure 15.10. The new scalar particles must be massless, so that this new amplitude has an imaginary part in the same physical region as the gg state. When the imaginary part of figure 15.10 is calculated in the usual way, it will involve contributions from the tree graph of figure 15.9, and these can be arranged to cancel the unphysical polarisation pieces like (15.27).

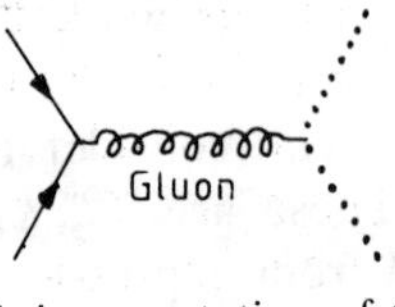

Figure 15.9 Tree graph interpretation of the expression (15.27).

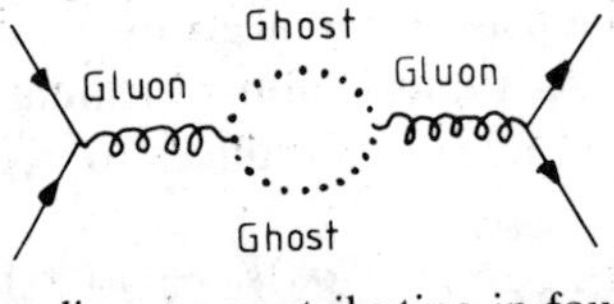

Figure 15.10 Ghost loop diagram contributing in fourth order to $f\bar{f} \rightarrow f\bar{f}$.

For this cancellation to work, the scalar particle loop graph of figure 15.10 must enter with the opposite sign from the three-gluon loop graph of figure 15.8, which in retrospect was the cause of all the trouble. Such a relative minus sign between single closed loop graphs would be expected if the scalar particles in figure 15.10 were in fact fermions! (See item (v) in the rules given at the start of this section.) Thus we appear to need scalar particles obeying Fermi statistics. Such particles are called 'ghosts'. We must emphasise that although we have introduced the tree graph of figure 15.9, which apparently involves ghosts as external lines, in reality the ghosts are always confined to loops, their function being to cancel unphysical contributions from intermediate gluons.

The preceding discussion has, of course, been entirely heuristic. In particular, Feynman recognised (Feynman 1963a, 1977) that unitarity alone was not a sufficient constraint to provide the prescription for treating more than one closed gluon loop. A compact derivation of the complete rules for forming unitarity-preserving amplitudes in non-Abelian gauge theories was given by Faddeev and Popov (1967) using a path-integral formalism. Their derivation shows that the ghosts, though spinless, obey Fermi–Dirac statistics. That the ghosts do restore unitarity was verified explicitly by 't Hooft (1971a, b). Ghosts are equally necessary in the 'hidden symmetry' case. We must admit that, in this matter of ghost loops, we have arrived at a part of the theory that cannot, so far as we can tell, be completely derived except by going to the appropriate field theory formalism (see e.g. Abers and Lee 1973). We shall not proceed any further along that path. Feynman rules for non-Abelian gauge field theories (of both manifest and hidden symmetry type) can be found elsewhere (Abers and Lee 1973): we do add the caution, however, that the form of the gluon–ghost coupling implied in (15.27) *depends on the choice of gauge chosen for the gluon fields*; in other gauges, other forms of coupling will be found. The rules for tree diagrams, for which there are no problems with ghosts, are given in Appendix F.

15.2 Grand unification

The fact that strong, weak and electromagnetic interactions all appear to be well described by gauge theories offers the remarkable prospect that perhaps all three can be understood as somehow different aspects of a single gauge theory. Partial unification (there are still two independent gauge coupling constants) of the weak and electromagnetic interactions has been achieved in the GSW theory. It is natural to try and include the strong (colour) interactions as well, in a 'grand unified' scheme. We shall give a brief introduction to this programme, but we stress that it is almost entirely speculative, and no agreed grand unified theory yet exists.

A truly unified theory of all three interactions would involve only one gauge coupling constant. Therefore, it will have to be the case that, in some sense to be understood, all three interactions must 'really' have the same strength. Now we have learned that the study of gauge theories themselves has profoundly altered the way we think about interaction strengths, in two quite distinct ways. 'Weak interactions are weak because M_W and M_Z are so large' summarises—perhaps slightly misleadingly—one aspect: interactions can be 'really' very similar, but the similarity can be hidden by the phenomenon of spontaneous symmetry breakdown (hidden gauge invariance). The $SU(2)_L \times U(1)$ symmetry only becomes apparent at $Q^2 \gg M_W^2, M_Z^2$. The idea of a Q^2-dependent coupling strength, introduced in §9.3, is equally revolutionary: apparently it makes little sense to talk of an 'inherent' strength at all—rather, the strength depends on the momentum (or length) scale under consideration. These two concepts suggest an economical possibility: perhaps the strong, weak and electromagnetic interactions are described by a single gauge group G, big enough to contain the $SU(3)_c$ and $SU(2)_L \times U(1)$ groups as subgroups (in the same way that SU(2) is a subgroup of SU(3)). G will have a single gauge coupling constant g_G; apparent differences in strengths among the interactions will be ironed out in one or other of the ways just mentioned.

Consider first the Q^2 dependence of the $SU(3)_c$, $SU(2)_L$ and U(1) coupling strengths, denoted respectively by α_3, α_2 and α_1. We saw in Chapter 9 that QCD was an asymptotically free theory, meaning that its coupling α_3 decreases (logarithmically) with Q^2. This is a property associated with non-Abelian gauge fields only—and thus the $SU(2)_L$ coupling α_2 has a similar behaviour, while the U(1) coupling α_1 *increases* with Q^2. At 'low' energies (O(1–10 GeV)) we know that the couplings are significantly different, and ordered as $\alpha_3 > \alpha_2 > \alpha_1$. As Q^2 increases, α_3 decreases, as does α_2 (but more slowly than α_3), while α_1 rises—thus the possibility exists that they may approach each other in magnitude for some large value of Q^2.

However, the following considerations show that the 'grand unified' symmetry associated with G must be (spontaneously) broken. In a world in which the three different interactions have the same intrinsic strength, the customary distinction between quarks and leptons, based on the idea that the former enjoy strong interactions while the latter do not, will no longer apply. Indeed, unification schemes naturally contemplate putting leptons and quarks into common multiplets of G. Symmetry transformations associated with certain generators of G will then exist, which turn leptons into quarks (and *vice versa*). If this seems disturbing, an analogy is provided by the $SU(2)_I$ and $U(1)_Y$ subgroups of $SU(3)_f$: there are generators of $SU(3)_f$ which turn non-strange particles into strange ones. When the symmetry of G is gauged, there will be a gauge quantum for every generator, as we have seen. Such quark $\leftrightarrow$ lepton transitions will almost inevitably lead to non-conservation of baryon number. In view of the present bound on the proton lifetime ($\tau_p \gtrsim 10^{32}$ yr), some mechanism for suppressing baryon-number-violating transitions is necessary. Its nature will depend on the precise model envisaged. An early proposal, by Georgi and Glashow (1974), took G to be SU(5), and suppressed proton decay by giving the quanta (generically called X) mediating it very high masses, M_X (by a spontaneous symmetry breakdown). A different mechanism operates in the model of Pati and Salam (1974). We shall discuss proton decay in a moment.

Returning to the Q^2 dependence of the coupling strengths, we now see that the α_i's are expected to be equal (up to possible group theoretic Clebsch–Gordan coefficients) only at Q^2 values well above the symmetry breakdown scale characterised by M_X^2. On the other hand, we know the empirical values of the α_i's at 'low' energies, and we can calculate (at least in some approximation) their dependence on energy. Hence we should be able to estimate the values of M_X, by requiring the known low-energy values of the α_i's to evolve to a common value at $Q^2 \geqslant M_X^2$. This was the principle behind the first such theoretical estimate of M_X by Georgi *et al* (1974), using formulae of the type (9.41), based on the 'summation of dominant logarithms' technique. The value of M_X will of course depend on G and details of the grand unified model, as well as on the approximations behind the assumed Q^2 dependence of the α_i's. A simple SU(5) estimate gives $M_X \sim 10^{15}$ GeV; more sophisticated calculations give a somewhat lower value, $M_X \sim$ 1–5 × 10^{14} GeV (Ross 1984).

Before continuing, we should pause for a moment to register what an enormous energy this is. The mathematical reason for it is, of course, that since the α_i's vary only logarithmically with energy, it takes many decades to change their values appreciably. The hypothesis that no new physics enters in the course of such a vast extrapolation (through the 'desert' from 10^{15} GeV to ~1 GeV) is almost absurdly bold. Other types of theory are possible, which populate this desert with phenomena occurring at new intermediate mass scales. But, within the classic grand unified framework, we are faced with a totally new scale of symmetry breaking, many orders of magnitude different in energy (or mass) from that which occurred in the GSW theory ($M_{W,Z} \sim 100$ GeV). No satisfactory way of understanding the origin of the very large ratio M_X/M_W, in a natural manner, has yet been found. Perhaps, indeed, this is an indication that intermediate scales do exist.

Is even such a value of M_X high enough to suppress proton decay sufficiently? A rough estimate of the decay rate can be obtained from dimensional reasoning. Since the decay must be mediated by an X boson, the amplitude will contain a factor M_X^{-2} arising from an X propagator evaluated at $q^2 \ll M_X^2$ (cf (10.91)–(10.93)), and hence the decay rate Γ_p will be proportional to M_X^{-4}. Since Γ_p must have dimension of mass (with $\hbar = c = 1$), a plausible estimate is

$$\Gamma_p \sim \alpha_G^2 m_p^5 / M_X^4 \tag{15.29}$$

where the proton mass has been taken to be the appropriate mass scale, and α_G is the fine structure constant of the group G. If we take $\alpha_G = \frac{1}{40}$ and $M_X \sim 10^{15}$, we find a proton lifetime

$$\tau_p = \Gamma_p^{-1} \sim 5 \times 10^{31} \text{ yr} \tag{15.30}$$

while a rather smaller M_X value will give a correspondingly shorter lifetime.

When Georgi *et al* (1974) wrote their paper, they quoted an experimental lower limit for τ_p of 10^{30} yr. The appearance of a theory with a prediction for the lifetime not very much longer than this added impetus to the unification idea, and brought a whole new type of high-energy experiment into being—the 'underground experiment'. Carried out deep underground to reduce cosmic ray background, these experiments typically search for the very rare proton decay by 'watching' a huge mass (100–1000 tons) of material, and hoping that an event identifiable as baryon decay will occur. Because the size of the 'detector' is so vast, it is not possible to get complete high-resolution instrumentation. Thus candidate decay events will not be individually unique, but have to be distinguished on a statistical basis against the dominant background, which is due to interactions of atmospheric neutrinos. Some idea of the difficulty of the enterprise can be gained from the fact that the relative rates of baryon decay and neutrino-induced processes are equal for a lifetime of order 10^{31} yr.

The experimental situation has been reviewed by Perkins (1984). The first detectors were designed to search for the SU(5)-favoured mode $p \to \pi^0 e^+$, and the current lower limit for the corresponding partial lifetime is around 3×10^{32} yr. This effectively rules out the simplest SU(5)-based prediction (cf (15.30) and Ross 1984). However, even within SU(5) it is possible to complicate the model so as to extend the lifetime, and of course other unified groups G are possible. In particular, supersymmetric theories (Ross 1984) tend to prefer modes such as $p \to \bar{\nu}_\tau K^+$, or $p \to \mu^+ K^0$. These are harder to see, and the lower bound for them is therefore weaker than for $p \to \pi^0 e^+$, standing currently at $\geqslant 5 \times 10^{31}$ yr for the partial lifetimes. From the theoretical side, none of the more elaborate and complicated models is able to make precise and unambiguous predictions. Hence 'the question of baryon stability, lifetime and decay modes has become an almost purely empirical matter' (Perkins 1984).

However, before passing on, we mention two points of a more general nature. Suppose that baryons are indeed stable—that is, baryon number is absolutely conserved. Then a thoroughgoing 'gauge principle' viewpoint would imply that—as with the absolute conservation of electromagnetic charge—this conservation law should be the result of a *local* symmetry, and be associated with the appearance of a long-range force field coupled to baryon number (Lee and Yang 1955). No evidence for such a field exists (Perkins 1984)—a statement equivalent to a suitably small upper limit on the magnitude of any associated coupling strength. The conventional preference is then to ignore any such potentially very weak interaction in favour of a grand unification of the known forces, within which baryons can decay.

The second point concerns the observed ratio of matter density (baryons) to radiation (photons), $n_{\rm B}/n_\gamma \sim 10^{-9}$, and the preponderance of matter over antimatter. It would take us too far afield to rehearse the arguments here, for which we must refer the reader to Perkins (1984) and Ross (1984). Suffice it to say that cosmological models incorporating grand unification and baryon decay, while at present unable to predict $n_{\rm B}/n_\gamma$ unambiguously, do offer the possibility that the observed value could arise as the universe evolved from an initially baryon-symmetric state. In the simplest SU(5) model, the predicted value is $n_{\rm B}/n_\gamma \leqslant 10^{-16}$, which is much too small—and is therefore further evidence against this theory. As with proton decay, however, more elaborate theories are possible which can, in principle, accommodate the desired value without predicting it.

The foregoing discussion concerning the phenomenological consequences of baryon decay is therefore somewhat inconclusive as regards the empirical status of grand unified theories. Fortunately, another important prediction can be made—that of $\sin^2\theta_{\rm W}$. We recall that the GSW theory did not fully unify electromagnetism and weak interactions, because two separate gauge groups are considered, $\mathrm{SU(2)_L}$ and $\mathrm{U(1)}_y$, each with its own independent gauge coupling constant (g and g' respectively). If there is indeed only one 'large' gauge group G, of which $\mathrm{SU(2)_L}$ and $\mathrm{U(1)}_y$ are subgroups, there can be only one independent coupling constant, and g' must be related to g, i.e. (cf (14.63a)) $\tan\theta_{\rm W}$ is known. Remarkably enough, it is possible to make a quantitative prediction on the basis of little more than this idea. In the $\mathrm{SU(2)_L}\times\mathrm{U(1)}_y$ theory, the covariant derivative had the form

$$\partial_\mu + \mathrm{i}g\boldsymbol{t}^{(t)}\cdot\boldsymbol{W}_\mu + \mathrm{i}g'(y/2)B_\mu \tag{15.31}$$

which we may write as

$$\partial_\mu + \mathrm{i}g\boldsymbol{t}^{(t)}\cdot\boldsymbol{W}_\mu + \mathrm{i}g\tan\theta_{\rm W}(Q - t_3)B_\mu \tag{15.32}$$

using (14.63a) and (14.17). If the $\mathrm{U(1)}_y$ part is to enter in a symmetrical way as compared to the $\mathrm{SU(2)_L}$ part, we must regard the quantity

$$t_0 = \tan\theta_{\rm W}(Q - t_3) \tag{15.33}$$

as the U(1) generator exactly analogous to the $\boldsymbol{t}$'s. The coupling of this to the B_μ is then equal to (i.e. unified with) that of the SU(2) generators to the $\boldsymbol{W}_\mu$. The crucial point (Georgi and Glashow 1974) is that the t_0 and $\boldsymbol{t}$'s, *if generators of a single group*, must satisfy common orthogonality and normalisation conditions. These take the form

$$\mathrm{Tr}(t_\alpha t_\beta) = N\delta_{\alpha\beta} \tag{15.34}$$

where N may depend on the representation, but not on α or β, and where the labels α, β run over all the generators of G. The relation (15.34) is a generalisation of the ones already adopted for SU(2) and SU(3). Using (15.34), we then obtain from (15.33) the result

$$\mathrm{Tr}(Q^2) = \mathrm{Tr}(t_3)^2 + \cot^2\theta_\mathrm{W}\mathrm{Tr}(t_0)^2 = (1 + \cot^2\theta_\mathrm{W})\mathrm{Tr}(t_3)^2 \tag{15.35}$$

$$\sin^2\theta_\mathrm{W} = \mathrm{Tr}(t_3)^2/\mathrm{Tr}(Q^2). \tag{15.36}$$

The normalisation factor N of (15.34) will cancel in the ratio (15.36) when we calculate the traces in any given representation of G. Hence it is sufficient to consider only one generation (say u, d, e and ν), and we may calculate (15.36) by assigning the particles to a single (doubtless reducible) representation of G. We then have three left-handed doublets of coloured quarks, one left-handed lepton doublet, and the corresponding singlets. These give $\mathrm{Tr}(t_3^2) = 2$ and $\mathrm{Tr}(Q^2) = \frac{16}{3}$, so that we obtain

$$\sin^2\theta_\mathrm{W} = \tfrac{3}{8} = 0.375. \tag{15.37}$$

This is the value predicted in the SU(5) model of Georgi and Glashow (1974).

At this point one may well be discouraged: (15.37) is far from the experimental value. However, we must remember that G is only supposed to be a good symmetry at energies $\gtrsim M_\mathrm{X}$. The quantity $\sin^2\theta_\mathrm{W}$ (or $\tan\theta_\mathrm{W}$) measures the relative strengths of the GSW coupling constants of g and g'—but, as stressed above, these (or the corresponding α_2 and α_1) vary with the energy scale. In fact, since $\tan\theta_\mathrm{W} = g'/g$, where g' is associated with the Abelian U(1) group while g is associated with the non-Abelian $\mathrm{SU(2)_L}$ group, g' will decrease as Q^2 decreases from values $\sim M_\mathrm{X}^2$ (where (15.37) is presumed to hold), whereas g will increase. Thus $\tan\theta_\mathrm{W}$ (and $\sin^2\theta_\mathrm{W}$) will certainly decrease, and we have the possibility of understanding how $\sin^2\theta_\mathrm{W}$ can be $\frac{3}{8}$ at the unification scale, but about 0.23 at 'low' energies.

Once again, precise predictions are only possible in specific models. The simple SU(5) estimate, based on the 'dominant logarithm' technique which yielded $M_\mathrm{X} \sim 10^{15}$ GeV, gives $\sin^2\theta_\mathrm{W} \approx 0.21$. This is obviously a most encouraging result—indeed it is probably the one tangible reason for continuing to have some sort of faith in grand unification. As in the case of M_X, more sophisticated calculations within SU(5) can be done, but they do not alter $\sin^2\theta_\mathrm{W}$ significantly. Once again, more complicated models, none particularly compelling, can be found which accommodate the observed value of $\sin^2\theta_\mathrm{W}$.

Another class of prediction characteristic of grand unified theories stems from the assignment of quarks and leptons to the *same* multiplets.

Typically, therefore, quark to lepton mass ratios will be predicted: for example, the simplest SU(5) model predicts $m_{\mathrm{d}} = m_{\mathrm{e}}$, $m_{\mathrm{s}} = m_{\mu}$ and $m_{\mathrm{b}} = m_{\tau}$. Here again we must remember that these predictions refer to energy scales beyond M_{X}: as we saw in Chapter 6, masses are also subject to renormalisation, and the idea of finite Q^2-dependent mass renormalisation effects must be invoked to obtain predictions at 'low' energies. The corrected value for m_{b} is ~ 5.3 GeV for six flavours (Ross 1984), which is a good result. The equivalent prediction for m_{s} is ~ 500 MeV, which is fair. The worst result is $m_{\mathrm{d}}/m_{\mathrm{s}} \approx \frac{1}{200}$, to be compared with the value obtained from current algebra, $\frac{1}{24}$. More complicated models contain more parameters and are less predictive.

There is an important technical point which should at least be mentioned in view of the emphasis of this section on renormalisation effects: the entire discussion presupposes that the basic grand unified gauge theory is renormalisable. One of the crucial tools in the proof of renormalisability is the use of Ward identities, which are an expression of the gauge invariance of the theory. However, in theories with axial currents, anomalous contributions to the Ward identities can occur, and these will in general spoil renormalisability. In fact it turns out that it is possible to arrange for these so-called 'anomalies' to cancel, and this is an important constraint in the construction of grand unified models (see e.g. Taylor 1976).

We should not leave the subject of grand unified theories without mentioning one more fundamental reason for being interested in them—namely, that they offer the hope of understanding charge quantisation. In §8.3 we pointed out that only one universal coupling strength, g_{G} say, is allowed in the case of a non-Abelian gauge symmetry group G. This implies that, in any representation p of the group, the 'generalised charges' $g_{\mathrm{G}} G_i^{(\mathrm{p})}$ will all be multiples of the basic g_{G}, the magnitudes depending on the allowed eigenvalues of the diagonalisable generators $G_i^{(\mathrm{p})}$ in that representation (a helpful analogy is that of angular momentum; all J_z values are quantised in units of $\hbar/2$, as follows from the commutation relations). No such constraint exists for an Abelian gauge group—roughly speaking, because no commutators set the scale. It is a theoretical possibility that particles could exist with charges incommensurate with that of the proton. At the level of $\mathrm{SU(2)_L} \times \mathrm{U(1)}_y$, the association $Q = (t_3 + y/2)$ was put in by hand: there is no *a priori* reason why the $\mathrm{U(1)}_y$ part should be quantised (note, on the other hand, that if y were absent, Q would necessarily be quantised in multiples of $e/2$, as determined by the eigenvalues of t_3). However, in a grand unified group with no U(1) part—such as SU(5), for example—Q will have to be a generator and its scale will be set, in terms of the basic group constant g_{G}, by the group commutation relations and associated eigenvalues. Charge quantisation will therefore automatically emerge from any such grand unified model which does not have a U(1) factor.

15.3 Confinement

In §1.2.3 we referred to the fact that, with one possibly significant exception (La Rue *et al* 1977, 1981), neither quarks nor gluons, nor any other particles apparently carrying colour quantum numbers, have been observed as free isolated particles, despite many careful searches. As we have seen, there is a large and impressive body of evidence to indicate the presence of quarks and gluons inside hadrons, but it seems impossible to prise an individual one loose; they are *confined* to the interior of hadrons. It may be that isolated particles carrying colour will one day be discovered; at present the prevailing view is that all coloured states are permanently confined, and only colour singlets are free.

The problem is then to explain the property of confinement in terms of the assumed dynamics—namely, QCD. We begin with a simple point. We have seen that in the non-Abelian gauge theory of QCD the effective coupling constant $\alpha_s(Q^2)$ decreases as Q^2 increases (provided $f \leqslant 16$). Correspondingly, α_s increases as Q^2 becomes smaller so that, although the approximate expression (9.41) will not then be valid, it is plausible to suppose that $\alpha_s(Q^2)$ may become very strong at sufficiently small Q^2. This would imply that as the particles carrying colour become more widely separated, the effective force holding them together increases ever more strongly. This is the kind of result we are looking for.

Nobody yet knows for certain the behaviour of $\alpha_s(Q^2)$ at low Q^2, because perturbation theory is inapplicable and calculational techniques which go beyond perturbation theory are not yet totally reliable. We shall describe one such technique in the following section—the idea of studying a field theory on a discrete space–time 'lattice' (Creutz 1979). Here we shall give a brief introduction to the conceptual aspects of one of the present lines of research on confinement—stressing that this is, as yet, an unsolved problem.

We have already encountered one distinctly non-perturbative aspect of the gauge theories apparently used by nature: the phenomenon of hidden symmetry. In Chapter 13 we emphasised that the nature of the superconducting ground state could not be arrived at in perturbation theory, and the same is true for the physical vacuum, in respect of the weak force. In the latter case, no attempt was made at a dynamical (BCS-type?) explanation of the phenomenon—the characteristic $\langle\tilde{0}|\hat{\phi}|\tilde{0}\rangle \neq 0$ was simply postulated. This dealt with the massive vector boson problem. Could some unusual features of a non-perturbative QCD vacuum be held responsible, in some analogous way, for confinement?

Remarkably enough, some promising ideas emerge (Mandelstam 1980) from further consideration of the 'relativistic superconductor' introduced in Chapter 13. An important clue is provided by a prominent feature of hadron spectroscopy: the hadronic resonance levels lie on linearly rising rotational sequences. For such sequences a plot of J, the angular momentum of each resonance, versus the square of the resonance mass yields a nearly straight line; for baryons, sequences exist up to perhaps $J = \frac{19}{2}$. This kind of sequence is well described, for the higher-angular-momentum states, by the 'string model' of hadrons, also called the 'dual-resonance model' (see e.g. Mandelstam 1974). This model describes hadrons as quantised relativistic one-dimensional structures—'strings'. Nielsen and Olesen (1973) and Tassie (1974) suggested an interpretation of these strings in terms of vortex line solutions to the particular U(1) gauge theory we called the relativistic superconductor.

The existence of such solutions is well known in the case of real superconductors. In fact, our discussion of superconductivity in Chapter 13 dealt with only one class of superconductors, that called type I; these remain superconducting throughout (exhibiting a complete Meissner effect) when an external magnetic field of less than a certain critical value is applied. There is a quite separate class—type II superconductors—which allows partial entry of the external field, in the form of thin filaments of flux. Within each filament the field is high and the material is not superconducting. Outside the core of the filaments the material is superconducting and the field dies off over the characteristic penetration length λ. Around each filament of magnetic flux there circulates a vortex of screening current; the filaments are often called vortex lines. It is as if numerous thin cylinders, each enclosing flux, had been drilled in a block of type I material, thereby producing a non-simply connected geometry.

In real superconductors, screening currents are associated with the macroscopic pair wavefunction (field) ψ. For type II behaviour to be possible, $|\psi|$ must vanish at the centre of a flux filament, and rise to the constant value appropriate to the superconducting state over a distance $\xi < \lambda$. ξ is called the 'coherence length'. According to the Ginzburg–Landau (GL) theory (which is the static version of the relativistic superconductor), a more precise criterion is that type II behaviour holds if $\xi < 2^{1/2}\lambda$; both ξ and λ are, of course, temperature-dependent. The behaviour of $|\psi|$ and B in the vicinity of a flux filament is shown in figure 15.11. Thus, whereas for simple type I superconductivity $|\psi|$ is simply set equal to a constant, in the type II case $|\psi|$ has the variation shown in this figure. Solutions of the coupled GL equations for $\boldsymbol{A}$ and ψ can be obtained which exhibit this behaviour.

An important result is that the flux through a vortex line is quantised. This fact was already mentioned in Chapter 13 (footnote after equation (13.76)). To see it explicitly, we write

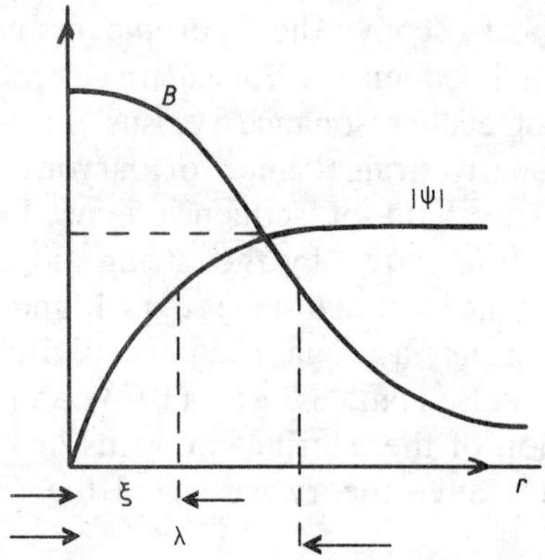

Figure 15.11 Magnetic field B and modulus of macroscopic (pair) wavefunction $|\psi|$ in the neighbourhood of a flux filament.

$$\psi = |\psi|e^{i\theta} \tag{15.38}$$

so that the expression for the current in the superconductor is

$$\boldsymbol{j} = \frac{q|\psi|^2}{m}(\boldsymbol{\nabla}\theta - q\boldsymbol{A}) \tag{15.39}$$

as usual. Rearranging, we have

$$\boldsymbol{A} = \frac{-m}{q^2|\psi|^2}\boldsymbol{j} + \frac{1}{q}\boldsymbol{\nabla}\theta. \tag{15.40}$$

Let us integrate equation (15.40) around any closed loop C in the type II superconductor, which includes a vortex line. Far enough away from the vortex, the screening currents $\boldsymbol{j}$ will have dropped to zero, and hence

$$\oint_C \boldsymbol{A}\cdot d\boldsymbol{s} = \frac{1}{q}\oint_C \boldsymbol{\nabla}\theta\cdot d\boldsymbol{s} = \frac{1}{q}[\theta]_C \tag{15.41}$$

where $[\theta]_C$ is the change in phase around C. If the wavefunction ψ is single-valued, the change in phase $[\theta]_C$ for a closed path can only be zero or an integer multiple of 2π. Transforming the left-hand side of (15.41) by Stokes' theorem, we obtain the result that the flux Φ through the surface spanning C is quantised:

$$\Phi = \int \boldsymbol{B}\cdot d\boldsymbol{S} = \frac{2n\pi}{q} = n\Phi_0 \tag{15.42}$$

where $\Phi_0 = 2\pi/q$ is the flux quantum (with $\hbar = c = 1$). It is not entirely self-evident why ψ should be single-valued, but experiments do indeed clearly demonstrate the phenomenon of flux quantisation, in units of Φ_0 with $q = 2e$ (which can be interpreted as the charge on the Cooper pair). The phenomenon is seen in non-simply connected specimens of type I superconductors (see footnote on page 414), and in the flux filaments of type II materials; in the latter case each filament carries a single quantum Φ_0.

We now take over these results to the 'relativistic superconductor' case. Suppose that the physical vacuum is such that it favours 'type II' behaviour, rather than the complete Meissner effect. As usual, the role of the pair wavefunction is played by the Higgs field ϕ. By a suitable choice of the parameters of the Higgs potential $V(\phi)$, the vortex lines can be made thin. In fact, λ corresponds to M_V^{-1}, as before, while ξ is just μ^{-1}, the Compton wavelength of the Higgs boson σ (see §14.7: the static solution of $(\Box + \mu^2)\sigma = 0$ gives an exponential fall-off distance of order μ^{-1}). These thin vortex lines will be interpreted as prototype hadronic strings. However, we do not want them to be infinitely extended. We can imagine closed loops of flux, or we can terminate strings on objects which absorb their flux. Such an object is a Dirac monopole (Dirac 1931). Dirac's quantisation condition

$$qg_{\text{m}} = n/2 \tag{15.43}$$

where q is the electric charge and g_{m} is the monopole strength, precisely guarantees that the flux $4\pi g_{\text{m}}$ out of any closed surface surrounding the monopole is quantised in units of Φ_0. Hence a flux filament can originate from, or be terminated by, a Dirac monopole (with the appropriate sign of g_{m}) as was first pointed out by Nambu (1974).

This is the basic model which, in one way or another, underlies many theoretical attempts to understand confinement. The monopole–antimonopole pair in a type II superconducting vacuum, joined by a quantised magnetic flux filament, provides a model of a meson. As the distance between the pair—the length of the filament—increases, so does the energy of the filament, at a rate proportional to its length, since the flux cannot spread out in directions transverse to the filament. This is exactly the kind of linearly rising potential energy required by hadron spectroscopy. The configuration is stable, because there is no way for the flux to leak away: it is a conserved quantised quantity.

For the eventual application to QCD, one will want (presumably) particles carrying non-zero values of the colour quantum numbers to be confined. These quantum numbers are the analogues of electric charge in the U(1) case, rather than of magnetic charge. We imagine, therefore, interchanging the roles of magnetism and electricity in all of the foregoing. Indeed, the Maxwell equations have such a symmetry when monopoles are present. The essential feature of the superconducting ground state was that it involved the coherent state formed by condensation of electrically charged bosonic fermion pairs. A vacuum which confined filaments of $\boldsymbol{E}$ rather than $\boldsymbol{B}$ may be formed as a coherent state of condensed magnetic monopoles (Mandelstam 1976, 't Hooft 1976). These $\boldsymbol{E}$ filaments would then terminate on electric charges. Now magnetic monopoles do not occur naturally as solutions of QED: they would have to be introduced by hand. Remarkably enough, however, solutions of the magnetic monopole type do occur in the case of non-Abelian gauge field theories, whose symmetry is spontaneously broken to an electromagnetic $U(1)_{em}$ gauge group. Just this circumstance can arise in a grand unified theory which contains $SU(3)_c$ and a residual $U(1)_{em}$. Incidentally, these monopole solutions provide an illuminating alternative way of thinking about the necessity of charge quantisation in such theories (discussed in the previous section). As Dirac (1931) pointed out, the existence of just one monopole implies, from his quantisation condition, (15.43), that charge is quantised.

When these ideas are applied to QCD, $\boldsymbol{E}$ and $\boldsymbol{B}$ must be understood as the appropriate colour fields (i.e. they carry an $SU(3)_c$ index). The group structure of SU(3) is also quite different from that of the U(1) models, and we do not want to be restricted just to static solutions (as in the GL theory used as an analogue). Whether in fact the real QCD vacuum (ground state) is formed as some such coherent plasma of monopoles, with confinement of electric charges and flux, is a subject of continuing research; other schemes are also possible. As so often stressed, the difficulty lies in the non-perturbative nature of the confinement problem.

Is there any way in which one can test the above ideas, and explore the long-distance, non-perturbative aspects of QCD? Quite apart from these intriguing speculations about quark confinement and the QCD vacuum, the non-perturbative regime of QCD presumably holds the key to the explanation of all low-energy hadronic physics—a subject which the perturbative techniques of this book have not touched. In our final section we shall describe one attempt to tackle these non-perturbative features, based on formulating (gauge) field theories on a discrete space–time lattice.

15.4 Lattice gauge theories

At first sight, formulating a field theory on a discrete lattice of space–time points seems to be a very drastic modification of the continuum version. For example, it is clear that such a lattice theory will not possess the familiar rotation and Lorentz invariance properties which are appropriate to continuous four-dimensional space. We shall have to hope that this will not matter, if the lattice spacing is small enough. There are, however, some significant advantages. Since we are primarily interested in non-perturbative aspects, it is most unlikely that we shall be able to perform calculations analytically—as is possible, of course, for at least the tree diagrams of perturbation theory. Even in perturbation theory, however, one soon encounters integrals that have to be evaluated numerically, and the standard way of doing this is to approximate the integral by some kind of discrete sum of the form

$$\int f(x)\,\mathrm{d}x = \sum_i f(x_i)\,\Delta x$$

in the simplest case. Thus a mesh of points (in general multidimensional) has to be chosen: perhaps it would make sense to formulate the theory on such a mesh in the first place.

But there is a more fundamental point involved here. As we have seen throughout this book, one of the triumphs of theoretical physics over the past 50 years has been the 'taming' of the ultraviolet divergences that arise in quantum field theory. This is done order by order in perturbation theory by first regulating each divergent graph so as to isolate the infinity in a well defined form. When all the infinities at a given order have been identified in this way, the renormalisation programme is able to sweep them all up into redefinitions of observable parameters like masses and coupling constants. But this immediately creates a problem for any non-perturbative approach: how can we regulate the ultraviolet divergences, and thereby define the theory, if we cannot get to grips with them via the specific (divergent) integrals supplied by perturbation theory? We need to be able to control the ultraviolet divergences and define the theory in an entirely non-perturbative fashion. In this context, approximating continuous space–time by a lattice appears as a rather natural way of regulating the divergences. The lattice of space–time points introduces a minimum distance—namely, the lattice spacing 'a' between neighbouring points. Since no two points can ever be closer than this, there is now a corresponding maximum momentum $\Lambda \sim \hbar/a$ in the lattice version of the theory. Thus the theory is automatically ultraviolet finite from the start, without presupposing the existence of any perturbative expansion (renormalisation questions will, of course, enter when we consider the a dependence of our parameters). This was Wilson's primary motivation for developing lattice gauge theory (Wilson 1974, 1975).

A further very significant advantage of the lattice formulation is that it enables direct contact to be made between quantum field theory and *statistical mechanics*, as we shall see below. This relationship allows physical insights and numerical techniques to pass from one subject to the other, in a way which has been extremely beneficial to both.

We are therefore looking for a formulation of quantum field theory which is going to be suitable for direct numerical computation. This at once seems to rule out any formalism based on non-commuting *operators*, as it is hard to see how they could be numerically simulated. Indeed, the same would be true of ordinary quantum mechanics: is there a formulation of that which avoids operators? The answer is provided by Feynman's *sum over paths* formulation of quantum theory, which is the essential starting point for the lattice approach.

To understand Feynman's formulation, we must go right back to first principles—because it is, in fact, a beautiful generalisation of the classical ideas briefly explained in §4.2.1. There we met a way of doing classical mechanics which seemed quite different from the traditional Newtonian one. Instead of the differential equations postulated in Newton's laws, which provide a local point-to-point account of trajectories (or paths), the action principle was based on a global concept: the contribution of a whole path to the action integral. The question naturally arises: if the differential equations of Schrödinger or Heisenberg are the quantum generalisations of Newtonian mechanics, what is the analogous generalisation of the global 'path-contribution' approach?

The answer was first hinted at by Dirac (1981, §32), but only fully worked out by Feynman. The book by Feynman and Hibbs (1965) presents a characteristically fascinating discussion—here we can only indicate the central idea. We ask: how does a particle get from the point $q(t_1)$ a time t_1 to the point $q(t_2)$ at t_2? Referring back to figure 4.4, in the classical case we imagined (infinitely) many possible paths $q_i(t)$, of which, however, only *one* was the actual path followed, namely the one which minimised the action integral

$$S = \int_{t_1}^{t_2} L(q(t), \dot{q}(t))\,\mathrm{d}t \qquad (15.44)$$

as a function of $q(t)$. In the quantum case, however, we noted in §4.2.2 that a particle does not follow a definite path because of quantum fluctuations. In conventional formulations of quantum theory, this fact is usually taken to imply the uselessness of the path concept *per se*. Feynman's insight was to appreciate that the 'opposite' viewpoint is also possible: since we are forbidden a unique path in the quantum theory, we should in principle include *all possible* paths! In other words, we take all the trajectories on figure 4.4 as physically possible (together with all the other infinitely many ways of accomplishing the trip).

But surely not all paths are equally *likely*: after all, we must presumably recover the classical trajectory as $\hbar \rightarrow 0$, in some sense. Thus we must find an appropriate weighting for the paths. Feynman's recipe is beautifully simple: weight each path by the factor

$$\mathrm{e}^{\mathrm{i}S/\hbar} \qquad (15.45)$$

where S is the action for that particular path. At first sight this is a rather strange proposal, since all paths—even the classical one—are 'weighted' by a quantity which is of unit modulus. But of course contributions of the form (15.45) from all paths have to be added coherently—just as we superposed the amplitudes in the 'two-slit' discussion in §2.5. What distinguishes the classical path is that it makes S stationary under small changes of path: thus in its vicinity paths have a strong tendency to add up constructively, while far from it the phase factors will tend to produce cancellations. The amount the particle can 'stray' from the classical path depends on the magnitude of the corresponding action relative to $\hbar$, the quantum of action. The scale of path coherence is set by $\hbar$—which is why we are here not setting it equal to 1.

In summary, then, the quantum mechanical amplitude to go from $q(t_1)$ to $q(t_2)$ is proportional to

$$\sum_{\text{all paths } q(t)} \exp\left(\frac{\mathrm{i}}{\hbar}\int_{t_1}^{t_2} L(q(t), \dot{q}(t))\, \mathrm{d}t\right). \tag{15.46}$$

The generalisation to quantum field theory is immediate. In §4.2.4 we saw how classical field theory could be formulated 'globally', the equation of motion for a field $\phi(\boldsymbol{x}, t)$ being derived from the requirement that ϕ minimise the action integral

$$S = \int \mathcal{L}(\phi, \dot{\phi}, \partial\phi/\partial x)\, \mathrm{d}x\, \mathrm{d}t \tag{15.47}$$

(with extension to four dimensions in §4.4). This suggests that the natural field generalisation of (15.46) is

$$\sum_{\text{all fields } \phi} \exp\left(\frac{\mathrm{i}}{\hbar}\int_{t_1}^{t_2} \mathrm{d}t \int_{\boldsymbol{x}_1}^{\boldsymbol{x}_2} \mathrm{d}^3x\mathcal{L}\right). \tag{15.48}$$

But what is this the amplitude *for*? To answer this, we must remember that we are no longer dealing with a single-particle system (with coordinate $q(t)$). Expression (15.46) is proportional to the amplitude for 'the system'—a particle with coordinate $q(t)$—to go from the initial configuration $q(t_1)$ at t_1 to the final configuration $q(t_2)$ at t_2. In the case of (15.48) the system is an entire quantum field, which is capable of containing arbitrarily many particles. However, in the absence of any external stimulus, the field is presumably, on energetic grounds, in its state of lowest energy—namely, its ground or vacuum state. Thus in (15.48) the 'system' which is passing from a certain field configuration at $(\boldsymbol{x}_1, t_1)$ to another at $(\boldsymbol{x}_2, t_2)$ is the vacuum state of the theory. The fact that the vacuum can make this evolution via (infinitely) many field 'trajectories' corresponds simply to the fact that arbitrary quantum fluctuations about the classical vacuum are possible. Amplitudes for propagating particles (excitations above the ground state) are represented by formulae analogous to (15.48), in which products of fields are weighted by $\exp(\mathrm{i}S/\hbar)$.

Feynman has therefore provided us with a version of quantum theory which employs only classical functions, not operators. Before we can submit it to the computer, however, one more step is necessary—which also bridges the gap to statistical mechanics. The quantum amplitudes of (15.46) and (15.48) are very difficult to handle numerically because of the rapidly oscillating complex exponentials (far from the classical trajectory). To deal with this, we make a bold move and rotate time into an imaginary direction:

$$t \rightarrow \mathrm{i}\tau. \tag{15.49}$$

This has the immediate effect of converting the oscillating factors $\sim \exp(\mathrm{i}S/\hbar)$ into (presumably) exponentially decreasing functions $\sim \exp(-S_{\mathrm{E}}/\hbar)$, so that they now do indeed appear much more like classical weighting functions, the most important path still being the one of least S_{E}. The subscript 'E' stands for Euclidean, since we have now effectively moved from ordinary space–time with an invariant squared length equal to $c^2t^2 - \boldsymbol{x}^2$ to a (four-dimensional) Euclidean space with squared length $-c^2\tau^2 - \boldsymbol{x}^2$.

This Euclidean version of quantum field theory, when put on a lattice, is in fact analogous to a statistical mechanical system, as we shall now see. The connection is made via the fundamental quantity of statistical mechanics, the *partition function* Z defined by

$$Z = \sum_{\text{configurations}} \exp(-H/kT) \tag{15.50}$$

which is simply the 'sum over states' or configurations (i.e. of the relevant degrees of freedom), with the Boltzmann weighting factor. H is the classical Hamiltonian evaluated for each configuration. Z already bears some resemblance to the Euclidean versions of (15.46) and (15.48). The parallel is much stronger when we 'discretise' space–time. Consider (15.46), for simplicity, with the single variable t (quantum mechanics, as we see by comparing (15.46) with (15.47), is like quantum field theory in zero space dimensions). We discretise t by, for example, writing $t = ja$, with j an integer. Then $q(t)$ becomes q_j, the value of q at 'site j', $\mathrm{d}q/\mathrm{d}t$ becomes $(q_{j+1} - q_j)/a$, and the integral over t in (15.46) becomes a sum over j. In particular, if L were essentially just $\dot{q}^2$, then the discrete (Euclidean) form of (15.46) would be

$$\sum_{\text{configurations } q_j} \exp\left(-\sum_j (q_{j+1} - q_j)^2/a^2\hbar\right). \tag{15.51}$$

Consider now the statistical mechanics of a system of spins on a lattice. In an Ising model, for example, the Hamiltonian has the form

$$H = J \sum_j S_j S_{j+1} \tag{15.52}$$

where J is a constant, the sum is over lattice points j, and the spin variables S_j take the values ± 1. Clearly, when (15.52) is inserted into (15.50), we arrive at something very similar to (15.51), with the q_j's playing the role of the spin variables. Naturally the 'effective Hamiltonian' is not quite the same (though the 'nearest neighbour interaction' is present in (15.51)); however, the basic analogy is plain.

Clearly this generalises straightforwardly to four dimensions—if we can imagine statistical mechanics in four Euclidean dimensions. We note that the statistical mechanical parameter kT in (15.50) is in some way analogous to $\hbar$ in (15.51). However, another parameter is also relevant. We have seen repeatedly that in gauge theories a fundamental role is played by the covariant derivative. In the U(1) case, for example, this is $D^\mu = \partial^\mu + ieA^\mu$. In lattice work it is conventional to redefine the gauge fields so as to make this quantity independent of the coupling parameter e; thus we introduce $\widetilde{A}^\mu = eA^\mu$. The free Maxwell Lagrangian then becomes

$$-\frac{1}{4e^2}\widetilde{F}^{\mu\nu}\widetilde{F}_{\mu\nu} \tag{15.53}$$

with similar expressions for the non-Abelian gauge field Lagrangians. Thus in general kT corresponds to $g^2\hbar$, and 'high (low) temperature' can also be interpreted as 'strong (weak) coupling'.

The benefit for quantum field theory of this connection with statistical mechanics lies in the fact that a whole variety of analytical and numerical tools has been developed, over the years, for dealing with the partition function. In particular, high-temperature/strong-coupling expansions, together with low-temperature/weak-coupling expansions and numerical simulations, can shed light on the detailed *phase transition structure* of field theories when put on a lattice.

We shall illustrate these ideas by considering the question of confinement in gauge field theories. Presumably it is essential to preserve, on the lattice, something like the local gauge invariance of the continuum theory. This is a basic new feature, peculiar to *gauge* theories on a lattice. Curiously, the essential steps were first explored in the context of the Ising model. Wegner (1971) was interested in constructing a theory with a local symmetry for which one could not characterise the phase structure in terms of a 'local order parameter'. For the ordinary Ising spin model, for example, the phase structure can be characterised by the magnetisation

$$\langle S\rangle = \frac{\Sigma_{\text{configs}} S_i \exp(-H/kT)}{Z} \tag{15.54}$$

which serves as a local order parameter to distinguish the two (ordered and disordered) phases. The Ising Hamiltonian, however, does not possess a local symmetry: rather, it has a global invariance under the replacement

$$S_i \to -S_i \qquad \text{all sites } i. \tag{15.55}$$

This may be called a 'global Z(2) symmetry': global—because the spins must be flipped everywhere at the same time; and Z(2)—because the possible spin values ± 1 are just the elements of the group Z(2) under multiplication. Now we know that local gauge invariance is concerned with the freedom to make *local* changes as opposed to *global*. To construct a spin theory with an invariance under a *local* spin transformation, Wegner formulated an Ising spin system in which the spins were associated with the *links* instead of the sites. The links can be labelled by the site n, and a direction vector μ and ν (for two dimensions), as shown in figure 15.12. We can now define a local spin transformation as the operation $G(n)$ which reverses all the spins on the links attached to the site n (figure 15.13). To construct a 'Hamiltonian' that is invariant under this local spin flip operation, the simplest 'most local' way is to make H depend only on 'plaquette' variables. An elementary plaquette spin interaction is just the product of spins around an elementary square of links on the lattice, as shown in figure 15.14. If we define the total 'energy' as

$$H = \sum_{\square} H_{\square} \tag{15.56}$$

with the plaquette energy given by

$$H_{\square} = -JS_1S_2S_3S_4 \tag{15.57}$$

for each plaquette, then this Hamiltonian is clearly invariant under local spin transformations of any site of the lattice, since *two* spins of any plaquette are always reversed. Thus we have defined a spin theory invariant under local Z(2) transformations—the so-called gauge Ising model.

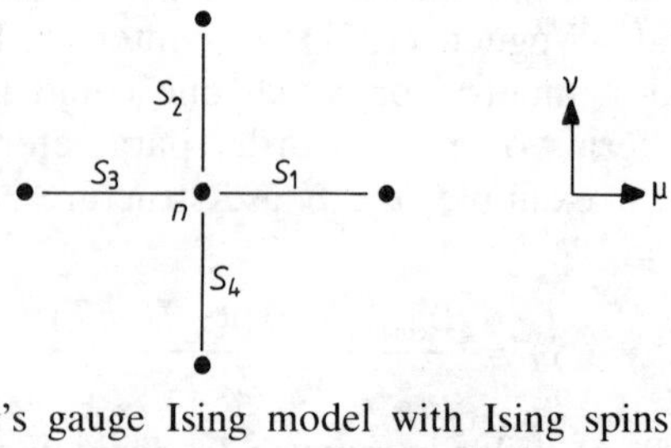

Figure 15.12 Wegner's gauge Ising model with Ising spins associated with the links from a site n.

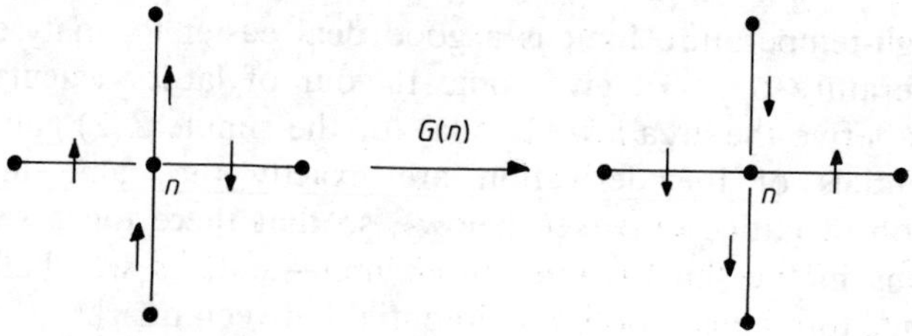

Figure 15.13 Local spin transformation for the gauge Ising model.

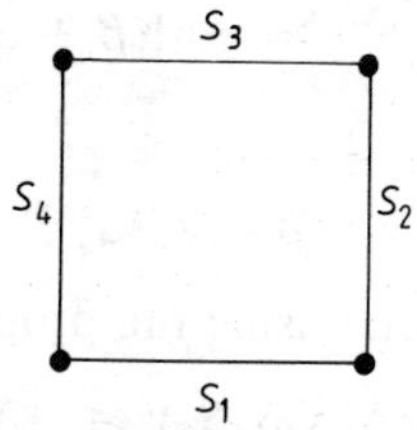

Figure 15.14 Elementary 'plaquette' of spins.

The model is interesting because one can prove there can be no non-zero local order parameter. Nevertheless, Wegner was able to show that there were still two phases by examining the behaviour of the product of link spins around some closed loop C (figure 15.15). This *loop parameter* $W(C)$ is locally invariant, and is the forerunner of the Wilson loop of lattice gauge theories. It can be shown that $W(C)$ has the following asymptotic behaviour:

$$\text{high temperature } T \qquad W(C) \sim \exp(-A) \tag{15.58a}$$

$$\text{low temperature } T \qquad W(C) \sim \exp(-P) \tag{15.58b}$$

where A is the area enclosed by the contour C, and P is its perimeter. Thus $W(C)$ serves to distinguish the two phases.

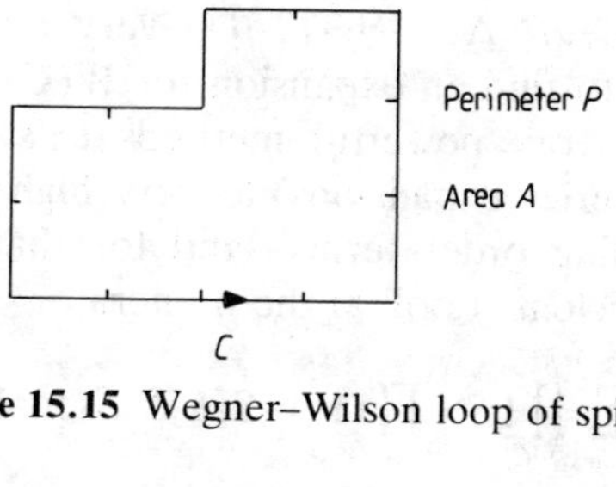

Figure 15.15 Wegner–Wilson loop of spins.

The high-temperature limit is a good deal easier to analyse than the low-temperature one. To give some flavour of lattice calculations, we shall now derive the area law (15.58a) for the simple Z(2) gauge theory. The essentials of the derivation are exactly the same in Wilson's formulation of lattice QCD (see below), so that there too a similar area law obtains in the 'high-temperature' limit—and, as we shall see, this corresponds to a linear confining potential between quarks.

We begin by rewriting the exponential in (15.54):

$$\begin{aligned}\exp(-H_\Box/kT) &= \exp((J/kT)S_1S_2S_3S_4) \\ &= \cosh\beta + S_1S_2S_3S_4\sinh\beta\end{aligned} \tag{15.59}$$

where we have defined

$$\beta = J/kT \tag{15.60}$$

and expanded the exponential using the property that

$$(S_1S_2S_3S_4)^2 = +1. \tag{15.61}$$

Now rewrite the exponential of a sum of terms as the product of exponentials, so that

$$\exp(-\textstyle\sum H_\Box/kT) = \Pi_\Box(\cosh\beta + S_1S_2S_3S_4\sinh\beta) \tag{15.62}$$

where $S_1S_2S_3S_4$ means the product of spins around the appropriate plaquette □. The expectation value of an operator Q in statistical mechanics is given by

$$\langle Q\rangle = \frac{\Sigma_{\text{configs}}\, Q\mathrm{e}^{-H/kT}}{\Sigma_{\text{configs}}\, \mathrm{e}^{-H/kT}} \tag{15.63}$$

which has the desirable property that the expectation value of the unit operator is one, as required. Applying this formula to the Wegner–Wilson loop operator W, we obtain

$$W(C) = \left\langle \prod_C S_i \right\rangle = \frac{\Sigma_{\text{configs}}(\Pi_C S_i)\Pi_\Box(1 + S_1S_2S_3S_4\tanh\beta)}{\Sigma_{\text{configs}}\Pi_\Box(1 + S_1S_2S_3S_4\tanh\beta)} \tag{15.64}$$

after dividing top and bottom by the same number of $(\cosh\beta)$ factors. How does all this help? At high T, β is very small and so is $(\tanh\beta)$. We want therefore to find an expansion for $W(C)$ in terms of powers of $(\tanh\beta)$. Although very powerful methods exist for performing such 'high-temperature series expansions' to very high orders, we only want to exhibit the leading order terms—and for that a little thought and brute force are sufficient. Look at the numerator

$$\sum_{\text{configs}}\left(\prod_C S_i\right)\prod_\Box(1 + S_1S_2S_3S_4\tanh\beta). \tag{15.65}$$

Clearly the leading order term will be the one with fewest factors of $(\tanh\beta)$ coming from the product over all plaquettes of the lattice. Why can we not just take the '1' term for all these plaquettes? We cannot because of the factors of S_i for each link on the contour C. Consider one such factor and a two-dimensional lattice for simplicity, so that this link contributes to two plaquettes (figure 15.16).

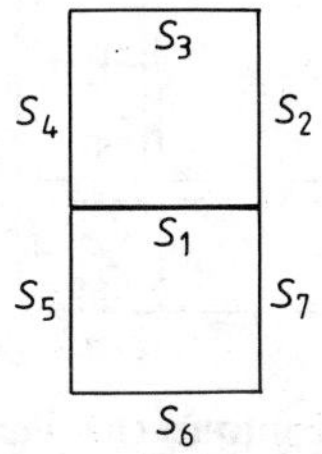

Figure 15.16 On a two-dimensional lattice, link spin S_1 contributes to two plaquettes in the action.

In the numerator we have a sum over spin configurations: this is a sum over the two possible values, ± 1, of an Ising (Z(2)) spin. Consider the sum over the two values of the link spin S_1:

$$\sum_{S_1=\pm 1} S_1(1 + S_1S_2S_3S_4 \tanh\beta)(1 + S_1S_5S_6S_7 \tanh\beta). \qquad (15.66)$$

If we take the 1 factors from both the plaquette terms, we are left with zero since

$$\sum_{S_1=\pm 1} S_1 = 0. \qquad (15.67)$$

We only get a non-zero contribution if we take one $(\tanh\beta)$ from one of the plaquette terms, since then

$$\sum_{S_1=\pm 1} (S_1)^2 = 2. \qquad (15.68)$$

Thus to avoid getting zero, we must have a $(\tanh\beta)$ term for all links on the contour C. Then to avoid getting zero from the spin sum over one of the other plaquette spins such as S_3, we must include the $(\tanh\beta)$ term from that plaquette and so on. In fact the leading order contribution with the minimum number of $(\tanh\beta)$ factors must come from 'tiling' the inside of the contour with the minimum number of plaquettes. In two dimensions this is of course necessarily flat, but in higher dimensions there are non-leading contributions coming from closed surfaces with bumps and so on. Thus, for a contour C, the leading high-temperature contribution corresponds to taking the $(\tanh\beta)$ factors for each of the plaquettes enclosed by C (figure 15.17). Finally then, for A tiles in area A, we obtain

$$W(C) \sim (\tanh\beta)^A \sim \exp[A \ln(\tanh\beta)] \tag{15.69}$$

and an area law dependence of the loop order parameter.

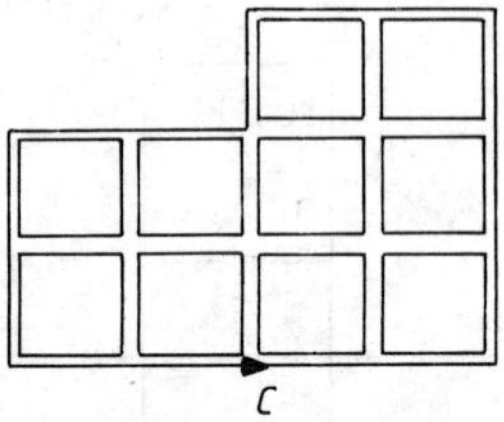

Figure 15.17 Minimal tiling of contour C giving the leading contribution to the Wegner–Wilson loop at high temperatures.

It is now not difficult to understand Wilson's formulation of lattice gauge theory. The Wilson action is similarly defined in terms of plaquette variables, and the gauge field variables are associated with the links of the lattice as for the Z(2) gauge theory. Now, however, instead of just the two elements ± 1 of Z(2), each link has associated with it an element of the corresponding gauge group:

$$\mathrm{U}(1) \quad U = \mathrm{e}^{\mathrm{i}\theta} \tag{15.70}$$

$$\mathrm{SU}(3) \quad \mathrm{U} = \mathrm{e}^{\mathrm{i}\boldsymbol{\lambda}\cdot\boldsymbol{\theta}/2}. \tag{15.71}$$

These lattices theories clearly have the required local invariance, but we must relate θ to the appropriate gauge field. Consider the case of U(1) which should correspond to QED. If we take

$$U = \mathrm{e}^{\mathrm{i}\theta} \tag{15.72}$$

and

$$S(U) = \sum_B S_B \tag{15.73}$$

with S_B the product of link variables around a plaquette, we can consider how this extrapolates to the familiar QED action. With

$$\theta = agA_\mu \tag{15.74}$$

and

$$\sum_{\Box} \rightarrow \int \frac{\mathrm{d}^4 x}{a^4} \tag{15.75}$$

we can show that S formally reduces to the familiar form as the lattice spacing a tends to zero:

$$S \to \tfrac{1}{4}\int d^4x\, F_{\mu\nu}F^{\mu\nu}. \tag{15.76}$$

As we have seen, the 'high-temperature' limit corresponds on the lattice to strong coupling, with the coupling g specific to a particular lattice spacing a. The high-temperature/strong-coupling limit is quite easy to obtain and, as in the gauge Z(2) case, both the U(1) and SU(3) lattice gauge theories display an area law behaviour:

$$\text{strong coupling} \quad W(C) \sim e^{-A}. \tag{15.77}$$

Wilson interpreted this area law behaviour for gauge theories as a signal for a confining phase. In the case of a long rectangular contour, labelled by t and R (figure 15.18), the effect of the loop operator is equivalent to the creation of a charged pair, rapidly separated to a distance R apart, followed by evolution of the system for a time t, and the subsequent annihilation of the charges. In this case $W(C)$ measures the static charge potential

$$W(C) \sim e^{-V(R)t}. \tag{15.78}$$

Thus an area law behaviour corresponds to a potential linear in R, the traditional signal for a confining force.

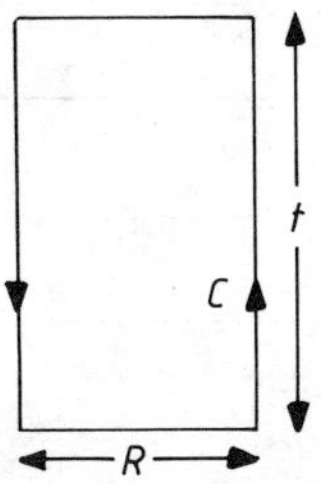

Figure 15.18 Wilson loop contour for quark–antiquark potential.

Such a confining phase is of course desired for SU(3), but not for U(1) (QED), since we know that electrons, for example, are not confined. How can we resolve this apparent problem? In the case of QCD, the effective coupling decreases with distance thanks to asymptotic freedom. To make connection with continuum physics, we must ensure that the lattice is not too coarse and there are enough discrete points inside a lattice hadron to approximate continuum QCD. Taking this continuum limit in QCD corresponds to taking g^2 on the lattice to zero. Thus, although we have a signal for confinement at large g^2, we must make sure this persists for all g^2 down to zero, and that we do not encounter a phase transition as g^2 is reduced (i.e. as the 'temperature' is lowered). By contrast, in QED, since electrons are not confined, there must be a phase transition as the coupling is varied and we approach the continuum limit. (Note that this may occur at a non-zero value of the coupling, since QED is not an asymptotically free theory.) This phase transition should be visible in the changeover from area to perimeter law dependence for the Wilson loop.

Although lattice QCD was initiated in the early 1970s by Wilson, it was not until Creutz (1979) performed some striking numerical simulations that it became an established technique for investigating the non-perturbative behaviour of field theories. Creutz used a simple Monte Carlo method to simulate the behaviour of both U(1) and SU(2) lattice theories. Figure 15.19 shows the variation of plaquette energy P with

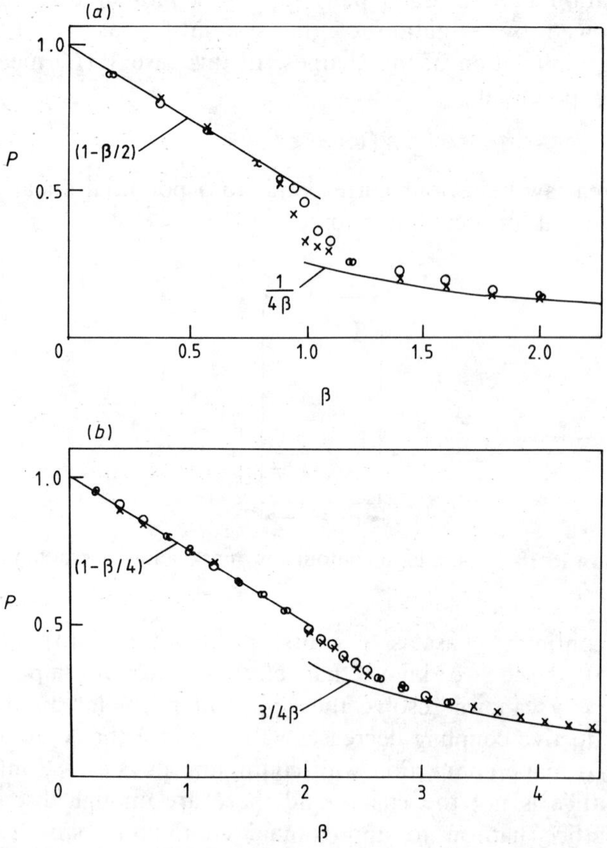

Figure 15.19 Numerical simulations (Creutz 1983) of thermal cycles for (*a*) U(1) and (*b*) SU(2) lattice gauge theories in four dimensions. The full curves are low-order high- and low-temperature predictions; X, heating; O, cooling.

$\beta(\sim 1/T) \sim 1/g^2$ for 'hysteresis' cycles for both these theories. In a hysteresis cycle, the system is equilibrated at one temperature, the energy measured, then the temperature changed a little and the system re-equilibrated. In this way the system can be started at high temperature (low β), cooled and then reheated. As can be seen for U(1) in four dimensions, there is a clear indication of a hysteresis loop in the middle of the range, suggestive of a phase transition. By contrast, SU(2) in four dimensions shows no such effect. This is exactly the behaviour we expect: a deconfining phase transition for the Abelian U(1) gauge theory—but for the non-Abelian SU(2) theory, confinement persists for all values of the coupling.

From this early numerical work, performed on lattices as small as 8^4, a whole industry has grown. The lattice theories we have so far discussed have only possessed gauge fields: a realistic simulation of both QED and QCD must include fermions. Much work in examining the hadron spectrum, glueballs, structure functions and form factors is still in progress, using more sophisticated simulation techniques and ever more powerful computers. Besides using the latest vector supercomputers, particle physicists have also built special-purpose parallel computers dedicated to realistic simulations of QCD. Although there are promising indications, a definitive solution for the spectrum of QCD is probably still many years away.

Summary

Necessity for ghost loops
- difference between imaginary part of Feynman graph, and unitarity relation, for intermediate states involving one photon;
- difference is due to unphysical ($\varepsilon \cdot k \neq 0$) polarisation states for internal photon in Feynman graph;
- difference vanishes due to Ward identity;
- no unphysical contributions at all in QED;
- unphysical contribution fails to vanish in non-Abelian case for gg intermediate state;
- offending remainder traced to gluon loop;
- *ad hoc* introduction of compensating term: 'ghost loop'.

Grand unification
- symmetry expected only for energies $\gg$ some M_X;
- prediction of M_X;
- proton lifetime estimates, and experiments;
- general arguments for baryon instability;
- prediction of $\sin^2 \theta_W$;
- renormalisation effects;

prediction of lepton masses;
charge quantisation.

Confinement
type II superconductor;
flux filament ending on magnetic monopoles, as model for meson; confines monopoles;
magnetic ↔ electric swop, to confine charges.

Lattice gauge theory
lattice as an ultraviolet regulator;
sum-over-paths formalism;
statistical mechanics analogy;
Wegner's Z(2) gauge Ising model, and loop parameter;
Wilson loop;
confining and deconfining phases in U(1) and SU(2) gauge theories.

Problem

15.1 Verify equation (15.10). (Hint: from covariance the second-rank tensor $P_{\mu\nu}$ must have the general form

$$P_{\mu\nu} = Ag_{\mu\nu} + Bk_\mu k_\nu + Ck_\mu t_\nu + Dk_\nu t_\mu + Et_\mu t_\nu$$

where A, B, C, D and E are invariant functions, since $g_{\mu\nu}$, $k_\mu k_\nu$, etc are the only second-rank tensors available. By contracting this with $\varepsilon^\mu(k, \lambda')$ and using (15.8), show that $A = -1$. Next, contract $P_{\mu\nu}$ with k^μ and k^ν to show that E must be zero. Using contractions with k and t, find $C(= D)$ and B.)

APPENDIX A

QUANTUM MECHANICS, SPECIAL RELATIVITY AND δ FUNCTION

Sections A.1 and A.2 are intended as very terse summaries of those elements of non-relativistic quantum mechanics and special relativity that are used throughout this book. Fuller accounts of these topics may be found in the books by Gasiorowicz (1974), Feynman (1963b) and Aitchison (1972). Section A.3 deals with a more advanced topic, the various 'pictures' which can be used in quantum mechanics, together with an application to time-dependent perturbation theory which generalises (A.17), (A.19) and (A.20). Section A.4 contains an elementary introduction to the δ function.

A.1 Non-relativistic quantum mechanics (NRQM)

Natural units $\hbar = c = 1$ (see Appendix B).
Fundamental postulate of QM:

$$[\hat{p}_i, \hat{x}_j] = -\mathrm{i}\delta_{ij}. \tag{A.1}$$

Coordinate representation:

$$\hat{\boldsymbol{p}} = -\mathrm{i}\boldsymbol{\nabla} \tag{A.2}$$

$$\hat{H}\psi(\boldsymbol{x}, t) = \mathrm{i}\frac{\partial\psi(\boldsymbol{x}, t)}{\partial t}. \tag{A.3}$$

A.1.1 Schrödinger equation for a spinless particle

$$\hat{H} = \frac{\hat{\boldsymbol{p}}^2}{2m} + \hat{V} \tag{A.4}$$

and so

$$\left(-\frac{1}{2m}\boldsymbol{\nabla}^2 + \hat{V}(\boldsymbol{x}, t)\right)\psi(\boldsymbol{x}, t) = \mathrm{i}\,\frac{\partial\psi(\boldsymbol{x}, t)}{\partial t}. \tag{A.5}$$

Probability current (see problem 3.1)

$$\rho = \psi^*\psi \tag{A.6}$$

$$= |\psi|^2 \geqslant 0 \tag{A.7}$$

$$\boldsymbol{j} = -\frac{\mathrm{i}}{2m}[\psi^*(\boldsymbol{\nabla}\psi) - (\boldsymbol{\nabla}\psi^*)\psi] \tag{A.8}$$

with

$$\frac{\partial \rho}{\partial t} + \nabla \cdot \boldsymbol{j} = 0. \tag{A.9}$$

Free particle solutions

$$\phi(\boldsymbol{x}, t) = u(\boldsymbol{x})\mathrm{e}^{-\mathrm{i}Et} \tag{A.10}$$

$$\hat{H}_0 u = Eu \tag{A.11}$$

where

$$\hat{H}_0 = \hat{H}(\hat{V} = 0). \tag{A.12}$$

Box normalisation:

$$\int_V u^*(\boldsymbol{x})u(\boldsymbol{x})\mathrm{d}^3x = 1. \tag{A.13}$$

Interaction with electromagnetic field
Particle of charge q in electromagnetic vector potential $\boldsymbol{A}$

$$\boxed{\hat{\boldsymbol{p}} \rightarrow \hat{\boldsymbol{p}} - q\boldsymbol{A}.} \tag{A.14}$$

Thus

$$\frac{1}{2m}(\hat{\boldsymbol{p}} - q\boldsymbol{A})^2\psi = \mathrm{i}\frac{\partial \psi}{\partial t} \tag{A.15}$$

and so

$$-\frac{1}{2m}\nabla^2\psi + \mathrm{i}\frac{q}{m}\boldsymbol{A}\cdot\nabla\psi + \frac{q^2}{2m}\boldsymbol{A}^2\psi = \mathrm{i}\frac{\partial \psi}{\partial t}. \tag{A.16}$$

Note: (a) chosen gauge $\nabla\cdot\boldsymbol{A} = 0$; (b) usually neglect q^2 term.

Time-dependent perturbation theory

$$\hat{H} = \hat{H}_0 + \hat{V}, \tag{A.17}$$

$$\hat{H}\psi = \mathrm{i}\partial\psi/\partial t. \tag{A.3}$$

Unperturbed problem:

$$\hat{H}_0 u_n = E_n u_n. \tag{A.18}$$

Completeness:

$$\psi(\boldsymbol{x}, t) = \sum_n a_n(t)u_n(\boldsymbol{x})\mathrm{e}^{-\mathrm{i}E_n t}. \tag{A.19}$$

First-order perturbation theory

$$a_{\mathrm{fi}} = \mathrm{i}^{-1} \int \int \underbrace{\mathrm{d}^3x\, \mathrm{d}t}_{\substack{\text{four-dimensional}\\ \text{volume element}}} \underbrace{u_{\mathrm{f}}^*(\boldsymbol{x})\mathrm{e}^{+\mathrm{i}E_{\mathrm{f}}t}}_{\text{final state}} \underbrace{\hat{V}(\boldsymbol{x}, t)}_{\substack{\text{perturbing}\\ \text{potential}}} \underbrace{u_{\mathrm{i}}(\boldsymbol{x})\mathrm{e}^{-\mathrm{i}E_{\mathrm{i}}t}}_{\text{initial state}}. \tag{A.20}$$

Important examples
(i) $\hat{V}$ independent of t:

$$a_{\mathrm{fi}} = \mathrm{i}^{-1} V_{\mathrm{fi}} 2\pi\delta(E_{\mathrm{f}} - E_{\mathrm{i}}) \tag{A.21}$$

where

$$V_{\mathrm{fi}} = \int \mathrm{d}^3x\, u_{\mathrm{f}}^*(\boldsymbol{x})\hat{V}(\boldsymbol{x})u_{\mathrm{i}}(\boldsymbol{x}). \tag{A.22}$$

(ii) Oscillating time-dependent potential:
(a) if $\hat{V} \sim \mathrm{e}^{-\mathrm{i}\omega t}$, time integral of a_{fi} is

$$\int \mathrm{d}t\, \mathrm{e}^{+\mathrm{i}E_{\mathrm{f}}t}\mathrm{e}^{-\mathrm{i}\omega t}\mathrm{e}^{-\mathrm{i}E_{\mathrm{i}}t} = 2\pi\delta(E_{\mathrm{f}} - E_{\mathrm{i}} - \omega) \tag{A.23}$$

i.e. system has absorbed energy from potential;
(b) if $\hat{V} \sim \mathrm{e}^{+\mathrm{i}\omega t}$, time integral of a_{fi} is

$$\int \mathrm{d}t\, \mathrm{e}^{+\mathrm{i}E_{\mathrm{f}}t}\mathrm{e}^{+\mathrm{i}\omega t}\mathrm{e}^{-\mathrm{i}E_{\mathrm{i}}t} = 2\pi\delta(E_{\mathrm{f}} + \omega - E_{\mathrm{i}}) \tag{A.24}$$

i.e. potential has absorbed energy from system.

Absorption and emission of photons
For electromagnetic radiation, far from sources, vector potential satisfies the wave equation

$$\boldsymbol{\nabla}^2\boldsymbol{A} - \frac{\partial^2\boldsymbol{A}}{\partial t^2} = 0. \tag{A.25}$$

Solution:

$$\boldsymbol{A}(\boldsymbol{x}, t) = \boldsymbol{A}_0 \exp(-\mathrm{i}\omega t + \mathrm{i}\boldsymbol{k}\cdot\boldsymbol{x}) + \boldsymbol{A}_0^* \exp(+\mathrm{i}\omega t - \mathrm{i}\boldsymbol{k}\cdot\boldsymbol{x}). \tag{A.26}$$

With gauge condition $\boldsymbol{\nabla}\cdot\boldsymbol{A} = 0$ we have

$$\boldsymbol{k}\cdot\boldsymbol{A}_0 = 0 \tag{A.27}$$

and there are two independent polarisation vectors for photons.
Treat interaction in first-order perturbation theory:

$$\hat{V}(\boldsymbol{x}, t) = (\mathrm{i}q/m)\boldsymbol{A}(\boldsymbol{x}, t)\cdot\boldsymbol{\nabla}. \tag{A.28}$$

Thus

$$\boldsymbol{A}_0 \exp(-\,\mathrm{i}\omega t + \mathrm{i}\boldsymbol{k}\cdot\boldsymbol{x}) \equiv \text{absorption of photon of energy } \omega$$

$$\boldsymbol{A}_0^* \exp(+\,\mathrm{i}\omega t - \mathrm{i}\boldsymbol{k}\cdot\boldsymbol{x}) \equiv \text{emission of photon of energy } \omega.$$

A.1.2 Pauli equation for a spin-$\frac{1}{2}$ particle

Wavefunction now two-component column matrix (spinor): general stationary state can be written as linear combination of separable wavefunctions of the form

$$\psi(\boldsymbol{x}, s) = \phi(\boldsymbol{x})\chi \tag{A.29}$$

with

$$\chi = \begin{pmatrix} a \\ b \end{pmatrix}. \tag{A.30}$$

Interaction with electromagnetic field

Electron in magnetic field $\boldsymbol{B}$ described by vector potential $\boldsymbol{A}$ is found to obey Pauli equation

$$\left(\frac{1}{2m}(\hat{\boldsymbol{p}} + \mathrm{e}\boldsymbol{A})^2 + \frac{ge}{2m}\boldsymbol{S}\cdot\boldsymbol{B}\right)\psi = E\psi \tag{A.31}$$

where the g factor is given by

$$g \simeq 2 \qquad \text{for spin angular momentum}$$

to be compared with

$$g = 1 \qquad \text{for classical angular momentum.}$$

Spin angular momentum

Operators $\boldsymbol{S}$ are matrices acting on χ's satisfying usual angular momentum commutation relations†

$$[S_i, S_j] = \mathrm{i}\varepsilon_{ijk}S_k. \tag{A.32}$$

Conventional choice is

$$\boldsymbol{S} = \tfrac{1}{2}\boldsymbol{\sigma} \tag{A.33}$$

where the Pauli matrices are

$$\sigma_x = \begin{pmatrix} 0 & 1 \\ 1 & 0 \end{pmatrix}, \qquad \sigma_y = \begin{pmatrix} 0 & -\mathrm{i} \\ \mathrm{i} & 0 \end{pmatrix}, \qquad \sigma_z = \begin{pmatrix} 1 & 0 \\ 0 & -1 \end{pmatrix} \tag{A.34}$$

†Note summation convention for repeated indices:

$$\varepsilon_{ijk}S_k \equiv \sum_{k=1}^{3} \varepsilon_{ijk}S_k.$$

This convention is used throughout this book, unless specifically indicated otherwise.

which satisfy (problem 3.2)

$$\sigma_i \sigma_j = \delta_{ij} + \mathrm{i}\varepsilon_{ijk}\sigma_k. \tag{A.35}$$

A.2 Special relativity: invariance and covariance

Under a rotation, the coordinates of a point P in the two systems are related by an orthogonal matrix $\mathbf{M}$

$$\boldsymbol{r}' = \mathbf{M}\boldsymbol{r} \tag{A.36a}$$

or

$$r_i' = M_{ij} r_j. \tag{A.36b}$$

Since

$$\mathbf{M}^{\mathrm{T}}\mathbf{M} = 1 \tag{A.37}$$

rotations leave length OP 'invariant':

$$\boldsymbol{r}'^2 = \boldsymbol{r}^2. \tag{A.38}$$

3-vectors
Objects with three components that transform like $\boldsymbol{r}$:

$$\boldsymbol{A}' = \mathbf{M}\boldsymbol{A}. \tag{A.39}$$

If $\boldsymbol{A}$ and $\boldsymbol{B}$ are 3-vectors,

$$\boldsymbol{A}'\cdot\boldsymbol{B}' = \boldsymbol{A}\cdot\boldsymbol{B}, \qquad \text{'invariant'}. \tag{A.40}$$

Tensors
Defined by generalising definition of 3-vector, e.g.

$$T_{ik}' = M_{ij} M_{kl} T_{jl}. \tag{A.41}$$

Each index transforms with $\mathbf{M}$.
Form invariance or covariance
An example is Newton's law:

$$\boldsymbol{F} = m\ddot{\boldsymbol{r}}. \tag{A.42}$$

In rotated frame it has the same form, i.e.

$$\boldsymbol{F}' = m\ddot{\boldsymbol{r}}', \qquad \text{'covariance'}. \tag{A.43}$$

Spinors and rotations
For an electron spin in a magnetic field, the Pauli equation effectively reduces to

$$\boldsymbol{\sigma}\cdot\boldsymbol{B}\chi = E\chi \tag{A.44}$$

where

$$\begin{aligned} \boldsymbol{\sigma}\cdot\boldsymbol{B} &= \sigma_x B_x + \sigma_y B_y + \sigma_z B_z \\ &= \begin{pmatrix} B_z & B_x - \mathrm{i}B_y \\ B_x + \mathrm{i}B_y & -B_z \end{pmatrix}. \end{aligned} \tag{A.45}$$

Under rotation of coordinates:

$$\boldsymbol{B} \rightarrow \boldsymbol{B}', \qquad \text{vector}$$

but

$$E \rightarrow E, \qquad \text{scalar} \rightarrow \text{unchanged.}$$

Covariance or *form invariance* of Pauli equation implies there must be χ' for rotated system such that

$$\boldsymbol{\sigma}\cdot\boldsymbol{B}'\chi' = E\chi' \tag{A.46}$$

where

$$\boldsymbol{B}' = \mathbf{M}\boldsymbol{B}. \tag{A.47}$$

Rewrite

$$\boldsymbol{\sigma}\cdot\boldsymbol{B}\chi = E\chi \tag{A.44}$$

by introducing matrix $\mathbf{U}$:

$$\mathbf{U}(\boldsymbol{B}\cdot\boldsymbol{\sigma})\mathbf{U}^{-1}\mathbf{U}\chi = E\mathbf{U}\chi \tag{A.48}$$

and construct $\mathbf{U}$ so that

$$\mathbf{U}(\boldsymbol{B}\cdot\boldsymbol{\sigma})\mathbf{U}^{-1} = \boldsymbol{B}'\cdot\boldsymbol{\sigma}; \tag{A.49}$$

then

$$\chi' = \mathbf{U}\chi \tag{A.50}$$

is the required form of the spinor in the rotated frame.

Example
For a rotation by an angle θ about axis $\boldsymbol{n}$, can show

$$\mathbf{U} = \exp(\mathrm{i}\boldsymbol{\sigma}\cdot\boldsymbol{n}\theta/2) \tag{A.51}$$

so that

$$\chi' = \exp(\mathrm{i}\boldsymbol{\sigma}\cdot\boldsymbol{n}\theta/2)\chi. \tag{A.52}$$

$\chi^\dagger\chi$ is invariant under rotations

$$\chi'^\dagger\chi' = \chi^\dagger\mathbf{U}^\dagger\mathbf{U}\chi = \chi^\dagger\chi \tag{A.53}$$

since

$$\mathbf{U}^\dagger\mathbf{U} = \mathbf{1}. \tag{A.54}$$

Hence $\chi^\dagger\chi$ is *invariant* under rotations: a *scalar*.

$\chi^\dagger \boldsymbol{\sigma} \chi$ is a vector under rotations

$$V = \chi^\dagger \boldsymbol{\sigma} \chi \tag{A.55}$$

and

$$\chi' = \mathbf{U}\chi. \tag{A.50}$$

$\mathbf{U}$ has explicit property (from (A.49))

$$\mathbf{U}^\dagger \sigma_i \mathbf{U} = M_{ij}\sigma_j. \tag{A.56}$$

Hence

$$V' = \mathbf{M} V \tag{A.57}$$

and $\chi^\dagger \boldsymbol{\sigma} \chi$ is a vector under rotations.

Lorentz invariance
Define $x^0 = ct$; $c = 1$. For Lorentz transformation in x^1 direction

$$\begin{aligned} x'^0 &= x^0 \cosh\omega - x^1 \sinh\omega \\ x'^1 &= -x^0 \sinh\omega + x^1 \cosh\omega \\ x'^2 &= x^2 \qquad x'^3 = x^3 \end{aligned} \tag{A.58}$$

where $\beta = v/c$ and $\cosh\omega = (1 - \beta^2)^{-1/2}$. Analogous to rotation through complex angle: Lorentz transformations leave invariant the quantity

$$x'^{02} - \boldsymbol{x}'^2 = x^{02} - \boldsymbol{x}^2. \tag{A.59}$$

4-vectors

$$x^\mu = (x^0, \boldsymbol{x}) \qquad (\mu = 0, 1, 2, 3). \tag{A.60}$$

Example:

$$p^\mu = (E, \boldsymbol{p}) \tag{A.61}$$

$$\partial^\mu = (\partial^0, -\boldsymbol{\nabla}). \tag{A.62}$$

Scalar products and the metric tensor
For two 4-vectors, A^μ and B^μ, the 'scalar product'

$$A \cdot B \equiv A^0 B^0 - \boldsymbol{A} \cdot \boldsymbol{B} \tag{A.63}$$

is invariant under Lorentz transformations. Convenient to introduce upper and lower indices via metric tensor

$$A_\mu = g_{\mu\nu} A^\nu \tag{A.64}$$

where

$$g_{\mu\nu} = g^{\mu\nu} = \begin{pmatrix} 1 & 0 & 0 & 0 \\ 0 & -1 & 0 & 0 \\ 0 & 0 & -1 & 0 \\ 0 & 0 & 0 & -1 \end{pmatrix}. \tag{A.65}$$

Thus

$$A^{\mu} = (A^0, \mathbf{A}) \tag{A.63}$$

but

$$A_{\mu} = (A^0, -\mathbf{A}). \tag{A.66}$$

The scalar product $A \cdot B$ may now be written as

$$A \cdot B = A_{\mu}B^{\mu} = g_{\mu\nu}A^{\nu}B^{\mu}. \tag{A.67}$$

A.3 Schrödinger, Heisenberg and interaction pictures: elementary theory of transitions in quantum field theory

The formulation of quantum mechanics briefly recapitulated in §A.1 is one in which the dynamical variables (such as $\boldsymbol{x}$ and $\hat{\boldsymbol{p}} = -\mathrm{i}\boldsymbol{\nabla}$) are independent of time, while the wavefunction ψ changes with time according to equation (A.3), which we rewrite here:

$$\hat{H}\psi(\boldsymbol{x}, t) = \mathrm{i}\frac{\partial\psi(\boldsymbol{x}, t)}{\partial t}. \tag{A.68}$$

Matrix elements of operators $\hat{A}$ depending on $\boldsymbol{x}$, $\hat{\boldsymbol{p}}$, . . . then have the form

$$\langle\phi|\hat{A}|\psi\rangle = \int \phi^*(\boldsymbol{x}, t)\,\hat{A}\,\psi(\boldsymbol{x}, t)\,\mathrm{d}^3\boldsymbol{x} \tag{A.69}$$

and will in general depend on time via the time dependences of ϕ and ψ. Although used almost universally in introductory courses on quantum mechanics, this formulation is not the only possible one, nor is it always the most convenient.

We may, for example, wish to bring out similarities (and differences) between the general dynamical frameworks of quantum and classical mechanics. The above formulation does not seem to be well adapted to this purpose, since in the classical case the dynamical variables depend on time ($\boldsymbol{x}(t)$, $\boldsymbol{p}(t)$, . . .) and obey equations of motion, while according to (A.68) the quantum variables are time-independent and the 'equation of motion' (A.68) is for the wavefunction ψ, which has no classical counterpart. In quantum mechanics, however, it is always possible to make unitary transformations of the state vectors or wavefunctions. We can make use of this possibility to obtain an alternative formulation of quantum mechanics, which is in some ways closer to the spirit of classical mechanics, as follows.

Equation (A.68) can be formally solved to give

$$\psi(\boldsymbol{x}, t) = \mathrm{e}^{-\mathrm{i}\hat{H}t}\,\psi(\boldsymbol{x}, 0) \tag{A.70}$$

where the exponential (of an operator!) can be defined by the corresponding power series, for example:

$$e^{-i\hat{H}t} = 1 - i\hat{H}t + \frac{1}{2!}(-i\hat{H}t)^2 + \ldots. \tag{A.71}$$

It is simple to check that (A.70), as defined by (A.71), does satisfy (A.68), and that the operator $\hat{U} = \exp(-i\hat{H}t)$ is unitary:

$$U^\dagger = [\exp(-i\hat{H}t)]^\dagger = \exp(i\hat{H}^\dagger t) = \exp(i\hat{H}t) = U^{-1} \tag{A.72}$$

where the Hermitian property $\hat{H}^\dagger = \hat{H}$ has been used. Thus (A.70) can be viewed as a unitary transformation from the time-dependent wavefunction $\psi(\boldsymbol{x}, t)$ to the time-independent one $\psi(\boldsymbol{x}, 0)$. Correspondingly the matrix element (A.69) is then

$$\langle\phi|\hat{A}|\psi\rangle = \int \phi^*(\boldsymbol{x}, 0)e^{i\hat{H}t}\hat{A}e^{-i\hat{H}t}\psi(\boldsymbol{x}, 0)\, d^3\boldsymbol{x} \tag{A.73}$$

which can be regarded as the matrix element of the *time-dependent* operator

$$\hat{A}(t) = e^{i\hat{H}t}\hat{A}e^{-i\hat{H}t} \tag{A.74}$$

between time-independent wavefunctions $\phi^*(\boldsymbol{x}, 0)$, $\psi(\boldsymbol{x}, 0)$.

Since (A.73) is perfectly general, it is clear that we can calculate amplitudes in quantum mechanics in either of the two ways outlined: (a) by using the time-dependent ψ's and time-independent $\hat{A}$'s, which is called the 'Schrödinger picture'; or (b) by using time-independent ψ's and time-dependent $\hat{A}$'s, which is called the 'Heisenberg picture'. The wavefunctions and operators in the two pictures are related by (A.70) and (A.74). We note that the pictures coincide at the (conventionally chosen) time $t = 0$.

Since $\hat{A}(t)$ is now time-dependent, we can ask for its equation of motion. Differentiating (A.74) carefully, we find (if $\hat{A}$ does not depend explicitly on t)

$$\frac{d\hat{A}(t)}{dt} = -i\,[\hat{A}(t), \hat{H}] \tag{A.75}$$

which is called the Heisenberg equation of motion for $\hat{A}(t)$. On the right-hand side of (A.75), $\hat{H}$ is the Schrödinger operator; however, if $\hat{H}$ is substituted for $\hat{A}$ in (A.74), one finds $\hat{H}(t) = \hat{H}$, so $\hat{H}$ can equally well be interpreted as the Heisenberg operator. For simple Hamiltonians $\hat{H}$, (A.75) leads to operator equations quite analogous to classical equations of motion, which can sometimes be solved explicitly (see §4.2.2).

The foregoing ideas apply equally well to the operators and state vectors of quantum field theory. In our introduction to field theory concepts in Chapter 4, we work in the Heisenberg picture. For the most part we discuss the formalism for free fields, and we only sketch the way in which interactions are handled. Thus, for example, the time dependence of the operators as given by the mode expansions (4.130) and (4.167) is that generated by the corresponding free particle Hamiltonians. In an interacting field theory, however, we cannot make much progress using this picture, since we do not then know the time dependence of the operators (which is generated by the full Hamiltonian). Instead we return to the Schrödinger picture, in which (adopting a more abstract notation) the states change with time according to

$$\hat{H}|\psi(t)\rangle = \mathrm{i}\frac{\mathrm{d}}{\mathrm{d}t}|\psi(t)\rangle \tag{A.76}$$

where

$$\hat{H} = \int \hat{\mathscr{H}}\, \mathrm{d}^3\boldsymbol{x}. \tag{A.77}$$

Note that although (A.76) is a 'Schrödinger equation', the Hamiltonian $\hat{H}$ will of course be fully relativistic. The field operators appearing in the density $\hat{\mathscr{H}}$ are all evaluated at fixed time $t = 0$, which is the time at which the Schrödinger and Heisenberg pictures coincide. At this fixed time, expansions of the form (4.130) and (4.167) with $t = 0$ are certainly possible, since the basis functions form complete sets.

In practice, of course, $\hat{H}$ is too complicated for (A.76) to be solved, and one resorts to perturbation theory, as in NRQM. We separate $\hat{H}$ into a part $\hat{H}_0$ which is the sum of all the free particle Hamiltonians for the relevant fields, and an interaction part $\hat{H}_{\text{int}}$ where

$$\hat{H}_{\text{int}} = \int \hat{\mathscr{H}}_{\text{int}}(\boldsymbol{x}, 0)\, \mathrm{d}^3\boldsymbol{x}. \tag{A.78}$$

We show in Chapter 4 that $\hat{H}_0$ is time-independent, and we have indicated explicitly in (A.78) that the field operators in $\hat{\mathscr{H}}_{\text{int}}$ are evaluated at $t = 0$. Thus for field theory in the Schrödinger picture, perturbation theory based on $\hat{H}_0 + \hat{H}_{\text{int}}$ is analogous to time-independent perturbation theory in NRQM. Indeed, to first order in $\hat{H}_{\text{int}}$ the amplitude for the transition between eigenstates $|\mathrm{i}\rangle$ and $|\mathrm{f}\rangle$ of $\hat{H}_0$ is

$$\mathscr{A}_{\mathrm{fi}}^{(1)} = -\mathrm{i}\int \mathrm{d}t\, \mathrm{e}^{\mathrm{i}E_{\mathrm{f}}t}\langle \mathrm{f}|\hat{H}_{\text{int}}|\mathrm{i}\rangle \mathrm{e}^{-\mathrm{i}E_{\mathrm{i}}t} \tag{A.79}$$

the only difference from (A.20) being that the matrix element is now understood to be between states specified by appropriate occupancies of free particle quanta. In general, of course, $\hat{H}_{\text{int}}$ will cause changes in the particle content, so that $|\mathrm{i}\rangle$ and $|\mathrm{f}\rangle$ need not contain the same particles.

Equation (A.79) can be written in various equivalent ways. We can integrate over t to obtain

$$\mathscr{A}_{\mathrm{fi}}^{(1)} = -\mathrm{i}\,\langle \mathrm{f}|\hat{H}_{\text{int}}|\mathrm{i}\rangle\, 2\pi\delta(E_{\mathrm{f}} - E_{\mathrm{i}}) \tag{A.80}$$

as in (A.21). Alternatively, we can observe that since $\hat{H}_0|\mathrm{i}\rangle = E_\mathrm{i}|\mathrm{i}\rangle$, and similarly for $|f\rangle$, the exponential factors can be rewritten in the form

$$\mathscr{A}_{\mathrm{fi}}^{(1)} = -\mathrm{i}\int \mathrm{d}t\,\langle \mathrm{f}|\mathrm{e}^{\mathrm{i}\hat{H}_0 t}\hat{H}_{\mathrm{int}}\mathrm{e}^{-\mathrm{i}\hat{H}_0 t}|\mathrm{i}\rangle. \tag{A.81}$$

Using (A.78), equation (A.81) becomes

$$\mathscr{A}_{\mathrm{fi}}^{(1)} = -\mathrm{i}\int \mathrm{d}^4x\,\langle \mathrm{f}|\hat{\mathscr{H}}_{\mathrm{int}}^{\mathrm{I}}(\boldsymbol{x}, t)|\mathrm{i}\rangle \tag{A.82}$$

where

$$\hat{\mathscr{H}}_{\mathrm{int}}^{\mathrm{I}}(\boldsymbol{x}, t) = \mathrm{e}^{\mathrm{i}\hat{H}_0 t}\hat{\mathscr{H}}_{\mathrm{int}}(\boldsymbol{x}, 0)\mathrm{e}^{-\mathrm{i}\hat{H}_0 t}. \tag{A.83}$$

Comparing (A.83) with (A.74), we see that the time-dependent operator $\hat{\mathscr{H}}_{\mathrm{int}}^{\mathrm{I}}(\boldsymbol{x}, t)$ is like a Heisenberg picture operator, except that its time dependence is generated (from the fixed t Schrödinger operator) by the *free* Hamiltonian $\hat{H}_0$, rather than the complete one $\hat{H}$. $\hat{\mathscr{H}}_{\mathrm{int}}^{\mathrm{I}}(\boldsymbol{x}, t)$ is called the 'interaction picture' operator corresponding to $\hat{\mathscr{H}}_{\mathrm{int}}(\boldsymbol{x}, 0)$. The fields in $\hat{\mathscr{H}}_{\mathrm{int}}^{\mathrm{I}}(\boldsymbol{x}, t)$ all have free particle time dependence and are therefore represented by mode expansions of the form (4.130) or (4.167), for example. Thus, in this formulation, the field expansions are fully relativistically covariant, and the volume element in (A.82) is invariant—hence $\mathscr{A}_{\mathrm{fi}}^{(1)}$ is invariant. Equation (A.82) forms the basis of our discussion in §5.2.

Second-order perturbation theory can also be considered. In the wave mechanical formalism, this will be cumbersome if $\hat{V}$ (cf §A.1) depends on time explicitly—and in fact we do not consider such cases in the text. In field theory, however, as stressed above, the perturbation $\hat{H}_{\mathrm{int}}$ is time-independent in the Schrödinger picture. Thus we can take over the formula of second-order transition theory for potentials constant in time (see e.g. Schiff 1955, §29). This means that the matrix element $\langle \mathrm{f}|\hat{H}_{\mathrm{int}}|\mathrm{i}\rangle$ in (A.80) is replaced by

$$\langle \mathrm{f}|\hat{H}_{\mathrm{int}}|\mathrm{i}\rangle + \sum_{n\neq \mathrm{i}}\langle \mathrm{f}|\hat{H}_{\mathrm{int}}|n\rangle\frac{1}{E_\mathrm{i}-E_n}\langle n|\hat{H}_{\mathrm{int}}|\mathrm{i}\rangle. \tag{A.84}$$

Since $\hat{H}_{\mathrm{int}}$ will, in general, not conserve the numbers of the various particles, the states $|n\rangle$ may contain different particles from those in $|\mathrm{i}\rangle$ or $|\mathrm{f}\rangle$; in particular, $|\mathrm{i}\rangle$ and $|\mathrm{f}\rangle$ may contain the same particles (in different momentum states), while $|n\rangle$ contains an additional quantum. This observation forms the basis of our treatment of single-quantum exchange processes, and forces, in Chapter 5 (§§5.6 and 5.8.2).

In contrast to the simple expression (A.82), which is 'manifestly' invariant for the reasons outlined above, it is *not* immediately apparent that the introduction of the second term in (A.84) into (A.80) will yield an invariant amplitude. Nevertheless, we demonstrate explicitly for a particular case in §5.6 that the second-order amplitude is invariant, provided we include contributions from *two* possible intermediate states $|n\rangle$. The single invariant amplitude corresponds in this case to the sum of two separate terms in the 'non-covariant' formalism of (A.84).

It is also possible to formulate directly a 'manifestly covariant' perturbation theory, extending (A.82) through use of the interaction picture operator $\mathcal{H}^{\mathrm{I}}_{\mathrm{int}}(\boldsymbol{x}, t)$. This is now one of the standard procedures explained in any professional quantum field theory book. The approach based on (A.84) is, however, quite sufficient for our purpose, and possibly more intuitively understandable.

A.4 The Dirac delta function

A.4.1. Introduction

Consider approximating an integral by a sum over strips Δx wide as shown in figure A.1:

$$\int_{x_1}^{x_2} f(x)\,\mathrm{d}x \simeq \sum_i f(x_i)\,\Delta x. \tag{A.85}$$

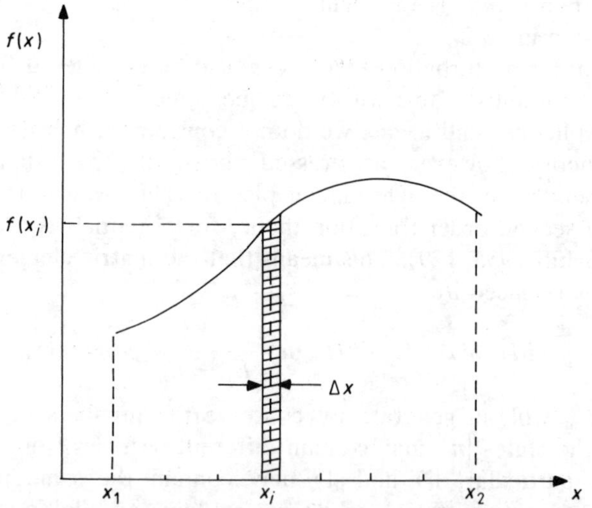

Figure A.1 Approximate evaluation of integral.

Consider the function $\delta(x - x_j)$ shown in figure A.2,

$$\delta(x - x_j) = \begin{cases} 1/\Delta x & \text{in } j\text{th interval} \quad (\text{A.86}) \\ 0 & \text{all others} \quad (\text{A.87}) \end{cases}.$$

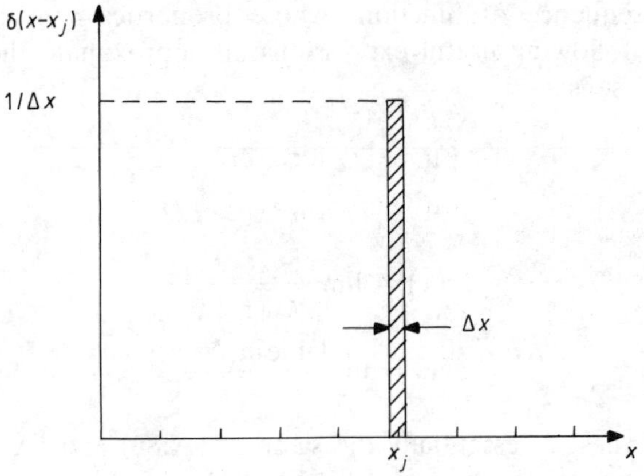

Figure A.2 The function $\delta(x - x_j)$.

Clearly this function has the properties

$$\sum_i f(x_i)\, \delta(x_i - x_j)\, \Delta x = f(x_j) \tag{A.88}$$

and

$$\sum_i \delta(x_i - x_j)\Delta x = 1. \tag{A.89}$$

In the limit as we pass to an integral form, we might expect (applying (A.85) to the left hand sides) that these equations to reduce to

$$\int_{x_1}^{x_2} f(x)\, \delta(x - x_j) = f(x_j) \tag{A.90}$$

and

$$\int_{x_1}^{x_2} \delta(x - x_j)\, \mathrm{d}x = 1 \tag{A.91}$$

provided that $x_1 < x_j < x_2$. Clearly such 'δ functions' can easily be generalised to more dimensions, e.g. 3D:

$$\mathrm{d}V = \mathrm{d}x\,\mathrm{d}y\,\mathrm{d}z, \qquad \delta(\boldsymbol{r} - \boldsymbol{r}_j) \equiv \delta(x - x_j)\delta(y - y_j)\delta(z - z_j). \tag{A.92}$$

Informally, therefore, we can think of the delta function as a function that is zero everywhere except where its argument vanishes—where it is infinite in such a way that its integral has unit area, and equations (A.90) and (A.91) hold. Do such amazing functions exist? In fact, the informal idea just given does not define a respectable mathematical function. More properly the use of the 'δ function' can be justified by introducing the notion of 'distributions' or 'generalised functions'. Roughly speaking, this means we can think of the 'δ function' as the limit of a sequence of functions whose properties converge to those above. The following useful expressions all approximate the δ function in the above sense:

$$\delta(x) = \begin{cases} \lim_{\varepsilon \to 0} \dfrac{1}{\varepsilon} & \text{for } -\varepsilon/2 \leqslant x \leqslant \varepsilon/2 \\ 0 & \text{for } |x| > \varepsilon/2 \end{cases} \tag{A.93}$$

$$\delta(x) = \lim_{\varepsilon \to 0} \frac{1}{\pi} \frac{\varepsilon}{x^2 + \varepsilon^2} \tag{A.94}$$

$$\delta(x) = \lim_{N \to \infty} \frac{1}{\pi} \frac{\sin(Nx)}{x}. \tag{A.95}$$

The first of these is essentially the same as (A.86) and (A.87), and the second is a 'smoother' version of the first. The third is sketched in figure A.3: as N tends to infinity, the peak becomes infinitely high and narrow, but it still preserves unit area.

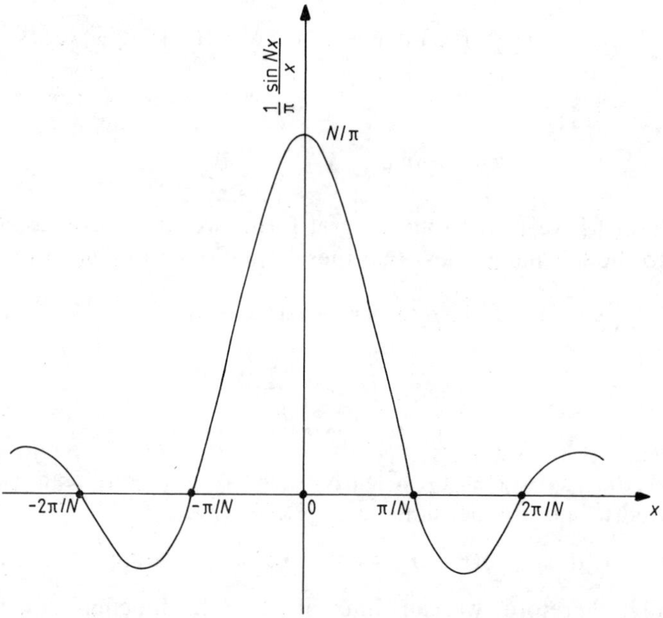

Figure A.3 The function (A.95) for finite N.

Usually, under integral signs, δ functions can be manipulated with no danger of obtaining a mathematically incorrect result. However, care must be taken when products of two such generalised functions are encountered.

A.4.2 Resumé of Fourier series and Fourier transforms

Fourier's theorem asserts that any suitably well behaved periodic function with period L can be expanded as follows:

$$f(x) = \sum_{n=-\infty}^{\infty} a_n e^{i2n\pi x/L}. \tag{A.96}$$

Using the orthonormality relation

$$\frac{1}{L}\int_{-L/2}^{L/2} e^{-2\pi imx/L} \cdot e^{2\pi inx/L} dx = \delta_{mn} \tag{A.97}$$

with the Krönecker delta symbol defined by

$$\delta_{mn} = \begin{cases} 1 & \text{if } m = n \\ 0 & \text{if } m \neq n \end{cases} \tag{A.98}$$

the coefficients in the expansion may be determined:

$$a_m = \frac{1}{L}\int_{-L/2}^{L/2} f(x) e^{-2\pi imx/L} dx. \tag{A.99}$$

Consider the limit of these expressions as $L \to \infty$. We may write

$$f(x) = \sum_{n=-\infty}^{\infty} F_n \Delta n \tag{A.100}$$

with

$$F_n = a_n e^{2\pi inx/L} \tag{A.101}$$

and the interval $\Delta n = 1$. Defining

$$2\pi n/L = k \tag{A.102}$$

and

$$La_n = g(k) \tag{A.103}$$

we can take the limit $L \to \infty$ to obtain

$$f(x) = \int_{-\infty}^{\infty} F_n dn \tag{A.104}$$

$$= \int_{-\infty}^{\infty} \frac{g(k)e^{ikx}}{L} \cdot \frac{L\,dk}{2\pi}. \tag{A.105}$$

Thus

$$f(x) = \frac{1}{2\pi}\int_{-\infty}^{\infty} g(k) e^{ikx} dk \tag{A.106}$$

and from (A.99)

$$g(k) = \int_{-\infty}^{\infty} f(x) e^{-ikx} dx. \tag{A.107}$$

These are the Fourier transform relations, and they lead us to an important representation of the Dirac delta function.

Substitute $g(k)$ from (A.107) into (A.106) to obtain

$$f(x) = \frac{1}{2\pi}\int_{-\infty}^{\infty} \mathrm{d}k\, \mathrm{e}^{\mathrm{i}kx}\int_{-\infty}^{\infty} \mathrm{d}x' \mathrm{e}^{-\mathrm{i}kx'} f(x'). \tag{A.108}$$

Reordering the integrals, we arrive at the result

$$f(x) = \int_{-\infty}^{\infty} \mathrm{d}x' f(x') \left(\frac{1}{2\pi}\int_{-\infty}^{\infty} \mathrm{e}^{\mathrm{i}k(x-x')}\mathrm{d}k\right) \tag{A.109}$$

valid for any function $f(x)$. Thus the expression

$$\frac{1}{2\pi}\int_{-\infty}^{\infty} \mathrm{e}^{\mathrm{i}k(x-x')}\mathrm{d}k \tag{A.110}$$

has the remarkable property of vanishing everywhere except at $x = x'$, and its integral with respect to x' over any interval including x is unity. In other words, (A.110) provides us with a new representation of the Dirac delta function:

$$\boxed{\delta(x) = \frac{1}{2\pi}\int_{-\infty}^{\infty} \mathrm{e}^{\mathrm{i}kx}\mathrm{d}k.} \tag{A.111}$$

Equation (A.111) is very important. It is the representation of the δ function which is most commonly used, and it occurs throughout this book. Note that if we replace the upper and lower limits of integration in (A.111) by N and $-N$, and consider the limit $N \to \infty$, we obtain exactly (A.95).

The integral in (A.111) represents the superposition, with identical uniform weight $(2\pi)^{-1}$, of plane waves of all wave numbers. Physically it may be thought of (cf (A.106)) as the Fourier transform of unity. Equation (A.111) asserts that the contributions from all these waves cancel completely, unless the phase parameter x is zero—in which case the integral manifestly diverges and '$\delta(0)$ is infinity', as expected. The fact that the Fourier transform of a constant is a δ function is an extreme case of the bandwidth theorem from Fourier transform theory, which states that if the (suitably defined) 'spread' in a function $g(k)$ is Δk, and that of its transform $f(x)$ is Δx, then $\Delta x \Delta k \geqslant \frac{1}{2}$. In the present case Δk is tending to infinity and Δx to zero.

One very common use of (A.111) refers to the normalisation of plane wave states. If we rewrite it in the form

$$\delta(k' - k) = \int_{-\infty}^{\infty} \frac{\mathrm{e}^{-\mathrm{i}k'x}}{(2\pi)^{1/2}} \cdot \frac{\mathrm{e}^{\mathrm{i}kx}}{(2\pi)^{1/2}}\, \mathrm{d}x \tag{A.112}$$

we can interpret it to mean that the wavefunctions $\mathrm{e}^{\mathrm{i}kx}/(2\pi)^{1/2}$ and $\mathrm{e}^{\mathrm{i}k'x}/(2\pi)^{1/2}$ are orthogonal on the real axis $-\infty \leqslant x \leqslant \infty$ for $k \neq k'$

(since the left-hand side is zero), while for $k = k'$ their overlap is infinite, in such a way that the *integral* of this overlap is unity. This is the continuum analogue of orthonormality for wavefunctions labelled by a discrete index, as in (A.97). We say that the plane waves in (A.112) are 'normalised to a δ function'. There is, however, a problem with this: plane waves are not square integrable and thus do not strictly belong to a Hilbert space. Mathematical physicists concerned with such matters have managed to deal with this by introducing 'rigged' Hilbert spaces in which such a normalisation is legitimate. Although we often, in the text, appear to be using 'box normalisation' (i.e. restricting space to a finite volume V), in practice when we evaluate integrals over plane waves the limits will be extended to infinity, and results like (A.112) will be used repeatedly.

A.4.3 Properties of the delta function

(i)
$$\delta(ax) = \frac{1}{|a|}\,\delta(x). \tag{A.113}$$

Proof

$$\text{For } a > 0, \qquad \int_{-\infty}^{\infty} \delta(ax)\mathrm{d}x = \int_{-\infty}^{\infty} \delta(y)\,\frac{\mathrm{d}y}{a} = \frac{1}{a}; \tag{A.114}$$

$$\text{for } a < 0, \qquad \int_{-\infty}^{\infty} \delta(ax)\mathrm{d}x = \int_{\infty}^{-\infty} \delta(y)\,\frac{\mathrm{d}y}{a} = \int_{-\infty}^{\infty} \delta(y)\,\frac{\mathrm{d}y}{|a|} = \frac{1}{|a|}. \tag{A.115}$$

(ii)
$$\delta(x) = \delta(-x), \qquad \text{i.e. even function.} \tag{A.116}$$

Proof

$$f(0) = \int \delta(x)\, f(x)\, \mathrm{d}x. \tag{A.117}$$

If $f(x)$ is an odd function, $f(0) = 0$. Thus $\delta(x)$ must be an even function.

(iii)
$$\delta(f(x)) = \sum_i \frac{1}{|\mathrm{d}f/\mathrm{d}x|_{x=a_i}}\,\delta(x - a_i) \tag{A.118}$$

where a_i are the roots of $f(x) = 0$.

Proof. The δ function is only non-zero when its argument vanishes. Thus we are concerned with the roots of $f(x) = 0$. In the vicinity of a root

$$f(a_i) = 0 \tag{A.119}$$

we can make a Taylor expansion

$$f(x) = f(a_i) + (x - a_i)\left(\frac{\mathrm{d}f}{\mathrm{d}x}\right)_{x=a_i} + \ldots. \tag{A.120}$$

Thus the delta function has non-zero contributions from each of the roots a_i of the form

$$\delta(f(x)) = \sum_i \delta\left[(x - a_i)\left(\frac{df}{dx}\right)_{x=a_i}\right]. \tag{A.121}$$

Hence (using property (i)) we have

$$\delta(f(x)) = \sum_i \frac{1}{|df/dx|_{x=a_i}} \delta(x - a_i). \tag{A.122}$$

Consider the example

$$\delta(x^2 - a^2). \tag{A.123}$$

Thus

$$f(x) = x^2 - a^2 = (x - a)(x + a) \tag{A.124}$$

with two roots $x = \pm a$ $(a > 0)$, and $df/dx = 2x$. Hence

$$\delta(x^2 - a^2) = \frac{1}{2a}[\delta(x - a) + \delta(x + a)]. \tag{A.125}$$

(iv)

$$x\delta(x) = 0. \tag{A.126}$$

This is to be understood as always occurring under an integral. It is obvious from the definition or from property (ii).

(v)

$$\int_{-\infty}^{\infty} f(x)\delta'(x)\, dx = -f'(0) \tag{A.127}$$

where

$$\delta'(x) = \frac{d}{dx}\delta(x). \tag{A.128}$$

Proof

$$\begin{aligned}\int_{-\infty}^{\infty} f(x)\, \delta'(x)dx &= -\int_{-\infty}^{\infty} f'(x)\delta(x)dx + [f(x)\delta(x)]_{-\infty}^{\infty} \\ &= -f'(0)\end{aligned} \tag{A.129}$$

since the surface term vanishes.

(vi)

$$\int_{-\infty}^{x} \delta(x' - a)dx' = \theta(x - a) \tag{A.130}$$

where

$$\theta(x) = \begin{cases} 0 & \text{for } x < 0 \\ 1 & \text{for } x > 0 \end{cases} \tag{A.131}$$

is the so-called 'theta function'.

Proof

$$\text{For } x > a, \qquad \int_{-\infty}^{x} \delta(x' - a)\mathrm{d}x' = 1; \tag{A.132}$$

$$\text{for } x < a, \qquad \int_{-\infty}^{x} \delta(x' - a)\mathrm{d}x' = 0. \tag{A.133}$$

By a simple extension it is easy to prove the result

$$\int_{x_1}^{x_2} \delta(x - a)\mathrm{d}x = \theta(x_2 - a) - \theta(x_1 - a). \tag{A.134}$$

Exercise. Use property (iii) plus the definition of the theta function to perform the p^0 integration and prove the useful phase space formula

$$\int \mathrm{d}^4 p\,\delta(p^2 - m^2)\theta(p^0) = \int \mathrm{d}^3\boldsymbol{p}/2E \tag{A.135}$$

where

$$p^2 = (p^0)^2 - \boldsymbol{p}^2 \tag{A.136}$$

and

$$E = +\,(\boldsymbol{p}^2 + m^2)^{1/2}. \tag{A.137}$$

APPENDIX B

NATURAL UNITS

In particle physics, a widely adopted convention is to work in a system of units, called natural units, in which

$$\boxed{\hbar = c = 1.} \tag{B.1}$$

This avoids having to keep track of untidy factors of $\hbar$ and c throughout a calculation: only at the end is it necessary to convert back to more usual units. Let us spell out the implications of this choice of c and $\hbar$.

(*i*) $c = 1$
In conventional MKS units c has the value

$$c \simeq 3 \times 10^{8}\ \mathrm{m\,s^{-1}}. \tag{B.2}$$

By choosing units such that

$$c = 1 \tag{B.3}$$

since a velocity has the dimensions

$$[c] = [\mathrm{L}][\mathrm{T}]^{-1} \tag{B.4}$$

we are implying that our unit of length is numerically equal to our unit of time. In this sense, length and time are equivalent dimensions:

$$[\mathrm{L}] = [\mathrm{T}]. \tag{B.5}$$

Similarly, from the energy–momentum relation of special relativity

$$E^2 = \boldsymbol{p}^2c^2 + m^2c^4 \tag{B.6}$$

we see that the choice of $c = 1$ also implies that energy, mass and momentum all have equivalent dimensions. In fact, it is customary to refer to momenta in units of 'MeV/c' or 'GeV/c' and, less frequently, to masses in units of 'MeV/c^2' or 'GeV/c^2'; these all become 'MeV' or 'GeV' when $c = 1$.

(*ii*) $\hbar = 1$
The numerical value of Planck's constant is

$$\hbar \simeq 6.6 \times 10^{-22}\ \mathrm{MeV\,s} \tag{B.7}$$

and $\hbar$ has dimensions of energy times time so that

$$[\hbar] = [\mathrm{M}][\mathrm{L}]^2[\mathrm{T}]^{-1}. \tag{B.8}$$

Setting $\hbar = 1$ therefore relates our units of [M], [L] and [T]. Since [L] and [T] are equivalent by our choice of $c = 1$, we can choose [M] as the single independent dimension for our natural units:

$$\boxed{[\mathrm{M}] = [\mathrm{L}]^{-1} = [\mathrm{T}]^{-1}.} \tag{B.9}$$

An example: the pion Compton wavelength
How do we convert from natural units to more conventional units? Consider the pion Compton wavelength

$$\lambda_\pi = \hbar/M_\pi c \tag{B.10}$$

evaluated in both natural and conventional units. In natural units

$$\lambda_\pi = 1/M_\pi \tag{B.11}$$

where $M_\pi \simeq 140\ \mathrm{MeV}/c^2$. In conventional units, using M_π, $\hbar$ (B.7) and c (B.2), we have the familiar result

$$\lambda_\pi \simeq 1.41\ \mathrm{fm} \tag{B.12}$$

where the 'fermi' or femtometre, fm, is defined as

$$1\ \mathrm{fm} = 10^{-15}\ \mathrm{m}.$$

We therefore have the correspondence

$$\boxed{\lambda_\pi = 1/M_\pi \simeq 1.41\ \mathrm{fm}.} \tag{B.13}$$

Practical cross section calculations
An easy-to-remember relation may be derived from the result

$$\boxed{\hbar c \simeq 200\ \mathrm{MeV\,fm}} \tag{B.14}$$

obtained directly from (B.2) and (B.7). Hence, in natural units, we have the relation

$$\boxed{1\ \mathrm{fm} \simeq \frac{1}{200\ \mathrm{MeV}} = 5\ (\mathrm{GeV})^{-1}.} \tag{B.15}$$

Cross sections are calculated without $\hbar$'s and c's and all masses, energies and momenta typically in MeV or GeV. To convert the result to an area, we merely remember the dimensions of a cross section:

$$[\sigma] = [\mathrm{L}]^2 = [\mathrm{M}]^{-2}. \tag{B.16}$$

If masses, momenta and energies have been specified in GeV, from

(B.15) we derive the useful result (from the more precise relation $\hbar c = 197.328\ \text{MeV fm}$)

$$\boxed{\left(\frac{1}{1\ \text{GeV}}\right)^2 = 1\ (\text{GeV})^{-2} = 0.38\,939\ \text{mb}} \tag{B.17}$$

where a millibarn, mb, is defined to be

$$1\ \text{mb} = 10^{-31}\ \text{m}^2.$$

Notice that a 'typical' hadronic cross section corresponds to an area of about λ_π^2 where

$$\lambda_\pi^2 = 1/M_\pi^2 \simeq 20\ \text{mb}.$$

Electromagnetic cross sections are an order of magnitude smaller: specifically for lowest-order $e^+e \rightarrow \mu^+\mu^-$

$$\sigma \approx \frac{88}{s}\ \text{nb}$$

where s is in $(\text{GeV})^2$.

APPENDIX C

MAXWELL'S EQUATIONS: CHOICE OF UNITS

In high-energy physics, it is not the convention to use the rationalised MKS system of units when treating Maxwell's equations. Since the discussion is always limited to field equations *in vacuo*, it is usually felt desirable to adopt a system of units in which these equations take their simplest possible form—in particular, one such that the constants ε_0 and μ_0, employed in the MKS system, do not appear. These two constants enter, of course, via the force laws of Coulomb and Ampère, respectively. These laws relate a mechanical quantity (force) to electrical ones (charge and current). The introduction of ε_0 in Coulomb's law

$$\boldsymbol{F} = \frac{q_1 q_2 \boldsymbol{r}}{4\pi\varepsilon_0 r^3} \qquad \text{(C.1)}$$

enables one to choose arbitrarily one of the electrical units and assign to it a dimension independent of those entering into mechanics (mass, length and time). If, for example, we use the coulomb as the basic electrical quantity (as in the MKS system), ε_0 has dimension (coulomb)^2 $[\mathrm{T}]^2/[\mathrm{M}][\mathrm{L}]^3$. Thus the common practical units (volt, ampere, coulomb, etc) can be employed in applications to both fields and circuits. However, for our purposes this advantage is irrelevant, since we are only concerned with the field equations, not with practical circuits. In our case, we prefer to define the electrical units in terms of mechanical ones in such a way as to reduce the field equations to their simplest form. The field equation correponding to (C.1) is

$$\boldsymbol{\nabla}\cdot\boldsymbol{E} = \rho/\varepsilon_0 \qquad \text{(Gauss's law; MKS)} \qquad \text{(C.2)}$$

and this may obviously be simplified if we choose the unit of charge such that ε_0 becomes unity. Such a system, in which CGS units are used for the mechanical quantities, is a variant of the electrostatic part of the 'Gaussian CGS' system. The original Gaussian system set $\varepsilon_0 \to 1/4\pi$, thereby simplifying the force law (C.1), but introducing a compensating 4π into the field equation (C.2). The field equation is, in fact, primary, and the 4π is a geometrical factor appropriate only to the specific case of three dimensions, so that it should not appear in a field equation of general validity. The system in which ε_0 in (C.2) may be replaced by unity is called the 'rationalised Gaussian CGS' or 'Heaviside–Lorentz' system:

$$\nabla \cdot \boldsymbol{E} = \rho \qquad \text{(Gauss's law; Heaviside–Lorentz).} \tag{C.3}$$

Generally, systems in which the 4π factors appear in the force equations rather than the field equations are called 'rationalised'.

Of course, (C.3) is only the first of the Maxwell equations in Heaviside–Lorentz units. In the Gaussian system, μ_0 in Ampère's force law

$$\boldsymbol{F} = \frac{\mu_0}{4\pi} \int \int \frac{\boldsymbol{j}_1 \times (\boldsymbol{j}_2 \times \boldsymbol{r}_{12})}{r_{12}^3} \, \mathrm{d}^3 r_1 \, \mathrm{d}^3 r_2 \tag{C.4}$$

was set equal to 4π, thereby defining a unit of current (the electromagnetic unit or Biot (Bi emu)). The unit of charge (the electrostatic unit or Franklin (Fr esu)) was already defined by the (Gaussian) choice $\varepsilon_0 = 1/4\pi$, and the ratio of these units is a fundamental constant, namely the velocity of light, c. The Gaussian system therefore defines charges via $\varepsilon_0 \to 1/4\pi$ and currents via $\mu_0 \to 4\pi$, and c appears explicitly in the equations. In the rationalised (Heaviside–Lorentz) form of this system, $\varepsilon_0 \to 1$ and $\mu_0 \to 1$, and the remaining Maxwell equations are

$$\nabla \times \boldsymbol{E} = -\frac{1}{c} \frac{\partial \boldsymbol{B}}{\partial t} \tag{C.5}$$

$$\nabla \cdot \boldsymbol{B} = 0 \tag{C.6}$$

$$\nabla \times \boldsymbol{B} = \boldsymbol{j} + \frac{1}{c} \frac{\partial \boldsymbol{E}}{\partial t}. \tag{C.7}$$

A further discussion of units in electromagnetic theory is given in Panofsky and Phillips (1962, Appendix I).

Finally, throughout this book we have used a particular choice of units for mass, length and time such that $\hbar = c = 1$ (see Appendix B). In that case, the Maxwell equations we use are as in (C.3), (C.5)–(C.7), but with c replaced by unity.

As an example of the relation between MKS and the system employed in this book (and universally in high-energy physics), we remark that the fine structure constant is written as

$$\alpha = \begin{cases} \dfrac{e^2}{4\pi\varepsilon_0 \hbar c} & \text{in MKS units} \qquad \text{(C.8)} \\[2ex] \dfrac{e^2}{4\pi} & \text{in Heaviside–Lorentz units with } \hbar = c = 1. \qquad \text{(C.9)} \end{cases}$$

Clearly the value of $\alpha(\simeq 1/137)$ is the same in both cases, but the numerical values of 'e' in (C.8) and in (C.9) are, of course, different.

The choice of rationalised MKS units for Maxwell's equations is a part of the SI system of units. In this system of units the numerical values of μ_0 and ε_0 are

$$\mu_0 = 4\pi \times 10^{-7} \; (\mathrm{kg\,m\,C^{-2} = H\,m^{-1}})$$

and, since $\mu_0 \varepsilon_0 = 1/c^2$,

$$\varepsilon_0 = \frac{10^7}{4\pi c^2} \simeq \frac{1}{36\pi \times 10^9} \; (\mathrm{C^2\,s^2\,kg^{-1}\,m^{-3} = F\,m^{-1}}).$$

APPENDIX D

DIRAC ALGEBRA AND TRACE THEOREMS

D.1 Dirac algebra

D.1.1 γ matrices

The fundamental anticommutator

$$\{\gamma^{\mu}, \gamma^{\nu}\} = 2g^{\mu\nu} \tag{D.1}$$

may be used to prove the following results:

$$\gamma_{\mu}\gamma^{\mu} = 4 \tag{D.2}$$

$$\gamma_{\mu}\not{a}\gamma^{\mu} = -2\not{a} \tag{D.3}$$

$$\gamma_{\mu}\not{a}\not{b}\gamma^{\mu} = 4a\cdot b \tag{D.4}$$

$$\gamma_{\mu}\not{a}\not{b}\not{c}\gamma^{\mu} = -2\not{c}\not{b}\not{a} \tag{D.5}$$

$$\not{a}\not{b} = -\not{b}\not{a} + 2a\cdot b. \tag{D.6}$$

As an example, we prove this last result:

$$\begin{aligned}\not{a}\not{b} &= a_{\mu}b_{\nu}\gamma^{\mu}\gamma^{\nu}\\ &= a_{\mu}b_{\nu}(-\gamma^{\nu}\gamma^{\mu} + 2g^{\mu\nu})\\ &= -\not{b}\not{a} + 2a\cdot b.\end{aligned}$$

D.1.2 γ_5 identities

Define

$$\gamma_5 = \mathrm{i}\gamma^0\gamma^1\gamma^2\gamma^3. \tag{D.7}$$

In the usual representation with (see footnote on p. 344)

$$\gamma^0 = \begin{pmatrix}1 & 0\\ 0 & -1\end{pmatrix} \quad \text{and} \quad \boldsymbol{\gamma} = \begin{pmatrix}0 & \boldsymbol{\sigma}\\ -\boldsymbol{\sigma} & 0\end{pmatrix} \tag{D.8}$$

γ_5 is the matrix

$$\gamma_5 = \begin{pmatrix}0 & 1\\ 1 & 0\end{pmatrix}. \tag{D.9}$$

Either from the definition or using this explicit form it is easy to prove that

$$\gamma_5^2 = 1 \tag{D.10}$$

and

$$\{\gamma_5, \gamma^\mu\} = 0 \tag{D.11}$$

i.e. γ_5 anticommutes with the other γ matrices. Defining the totally antisymmetric tensor

$$\varepsilon_{\mu\nu\rho\sigma} = \begin{cases} +1 & \text{for an even permutation of } 0, 1, 2, 3 \\ -1 & \text{for an odd permutation of } 0, 1, 2, 3 \\ 0 & \text{if two or more indices are the same,} \end{cases} \tag{D.12}$$

we may write

$$\gamma_5 = \frac{\mathrm{i}}{4!}\,\varepsilon_{\mu\nu\rho\sigma}\gamma^\mu\gamma^\nu\gamma^\rho\gamma^\sigma. \tag{D.13}$$

With this form it is possible to prove

$$\gamma_5\gamma_\sigma = \frac{\mathrm{i}}{3!}\,\varepsilon_{\mu\nu\rho\sigma}\gamma^\mu\gamma^\nu\gamma^\rho \tag{D.14}$$

and the identity

$$\gamma^\mu\gamma^\nu\gamma^\rho = g^{\mu\nu}\gamma^\rho - g^{\mu\rho}\gamma^\nu + g^{\nu\rho}\gamma^\mu + \mathrm{i}\gamma_5\varepsilon^{\mu\nu\rho\sigma}\gamma_\sigma. \tag{D.15}$$

D.1.3 Hermitian conjugate of spinor matrix elements

$$[\bar{u}(p', s')\Gamma u(p, s)]^\dagger = \bar{u}(p, s)\bar{\Gamma}u(p', s') \tag{D.16}$$

where Γ is any collection of γ matrices and

$$\bar{\Gamma} \equiv \gamma^0\Gamma^\dagger\gamma^0. \tag{D.17}$$

For example

$$\overline{\gamma^\mu} = \gamma^\mu \tag{D.18}$$

and

$$\overline{\gamma^\mu\gamma_5} = \gamma^\mu\gamma_5. \tag{D.19}$$

D.1.4 Spin sums and projection operators

Positive-energy projection operator:

$$[\Lambda_+(p)]_{\alpha\beta} \equiv \sum_s u_\alpha(p, s)\bar{u}_\beta(p, s) = (\not{p} + m)_{\alpha\beta}. \tag{D.20}$$

Negative-energy projection operator:

$$[\Lambda_-(p)]_{\alpha\beta} \equiv -\sum_s v_\alpha(p, s)\bar{v}_\beta(p, s) = (-\not{p} + m)_{\alpha\beta}. \qquad \text{(D.21)}$$

Note that these forms are specific to the normalisations

$$\bar{u}u = 2m, \qquad \bar{v}v = -2m \qquad \text{(D.22)}$$

for the spinors.

D.2 Trace theorems

(1) $$\mathrm{Tr}\,\mathbf{1} = 4. \qquad \text{(D.23)}$$

(2) $$\mathrm{Tr}\,\gamma_5 = 0. \qquad \text{(D.24)}$$

(3) $$\mathrm{Tr}\,(\text{odd number of } \gamma\text{'s}) = 0 \qquad \text{(D.25)}$$

Proof
Consider

$$T \equiv \mathrm{Tr}(\not{a}_1\not{a}_2 \ldots \not{a}_n) \qquad \text{(D.26)}$$

where n is odd. Now insert $1 = (\gamma_5)^2$ into T, so that

$$T = \mathrm{Tr}(\not{a}_1\not{a}_2 \ldots \not{a}_n\gamma_5\gamma_5). \qquad \text{(D.27)}$$

Move the first γ_5 to the front of T by repeatedly using the result

$$\not{a}\gamma_5 = -\gamma_5\not{a}. \qquad \text{(D.28)}$$

We therefore pick up n minus signs:

$$\begin{aligned} T = \mathrm{Tr}(\not{a}_1 \ldots \not{a}_n) &= (-1)^n\, \mathrm{Tr}(\gamma_5\not{a}_1 \ldots \not{a}_n\gamma_5) \\ &= (-1)^n \mathrm{Tr}(\not{a}_1 \ldots \not{a}_n\gamma_5\gamma_5) \qquad \text{(cyclic property of trace)} \\ &= -\,\mathrm{Tr}(\not{a}_1 \ldots \not{a}_n) \qquad \text{for } n \text{ odd.} \qquad \text{(D.29)} \end{aligned}$$

Thus, for n odd, T must vanish.

(4) $$\mathrm{Tr}(\not{a}\not{b}) = 4a\cdot b. \qquad \text{(D.30)}$$

Proof

$$\begin{aligned} \mathrm{Tr}(\not{a}\not{b}) &= \tfrac{1}{2}\mathrm{Tr}(\not{a}\not{b} + \not{b}\not{a}) \\ &= \tfrac{1}{2}a_\mu b_\nu \mathrm{Tr}(\mathbf{1}\cdot 2g^{\mu\nu}) \\ &= 4a\cdot b. \end{aligned}$$

(5) $$\mathrm{Tr}(\not{a}\not{b}\not{c}\not{d}) = 4[(a\cdot b)(c\cdot d) + (a\cdot d)(b\cdot c) - (a\cdot c)(b\cdot d)]. \qquad \text{(D.31)}$$

Proof

$$\mathrm{Tr}(\not{a}\not{b}\not{c}\not{d}) = 2(a\cdot b)\mathrm{Tr}(\not{c}\not{d}) - \mathrm{Tr}(\not{b}\not{a}\not{c}\not{d}) \tag{D.32}$$

using the result of (D.6). We continue taking $\not{a}$ through the trace in this manner and use (D.30) to obtain

$$\begin{aligned}\mathrm{Tr}(\not{a}\not{b}\not{c}\not{d}) &= 2(a\cdot b)4(c\cdot d) - 2(a\cdot c)\mathrm{Tr}(\not{b}\not{d}) + \mathrm{Tr}(\not{b}\not{c}\not{a}\not{d}) \\ &= 8(a\cdot b)(c\cdot d) - 8(a\cdot c)(b\cdot d) + 8(b\cdot c)(a\cdot d) - \mathrm{Tr}(\not{b}\not{c}\not{d}\not{a})\end{aligned} \tag{D.33}$$

and, since we can bring $\not{a}$ to the front of the trace, we have proved the theorem.

(6) $$\mathrm{Tr}[\gamma_5\not{a}] = 0. \tag{D.34}$$

This is a special case of (3) since γ_5 contains four γ matrices.

(7) $$\mathrm{Tr}[\gamma_5\not{a}\not{b}] = 0. \tag{D.35}$$

This is not so obvious; it may be proved by writing out all the possible products of γ matrices that arise.

(8) $$\mathrm{Tr}[\gamma_5\not{a}\not{b}\not{c}] = 0. \tag{D.36}$$

Again this is a special case of (3).

(9) $$\mathrm{Tr}[\gamma_5\not{a}\not{b}\not{c}\not{d}] = 4\mathrm{i}\varepsilon_{\alpha\beta\gamma\delta}a^\alpha b^\beta c^\gamma d^\delta. \tag{D.37}$$

This theorem follows by looking at components: the ε tensor just gives the correct sign of the permutation.

The ε tensor is the four-dimensional generalisation of the three-dimensional antisymmetric tensor ε_{ijk}. In the three-dimensional case we have the well known results

$$(\boldsymbol{b} \times \boldsymbol{c})_i = \varepsilon_{ijk}b_jc_k \tag{D.38}$$

and

$$\boldsymbol{a}\cdot(\boldsymbol{b} \times \boldsymbol{c}) = \varepsilon_{ijk}a_ib_jc_k \tag{D.39}$$

for the triple scalar product. See problem 10.3 for further results on such antisymmetric tensors.

APPENDIX E

EXAMPLE OF A CROSS SECTION CALCULATION

In this appendix we outline in more detail the calculation of the π^+e^- elastic scattering cross section in §§6.3–6.5. The standard factors for the unpolarised cross section lead to the expression

$$d\bar{\sigma} = \frac{1}{4E\omega|\boldsymbol{v}|} \tfrac{1}{2} \sum_{ss'} |F_{ss'}|^2 \, \mathrm{dLips}(s; k', p') \tag{E.1}$$

$$= \frac{1}{4[(k\cdot p)^2 - m^2M^2]^{1/2}} \tfrac{1}{2} \sum_{ss'} |F_{ss'}|^2 \, \mathrm{dLips}(s; k', p') \tag{E.2}$$

using the result of problem 5.4, and the definition of Lorentz invariant phase space:

$$\mathrm{dLips}(s; k', p') \equiv (2\pi)^4 \delta^4(k' + p' - k - p) \frac{d^3p'}{(2\pi)^3 2E'} \frac{d^3k'}{(2\pi)^3 2\omega'}. \tag{E.3}$$

Instead of evaluating the matrix element and phase space integral in the CM frame, or writing the result in invariant form, we shall perform the calculation entirely in the 'laboratory' frame, defined as the frame in which

$$\boxed{p^\mu = (M, 0).} \tag{E.4}$$

Let us look in some detail at the 'laboratory' frame kinematics for elastic scattering (figure E.1). Conservation of energy and momentum in the form

$$p'^2 = (p + q)^2 \tag{E.5}$$

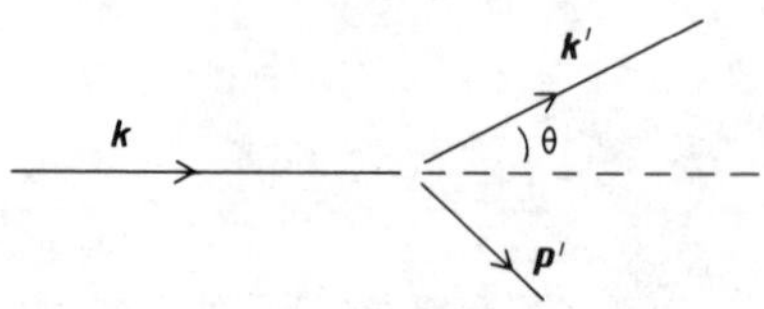

Figure E.1

allows us to eliminate p' to obtain the elastic scattering condition

$$2p \cdot q + q^2 = 0 \tag{E.6}$$

or

$$\boxed{2p \cdot q = Q^2} \tag{E.7}$$

if we introduce the positive quantity

$$Q^2 = -q^2 \tag{E.8}$$

for a scattering process.

In all the applications with which we are concerned it will be a good approximation to neglect electron mass effects for high-energy electrons. We therefore set

$$k^2 = k'^2 \simeq 0 \tag{E.9}$$

so that

$$s + t + u \simeq 2M^2 \tag{E.10}$$

where

$$s = (k + p)^2 = (k' + p')^2 \tag{E.11}$$

$$t = (k - k')^2 = (p' - p)^2 = q^2 \tag{E.12}$$

$$u = (k - p')^2 = (k' - p)^2 \tag{E.13}$$

are the usual Mandelstam variables. For the electron 4-vectors

$$k^\mu = (\omega, \boldsymbol{k}) \tag{E.14}$$

$$k'^\mu = (\omega', \boldsymbol{k}') \tag{E.15}$$

we can neglect the difference between the magnitude of the 3-momentum and the energy,

$$\omega \simeq |\boldsymbol{k}| \equiv k \tag{E.16}$$

$$\omega' \simeq |\boldsymbol{k}'| \equiv k' \tag{E.17}$$

and in this approximation

$$q^2 = -2kk'(1 - \cos\theta) \tag{E.18}$$

or

$$\boxed{q^2 = -4kk' \sin^2(\theta/2).} \tag{E.19}$$

The elastic scattering condition (E.7) gives the following relation between k, k' and θ:

$$(k/k') = 1 + (2k/M)\sin^2(\theta/2). \tag{E.20}$$

It is important to realise that this relation is *only* true for elastic scattering; for inclusive inelastic electron scattering k, k' and θ are independent variables.

The first element of the cross section, the flux factor, is easy to evaluate:

$$4[(k\cdot p)^2 - m^2M^2]^{1/2} \simeq 4Mk \tag{E.21}$$

in the approximation of neglecting the electron mass m. We now consider the calculation of the spin-averaged matrix element and the phase space integral in turn.

E.1 The spin-averaged squared matrix element

The Feynman rules for $e\pi$ scattering enable us to write the spin sum in the form

$$\tfrac{1}{2}\sum_{s,s'}|F_{ss'}|^2 = \left(\frac{4\pi\alpha}{q^2}\right)^2 L_{\mu\nu}T^{\mu\nu} \tag{E.22}$$

where $L_{\mu\nu}$ is the lepton tensor, $T^{\mu\nu}$ the pion tensor, and the one-photon exchange approximation has been assumed. From problems 6.4 and 6.5 we find the result

$$L_{\mu\nu}T^{\mu\nu} = 8[2(k\cdot p)(k'\cdot p) + (q^2/2)M^2]. \tag{E.23}$$

In the 'laboratory' frame, neglecting the electron mass, this becomes

$$L_{\mu\nu}T^{\mu\nu} = 16M^2kk'\cos^2(\theta/2). \tag{E.24}$$

E.2 Evaluation of two-body Lorentz invariant phase space in 'laboratory' variables

We must evaluate

$$\mathrm{dLips}(s; k', p') \equiv \frac{1}{(4\pi)^2}\,\delta^4(k' + p' - k - p)\,\frac{\mathrm{d}^3p'}{E'}\,\frac{\mathrm{d}^3k'}{\omega'} \tag{E.25}$$

in terms of 'laboratory' variables. This is in fact rather tricky and requires some care. There are several ways it can be done:

(a) Use CM variables, put the cross section into invariant form, and then translate to the 'laboratory' frame. This involves relating $\mathrm{d}q^2$ to $\mathrm{d}(\cos\theta)$ which we shall do as an exercise at the end of this appendix.

(b) Alternatively, we can work directly in terms of 'laboratory' variables and write

$$d^3p'/2E' = d^4p'\delta(p'^2 - M^2)\theta(p'^0). \qquad \text{(E.26)}$$

The four-dimensional δ function then removes the integration over d^4p' leaving us only with an integration over the single δ function $\delta(p'^2 - M^2)$, in which p' is understood to be replaced by $k + p - k'$. For details of this last integration, see Bjorken and Drell (1964, p 114).

(c) We shall evaluate the phase space integral in a more direct manner. We begin by performing the integral over d^3p' using the three-dimensional δ function from $\delta^4(k' + p' - k - p)$. In the 'laboratory' frame $\boldsymbol{p} = 0$, so we have

$$\int d^3\boldsymbol{p}'\delta^3(\boldsymbol{k}' + \boldsymbol{p}' - \boldsymbol{k})f(\boldsymbol{p}', \boldsymbol{k}', \boldsymbol{k}) = f(\boldsymbol{p}', \boldsymbol{k}', \boldsymbol{k})\big|_{\boldsymbol{p}'=\boldsymbol{k}-\boldsymbol{k}'}. \qquad \text{(E.27)}$$

In the particular function $f(\boldsymbol{p}', \boldsymbol{k}', \boldsymbol{k})$ that we require, $\boldsymbol{p}'$ only appears via E', since

$$E'^2 = \boldsymbol{p}'^2 + M^2 \qquad \text{(E.28)}$$

and

$$\boldsymbol{p}'^2 = k^2 + k'^2 - 2kk'\cos\theta \qquad \text{(E.29)}$$

(setting the electron mass m to zero). We now change d^3k' to angular variables:

$$d^3k'/\omega' \simeq k'dk'd\Omega \qquad \text{(E.30)}$$

leading to

$$\text{dLips}(s; k', p') = \frac{1}{(4\pi)^2}\, d\Omega\, dk'\, \frac{k'}{E'}\, \delta(E' + k' - k - M). \qquad \text{(E.31)}$$

Since E' is a function of k' and θ for a given k (cf (E.28) and (E.29)), the δ function relates k' and θ as required for elastic scattering (cf (E.20)), but *until* the δ function integration is performed they must be regarded as independent variables. We have the integral

$$\frac{1}{(4\pi)^2}\int d\Omega\, dk'\, \frac{k'}{E'}\, \delta(f(k', \cos\theta)) \qquad \text{(E.32)}$$

where

$$f(k', \cos\theta) = [(k^2 + k'^2 - 2kk'\cos\theta) + M^2]^{1/2} + k' - k - M \qquad \text{(E.33)}$$

remaining to be evaluated. In order to obtain a differential cross section, we wish to integrate over k': for this k' integration we must regard $\cos\theta$ in $f(k', \cos\theta)$ as a constant, and use the result (A.118):

$$\delta(f(x)) = \frac{1}{|f'(x)|_{x=x_0}}\delta(x - x_0) \qquad \text{(E.34)}$$

where $f(x_0) = 0$. The derivative

$$\left.\frac{\mathrm{d}f}{\mathrm{d}k'}\right|_{\substack{\text{constant}\\ \cos\theta}} = \frac{1}{E'}(E' + k' - k\cos\theta) \tag{E.35}$$

and the δ function requires that k' is determined from k and θ by the elastic scattering condition

$$k' = \frac{k}{1 + (2k/M)\sin^2(\theta/2)} \equiv k'(\cos\theta). \tag{E.36}$$

We have

$$\frac{1}{(4\pi)^2}\int \mathrm{d}\Omega\,\mathrm{d}k'\,\frac{k'}{E'}\,\frac{1}{|\mathrm{d}f/\mathrm{d}k'|_{k'=k'(\cos\theta)}}\,\delta[k' - k'(\cos\theta)] \tag{E.37}$$

and, after some juggling, $\mathrm{d}f/\mathrm{d}k'$ evaluated at $k' = k'(\cos\theta)$ may be written as

$$\left.\frac{\mathrm{d}f}{\mathrm{d}k'}\right|_{k'=k'(\cos\theta)} = \frac{Mk}{E'k'}. \tag{E.38}$$

Thus we obtain finally the result

$$\boxed{\mathrm{dLips}(s;\,k',\,p') = \frac{1}{(4\pi)^2}\,\frac{k'^2}{Mk}\,\mathrm{d}\Omega} \tag{E.39}$$

for two-body elastic scattering in terms of 'laboratory' variables, neglecting lepton masses.

Putting all these elements together yields the advertised result

$$\left(\frac{\mathrm{d}\sigma}{\mathrm{d}\Omega}\right)_{\mathrm{ns}} \equiv \frac{\mathrm{d}\bar{\sigma}}{\mathrm{d}\Omega} = \frac{\alpha^2}{4k^2\sin^4(\theta/2)}\,\frac{k'}{k}\cos^2(\theta/2). \tag{E.40}$$

As a final twist to this calculation let us consider the change of variables from $\mathrm{d}\Omega$ to $\mathrm{d}q^2$ in this elastic scattering example. In the unpolarised case

$$\mathrm{d}\Omega = 2\pi\,\mathrm{d}(\cos\theta) \tag{E.41}$$

and

$$q^2 = -2kk'(1 - \cos\theta) \tag{E.18}$$

where

$$k' = \frac{k}{1 + (2k/M)\sin^2(\theta/2)}. \tag{E.20$'$}$$

Thus, since k' and $\cos\theta$ are not independent variables, we have

$$\mathrm{d}q^2 = 2kk'\mathrm{d}(\cos\theta) + (1 - \cos\theta)(-2k)\,\frac{\mathrm{d}k'}{\mathrm{d}(\cos\theta)}\,\mathrm{d}(\cos\theta). \tag{E.42}$$

From (E.20) we find

$$\frac{dk'}{d(\cos\theta)} = \frac{k'^2}{M} \tag{E.43}$$

and, after some routine juggling, arrive at the result

$$\boxed{dq^2 = 2k'^2\, d(\cos\theta).} \tag{E.44}$$

If we introduce the variable ν defined, for elastic scattering, by

$$2p\cdot q \equiv 2M\nu = -q^2 \tag{E.45}$$

we have immediately

$$d\nu = \frac{k'^2}{M}\, d(\cos\theta). \tag{E.46}$$

Similarly, if we introduce the variable y defined by

$$y = \nu/k \tag{E.47}$$

we find

$$dy = \frac{k'^2}{2\pi kM}\, d\Omega \tag{E.48}$$

for elastic scattering.

APPENDIX F

FEYNMAN RULES FOR TREE GRAPHS IN QED, QCD AND THE ELECTROWEAK THEORY

$2 \rightarrow 2$ *cross section formula*

$$\mathrm{d}\sigma = \frac{1}{4[(p_1 \cdot p_2)^2 - m_1^2 m_2^2]^{1/2}} |F|^2 \,\mathrm{dLips}\,(s; p_3\, p_4).$$

$1 \rightarrow 2$ *decay formula*

$$\mathrm{d}\Gamma = \frac{1}{2m_1} |F|^2 \,\mathrm{dLips}\,(m_1^2; p_2, p_3).$$

Note that for two identical particles in the final state an extra factor of $\frac{1}{2}$ must be included in these formulae.

The amplitude F is the invariant matrix element for the process under consideration, and is given by the Feynman rules of the relevant theory. For particles with non-zero spin, unpolarised cross sections are formed by averaging over initial spin components and summing over final.

F.1 QED: rules for tree graphs

F.1.1 External particles

Spin-$\frac{1}{2}$. For each fermion or antifermion line entering the graph include the spinor

$$u(p, s) \qquad \text{or} \qquad v(p, s)$$

and for spin-$\frac{1}{2}$ particles leaving the graph the spinor

$$\bar{u}(p', s') \qquad \text{or} \qquad \bar{v}(p', s').$$

Photons. For each photon line entering the graph include a polarisation vector

$$\varepsilon_\mu(k, \lambda)$$

and for photons leaving the graph the vector

$$\varepsilon_\mu^*(k', \lambda').$$

F.1.2 Propagators

Spin-0

$$= \frac{\mathrm{i}}{p^2 - m^2}.$$

Spin-$\frac{1}{2}$

$$= \frac{\mathrm{i}}{\not{p} - m} = \mathrm{i}\,\frac{\not{p} + m}{p^2 - m^2}.$$

Photon

$$= \frac{\mathrm{i}}{k^2}\left(-\,g^{\mu\nu} + (1 - \xi)\,\frac{k^\mu k^\nu}{k^2}\right)$$

for a general ξ gauge. Calculations are usually performed in Lorentz or Feynman gauge with $\xi = 1$ and photon propagator

$$= \mathrm{i}\,\frac{(-\,g^{\mu\nu})}{k^2}.$$

F.1.3 Vertices

Spin-0

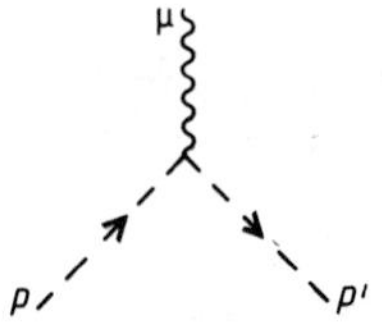

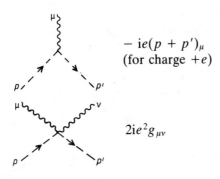

$-\,\mathrm{i}e(p + p')_\mu$
(for charge $+e$)

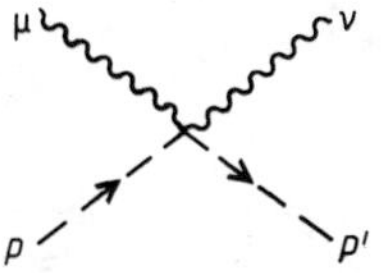

$2\mathrm{i}e^2 g_{\mu\nu}$

Spin-$\frac{1}{2}$

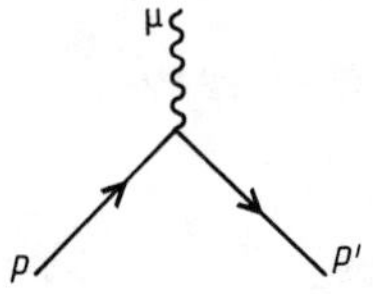

$-\,\mathrm{i}e\gamma_\mu$
(for charge $+e$)

F.2 QCD: rules for tree graphs

F.2.1 External particles

Quarks. The SU(3) colour degree of freedom is not written explicitly;

the spinors have 3(colour) × 4(Dirac) components:

$$\text{ingoing} \quad u(p, s) \quad \text{or} \quad v(p, s)$$
$$\text{outgoing} \quad \bar{u}(p', s') \quad \text{or} \quad \bar{v}(p', s')$$

as for QED.

Gluons. Besides the spin-1 polarisation vector, external gluons also have a 'colour polarisation' vector $a^\alpha(\alpha = 1, 2, \ldots, 8)$ specifying the particular colour state involved:

$$\text{ingoing} \quad \varepsilon_\mu(k, \lambda)a^\alpha$$
$$\text{outgoing} \quad \varepsilon^*_\mu(k', \lambda')a^{*\alpha}.$$

F.2.2 Propagators

Quark

$$\longrightarrow \; = \frac{\mathrm{i}}{\not{p} - m} = \mathrm{i}\,\frac{\not{p} + m}{p^2 - m^2}.$$

Gluon

$$= \frac{\mathrm{i}}{q^2}\left(-\, g^{\mu\nu} + (1 - \xi)\,\frac{q^\mu q^\nu}{q^2}\right)\delta^{\alpha\beta}$$

for a general ξ gauge. In Feynman gauge this reduces to

$$= \frac{\mathrm{i}}{q^2}\,(-\, g^{\mu\nu})\delta^{\alpha\beta}$$

which is usually the most convenient form.

F.2.3 Vertices

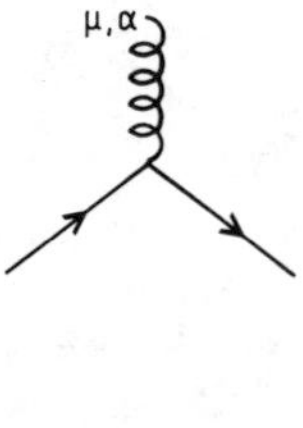

$$-\,\mathrm{i}g_{\mathrm{s}}\,\frac{\lambda^\alpha}{2}\,\gamma_\mu$$

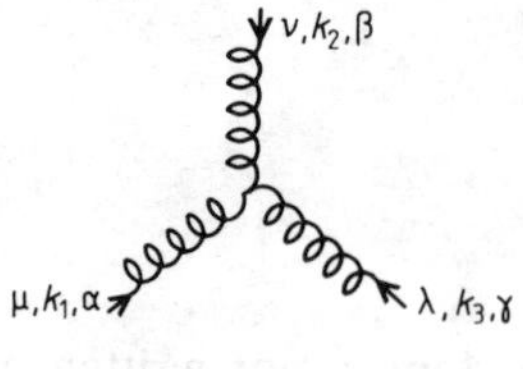

$$-\, g_{\mathrm{s}}f_{\alpha\beta\gamma}[g_{\mu\nu}(k_1 - k_2)_\lambda + g_{\nu\lambda}(k_2 - k_3)_\mu + g_{\lambda\mu}(k_3 - k_1)_\nu]$$

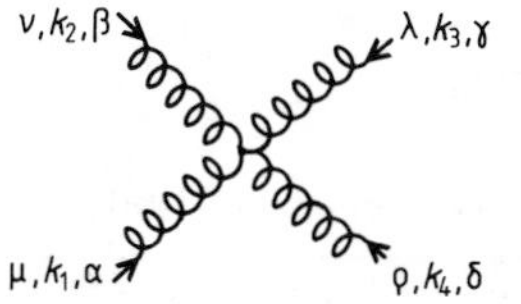

$$
\begin{aligned}
- \mathrm{i}g_s^2[&f_{\alpha\beta\eta}f_{\gamma\delta\eta}(g_{\mu\lambda}g_{\nu\rho} - g_{\mu\rho}g_{\nu\lambda}) \\
&+ f_{\alpha\delta\eta}f_{\beta\gamma\eta}(g_{\mu\nu}g_{\lambda\rho} - g_{\mu\lambda}g_{\nu\rho}) \\
&+ f_{\alpha\gamma\eta}f_{\delta\beta\eta}(g_{\mu\rho}g_{\nu\lambda} - g_{\mu\nu}g_{\lambda\rho})]
\end{aligned}
$$

It is important to remember that the rules given above are only adequate for tree diagram calculations in QCD (see §15.1).

F.3 The standard model of electroweak interactions: rules for tree graphs

For tree graph calculations, it is most convenient to use the U gauge Feynman rules (Chapters 13, 14) in which no unphysical particles appear. These U gauge rules are given below for the (e, ν_e), (μ, ν_μ), (u, d_c) and (c, s_c) doublets of leptons and quarks. Since the electroweak interactions are not sensitive to colour, there are three equal contributions for each quark transition corresponding to the different quark colours.

F.3.1 External particles

Leptons and quarks

ingoing	$u(p, s)$	or	$v(p, s)$
outgoing	$\bar{u}(p', s')$	or	$\bar{v}(p', s')$.

Vector bosons

ingoing	$\varepsilon_\mu(k, \lambda)$
outgoing	$\varepsilon_\mu^*(k', \lambda')$.

F.3.2 Propagators

Leptons and quarks

$$= \frac{\mathrm{i}}{\not{p} - m} = \mathrm{i}\,\frac{\not{p} + m}{p^2 - m^2}.$$

(Neutrinos are generally assumed to be massless.)
Vector bosons (U gauge)

$W^\pm$, Z^0  $\dfrac{\mathrm{i}}{k^2 - M_V^2}(-\, g^{\mu\nu} + k^\mu k^\nu/M_V^2)$

where the mass M_W of the charged W bosons is given by

$$\frac{G_F}{2^{1/2}} = \frac{g^2}{8M_W^2}$$

with $g \sin\theta_W = e$ (where, in our convention, $e > 0$), so that

$$M_W = \frac{e}{2^{5/4}G_F^{1/2}\sin\theta_W} \simeq \left(\frac{37.3}{\sin\theta_W}\right)\ \mathrm{GeV}/c^2.$$

The mass of the neutral Z boson is related to that of the charged W bosons by

$$M_Z = M_W/\cos\theta_W.$$

Higgs scalar

$$- - - \rightarrow - - - = \frac{i}{p^2 - \mu^2}.$$

F.3.3 Vertices

Charged current weak interactions

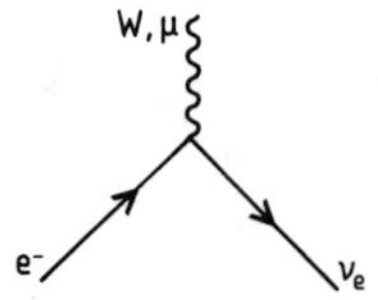

$$-i\frac{g}{2^{1/2}}\gamma_\mu\frac{1-\gamma_5}{2}$$

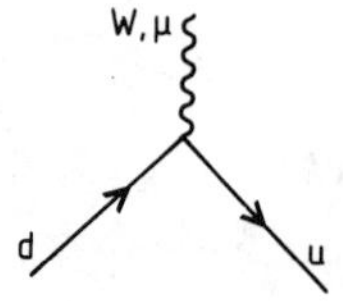

$$-i\frac{g}{2^{1/2}}\cos\theta_C\gamma_\mu\frac{1-\gamma_5}{2}$$

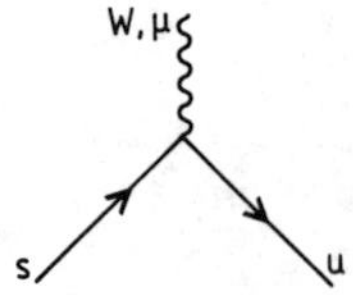

$$-i\frac{g}{2^{1/2}}\sin\theta_C\gamma_\mu\frac{1-\gamma_5}{2}$$

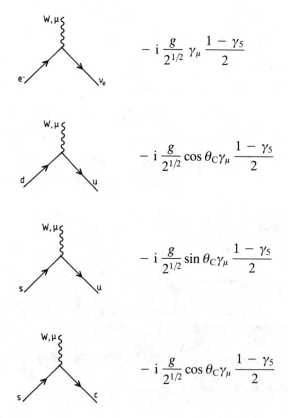

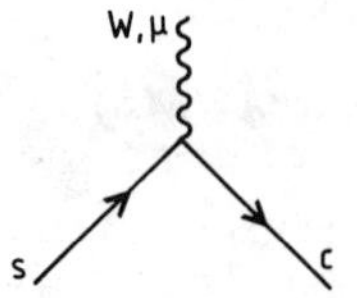

$$-i\frac{g}{2^{1/2}}\cos\theta_C\gamma_\mu\frac{1-\gamma_5}{2}$$

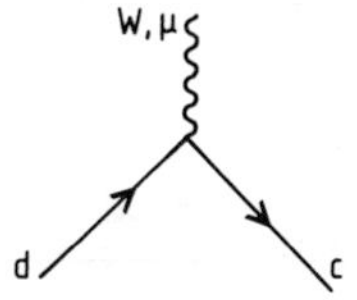

$$-\,\mathrm{i}\,\frac{g}{2^{1/2}}\,(-\sin\theta_{\mathrm{C}})\gamma_\mu\,\frac{1-\gamma_5}{2}$$

Neutral current weak interactions

Massless neutrinos:

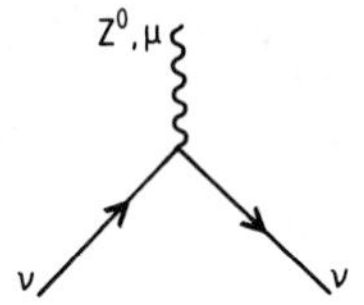

$$\frac{-\,\mathrm{i}e}{\sin\theta_{\mathrm{W}}\cos\theta_{\mathrm{W}}}\,\frac{1}{2}\,\gamma_\mu\,\frac{1-\gamma_5}{2}$$

Massive fermions:

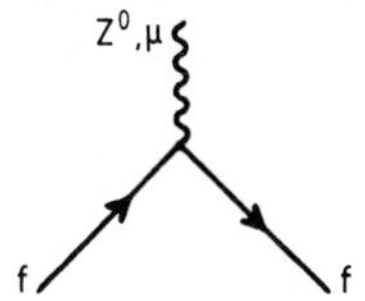

$$\frac{-\,\mathrm{i}e}{\sin\theta_{\mathrm{W}}\cos\theta_{\mathrm{W}}}\,\gamma_\mu\left(c_{\mathrm{L}}^f\,\frac{1-\gamma_5}{2}+c_{\mathrm{R}}^f\,\frac{1+\gamma_5}{2}\right)$$

where

$$c_{\mathrm{L}} = -\tfrac{1}{2}+\sin^2\theta_{\mathrm{W}}, \quad c_{\mathrm{R}} = \sin^2\theta_{\mathrm{W}}, \qquad \text{for } \mathrm{e}^-, \mu^-$$

$$c_{\mathrm{L}} = +\tfrac{1}{2}-\tfrac{2}{3}\sin^2\theta_{\mathrm{W}}, \quad c_{\mathrm{R}} = -\tfrac{2}{3}\sin^2\theta_{\mathrm{W}}, \qquad \text{for u, c}$$

$$c_{\mathrm{L}} = -\tfrac{1}{2}+\tfrac{1}{3}\sin^2\theta_{\mathrm{W}}, \quad c_{\mathrm{R}} = \tfrac{1}{3}\sin^2\theta_{\mathrm{W}}, \qquad \text{for d, s.}$$

(Massless neutrinos have $c_{\mathrm{L}} = \frac{1}{2}$; $c_{\mathrm{R}} = 0$.)

Vector boson couplings

(a) Trilinear couplings:

γW^+W^- vertex

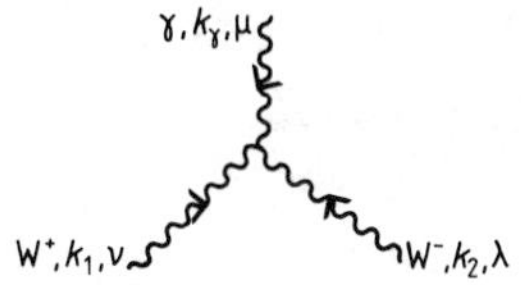

$$\mathrm{i}e[g_{\nu\lambda}(k_1-k_2)_\mu + g_{\lambda\mu}(k_2-k_\gamma)_\nu + g_{\mu\nu}(k_\gamma-k_1)_\lambda]$$

$Z^0W^+W^-$ vertex

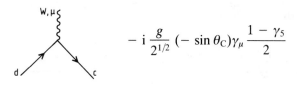

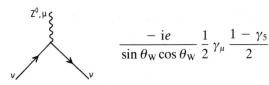

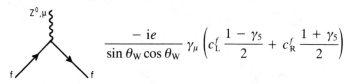

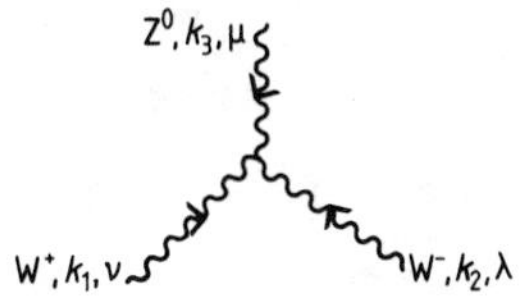

$$\mathrm{i}\,\frac{e\cos\theta_{\mathrm{W}}}{\sin\theta_{\mathrm{W}}}\,[g_{\nu\lambda}(k_1-k_2)_\mu + g_{\lambda\mu}(k_2-k_3)_\nu + g_{\mu\nu}(k_3-k_1)_\lambda]$$

(b) Quadrilinear couplings:

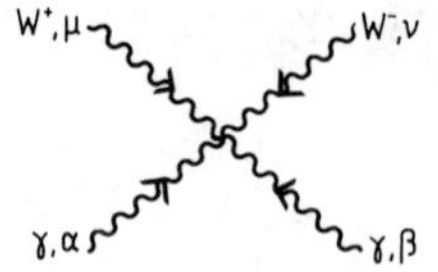

$$-\,\mathrm{i}e^2(2g_{\alpha\beta}g_{\mu\nu} - g_{\alpha\mu}g_{\beta\nu} - g_{\alpha\nu}g_{\beta\mu})$$

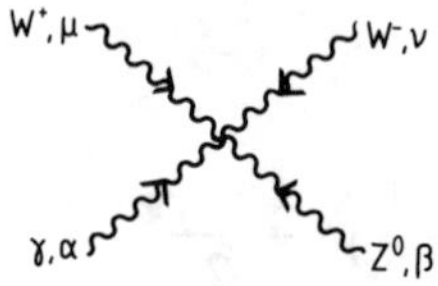

$$-\,\mathrm{i}e^2 \cot\theta_{\mathrm{W}}(2g_{\alpha\beta}g_{\mu\nu} - g_{\alpha\mu}g_{\beta\nu} - g_{\alpha\nu}g_{\beta\mu})$$

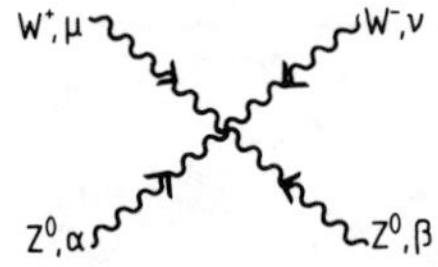

$$-\,\mathrm{i}e^2 \cot^2\theta_{\mathrm{W}}(2g_{\alpha\beta}g_{\mu\nu} - g_{\alpha\mu}g_{\beta\nu} - g_{\alpha\nu}g_{\beta\mu})$$

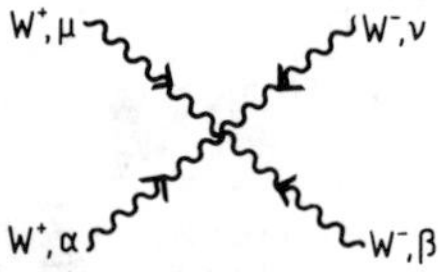

$$\frac{\mathrm{i}e^2}{\sin^2\theta_{\mathrm{W}}}\,(2g_{\mu\alpha}g_{\nu\beta} - g_{\mu\beta}g_{\alpha\nu} - g_{\mu\nu}g_{\alpha\beta})$$

Higgs couplings

(a) Trilinear couplings:

σW^+W^- vertex

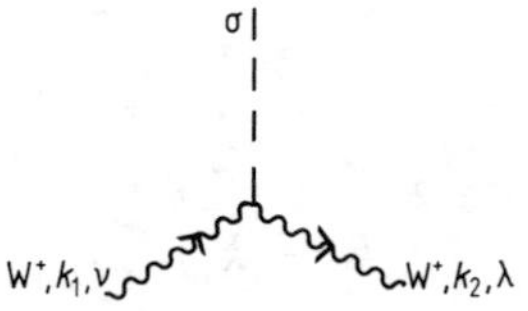

$$\frac{\mathrm{i}e}{\sin\theta_{\mathrm{W}}}\,M_{\mathrm{W}}g_{\nu\lambda}$$

σZ^0Z^0 vertex

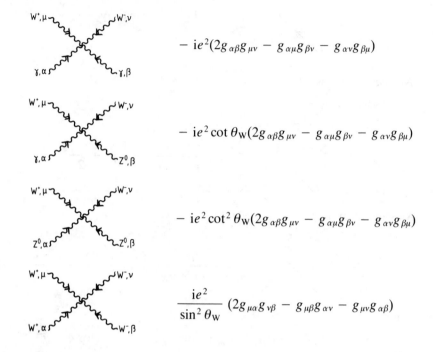

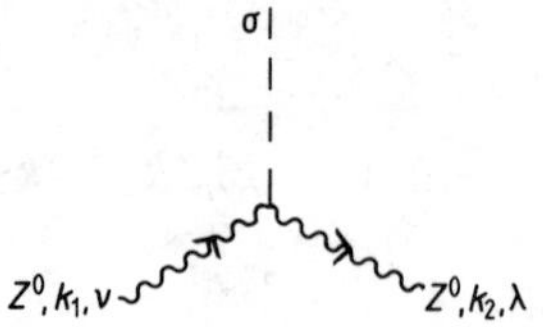

$$\frac{\mathrm{i}2e}{\sin 2\theta_{\mathrm{W}}}\,M_{\mathrm{Z}}g_{\nu\lambda}$$

Fermion Yukawa couplings (massive fermions, mass m_{f})

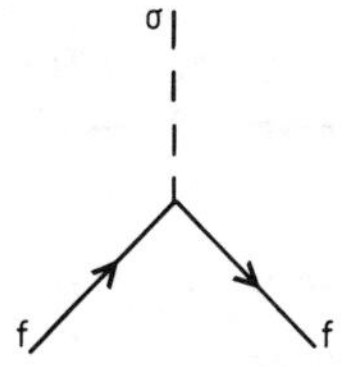

$$-\frac{\mathrm{i}e}{2\sin\theta_{\mathrm{W}}}\frac{m_{\mathrm{f}}}{M_{\mathrm{W}}}$$

Trilinear self-coupling

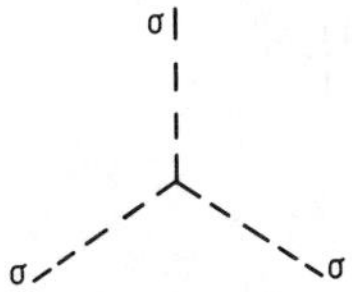

$$-\mathrm{i}\frac{3\mu^2 e}{2M_{\mathrm{W}}\sin\theta_{\mathrm{W}}}$$

(b) Quadrilinear couplings:

$\sigma\sigma\mathrm{W}^+\mathrm{W}^-$ vertex

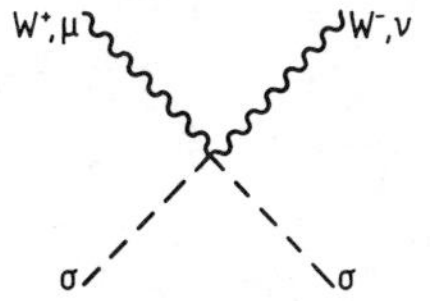

$$\frac{\mathrm{i}e^2}{2\sin^2\theta_{\mathrm{W}}}g_{\mu\nu}$$

$\sigma\sigma\mathrm{ZZ}$ vertex

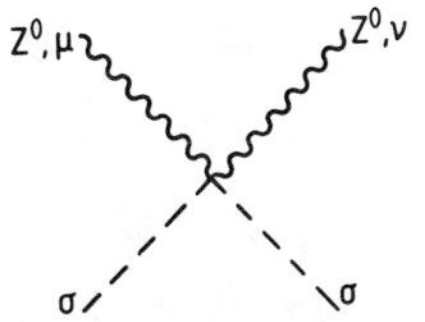

$$\frac{\mathrm{i}2e^2}{\sin^2 2\theta_{\mathrm{W}}}g_{\mu\nu}$$

Quadrilinear self-coupling

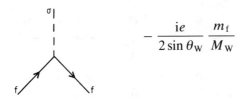

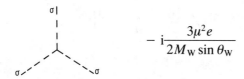

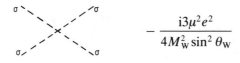

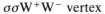

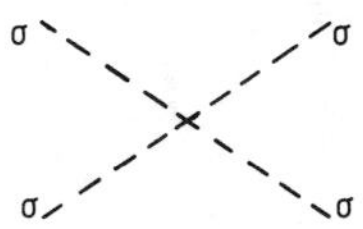

$$-\frac{\mathrm{i}3\mu^2 e^2}{4M_{\mathrm{W}}^2\sin^2\theta_{\mathrm{W}}}$$

(This is obtained from (14.156) with $\rho \to \sigma$, allowing for a combinatorial factor of 6 due to the different ways the σ^3 term can contribute, analogous to the 2 from an $A_\mu A^\mu$ term.)

Finally, we list here the λ matrices of SU(3), which appear in §F2.3 for example:

$$\lambda_1 = \begin{pmatrix} 0 & 1 & 0 \\ 1 & 0 & 0 \\ 0 & 0 & 0 \end{pmatrix}, \quad \lambda_2 = \begin{pmatrix} 0 & -\mathrm{i} & 0 \\ \mathrm{i} & 0 & 0 \\ 0 & 0 & 0 \end{pmatrix}, \quad \lambda_3 = \begin{pmatrix} 1 & 0 & 0 \\ 0 & -1 & 0 \\ 0 & 0 & 0 \end{pmatrix},$$

$$\lambda_4 = \begin{pmatrix} 0 & 0 & 1 \\ 0 & 0 & 0 \\ 1 & 0 & 0 \end{pmatrix}, \quad \lambda_5 = \begin{pmatrix} 0 & 0 & -\mathrm{i} \\ 0 & 0 & 0 \\ \mathrm{i} & 0 & 0 \end{pmatrix}, \quad \lambda_6 = \begin{pmatrix} 0 & 0 & 0 \\ 0 & 0 & 1 \\ 0 & 1 & 0 \end{pmatrix},$$

$$\lambda_7 = \begin{pmatrix} 0 & 0 & 0 \\ 0 & 0 & -\mathrm{i} \\ 0 & \mathrm{i} & 0 \end{pmatrix}, \quad \lambda_8 = \frac{1}{\sqrt{3}} \begin{pmatrix} 1 & 0 & 0 \\ 0 & 1 & 0 \\ 0 & 0 & -2 \end{pmatrix}.$$

REFERENCES

Abers E S and Lee B W 1973 *Phys. Rep.* **9C** 1
Abramowicz H *et al* 1982a *Z. Phys.* C **12** 289
—— 1982b *Z. Phys.* C **13** 199
—— 1983 *Z. Phys.* C **17** 283
Adeva B *et al* 1985 *Phys. Rev. Lett.* **55** 665
Adler S and Dashen R F 1968 *Current Algebras* (New York: Benjamin)
Aharonov Y and Bohm D 1959 *Phys. Rev.* **115** 485
Aitchison I J R 1972 *Relativistic Quantum Mechanics* (London: Macmillan)
—— 1982 *An Informal Introduction to Gauge Field Theories* (Cambridge: Cambridge University Press)
—— 1985 *Contemp. Phys.* **26** 333
Alles W, Boyer C and Buras A J 1977 *Nucl. Phys.* B **119** 125
Alper B *et al* 1973 *Phys. Lett.* **44B** 521
Altarelli G 1982 *Phys. Rep.* **81** 1
Altarelli G and Parisi G 1977 *Nucl. Phys.* B **126** 298
Altarelli G *et al* 1978a *Nucl. Phys.* B **143** 521
—— 1978b *Nucl. Phys.* B **146** 544 (E)
—— 1979 *Nucl. Phys.* B **157** 461
—— 1984 *Nucl. Phys.* B **246** 12
—— 1985 *Z. Phys.* C **27** 617
Althoff M *et al* 1984 *Z. Phys.* C **22** 307
Anderson C D 1932 *Science* **76** 238
Anderson C D and Neddermeyer S 1937 *Phys. Rev.* **51** 884
Anderson P W 1963 *Phys. Rev.* **130** 439
Antonelli F and Maiani L 1981 *Nucl. Phys.* B **186** 269
Appel J A *et al* 1986 *Z. Phys.* C **30** 341
Arnison G *et al* 1982 *Phys. Lett.* **118B** 167
—— 1983a *Phys. Lett.* **122B** 103
—— 1983b *Phys. Lett.* **126B** 398
—— 1984 *Phys. Lett.* **136B** 294
—— 1985 *Phys. Rev. Lett.* **158B** 494
—— 1986 *Phys. Lett.* **166B** 484
Attwood W B 1980 in *Proc. 1979 SLAC Summer Institute on Particle Physics* (SLAC-224) ed. A Mosher, vol 3
Aubert J J *et al* 1974 *Phys. Rev. Lett.* **33** 1404
Augustin J E *et al* 1974 *Phys. Rev. Lett.* **33** 1406
Bagnaia P *et al* 1983 *Phys. Lett.* **129B** 130
—— 1984 *Phys. Lett.* **144B** 283
Bailin D 1982 *Weak Interactions* (Bristol: Adam Hilger)
Banner M *et al* 1973 *Phys. Lett.* **44B** 537
—— 1982 *Phys. Lett.* **118B** 203
—— 1983a *Phys. Lett.* **122B** 322
—— 1983b *Phys. Lett.* **122B** 476
Barber D P *et al* 1979 *Phys. Rev. Lett.* **43** 830

Bardeen J 1957 *Nuovo Cimento* **5** 1766
Barger V, Martin A D and Phillips R J N 1983 *Z. Phys.* C **21** 99
Bartel W B *et al* 1968 *Phys. Lett.* **28B** 148
Behrend H-J *et al* 1982 *Phys. Lett.* **110B** 329
—— 1987 *Phys. Lett.* **183B** 400
Berends F A *et al* 1981 *Phys. Lett.* **103B** 124
Berends F A, Kleiss R and Jadach S 1983 *Nucl. Phys.* B **202** 63
Berman S M, Bjorken J D and Kogut J B 1971 *Phys. Rev.* D **4** 3388
Bernstein J 1968 *Elementary Particles and their Currents* (San Francisco: Freeman)
Berry M V 1984 *Proc. R. Soc.* A **392** 45
Bjorken J D 1969 *Phys. Rev.* **179** 1547
—— 1973 *Phys. Rev.* D **8** 4098
Bjorken J D and Drell S D 1964 *Relativistic Quantum Mechanics* (New York: McGraw-Hill)
Bludman S A 1958 *Nuovo Cimento* **9** 443
Bodek A *et al* 1973 *Phys. Rev. Lett.* **30** 1087
—— 1974 *Phys. Lett.* **51B** 417
Bogoliubov N N and Shirkov D V 1980 *Introduction to the Theory of Quantised Fields* (Chichester: Wiley-Interscience) 3rd edn, translated by S Chomet
Böhm M and Hollik W 1982 *Nucl. Phys.* B **204** 45
—— 1984 *Phys. Lett.* **139B** 213
Bohr N 1913 Phil Mag. **26** 1, 476, 857
Born M, Heisenberg W and Jordan P 1926 *Z. Phys.* **35** 557
Bosetti P C *et al* 1978 *Nucl. Phys.* B **142** 1
Bracci L and Fiorentini G 1982 *Phys. Rep.* **86** 169
Brandelik R *et al* 1979 *Phys. Lett.* **86B** 243
—— 1980 *Phys. Lett.* **97B** 453
Budny R 1975 *Phys. Lett.* **55B** 227
Büsser F W *et al* 1972 *Proc. XVI Int. Conf. on High Energy Physics, Chicago* (Batavia: FNAL) vol 3
—— 1973 *Phys. Lett.* **46B** 471
Cabibbo N 1963 *Phys. Rev. Lett.* **10** 531
Callan C G and Gross D 1969 *Phys. Rev. Lett.* **22** 156
Cashmore R J, Hawkes C M, Lynn B W and Stuart R G 1986 *Z. Phys.* C **30** 125
Casimir H B G 1948 *Koninkl. Ned. Akad. Wetenschap. Proc.* **51** 793
Chambers R G 1960 *Phys. Rev. Lett.* **5** 3
CHARM Collaboration 1981 *Phys. Lett.* **99B** 265
Chen M-S and Zerwas P 1975 *Phys. Rev.* D **12** 187
Close F E 1979 *An Introduction to Quarks and Partons* (London: Academic)
Coleman S 1975 *Laws of Hadronic Matter* ed. A Zichichi (London: Academic) pp 141–2
Coleman S and Gross D J 1973 *Phys. Rev. Lett.* **31** 851
Collins J C 1984 *Renormalization* (Cambridge: Cambridge University Press)
Collins P D B and Martin A D 1984 *Hadron Interactions* (Bristol: Adam Hilger)
Combridge B L *et al* 1977 *Phys. Lett.* **70B** 234
Conversi M E *et al* 1947 *Phys. Rev.* **71** 209

Cooper L N 1956 *Phys. Rev.* **104** 1189
Cottingham W N and Greenwood D A 1986 *An Introduction to Nuclear Physics* (Cambridge: Cambridge University Press)
Creutz M 1979 *Phys. Rev. Lett.* **45** 313
—— 1983 *Quarks, Gluons and Lattices* (Cambridge: Cambridge University Press)
Cundy D C 1974 *Proc. XVII Int. Conf. on High Energy Physics, London, July 1974* (Didcot, Berks: Science Research Council) pp IV-131–IV-148
Dalitz R H 1953 *Phil. Mag.* **44** 1068
—— 1965 in *High Energy Physics* ed. C de Witt and M Jacob (New York: Gordon and Breach)
—— 1967 in *Proc. Second Hawaii Topical Conf. in Particle Physics,* ed. S Pakvasa and S F Tuan (Honolulu: University of Hawaii Press)
Danby G *et al* 1962 *Phys. Rev. Lett.* **9** 36
Dawson S, Hagelin J S and Hall L 1981 *Phys. Rev.* D **23** 2666
Decker R, Paschos E A and Brown R W 1984 *Phys. Rev. Lett.* **52** 1192
de Groot J G H *et al* 1979 *Z. Phys.* C **1** 143
di Lella L 1985 *Ann. Rev. Nucl. Sci.* **35** 107
—— 1986 in *Proc. Int. Europhysics Conf. on High Energy Physics, Bari, Italy, July 1985* ed. L Nitti and G Preparata (Bari: Laterza) pp 761ff
Dirac P A M 1927 *Proc. R. Soc.* A **114** 243
—— 1928 *Proc. R. Soc.* A **117** 610
—— 1931 *Proc. R. Soc.* A **133** 60
—— 1981 *The Principles of Quantum Mechanics* (Oxford: Oxford University Press) 4th edn (reprinted)
Drell S D and Yan T M 1970 *Phys. Rev. Lett.* **25** 316
Duke D W and Roberts R G 1985 *Phys. Rep.* **120** 275
Eden R J, Landshoff P V, Olive D I and Polkinghorne J C 1966 *The Analytic S-matrix* (Cambridge: Cambridge University Press)
Ehrenburg W and Siday R E 1949 *Proc. Phys. Soc.* B **62** 8
Eisberg R M 1961 *Fundamentals of Modern Physics* (New York: Wiley)
Ellis J 1979 *Proc. LEP Summer Study, Geneva, 1979* CERN 79-01, pp 615ff
—— 1981 in *Proc. 1981 SLAC Summer Institute on Particle Physics* (SLAC-245) ed. A Mosher, pp 621ff
Ellis J and Karliner I 1979 *Nucl. Phys.* B **148** 141
Ellis J *et al* 1976a *Nucl. Phys.* B **106** 292
—— 1976b *Nucl. Phys.* B **111** 253
Englert F and Brout R 1964 *Phys. Rev. Lett.* **13** 321
Faddeev L D and Popov V N 1967 *Phys. Lett.* **25B** 29
Fermi E 1934a *Nuovo Cimento* **11** 1
—— 1934b *Z. Phys.* **88** 161
Feynman R P 1962 *The Theory of Fundamental Processes* (New York: Benjamin) Ch 5
—— 1963a *Acta Phys. Polon.* **26** 697
—— 1963b *The Feynman Lectures on Physics* vol 1 (Reading, MA: Addison-Wesley) ch 15
—— 1964 *The Feynman Lectures on Physics* vol 2 (Reading: MA: Addison-Wesley)

—— 1965 *The Feynman Lectures on Physics* vol 3 (Reading; MA: Addison-Wesley)
—— 1972 *Statistical Mechanics* (Reading, MA: Benjamin)
—— 1977 in *Weak and Electromagnetic Interactions at High Energy* ed. R Balian and C H Llewellyn Smith (Amsterdam: North-Holland) p 121
Feynman R P and Hibbs A R 1965 *Quantum Mechanics and Path Integrals* (New York: McGraw-Hill)
Franck J and Hertz G 1914 *Verh. Deutschen Phys. Ges.* **16** 457
Fregeau J H and Hofstadter R 1955 *Phys. Rev.* **99** 1503
French A P and Taylor E F 1978 *An Introduction to Quantum Physics* (New York: Norton)
Friedman J I and Kendall H W 1972 *Ann. Rev. Nucl. Sci.* **22** 203
Frishman Y 1974 *Phys. Rep.* **13C** 1
Fritzsch H and Gell-Mann M 1972 in *Proc. XVI Int. Conf. on High Energy Physics, Chicago* (Batavia: FNAL) vol 2
Furman M 1982 *Nucl. Phys.* B **197** 413
Gasiorowicz S 1966 *Elementary Particle Physics* (New York: Wiley)
—— 1974 *Quantum Physics* (New York: Wiley)
Gastmans R 1975 *Weak and Electromagnetic Interactions at High Energies, Cargese 1975* ed. M Levy, J L Basdevant, D Speiser and R Gastmans (New York: Plenum) pp 109ff
Geer S 1986 in *High Energy Physics 1985, Proc. Yale Theoretical Advanced Study Institute* (Singapore: World Scientific) vol 2
Geiger H and Marsden E 1909 *Proc. R. Soc.* **82** 495
—— 1913 *Phil. Mag.* **25** 604
Gell-Mann M 1953 *Phys. Rev.* **92** 833
—— 1961 *California Institute of Technology Report* CTSL-20 (reprinted in Gell-Mann and Ne'eman 1964)
—— 1964 *Phys. Lett.* **8** 214
Gell-Mann M and Ne'eman Y 1964 *The Eightfold Way* (New York: Benjamin)
Georgi H and Glashow S L 1974 *Phys. Rev. Lett.* **32** 438
Georgi H, Quinn H R and Weinberg S 1974 *Phys. Rev. Lett.* **33** 451
Gibson W M and Pollard B R 1976 *Symmetry Principles in Elementary Particle Physics* (Cambridge: Cambridge University Press)
Glashow S L 1961 *Nucl. Phys.* **22** 579
Glashow S L, Iliopoulos J and Maiani L 1970 *Phys. Rev.* D **2** 1285
Goldstone J 1961 *Nuovo Cimento* **19** 154
Gottschalk T and Sivers D 1980 *Phys. Rev.* D **21** 102
Green M B, Schwarz J H and Witten E 1987 *Superstring Theory* (Cambridge: Cambridge University Press) vols I and II
Greenberg O W 1964 *Phys. Rev. Lett.* **13** 598
Gross D J and Llewellyn Smith C H 1969 *Nucl. Phys.* B **14** 337
Gross D J and Wilczek F 1973 *Phys. Rev. Lett.* **30** 1343
Guralnick G S, Hagen C R and Kibble T W B 1964 *Phys. Rev. Lett.* **13** 585
Halzen F and Martin A D 1984 *Quarks and Leptons* (New York: Wiley)
Herb S W *et al* 1977 *Phys. Rev. Lett.* **39** 252
Heisenberg W 1932 *Z. Phys.* **77** 1

Higgs P W 1964 *Phys. Rev. Lett.* **13** 508
—— 1966 *Phys. Rev.* **145** 1156
Hofstadter R 1958 *Ann. Rev. Nucl. Sci.* **7** 231
't Hooft G 1971a *Nucl. Phys.* B **33** 173
—— 1971b *Nucl. Phys.* B **35** 167
—— 1971c *Phys. Lett.* **37B** 195
—— 1976 *High Energy Physics, Proc. European Physical Society Int. Conf.* ed. A Zichichi (Bologna: Editrice Composition) p 1225
—— 1980 *Sci. Am.* **242** (6) 90–116
't Hooft G and Veltman M 1972 *Nucl. Phys.* B **44** 189
Horgan R and Jacob M 1981 *Nucl. Phys.* B **179** 441
Hughes R J 1981 *Nucl. Phys.* B **186** 376
Innes W R *et al* 1977 *Phys. Rev. Lett.* **39** 1240
Isgur N and Karl G 1983 *Phys. Today* **36** (11) 36
Itzykson C and Zuber J-B 1980 *Quantum Field Theory* (New York: McGraw-Hill)
Jacob M and Landshoff P V 1978 *Phys. Rep.* **48C** 285
Jones S E 1986 *Nature* **321** 127
Kibble T W B 1967 *Phys. Rev.* **155** 1554
Kim K J and Schilcher K 1978 *Phys. Rev.* D **17** 2800
Kinoshita T 1962 *J. Math. Phys.* **3** 650
Kinoshita T and Sapirstein J 1984 in *Atomic Physics Nine*, ed. R S Van Dyck Jr and E N Fortson (Singapore: World Scientific)
Kitchener J A and Prosser A P 1957 *Proc. R. Soc.* A **242** 403
Kobayashi M and Maskawa K 1973 *Prog. Theor. Phys.* **49** 652
Kunszt Z and Piétarinen E 1980 *Nucl. Phys.* B **164** 45
La Rue G, Fairbank W M and Hebard A F 1977 *Phys. Rev. Lett.* **38** 1011
La Rue G, Phillips J D and Fairbank W M 1981 *Phys. Rev. Lett.* **46** 967
Lattes C M G *et al* 1947 *Nature* **159** 694
Lautrup B 1976 *Weak and Electromagnetic Interactions at High Energies, Cargese 1975* ed. M Levy, J L Basdevant, D Speiser and R Gastmans (New York: Plenum) p 40
Lee B W, Quigg C and Thacker H B 1977 *Phys. Rev.* D **16** 1519
Lee T D 1981 *Particle Physics and Introduction to Field Theory* (Chur: Harwood)
Lee T D and Nauenberg M 1964 *Phys. Rev.* B **133** 1549
Lee T D and Yang C N 1955 *Phys. Rev.* **98** 1501
—— 1956 *Phys. Rev.* **104** 254
—— 1962 *Phys. Rev.* **128** 855
LEP 1979 *Proc. LEP Summer Study, Geneva, 1979* CERN 79-01
Lichtenberg D B 1978 *Unitary Symmetry and Elementary Particles* (London: Academic) 2nd edn
Linde A D 1976 *JETP Lett.* **23** 73
Llewellyn Smith C H 1974 *Phenomenology of Particles at High Energies* ed. R L Crawford and R Jennings (London: Academic) pp 459–528
—— 1979 *Proc. LEP Summer Study, Geneva, 1979* CERN 79-01, pp 35ff
Llewellyn Smith C H and Wheater J F 1982 *Nucl. Phys.* B **208** 27

London F 1950 *Superfluids Vol 1, Microscopic Theory of Superconductivity* (New York: Wiley)
Lynn B W and Stuart R G 1985 *Nucl. Phys.* B **253** 216
Mandelstam S 1974 *Phys. Rep.* **13C** 259
—— 1976 *Phys. Rep.* **23C** 245
—— 1980 *Phys. Rep.* **67C** 109
Mandl F and Shaw G 1984 *Quantum Field Theory* (Chichester: Wiley)
Marciano W J 1979 *Phys. Rev.* D **20** 274
Marciano W J and Sirlin A 1984 *Phys. Rev.* D **29** 945
Maxwell J C 1864 *Phil. Trans. R. Soc.* **155** 459
Nambu Y 1960 *Phys. Rev. Lett.* **4** 380
—— 1974 *Phys. Rev.* D **10** 4262
Ne'eman Y 1961 *Nucl. Phys.* **26** 222
Nielsen H B and Olesen P 1973 *Nucl. Phys.* B **61** 45
Nielsen N K 1981 *Am. J. Phys.* **49** 1171
Nishijima K 1955 *Prog. Theor. Phys.* **13** 285
—— 1969 *Fields and Particles* (New York: W A Benjamin)
Noether E 1918 *Nachr. Ges. Wiss. Göttingen* 171
Occhialini G P S and Powell C F 1947 *Nature* **159** 186
Pais A 1952 *Phys. Rev.* **86** 663
Panofsky W K H and Phillips M 1962 *Classical Electricity and Magnetism* (Reading, MA: Addison-Wesley) 2nd edn
Parisi G 1980 *Phys. Lett.* **90B** 295
Pati J C and Salam A 1974 *Phys. Rev.* D **10** 275
Pauli W 1933 *Handb. Phys.* **24** 246
Pennington M R 1983 *Rep. Prog. Phys.* **46** 393
Perkins D H 1947 *Nature* **159** 126
—— 1972 *Introduction to High Energy Physics* (Reading, MA: Addison-Wesley)
—— 1982 *Introduction to High Energy Physics* (Reading, MA: Addison-Wesley) 2nd edn
—— 1984 *Ann. Rev. Nucl. Part. Sci.* **34** 1
—— 1987 *Introduction to High Energy Physics* (Reading, MA: Addison-Wesley) 3rd edn
Perl M *et al* 1975 *Phys. Rev. Lett.* **35** 1489
Politzer H D 1973 *Phys. Rev. Lett.* **30** 1346
Prescott C Y *et al* 1978 *Phys. Lett.* **77B** 347
Quigg C 1977 *Rev. Mod. Phys.* **49** 297
Review of Particle Properties 1984 *Rev. Mod. Phys.* **56** S1-S304
Ross G G 1984 *Grand Unified Theories* (Menlo Park, CA: Benjamin/Cummings)
Rubbia C, McIntyre P and Cline D 1977 *Proc. Int. Neutrino Conf., Aachen, 1976* (Braunschweig: Vieweg) p 683
Rutherford E 1911 *Phil. Mag.* **21** 669
Ryder L H 1985 *Quantum Field Theory* (Cambridge: Cambridge University Press)
Sachrajda C T C 1983 in *Gauge Theories in High Energy Physics, Les Houches Lectures, Session XXXVII* ed. M K Gaillard and R Stora (Amsterdam: North-Holland)

Sagawa H *et al* 1988 *Phys. Rev. Lett.* **60** 93
Saint-James D, Sarma G and Thomas E J 1969 *Type II Superconductivity* (Oxford: Pergamon)
Sakurai J J 1960 *Ann. Phys., NY* **11** 1
Salam A 1968 *Elementary Particle Theory* ed. N Svartholm (Stockholm: Almqvist) p 367
Salam A and Ward J C 1964 *Phys. Lett.* **13** 168
Schiff L I 1955 *Quantum Mechanics* (New York: McGraw-Hill) 2nd edn
Schrieffer J R 1964 *Theory of Superconductivity* (New York: Benjamin)
Schwinger J 1957 *Ann. Phys, NY* **2** 407
—— 1962 *Phys. Rev.* **125** 397
Scott D M 1985 in *Proc. School for Young High Energy Physicists, Rutherford Appleton Laboratory, 1984* (RAL-85-010) ed. J B Dainton
Shaw R 1955 *The Problem of Particle Types and Other Contributions to the Theory of Elementary Particles: PhD Thesis* University of Cambridge
Sirlin A and Marciano W J 1981 *Nucl. Phys.* B **189** 442
Sparnaay M J 1958 *Physica* **24** 751
Staff of the CERN $\bar{p}p$ project 1981 *Phys. Lett.* **107B** 306
Street J C and Stevenson E C 1937 *Phys. Rev.* **52** 1003
Sushkov O P, Flambaum V V and Khriplovich I B 1975 *Sov. J. Nucl. Phys.* **20** 537
Tassie L J 1974 *Int. J. Theor. Phys.* **9** 167
Taylor J C 1976 *Gauge Theories of Weak Interactions* (Cambridge: Cambridge University Press)
Taylor J R 1972 *Scattering Theory* (New York: Wiley)
Tilley D R and Tilley J 1986 *Superfluidity and Superconductivity* (Bristol: Adam Hilger)
Uehling E A 1935 *Phys. Rev.* **48** 55
Utiyama R 1956 *Phys. Rev.* **101** 1597
Van Dyck R S Jr, Schwinberg P B and Dehmelt H G 1987 *Phys. Rev. Lett.* **59** 26
Veltman M 1977 *Act. Phys. Polon.* **B8** 475
Wegner F J 1971 *J. Math. Phys.* **12** 2259
Weinberg S 1967 *Phys. Rev. Lett.* **19** 1264
—— 1976 *Phys. Rev. Lett.* **36** 294
—— 1980 *Rev. Mod. Phys.* **52** 513
von Weizsäcker C F 1934 *Z. Phys.* **88** 612
Wetzel W 1983 *Nucl. Phys.* B **227** 1
Wick G C 1938 *Nature* **142** 993
—— 1950 *Phys. Rev.* **80** 268
Wigner E P 1949 *Proc. Am. Phil. Soc.* **93** 521; also reprinted in *Symmetries and Reflections* (Bloomington: Indiana University Press) pp 3–13
Wilczek F 1977 *Phys. Rev. Lett.* **39** 1304
Williams E J 1934 *Phys. Rev.* **45** 729
Wilson K G 1969 *Phys. Rev.* **179** 1499
—— 1974 *Phys. Rev.* D **14** 2455
—— 1975 in *New Phenomena in Subnuclear Physics, Proc. 1975 Int. School on Subnuclear Physics 'Ettore Majorana'* ed. A Zichichi (New York: Plenum)

Wu C S, Ambler E, Hayward R W, Hoppes D D and Hudson R P 1957 *Phys. Rev.* **105** 1413
Yang C N and Mills R L 1954 *Phys. Rev.* **96** 191
Yukawa H 1935 *Proc. Phys. Math. Soc. Japan* **17** 48
Zweig G 1964 *CERN Preprint* TH-401

Index